What is the Universe made of? How old is it? How does a supernovae explode? Can we detect black holes? And where do cosmic rays originate? This volume provides a comprehensive and pedagogical introduction to modern ideas and challenging problems in nuclear and particle astrophysics.

Based on a graduate school, specially written articles by eight leading experts cover a wealth of exciting topics, including the search for black holes, nucleosynthesis and neutrino transport in supernovae, the physics of neutron stars, massive neutrinos, cosmic ray physics and astrophysics, and physical cosmology. Together, they present the Universe as a laboratory for testing cutting-edge physics and bridge the gap between conference proceedings and specialised monographs.

This volume provides an invaluable resource for graduate students and active researchers in nuclear and particle physics, astrophysics and cosmology.

CAMBRIDGE CONTEMPORARY ASTROPHYSICS

Nuclear and Particle Astrophysics

CAMBRIDGE CONTEMPORARY ASTROPHYSICS

Series editors
José Franco, Steven M. Kahn, Andrew R. King and Barry F. Madore

Titles available in this series

Gravitational Dynamics, *edited by O. Lahav, E. Terlevich and R. J. Terlevich* (ISBN 0 521 56327 5)

High-sensitivity Radio Astronomy, *edited by N. Jackson and R. J. Davis* (ISBN 0 521 57350 5)

Relativistic Astrophysics, *edited by B. J. T. Jones and D. Marković* (ISBN 0 521 62113 5)

Advances in Stellar Evolution, *edited by R. T. Rood and A. Renzini* (ISBN 0 521 59184 8)

Relativistic Gravitation and Gravitational Radiation, *edited by J.-A. Marck and J.-P. Lasota* (ISBN 0 521 59065 5)

Instrumentation for Large Telescopes, *edited by J. M. Rodríguez Espinosa, A. Herrero and F. Sánchez* (ISBN 0 521 58291 1)

Stellar Astrophysics for the Local Group, *edited by A. Aparicio, A. Herrero and F. Sánchez* (ISBN 0 521 63255 2)

Nuclear and Particle Astrophysics, *edited by J. G. Hirsch and D. Page* (ISBN 0 521 63010 X)

Nuclear and Particle Astrophysics

Proceedings of the Mexican School on Nuclear Astrophysics,
held in Guanajuato, México
August 13–20, 1997

Edited by
J. G. HIRSCH
Physics Department, CINVESTAV del IPN, México

D. PAGE
Astronomy Institute, UNAM, México

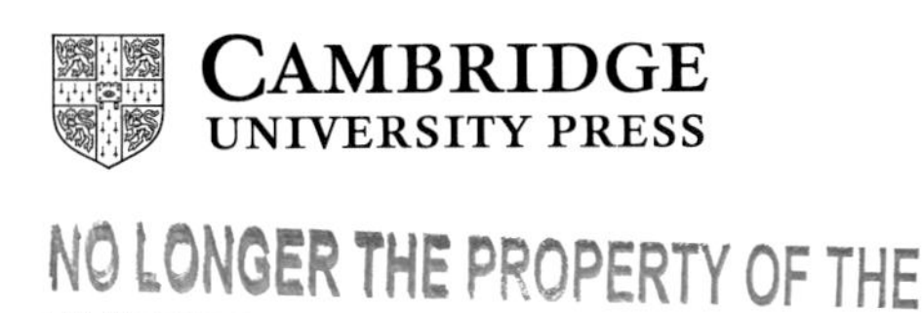
CAMBRIDGE
UNIVERSITY PRESS

PUBLISHED BY THE PRESS SYNDICATE OF THE UNIVERSITY OF CAMBRIDGE
The Pitt Building, Trumpington Street, Cambridge CB2 1RP, United Kingdom

CAMBRIDGE UNIVERSITY PRESS
The Edinburgh Building, Cambridge CB2 2RU, United Kingdom
40 West 20th Street, New York, NY 10011–4211, USA
10 Stamford Road, Oakleigh, Melbourne 3166, Australia

First published 1998

Printed in the United Kingdom at the University Press, Cambridge

A catalogue record for this book is available from the British Library

ISBN 0 521 63010 X hardback

Contents

Participants

Instituto de Astronomía, UNAM, México

Carlos Miguel Avendaño Villela	carlos@astroscu.unam.mx
Vladimir Avila Reese	avila@astroscu.unam.mx
Wilder Chicana Nuncebay	wilder@astroscu.unam.mx
Claudio Firmani	claudio@astroscu.unam.mx
Héctor M. Hernández Toledo	hector@astroscu.unam.mx
Luis López Martín	luislm@astroscu.unam.mx
Dany Page	page@astroscu.unam.mx
Miriam del Carmen Peña Cárdenas	miriam@astroscu.unam.mx
Carlos Pérez Torres	capeto@astroscu.unam.mx
Gabriella Piccinelli Bacchi	gabriela@astroscu.unam.mx
Luis Felipe Rodríguez Jorge	luisfr@astrosmo.unam.mx
Yolanda de Rodríguez	gocy@astrosmo.unam.mx
William J. Schuster B.	schuster@bufadora.astrosen.unam.mx
Octavio Valenzuela	octavio@astroscu.unam.mx

Instituto de Ciencias Nucleares, UNAM, México

Fernando Benítez Martínez	benitez@nuclecu.unam.mx
Alejandro Coca Santillana	
Octavio Castaños Garza	ocasta@nuclecu.unam.mx
Enrique Chávez Carrasco	echavez@xochitl.nuclecu.unam.mx
Juan Carlos D'Olivo	dolivo@nuclecu.unam.mx
Peter Otto Hess R.	hess@nuclecu.unam.mx
Ernesto Kirchuk	kirchuk@nuclecu.unam.mx
Román Linares Romero	linares@roxanne.nuclecu.unam.mx
Juan Carlos López Vieyra	vieyra@nuclecu.unam.mx
Enrique Lozano Ochoa	lozano@nuclecu.unam.mx
Elizabeth Padilla Rodal	padilla@sphinxm.nuclecu.unam.mx
Daniel Sudarsky	sudarsky@xochitl.nuclecu.unam.mx
Luis F. Urrutia	me@aurora.nuclear.unam.mx

Instituto de Geofísica, UNAM, México

Alejandro Lara Sánchez	alara@tonatiuh.igeocu.unam.mx
Raúl Meléndez Venancio	raulm@quetzalcoatl.igeofcu.unam.mx
Miguel Rodríguez Rojas	mrojas@ariel.igeofcu.unam.mx
José Francisco Valdéz	jfvaldez@igeofcu.unam.mx

Instituto de Física, UNAM, México

José Rubén Alfaro Molina	ruben@arrayan.ifisicacu.unam.mx
Mónica Borunda Pacheco	monica@ft.ifisicacu.unam.mx
Efraín R. Chávez L.	chavez@fenix.ifisicacu.unam.mx
Ma. Catalina Espinoza Hernández	catalina@ft.ifisicacu.unam.mx
Benjamín Gutiérrez García	benjamin@fenix.ifisicacu.unam.mx
Leandro Hernández De la Peña	leandro@feynmann.ifisicacu.unam.mx
Enriqueta Hernández Saldaña	queta@ft.ifisicacu.unam.mx
Ma. Esther Ortiz	ortiz@fenix.ifisicacu.unam.mx
Armando Perdomo Almeida	perdomo@feynmann.ifisicacu.unam.mx
Juan Suro Pérez	jsuro@fenix.ifisicacu.unam.mx

Manuel Torres
Liliana Esther Velazco Sevilla liliana@ft.ifisicacu.unam.mx
Erica Ivonne Zavala Carrasco yvonne@ft.ifisicacu.unam.mx

Instituto Nacional de Investigaciones Nucleares, México
Juan Jaime Vega Castro jvega@servidor.unam.mx

Universidad Michoacana de San Nicolás de Hidalgo, México
Francisco Astorga fastorga@zeus.ccu.umich.mx
José Edgar Madriz Aguilar jmadriz@scfic.eie.umich.mx
Gilberto Jerónimo Moreno gilberto@ginette.ifm.umich.mx
Maribel Ríos Cruz maribel@ifm1.ifm.umich.mx
Thomas Zannias zannias@2fm1.2fm.umich.mx

Instituto Nacional de Astrofísica, Optica y Electrónica, México
Gerardo Alarcón F. galarcon@inaoep.mx
César Alvarez Ochoa calvarez@ianoep.mx
Alberto Carramiñana alberto@inaoep.mx
Luis Antonio García Trujillo reyleon@ianoep.mx
Ma. Luisa Ortega Cruz ecruz@inaoep.mx
Elsa Recillas Pishmish elsare@inaoep.mx
Emilio Romano Díaz eromano@inaoep.mx
Miguel Angel Trinidad Hernández mtrini@inaoep.mx

CINVESTAV del IPN, México
Said Aranda Espinoza said@fis.cinvestav.mx
Guillermo Arreaga García garreaga@fis.cinvestav.mx
Victoria Cerón Angeles vceron@fis.cinvestav.mx
Rubén Cordero Elizalde cordero@fis.cinvestav.mx
Luz del Carmen Cortés Cuautli lucero@fis.cinvestav.mx
Enrique García Reyes engarcia@fis.cinvestav.mx
Francisco S. Guzmán Murillo siddh@fis.cinvestav.mx
Jorge Hirsch Ganievich hirsch@fis.cinvestav.mx
Tonatiuh Matos Chassín tmatos@fis.cinvestav.mx
Claudia Moreno González yaya@fis.cinvestav.mx
Ulises Nucamendi Gómez ulises@fis.cinvestav.mx
Abdel Pérez Lorenzana abdel@fis.cinvestav.mx
Juan Reyes Rivera mreyes@fis.cinvestav.mx
Efran Rojas Marcial efrain@fis.cinvestav.mx
Rubén Sánchez Sánchez rsanchez@fis.cinvestav.mx
Héctor Uriarte Rivera huriarte@fis.cinvestav.mx
Carlos Vargas Madrazo cvargas@fis.cinvestav.mx
Víctor Velázquez vvelaz@fis.cinvestav.mx
Hugo Villegas Brena brena@fis.cinvestav.mx
Arnulfo Zepeda zepeda@fis.cinvestav.mx

Universidad Autónoma de Puebla, México
Umberto Cotti ucotti@sirio.ifuap.buap.mx
Bolivia Martha Pérez Ramírez bolivia@fcfm.buap.mx
Enrique Zeleny Vázquez ezelenyv@sirio.ifuap.buap.mx

Universidad Autónoma de Zacatecas, México
Valeri V. Dvoeglazov valeri@camtera.rediaz.mx

Universidad de Guanajuato, México
Octavio Obregón
Laura Muñoz Salazar

Universidad Autónoma Metropolitana Iztapalapa,México
Eva Hernández Telles

Universidad de Santiago de Compostela, España
Jaime Alvarez Muñiz blanco@fpaxp1.usc.es
José Juan Blanco Pillado jaime@gaes.usc.es

Universidad Interamericana, Puerto Rico
Carlos Oliveras Martnez colivera@ponce.inter.edu
Pablo I. Rivera Díaz privera@ponce.inter.edu

State University of New York, Stony Brook, U. S. A.
Madappa Prakash prakash@nuclear.physics.sunysb.edu
Sanjay Reddy reddy@nuclear.physics.sunysb.edu

Universidad de Costa Rica, Costa Rica
Marcelo Magallón Gherardelli mmagallo@efis.ucr.ac.cr

IVE Observatory, VILSPA, Spain
Nora Loiseau mll@vilspa.esa.es

Astronomisches Institut Postdam, Germany
Ulrich R.M.E. Geppert urme@aip.de

Universidad de Sao Paulo, Brasil
Helio Dias heliodias@if.usp.br

Observatorio Interamericano de Cello Tololo, Chile
Ricardo Covarrubias Carreño riccov@ctiowb.ctio.noao.edu

Universidad Nacional de la Plata, Argentina
Luis Alfredo Anchordoqui doqui@venus.fisica.unlp.edu.ar

University of Chicago, U. S. A.
David Schramm dns@ddjob.uchicago.edu

University of Delaware, U. S. A.
Tom Gaisser gaisser@bartol.udel.edu

California Institute of Technology, U. S. A.
Peter Vogel vogel@lamppost.caltech.edu

ORNL University of Tennessee, U. S. A.
Mike Guidry guidry@csep4.rmt.vtk.edu

University of Notre Dame, U. S. A.
Michael Wiescher michael.c.wiescher.1@nd.edu

Institut fur Physik der Universitat Basel, Switzerland
F. K. Thielemann fkt@quasar.physik.unibas.ch

Preface

The Mexican School on Nuclear Astrophysics was held in the Hotel Castillo Santa Cecilia, Guanajuato, México, from August 13 to August 20, 1997. The goal of the school was to gather together researchers and graduate students working on related problems in astrophysics – to present areas of current research, to discuss some important open problems, and to establish and strengthen links between researchers. The school consisted of eight courses and material presented in these forms the basis of this book.

Non-stop interaction between the participants, through both formal and informal discussions, gave the school a relaxed and productive atmosphere. It provided the opportunity for researchers from a wide range of backgrounds to share their interests in and different perspectives of the latest developments in astrophysics.

The productivity of the meeting reflected the strong interest of the Mexican and Latin American scientific communities in the subjects covered, Indeed, a second school is planned for 1999.

Professor David Schramm very sadly died not long after the conference, in December 1997. His lectures at the School were fascinating. He will be sorely missed by us and the rest of the astrophysics community.

Jorge G. Hirsch and Dany Page
Guanajuato, México 1997

Acknowledgements

We would like to acknowledge the support of the following institutions:

Departamento de Física, CINVESTAV del IPN.
Instituto de Astronomía, UNAM.
Instituto de Ciencias Nucleares, UNAM.
Instituto de Física, UNAM.
Instituto de Geofísica, UNAM.
Instituto Nacional de Astrofísica, Optica y Electrónica (INAOE).
Instituto Nacional de Investigaciones Nucleares (ININ).
Universidad de Guanajuato.
Gobierno del Estado de Guanajuato.
Experimento Auger (México).
Consejo Nacional de Ciencia y Tecnología (CONACyT).
International Center for Theoretical Physics (ICTP), Trieste.
Centro Latinoamericano de Física - México (CLAF-México).
Centro Latinoamericano de Física - Brasil (CLAF-Brasil).
Academia Mexicana de Ciencias.
United States - México Foundation for Science.

Observational Astronomy: The Search for Black Holes

By LUIS F. RODRIGUEZ

Instituto de Astronomía, UNAM, Apdo. Postal 70-264, México, DF, 04510, MEXICO

A brief review of key concepts in multifrequency observational astronomy is presented. The basic physical scales in astronomy as well as the concept of stellar evolution are also introduced. As examples of the application of multifrequency astronomy, recent results related to the observational search for black holes in binary systems in our Galaxy and in the centers of other galaxies is described. Finally, the recently discovered microquasars are discussed. These are galactic sources that mimic in a smaller scale the remarkable relativistic phenomena observed in distant quasars.

---

1. Introduction

There have been many outstanding observational and theoretical discoveries made in astronomy during the twentieth century. However, in the future this ending century will most probably be remembered not by these achievements, but by being the time when astronomers started observing the Cosmos with a variety of techniques and in particular when we started to use all the "windows" in the electromagnetic spectrum.

During our century we started to investigate systematically the Universe using:

• The whole electromagnetic spectrum. At the beginning of the century, practically all the data was coming from the visible photons (that is, those detected by the human eye) only.

• Cosmic rays. These charged particles hit the Earth's atmosphere and can be detected by the air showers they produce. The origin of the most energetic cosmic rays (10^{19} ergs or more) remains a mystery.

• Neutrinos. These elusive particles have been detected from the Sun by various teams and from SN1987 A, the supernova that exploded in the Large Magellanic Cloud (Hirata *et al.* 1987; Svoboda *et al.* 1987).

• Gravitational waves. Although no direct detections exist, the expected decay produced by gravitational radiation damping has been observed in the orbit of the binary pulsar PSR 1913+16 by Taylor & Weisberg (1982). Direct detection of gravitational waves will quite likely be achieved in the XXI century, opening remarkable new possibilities of research.

• Radar astronomy. Although the possibility of bouncing and detecting radio waves can be applied only to the Moon and nearby asteroids and planets, this technique has provided valuable information on the nature of their surfaces. One of the great successes of this technique was the measure of the delays in the echoes from Mercury (Shapiro *et al.* 1971) predicted by general relativity for radiation passing close to a massive object, in this case the Sun.

• Direct exploration. As in the case of radar astronomy, this technique can only be applied to relatively near objects, but has allowed the detailed study of the characteristics and chemical composition of the Moon and some of the planets in the Solar System.

In this chapter I will concentrate on multifrequency astronomy (i.e., the study of objects and phenomena in all or al least several windows of the electromagnetic spectrum). Neutrinos and cosmic rays are discussed in the chapters of Vogel and Gaisser in this book. The multifrequency approach seems to be here to stay and it is becoming increasingly

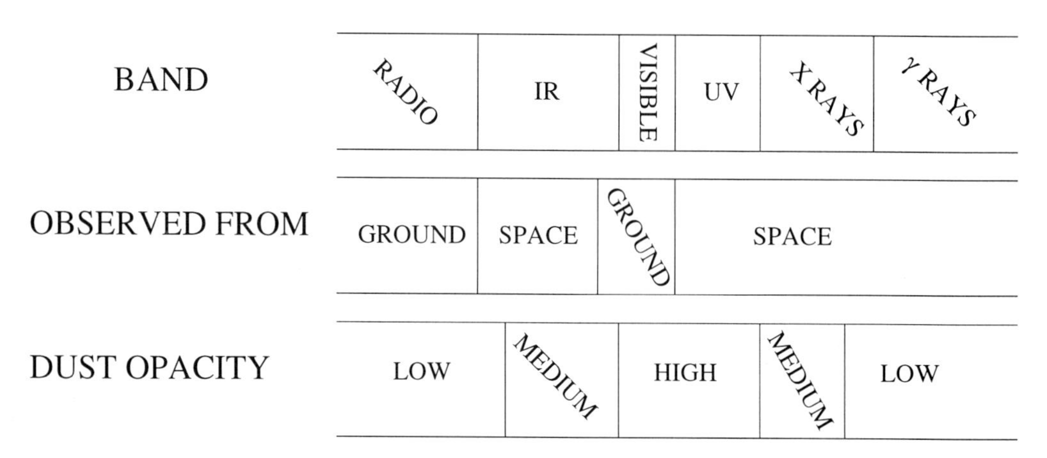

FIGURE 1. The bands of the electromagnetic spectrum, from where they are observed astronomically, and the opacity of cosmic dust to these bands.

difficult to reach solid conclusions about a phenomenon or a source if we do not combine and are conversant, at least to some degree, with all the available information.

I will first review several basic aspects of observational multifrequency astronomy, and will also give a brief discussion on scales in astronomy and on the powerful concept of stellar evolution.

Later on, I will discuss from an observer's point of view the efforts to test the existence of black holes both in binary systems in our own Galaxy, as well as in the centers of other galaxies. At the end of the chapter, I discuss the recently discovered microquasars.

2. Multifrequency observational astronomy

Electromagnetic radiation can be characterized by its wavelength (λ) or frequency (ν), that are related by $c = \lambda\nu$, with $c = 3.00 \times 10^{10}$ $cm\ sec^{-1}$ being the speed of light. The energy of a photon is $E = h\nu$ and the momentum is $p = h\nu/c$, with $h = 6.63 \times 10^{-27}$ $erg\ sec$ being Planck's constant. Besides the possibility of using wavelength, frequency, or energy one can also use an equivalent temperature, given by $T = h\nu/k$, with $k = 1.38 \times 10^{-16}$ $erg\ K^{-1}$ being Boltzmann's constant. An excellent introduction to the study of radiative processes in astrophysics is given in the textbook by Rybicki & Lightman (1979).

The astronomers accostume to divide the electromagnetic spectrum in six bands, that in order of increasing frequency are: radio, infrared (IR), visible, ultraviolet (UV), X rays, and γ rays. In figure 1 I show a qualitative description of the electromagnetic spectrum. The Earth's atmosphere is opaque to all radiations, with the exception of the radio, visible, and the near-IR. Correspondingly, a large fraction of astronomical research is now made from observatories in space (or at least from planes or balloons that can rise above most of the atmosphere). In the long wavelength end ($\lambda \geq 30\ m$), the ionosphere becomes opaque and the study of the Universe beyond those wavelengths also needs to be done from space. Besides the opacity of our own atmosphere, we have to consider the opacity of outer space. In the interstellar medium, in addition to ions, atoms and molecules, there are solid particles with typical dimensions of order 0.1 μm that are referred to as "dust" and are believed to be constituted by silicates, hydrogenated carbons, and graphite. The cosmic dust is particularly opaque to visible and UV radiation (see Figure 1) and at these wavelengths one cannot observe very far in our Galaxy. Regions such as the galactic center can be studied only at the other bands, where the opacity of dust is low or medium.

The electromagnetic bands can be defined by the wavelength (or energy) range they cover and although, as mentioned before, the tendency is to study cosmic objects at many wavelenghts, there are sources that are very bright or were discovered at a given band and that can be associated with it. In what follows, I list the wavelength (or energy) range covered and some of the sources or phenomena characteristic of each band.

• Radio ($\lambda \geq 0.5\ mm$): radio galaxies, supernova remnants, pulsars, molecular clouds, cosmic microwave background.

• Infrared ($500\ \mu m \geq \lambda \geq 0.8\ \mu m$): heated dust, young stars, evolved stars, star-burst galaxies.

• Visible (8000 Å $\leq \lambda \leq 4000$ Å): main-sequence stars, emission and reflection nebulae, normal galaxies.

• Ultraviolet (4000 Å $\leq \lambda \leq 80$ Å): hot stars, gas and molecules in absorption.

• X rays ($150\ eV \leq E \leq\ 100\ keV$): binary systems with compact component, gas in clusters of galaxies.

• γ rays ($100\ keV\ \leq E$): γ-ray bursts, binary systems with compact component, nuclear processes.

To a crude first approximation using Wien's displacement law for blackbody radiation ($h\nu_{max} = 2.8kT$, with ν_{max} being the frequency of maximum emission of the blackbody at temperature T), one could think that with increasing frequency (or energy) one is observing hotter bodies. This is true to some extent, with the cosmic microwave background being much cooler than the accretion disk around a compact object in a binary system. However, there are many counterexamples to this notion: synchrotron emission in astrophysical contexts is emitted by very relativistic electrons (that you could think of as being hot). Nevertheless, the bulk of the emission comes out in the radio regime.

The observational astronomer spends a good part of his time designing, building and observing with telescopes that operate in a given range of the electromagnetic spectrum. In addition of the wavelength range of the telescope, one has to think in several other parameters such as the resolution (that could be in angle, frequency, or time), and the sensitivity (usually related to bigger collecting areas and/or to better detectors). Experience indicates that the most successful instruments have been those that were designed to occupy an unexplored niche in the parameter space (frequency, resolution, sensitivity, etc.) of observational astronomy.

One of the crucial parameters of any telescope is its angular resolution, that is, how accurately can the direction in the sky from where the photons originated be determined. Alternatively, one can see the angular resolution as the smallest separation in the sky that two sources of emission can have before been detected by the instrument as a single, unresolved source. Consider a parabolic mirror with an aperture of diameter D (see Figure 2). It is impossible to say where in the surface the photons bounced before being concentrated in the focus. Then, they have a position uncertainty of $\Delta x = D$. On the other hand, the uncertainty in the angle of origin of the photon in the sky is given by $\theta = \Delta p/p$, where p and Δp are the momentum and the uncertainty in the momentum of the photon. Using Heisenberg's uncertainty principle, $\Delta x \Delta p \sim h$, and that $p = h/\lambda$, we find that $\theta = \lambda/D$. In other words, the smaller the wavelength or the larger the aperture, the better (smaller) the angular resolution. The equation of the angular resolution appears to condemn radio astronomy to have the poorest angular

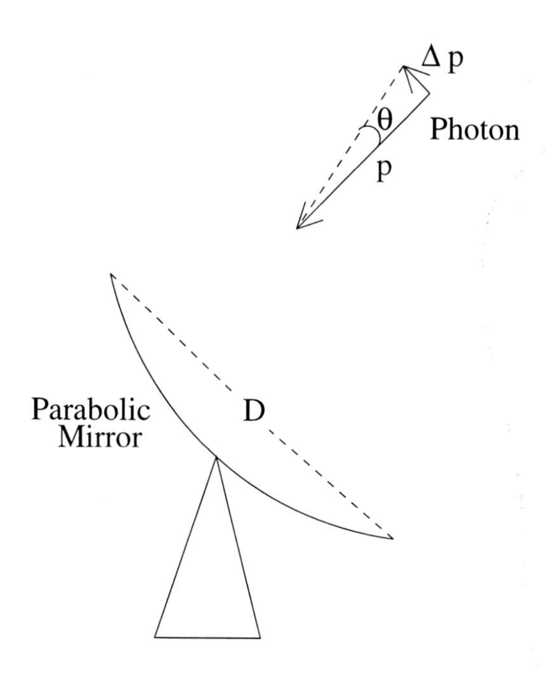

FIGURE 2. The angular resolution of a telescope is determined by the diameter of the telescope and the wavelength of the radiation observed (see text).

resolutions. However, in the case of an interferometer (two or more telescopes that combine the received signals), the diameter D is substituted by the maximum separation between the elements of the interferometer and much better angular resolutions can be obtained. In radio astronomy there are radio telescopes operating simultaneously as interferometers in different continents and now even in orbit (the project VSOP) and angular resolutions of $0''.001$ are routinely achieved. It should be emphazised, however, than when an instrument has very high angular resolution, a smaller number of photons is received inside its beam, and the source has to be very bright for these studies.

3. Astronomical scales

Another concept that, although in a sense trivial, has not always been fully appreciated, is that astronomical research is made on a very large diversity of physical scales.

In an attempt to ubicate the reader, although only approximately and somewhat arbitrarily, I have divided the typical astronomical scales in interplanetary, interstellar, galactic, and cosmological (see Figure 3).

• The interplanetary scale is of order 10^{13} cm and is associated to the distances between planets. A more accurate unit is the astronomical unit (1 AU $= 1.5 \times 10^{13}$ cm), the mean distance between the Earth and the Sun. Light takes minutes to hours to travel these distances.

• The interstellar scale is of order 10^{18} cm and can be associated to the typical separation between stars in the solar neighborhood. Of course, in other regions of the Universe the typical separation between stars can be much larger or much smaller. Although light-years (1 l-y $= 9.5 \times 10^{17}$ cm) are the unit used in popular literature (because light travels

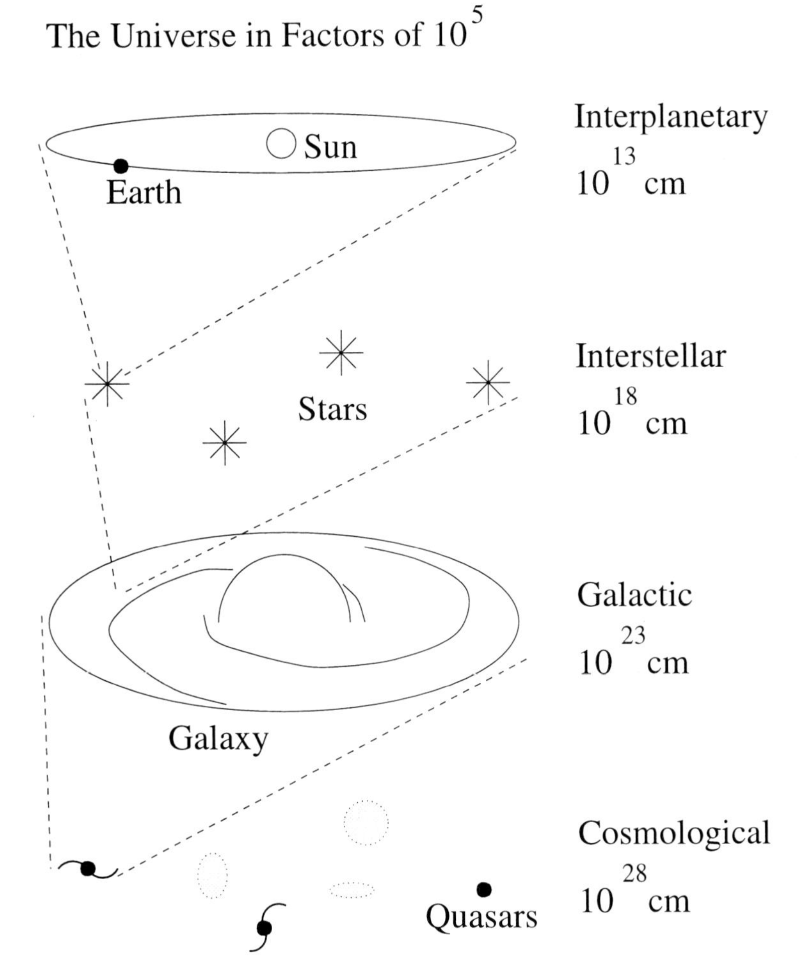

FIGURE 3. The physical scales over which astronomers work can be approximately divided in factors of 10^5, as shown in this cartoon.

these distances in times of years), the astronomer uses the parsec (1 pc $= 3.1 \times 10^{18}$ cm), which is the distance at which 1 AU would subtend 1 arc second in the sky.

• The galactic scale is of order 10^{23} cm and refers to the size of a large galaxy. Our Sun is one of about 10^{11} stars that form our Galaxy, the Milky Way. The astronomer would use the kiloparsec (1 kpc $= 10^3$ pc) in referring to these scales that are traveled by light in timescales of hundreds of thousands of years. If we consider that a star like the Sun lives for about 10^{10} years, when we observe objects in other places of our Galaxy we are observing them pretty much as they actually are "now".

• The cosmological scale is of order 10^{28} cm and is about the size of the observable Universe. It was only until the first decades of the XXth century that it became clear that our Galaxy is not all the Universe, but that outside it there is an enormous number of other galaxies. Galaxies come in three shapes: spiral (like our own, shaped like a disk of stars and gas with a big central bulge made mostly of stars; the spiral term comes from the fact that bright, young stars are born in a two-armed spiral pattern), elliptical (huge spherical or ellipsoidal configurations of star with little gas in them), and irregulars (everything else that cannot be classified as spiral or elliptical). Light takes in the order of 10^{10} years (that is, about the age of the Universe) to travel cosmological scales and it is here where we can expect important evolutionary effects (sources do change over cosmological timescales and very remote things are not necessarily as we see them now). In particular, far from our Galaxy, we observe the enigmatic quasars, sources of electromagnetic luminosities orders of magnitude in excess those of normal galaxies. In is believed that quasars are the nuclei of young galaxies, undergoing an evolutionary stage that disappears with time. Perhaps even the center of our own Galaxy went in the past through a quasar stage.

4. Stellar evolution

I can think of few other concepts in science that unify so much knowledge as that of stellar evolution. The stars are formed from gravitational contraction of dense gaseous clumps embedded in large clouds of molecules and dust that are known to exist in the interstellar space. These molecular clouds can be very massive (reaching 10^6 solar masses in some cases), so many stars can form from them. As expected, astronomers like to discuss the parameters of stars in solar units (one solar mass = 1 $M_\odot$ = 2.0×10^{33} g; one solar luminosity = 1 $L_\odot$ = 3.8×10^{33} erg s^{-1}; and one solar radius = 1 $R_\odot$ = 7.0×10^{10} cm).

The observed masses of stars are between 0.08 and $\sim$100 $M_\odot$. Objects with masses below 0.08 $M_\odot$ will not be able to ignite hydrogen fusion in their interiors. They are called brown dwarfs and are expected to be very dim ($L_\star \leq 10^{-3}$ $L_\odot$); some examples may have been detected in recent studies (Rebolo *et al.* 1996). Objects more massive than about 100 $M_\odot$ are believed to undergo instabilities that lead to the ejection of its outer envelope.

It is also known (although not understood) that Nature forms much more low-mass than high-mass stars, a fact first noted by Salpeter (1955). The so-called initial-mass-function (the number of just-formed stars as a function of the stellar mass) can be roughly approximated by the function $n(M)dM \propto M^{-2.5}$, where $n(M)dM$ is the number of stars in the interval M to $M + dM$. Then, about 30 times more stars are formed in the mass interval of 1 to 2 $M_\odot$ with respect to the mass interval of 10 to 20 $M_\odot$. Furthermore, high-mass stars live much shorter lifes than low-mass stars. The "life" of a star can be defined as the time they spend transforming hydrogen into helium (when they are in the so-called main sequence). The fraction of hydrogen that can be transformed into helium in a star is of order 0.1, so the "fuel" available goes as M. However, the luminosity of stars (that is, the rate at which they burn hydrogen) goes as $M^{3.5}$. Then, the time they spend in the main sequence will go as (fuel available/rate of burning), that is, $t \propto M^{-2.5}$. Finally, assuming a stationary state, the number of stars one expects to find at a given time goes as $N \propto n(M)dM\ t \propto M^{-5}dM$. Then, there are about 10^4 more stars in the mass interval of 1 to 2 $M_\odot$ with respect to the mass interval of 10 to 20 $M_\odot$.

After their life in the main sequence (of order 10^{10} years for a star with 1 $M_\odot$), the stars can start burning heavier elements for shorter timescales until they do not have an internal source of energy and "die". The endpoints of stellar evolution are three. If the star originally had "low" mass (loosely defined as less than $\sim 8M_\odot$), it will go first through a red giant phase, eject part of its outer layers in a planetary nebulae (these nebulae have no relation to planets, they got the name because in small telescopes they look like bright disks, similar to the appearance of planets), and end as a white dwarf

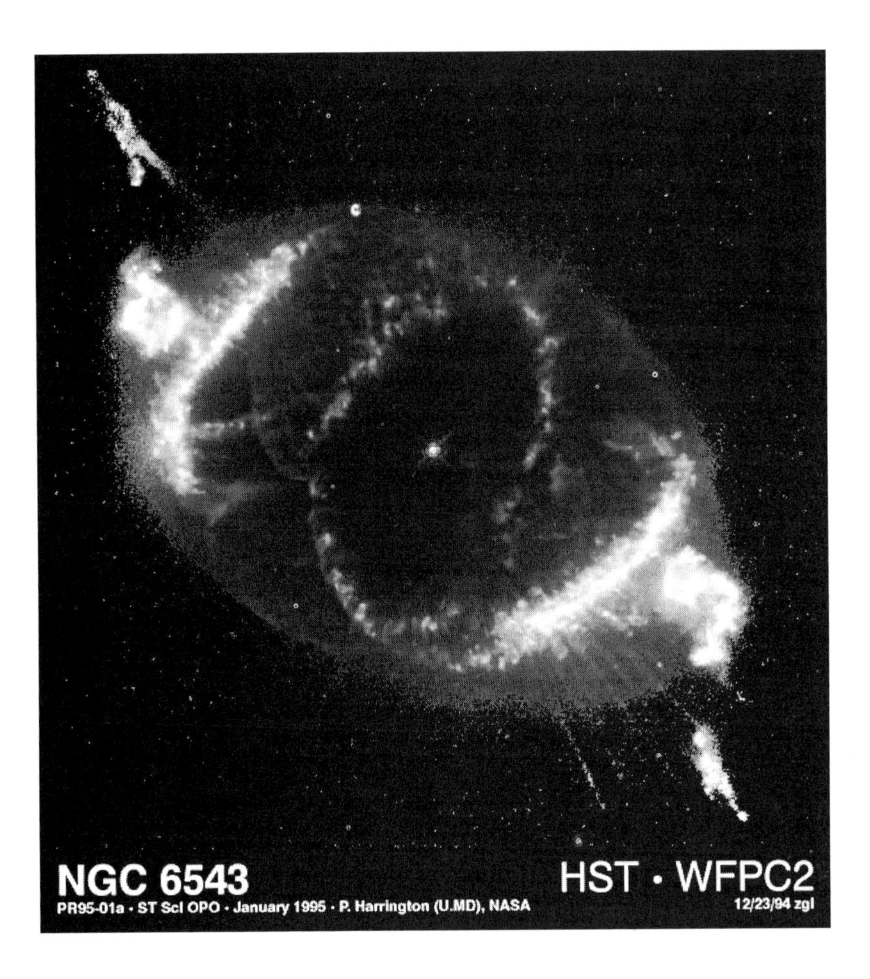

FIGURE 4. Hubble Space Telescope visible image of the planetary nebula NGC 6543. The star at the center of the nebula is in its way to becoming a white dwarf. Credit: J.P. Harrington and K.J. Borkowski (University of Maryland), and NASA.

(see Figure 4). White dwarfs have masses typically between 0.4 and 1 $M_\odot$. You may ask what happened to the remaining few solar masses that formed the original star. The answer lies in that stars loose significant amounts of gas in stellar winds and ejecta, in particular at the end of their lifes. White dwarfs are quite small with radii of order 10^9 cm and are supported against gravity by the pressure of degenerate electrons.

If the star is "massive" (more than $\sim 8M_\odot$), it is believed that it will end its life in a supernova explosion (see Figure 5). The outer layers of the star will be expelled at velocities of order 0.1c. The core will collapse and having a mass in excess of Chandrasekhar's limit, it will continue contracting below the white dwarf dimensions. In a landmark paper, Chandrasekhar (1931) showed that if the mass of a white dwarf exceeded 1.4 $M_\odot$ (the famous Chandrasekhar limit), pressure from the degenerate electrons will not be able to balance the force of gravity and the object would continue contracting. Two possibilities appear to be left. If the pressure from degenerate neutrons stops the

FIGURE 5. Visible image of the Crab supernova remnant. This supernova exploded in 1054 and at its center a pulsar is located. Credit: T. Credner and S. Kohle (Astronomical Institutes of the University of Bonn).

collapse (which should happen for $M_{core} \leq 3\ M_\odot$), a neutron star is formed (see chapter by Prakash). If the core is more massive that $3\ M_\odot$, it will continue contracting to form a black hole. A neutron star is a remarkable object; its mass should be between 1.4 and $3\ M_\odot$ and its radius of order 10 km. When radio emission is produced at its surface, the rotating neutron star is detected as a pulsar (Hewish *et al.* 1968). Thermal soft X-rays from the surface of neutron stars have also been detected and studied in a number of sources (e. g. Page 1994). If the mass of the core exceeds $\sim 3\ M_\odot$ no known mechanism can stop its collapse and a black hole should form.

We will dedicate the rest of this chapter to review the observational evidence in favor of the existence of black holes in the astrophysical context.

5. Historical background

In 1915, within months of the publication of Einstein's papers on general relativity, Karl Schwarzschild found the solution of the field equations for the case of a non-spinning spherical mass. Far from the mass, the familiar newtonian results appeared as a limit. However, close to the mass M, things were very different. In particular no particle or radiation could escape from inside the Schwarzschild radius to the exterior. The Schwarzschild radius is given by:

$$r_s = 2GM/c^2 \simeq 3\ km\ (M/M_\odot), \tag{5.1}$$

where G $= 6.7 \times 10^{-8}$ dynes cm^2 g^{-2} is the gravitational constant. Then, the "interior" of these bodies cannot be seen from the outside, hence the name black holes.

In 1931, as already mentioned, Chandrasekhar showed that if the mass of a white dwarf exceeded 1.4 $M_\odot$, pressure from the degenerate electrons will not be able to balance the force of gravity and the object would continue contracting. Then, it would appear that dead stars above Chandrasekhar's mass limit would be candidates for black holes. However, in 1932 Chadwick discovered the neutron and astronomers like Fritz Zwicky started to especulate on the possibility that neutron stars could exist. Perhaps these neutron stars could be the endpoint for objects above Chandrasekhar's limit. In the next few decades, it became clear through the work of physicists like Oppenheimer, Landau, Wheeler and others, that during a supernova explosion a neutron core could form. The pressure of degenerate neutrons can stabilize this core. However, if the core has a mass in excess of $\sim 3\ M_\odot$, in analogy with Chandrasekhar's calculations, collapse will continue to a black hole.

How to search for these cosmic black holes? No intrinsic radiation was expected from them (Hawking radiation from black holes was discovered later and anyway it is too weak to be detected for the masses associated with astronomical black holes).

For a detailed description of the history of black holes, the remarkable popularization books by Thorne (1994) and Begelman & Rees (1996) are highly recommended.

6. X-ray binaries

In the early 1960's, the astrophysicists Shklovskii, Salpeter, and Zel'dovich conceived independently an idea that proved to be crucial in the future searches for black holes. If significant amounts of gas fell from a large distance into a black hole, the gas would be accelerated close to the speed of light before entering the black hole, transforming large amounts of gravitational energy into kinetic energy. If even a small fraction of this kinetic energy could be transformed into radiation, the surroundings of an accreting black hole could be strong emitters of radiation.

In those same years, X-ray astronomy was developing rapidly and it soon became clear that there were bright X-ray sources in our Galaxy (Giacconi *et al.* 1962). These sources could reach very high luminosities, $L_X \simeq 10^6\ L_\odot$. Complementary studies in the visible (Webster & Murdin 1972; Bolton 1972) showed that the X-ray emission was coming from binary stellar systems.

Unlike the Sun that is a single star, most of the stars ($\geq 75\%$) are born in multiple systems, preferentially in pairs (a binary system). The two stars that form binary systems can have separations that go from $10^4\ R_\odot$ to having their surfaces touching (these last binaries are called contact binaries). When the separation is large, the two stars evolve practically independently, as two single stars. However, when the two stars are near each other, they form a close binary system (defined as one where the mean separation between the stars is less than a few times the sum of the radii of the two stars), and mass transfer from one star to the other can take place. Since the evolution of a star depends on its mass, this transfer alters the evolution of the component stars, making it very different to what it would have been for separate objects.

The stars in a close binary system are so close to each other that there is no way at present to make an image of the region showing two separate sources (we lack sufficient angular resolution). Their binary nature is usually established by spectroscopic means.

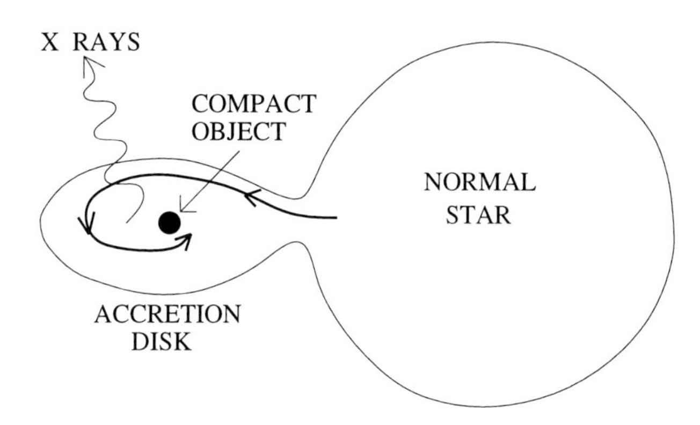

FIGURE 6. An X-ray binary is a double system with a "normal" star and a dead compact companion. Mass from the star can be transferred, via an accretion disk, to the compact companion. As the accreted gas spirals into the compact companion, viscous friction heats the disk to large temperatures, producing X ray emission.

The light from the stars is analyzed to search for spectral lines. These lines (that can appear in absorption or in emission, depending on the nature of the star) are produced by the atoms in the surface of the star. When the stars are in a close binary system, they translate in their orbits at velocities of the order of hundreds of km s^{-1}. Observing the spectrum of the system at different epochs one can detect the lines shifting sinusoidally in frequency with time as result of the Doppler effect (see Figure 8).

7. Accretion disks

In the case of the X-ray sources, only one star is observed, orbiting around an invisible companion. Furthermore, the visible star, emitting most of its radiation in the visible and UV, cannot be the source of the great X-ray luminosities observed. With the combined effort of several distinguished astrophysicists all these puzzles found an answer in the model for an X-ray binary system in which one of the stars was dead, transformed in a compact object (either a neutron star or a black hole). Mass from the "normal" star was being transferred to the compact companion. However, this transfer was not taking place radially, but via an accretion disk rotating around the compact object (see Figure 6).

On its way to the surface of a spherical compact object with mass M_c and radius R_c, accreting gas with mass m will gain energy given by $\Delta E_{acc} = GM_c m/R_c$. If the compact object is a neutron star with $R_c = 10$ km and $M_c = 1.4\ M_\odot$, we find that $\Delta E_{acc} \simeq 0.2\, mc^2$ (this calculation does not take into account that the radiation will suffer gravitational redshift, so that the actual luminosity detected by an external observer will be smaller). For comparison, the thermonuclear burning of hydrogen into helium yields $\Delta E_{nuc} = 0.007\ mc^2$, where m is the mass involved in the process. Then, accretion power is very important in the astrophysical context (Frank, King, & Raine 1992).

In the case of a black hole if the accretion were radial there would be very little radiation produced, with the accelerated gas disappearing quietly into the event horizon (i. e. the Schwarzschild's radius for a non-rotating black hole). The presence of the accretion disk

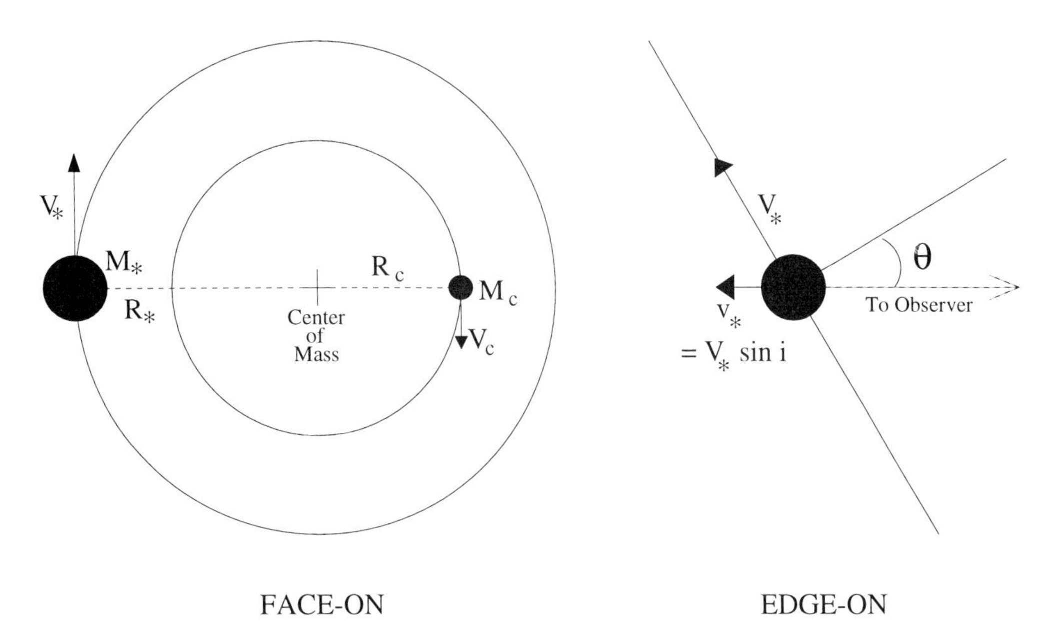

FIGURE 7. The diagram depicts a binary system with the two stars moving in circular orbits around the center of mass. On the left the orbital plane is presented face-on and on the right it is seen edge-on. The parameters of the stars (see text) are shown in the figure. In the case of X ray binaries one can obtain information on the normal star, and from that set a lower limit to the mass of the compact companion.

changes this completely, with the disk acting as an intermediator between the normal star and the black hole. As noted before, the energy gained during gravitational fall to the surface of a spherical compact object is $\Delta E_{acc} = GM_c m/R_c$. However, the kinetic energy of a mass m in Keplerian orbit near the surface of the object is (neglecting relativity) $GM_c m/2R_c$, exactly one half of the energy gained during the gravitational fall. Where has the other half of the energy gone? The answer is that it must have been lost in heating the accretion disk. Indeed, as the falling gas spirals down to the black hole, viscous friction between adjacent "rings" in the disk produces heating. It is relevant to remember that as the disk is believed to be approximately in Keplerian rotation, the gas in the inner orbits moves faster than that in the outer orbits and friction appears. This heated accretion disk is the one that produces the X-ray radiation seen in binaries with a compact companion (see Figure 6).

8. Determining the mass of the compact companion

Clearly, we want to determine the mass of these compact objects in binary systems. If the mass exceeds 3 $M_\odot$, a claim for a black hole can be made. Interestingly, these crucial mass determinations use an ancient astronomical technique, based on newtonian mechanics.

For simplicity we consider that the two stars are in circular orbits around the center of mass (CM) of the binary system. The visible star has mass, orbital velocity, and orbital radius given by $M_\star$, $V_\star$, and $R_\star$, while for the compact object these parameters

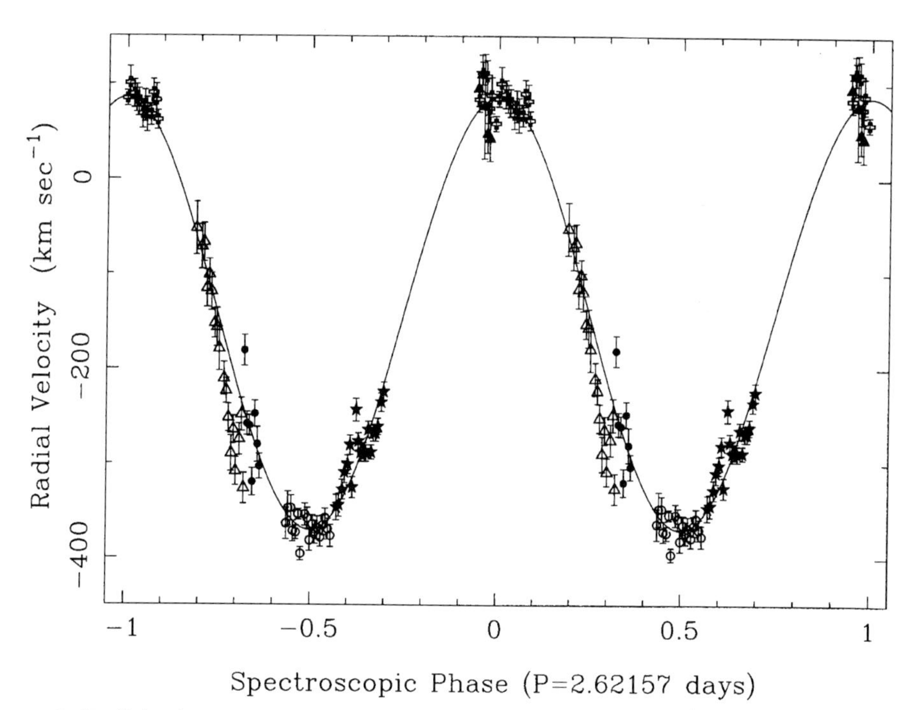

FIGURE 8. Radial velocities of the normal star in the X ray binary GRO J1655-40 as a function of time. The best fitting sinusoid is also shown. Credit: J. A. Orosz and C. D. Bailyn (Yale University).

are M_c, V_c, and R_c (see Figure 7). Kepler's law relates the period of the system with the separation of the stars by

$$\frac{G(M_\star + M_c)P^2}{(2\pi)^2} = (R_\star + R_c)^3, \tag{8.2}$$

where P is the period of the binary. The parameters of the stars are related by

$$\frac{M_\star}{M_c} = \frac{R_c}{R_\star} = \frac{V_c}{V_\star} \tag{8.3}$$

Finally, since the plane of the orbits is not necessarily parallel to the line of sight, the (observed) maximum radial velocity, $v_\star$, will be related to the true orbital velocity, $V_\star$, by

$$v_\star = V_\star \; sin \; i, \tag{8.4}$$

where i is the orbital inclination angle (the angle between the line of sight and a perpendicular to the plane of the orbits; see Figure 7). From the three formulae above, one can derive the so-called mass function:

$$f(M) = \frac{M_c^3 sin^3 i}{(M_\star + M_c)^2} = \frac{P v_\star^3}{2\pi G} \tag{8.5}$$

From spectroscopic observations of the visible star, we can find the right hand term in the mass function. Furthermore, from the spectrum of the visible star we can determine

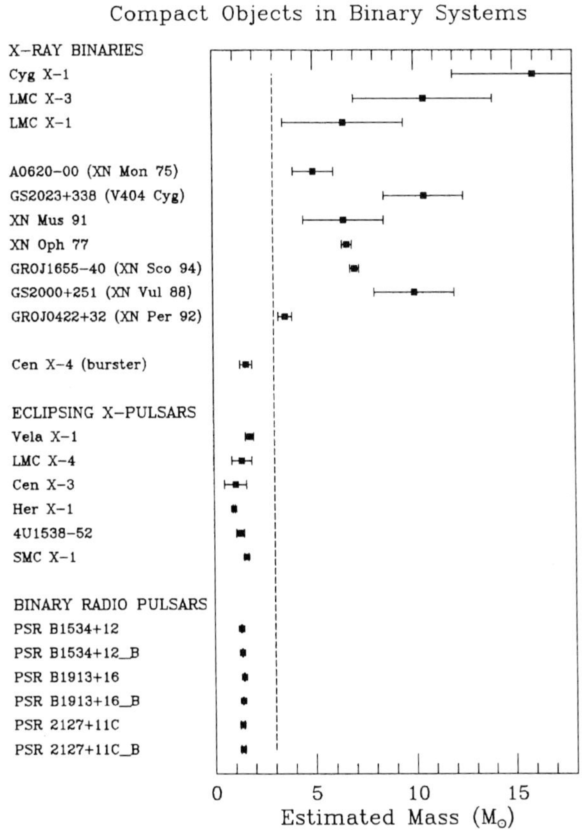

Compact Objects in Binary Systems

X-RAY BINARIES
Cyg X-1
LMC X-3
LMC X-1

A0620-00 (XN Mon 75)
GS2023+338 (V404 Cyg)
XN Mus 91
XN Oph 77
GROJ1655-40 (XN Sco 94)
GS2000+251 (XN Vul 88)
GROJ0422+32 (XN Per 92)

Cen X-4 (burster)

ECLIPSING X-PULSARS
Vela X-1
LMC X-4
Cen X-3
Her X-1
4U1538-52
SMC X-1

BINARY RADIO PULSARS
PSR B1534+12
PSR B1534+12_B
PSR B1913+16
PSR B1913+16_B
PSR 2127+11C
PSR 2127+11C_B

0 5 10 15
Estimated Mass ($M_\odot$)

FIGURE 9. A summary of mass determinations for compact objects in binary systems. The vertical dashed line is the 3 $M_\odot$ limit, and objects to its right are black hole candidates. Objects to the left of the dashed line are neutron stars. Credit: R. Remillard (MIT).

its mass, $M_\star$. We then obtain a lower limit to M_c. Consider for example, the case of the X-ray binary GRO J1655-40 (Orosz & Bailyn 1997). This X ray binary system has $P =$ 2.62 days, $v_\star = 228$ km s^{-1} (see Figure 8), from which we find $f(M) = 3.2\ M_\odot$. Since it can be established from spectroscopy that $M_\star = 2.3\ M_\odot$, we obtain $M_c \geq\ 6\ M_\odot$. The conclusion is that the compact companion in this binary is a black hole, or more conservatively (as many astronomers like to put it), a black hole candidate. In the case of GRO J1655-40, Orosz & Bailyn (1997) model the visible and X ray emissions from the system and can estimate the orbital inclination angle i, producing a mass determination, $M_c = 7.0 \pm 0.2 M_\odot$.

Over the years, it has been possible to estimate the mass of several compact objects in binary systems. A summary of these mass determinations is given in Figure 9. The compact objects with mass in excess of 3 $M_\odot$ are black hole candidates.

9. Black holes in the nuclei of quasars and radio galaxies

Also in the decade of the 1960s, astronomers became convinced that something strange was going on in the central regions of galaxies, in particular, in those of "active" galaxies, such as the quasars and the radio galaxies. Enormous amounts of visible and UV radi-

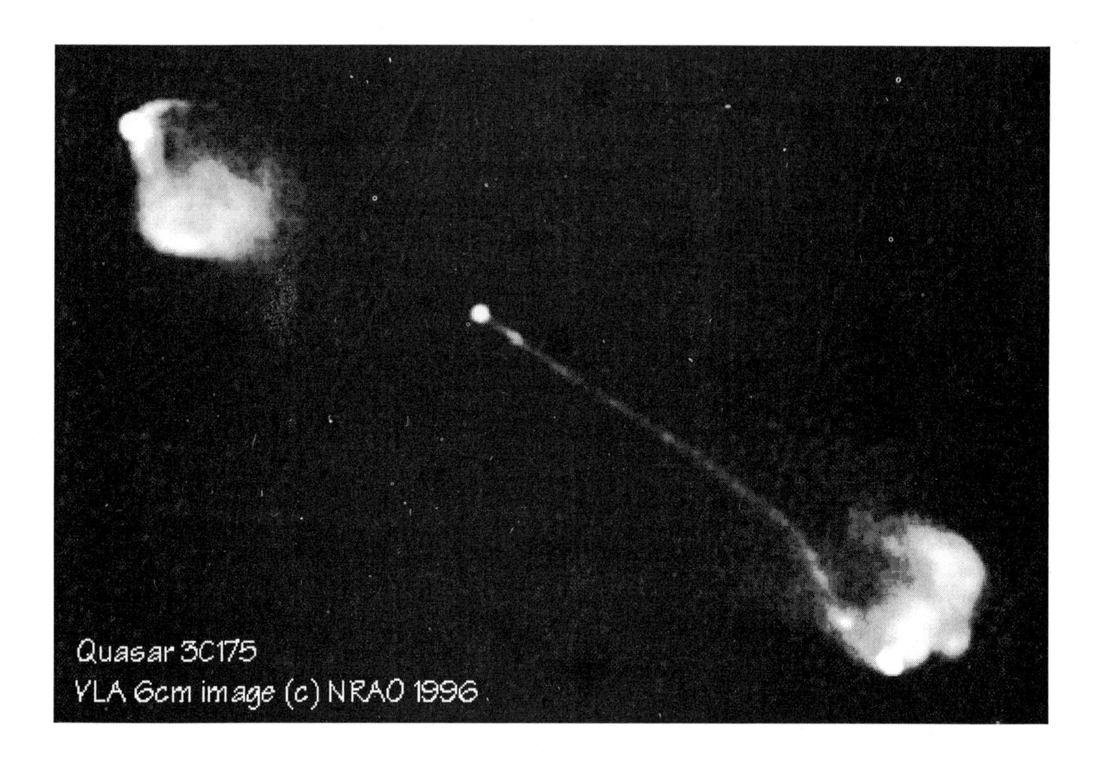

FIGURE 10. Radio image made at 6 cm wavelength of the quasar 3C175. The quasar is the bright spot located near the center of the image. The radio emission has an overall size (from lobe to lobe) of about 200 kpc, about 8 times the size of our Galaxy. The quasar is connected with one of the lobes by a narrow jet. The double lobes have prominent 'hot spots', bright regions where the energy of the jets from the quasar dissipates. Credit: A. Bridle (NRAO).

ation (reaching luminosities hundreds of times larger than those produced by ordinary galaxies) were produced in a small region (about the size of the Solar System) at the center of these extragalactic objects. In many of them thin synchrotron jets observable in the radio regime emanated from the core and traveled for Mpc in the intergalactic medium before ending in a bow shock larger than our Galaxy (see Figure 10).

As in the case of the X ray binaries, the large luminosities involved suggested the possibility of an accreting black hole. There are, however, important differences between a quasar and an X-ray binary. One difference, of course, is that the luminosities of quasars can be hundreds of millions of times that of X-ray binaries. This could be accounted for if the black hole in the quasars was supermassive, with masses in the range of 10^6–10^9 $M_\odot$. If we define $\dot{m}$ as the accretion rate (g s^{-1}) into an object, the accretion luminosity will be given by $L_{acc} = \eta G M_c \dot{m}/2R_c$, so the more massive the black hole, the more luminous a source is expected. In this last equation η is the efficiency of the process, that for an accretion disk around a black hole is estimated to be ~ 0.1. One could think that the luminosity of an accretion disk around a black hole could grow without limits, as long as $\dot{m}$ grows. There is, however, a limit for the luminosity of an object, called the Eddington limit, that cannot be exceeded because at greater luminosities the tremendous outward pressure of radiation will exceed the inward gravitational attraction and accretion would be halted. Eddington's luminosity is given by

$$\left[\frac{L_{Edd}}{10^5\ L_\odot}\right] = \frac{4\pi G M_c m_p c}{\sigma_T} \simeq 0.34 \left[\frac{M_c}{M_\odot}\right], \tag{9.6}$$

where $m_p = 1.7 \times 10^{-24}$ g is the mass of the proton, and $\sigma_T = 6.7 \times 10^{-25}\ cm^2$ is the Thomson cross section. Although the concept of the Eddington limit is valid only for spherical accretion, it can give a crude idea of the central mass required to counteract the pressure of radiation. For the luminosities of quasars ($10^{12}\ L_\odot$ and larger), the black hole masses required are in excess of $10^6\ M_\odot$. The other apparent major difference between quasars and X ray binaries is that while the former radiate mostly in the visible and UV, the latter do so in the X rays. Remarkably, this difference is expected from models of accretion disks, where the characteristic temperature of a disk accreting near the Eddington limit is proportional to $M_c^{-5/8}$ (Shakura & Sunyaev 1973). Then, for disks accreting near the Eddington limit, the more massive the black hole, the cooler the radiation emitted, consistent with the difference observed between quasars and X ray binaries.

I will present now recent results on three galactic nuclei, where supermassive black holes may exist: our own Galaxy, M87, and NGC 4258. A review on black holes models for active galactic nuclei has been given by Rees (1984).

9.1. *The nucleus of our galaxy*

The nucleus of our Galaxy seems like a bad place to search for a supermassive black hole. No great source of luminosity exists at its center (although a mysterious radio source that could mark the position of a black hole, Sgr A* does reside there, see Genzel & Townes 1987). However, we know that the surroundings of black holes become sources of radiation only if they are accreting considerable amounts of mass, so it can be argued that a quiescent black hole could be present at the center of the Milky Way. Under these conditions we have to detect the black hole by its gravitational influence on surrounding bodies.

Since the late 1970s several studies IR and radio studies (remember that because of dust absorption, the galactic center cannot be observed in the visible) indicated large velocity dispersions in a cloud of ionized gas that surrounds the galactic center (Wollman *et al.* 1976; Rodríguez & Chaisson 1979). Why did these large dispersions suggested the presence of a black hole? From infrared studies it was known that the mass density in stars at the center of the Galaxy goes approximately as $\rho \propto R_G^{-2}$, where R_G is the radius from the galactic center. Assuming virialization, the velocity dispersion of matter (gas and stars) inside a radius R_G will go as $\Delta v \propto R_G$. Then, if the stars were the only component of matter (the mass in gas is negligible) in the galactic center it was expected that the smaller the region observed, the smaller the velocity dispersion. However, the expected trend was observed only for $R_G \geq 3\ pc$. Inside this radius the dispersion increased with decreasing radius, pointing to the need of additional, dark matter. A black hole with a mass of a few million solar masses was proposed to exist at the center of the Milky Way to account for this anomalous increase in velocity dispersion.

However, gas can be accelerated relatively easy (in comparison with the much more compact and dense stars) and there was always the possibility that the observed large motions inside $R_G \simeq 3\ pc$ could be due to explosions or ejections of gas. Measuring the velocity dispersion of stars would give a more reliable estimate of the 'excess' mass, if any. Recent efforts by several groups (i. e. McGinn *et al.* 1989; Rieke & Rieke 1988; Lindqvist *et al.* 1992; Eckart & Genzel 1996) have produced measurements of the radial velocities (the motion along the line of sight measured by the Doppler effect) as well of the velocities in the plane of the sky (by measuring their proper motion and knowing the distance to the galactic center, 8.5 kpc) of stars in the galactic center. These two measurements give the complete velocity vector of the stars and allow a very reliable determination of their velocity dispersions, and thus of the gravitational potential in the region. The proper motion, (μ), of a star is its angular motion in the plane of the sky, that combined with the distance to the object, D, gives the transversal (that is, in the plane of the sky) velocity, v_t, since for small angles $v_t = \mu D$.

The most recent studies of Eckart & Genzel (1996; 1997) were based on 39 stars near the galactic center. They conclude that a central dark mass of $2.5 \pm 0.5 \times 10^6\ M_\odot$ coincides within $\leq$0.015 pc with Sgr A*. The stars that they investigate are as close as 0.04 pc from the galactic center. Even when these dimensions are relatively small, they still are $\sim 10^5$ times larger than the Schwarzschild radius of the suspected black hole. To compare with other astrophysical environments, these authors place a lower limit of $6.5 \times 10^9\ M_\odot\ pc^{-3}$ to the average mass density of the central region around Sgr A*. For example, the cores of stellar globular clusters have many stars in a small region and are one of the most dense astrophysical environments, with densities $\leq\ 10^5\ M_\odot\ pc^{-3}$. Clearly, the density derived for the surroundings of Sgr A* is very large and it seems difficult that something else that a central black hole could account for it. Other possibilities, such as the existence of a cluster of black holes with smaller masses cannot be ruled out, but they appear to be even more exotic than the hypothesis of a single massive black hole.

9.2. *M87*

M87 is a nearby giant elliptical galaxy located at a distance of about 15 Mpc. It is the host of the powerful radio galaxy Virgo A. A remarkable jet that is detectable from the radio to the X rays emanates from its center. Two decades ago (Sargent *et al.* 1978) proposed the existence of a supermassive black hole with mass of $5 \times 10^9\ M_\odot$ on the basis of the increase in the dispersion of stellar radial velocities at the nucleus of this galaxy. Of course, in contrast with our galactic center where the motions of individual stars can be analyzed, in the case of M87 it was the combined light of millions of stars the one being studied. Recent results obtained with the Hubble Space Telescope (Ford *et al.* 1994; Harms *et al.* 1994) reveal the presence at the center of M87 of a disk of ionized gas with radius of about 18 pc rotating at a velocity of $\sim$500 $km\ s^{-1}$. The mass of the disk itself is small, $\sim 4 \times 10^3\ M_\odot$. However, assuming that the disk is in Keplerian rotation, an inner mass of $2.4 \pm 0.7 \times 10^9\ M_\odot$ is required. The visible stars inside a radius of 18 pc in the center of M87 can account only for a fraction of the required mass, and a supermassive black hole has been proposed to provide for the rest.

9.3. *NGC 4258*

The evidence for a black hole in this spiral galaxy located at a distance of 6.4 Mpc is widely considered as the more convincing case for supermassive black holes in the extragalactic realm.

Since two decades ago it has been known that galaxies can emit powerfully in the radio lines of the molecules of hydroxyl (OH) at 18 cm wavelength (Baan *et al.* 1982) and water vapor (H_2O) at 1.3 cm wavelength (Dos Santos & Lepine 1979). Since these lines originate from levels whose populations are inverted, they emit as masers, and thus their intense brightness. The same maser emissions have been detected in countless star-forming regions in our Galaxy (Reid & Moran 1988). The most powerful of the extragalactic masers, however, are much more luminous than the galactic counterparts and are called megamasers.

The H_2O megamaser in NGC 4258 was detected by Claussen *et al.* (1984). The emis-

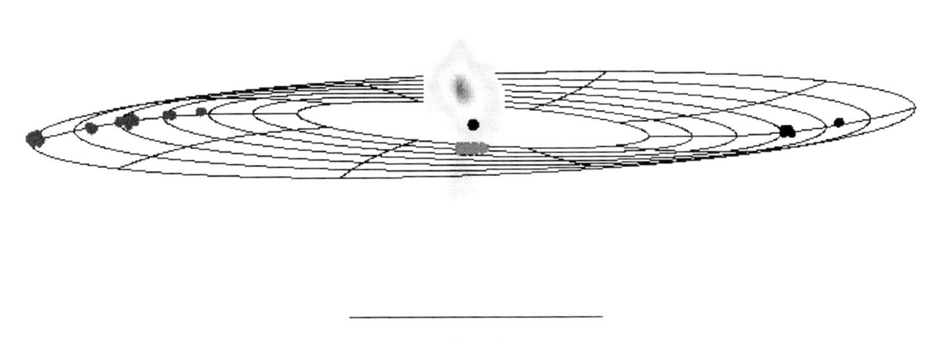

FIGURE 11. This graphic shows the geometric relationships between the H_2O maser emission (redshifted emission to the left, systemic velocity at the center, and blueshifted velocity to the right) in NGC 4258. These maser spots (marked as black dots in the graphic) are interpreted to be part of a warped disk in Keplerian rotation around a 3.7×10^7 $M_\odot$ black hole. In addition, at the center it is shown in grey tones the synchrotron jet that is believed to emanate from the surroundings of the black hole. Credit: J. Herrnstein *et al.* (Harvard-Smithsonian Center for Astrophysics) and NRAO.

sion was found, as in other megamasers, to originate from the center of that galaxy and to have velocities similar to the average velocity of the galaxy (the systemic velocity). While testing a very wide band spectrometer for the Nobeyama radio telescope, Nakai *et al.* (1993) found extremely high-velocity features symmetrically offset from the systemic radial velocity by $\sim$1000 km s^{-1}. In other words, there were three distinct velocity groups of H_2O maser features in NGC 4258. One set (the brighter one) was at the systemic velocity of the galaxy and the other two at $\pm$1000 km s^{-1} from the systemic velocity. These high velocity lines have not been detected previously because the spectrometers then available were not wide enough.

The mapping of the H_2O maser emission using the powerful technique of Very Long Baseline Interferometry (Miyoshi *et al.* 1995) revealed that the three velocity groups arise from different regions in space (see Figure 11). Their velocities and positions in the sky of the maser spots can be modeled accurately in terms of a warped disk in Keplerian rotation around a black hole with mass of 3.7×10^7 $M_\odot$. The explanation of why the maser emission is seen preferentially from three zones in the disk (and not from all the disk) is very well understood in terms of cosmic maser radiation transfer. At the center of the disk a synchrotron jet approximately perpendicular to the disk has been detected (Herrstein *et al.* 1997). The inner edge of this molecular disk is found to have a radius of $\sim$0.13 pc and the density inside this radius is $\geq 4 \times 10^9$ $M_\odot$ pc^{-3}.

In Table 1 we summarize the parameters of the suspected black holes at the center of the three galaxies discussed. In this Table the minimum size refers to the smallest scales resolved by the observations (in all cases many orders of magnitude larger than the Schwarzschild radius) and the mass density is the average value estimated inside that minimum size.

Galaxy	Distance (Mpc)	Black Hole Mass ($M_\odot$)	Minimum Size (pc)	Density ($M_\odot\ pc^{-3}$)
Milky Way	0.0085	2.5×10^6	0.04	$\geq 6.5 \times 10^9$
M87	15	2.4×10^9	18	$\geq 1.0 \times 10^5$
NGC 4258	6.4	3.6×10^7	0.13	$\geq 4.0 \times 10^9$

TABLE 1. Parameters of suspected black holes at the nucleus of selected galaxies

10. Microquasars in our Galaxy

In our review of the characteristics of the galactic X ray binaries and the active galactic nuclei, we noted that two key elements were believed to be present in both objects: a black hole and an accretion disk around it. A third element, the presence of relativistic, radio-emitting synchrotron jets was quite common in active galactic nuclei. However, until recently there was only one solid, well-studied case of relativistic jets in an X ray binary: the remarkable source SS 433.

SS 433 was first catalogued in a survey as one of many emission-line objects with strong Hα in our Galaxy (Stephenson & Sanduleak 1977). However, in 1979 it was found that the emission lines drifted across the spectrum with a period of about 164 days (Margon *et al.* 1979; Mammano *et al.* 1980). The variable Doppler shifts achieved very large magnitudes, of order 50,000 km s^{-1}. Two of the earliest theoretical papers (Fabian & Rees 1979; Milgrom 1979) contain the acccepted basic elements of the kinematical model of SS 433, that we describe in what follows. SS 433 is an X ray binary system (the X rays were detected later). The compact component (it is still unclear if it is a neutron star or a black hole) is surrounded by an accretion disk. Perpendicular to this disk emanate two collimated, oppositely aligned jets. The ejection velocity is 0.26c and the axis of the jets precesses with a period of 164 days. Is is the gas in the precessing jets the one that produces the drifting lines in the optical spectrum. The high angular resolution radio studies of this source (Hjellming & Johnston 1981) were able to actually follow the corkscrew motions of the ejecta and fully confirmed the kinematical model.

The discovery of SS 433 clearly indicated that the relativistic ejections observed in remote active galactic nuclei and in quasars could also be present in nearby X ray binaries. However, no other similar case was detected for more than a decade. Margon (1984) has reviewed the early years in the history of SS 433.

In the early 1990s, with the advent of the french-russian space telescope SIGMA-GRANAT for the hard X rays, it became possible to localize these sources with a positional accuracy of one arc minute. Radio observations of these regions soon revealed the presence of double jet sources associated with the hard X ray binary under study (Mirabel *et al.* 1992; Rodríguez *et al.* 1992). A contour map of the radio continuum emission from 1E1740.7-2942 (Mirabel *et al.* 1992) is shown in Figure 12. However, these jets are of relatively large dimensions and do not appear to show significant changes over periods of a few years. Finally, in 1994 the actual ejection of matter at relativistic speeds was observed in repeated ocassions by Mirabel & Rodríguez (1994) in the remarkable hard X ray binary GRS 1915+105. Furthermore, the plasma clouds ejected by this source appeared to move in the plane of the sky at speeds in excess of that of light, the so-called superluminal motions, first observed in quasars in the early 1970s (Whitney *et al.* 1971; Cohen *et al.* 1971).

The source GRS 1915+105 was discovered by the satellite GRANAT in 1992 (Castro-Tirado *et al.* 1994). It is located in the galactic plane and from studies of the λ21 cm

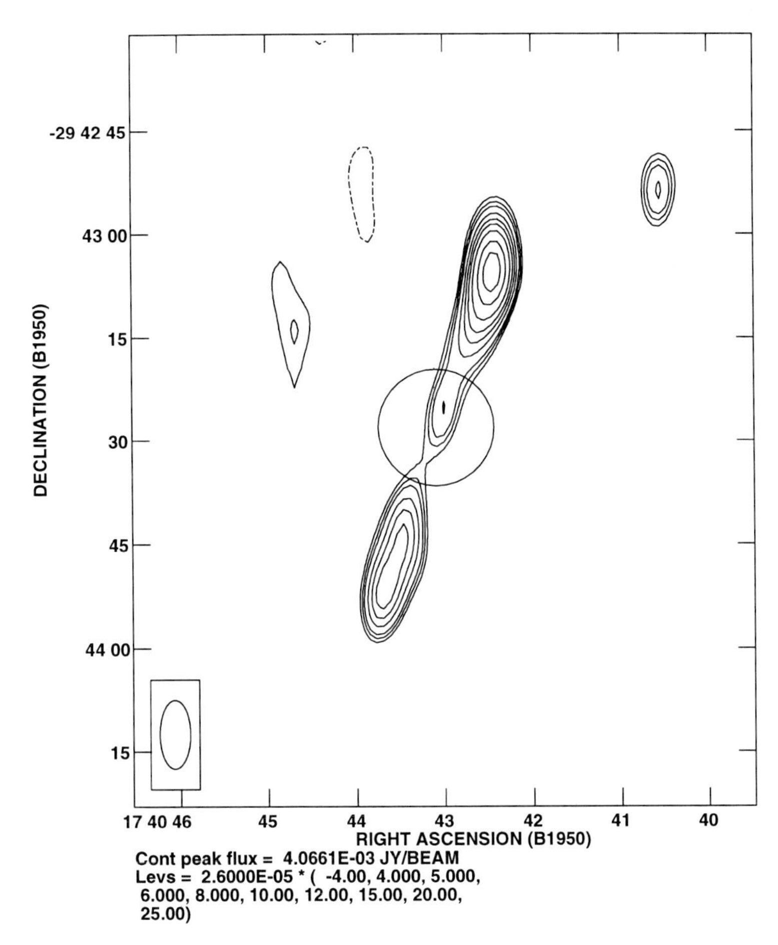

FIGURE 12. Contour map of the radio continuum emission at 6 cm wavelength from the hard X ray binary 1E1740.7-2942 (Mirabel *et al.* 1992). Two synchrotron-emitting jets emanate from the central source in opposite directions. The jets extend over a scale of 1 pc. The circle marks the position of the X-ray binary obtained by the satellite ROSAT.

line of HI in absorption a kinematic distance of 12.5±1.5 kpc is derived (Rodríguez *et al.* 1995). Given this large distance, the source is heavily obscured by dust and has been detected only in the radio, infrared, X rays, and γ rays. Radio observations made with the Very Large Array by Mirabel & Rodríguez (1994) in early 1994 revealed the ejection of radio-emitting clouds from the expected position of the binary star.

Figure 13 shows the time evolution of a pair of bright radio clouds emerging in opposite directions from the compact, variable core (the location of the binary system). Knowing the proper motions of the clouds and its distance, it was possible to determine that the apparent velocities in the plane of the sky were 1.25c and 0.65c for the clouds in the left and right side of Figure 13. The cloud on the left side not only appeared to be moving faster than the other cloud, but was also brighter. These apparently puzzling characteristics find a straightforward interpretation in terms of the "illusions" that appear

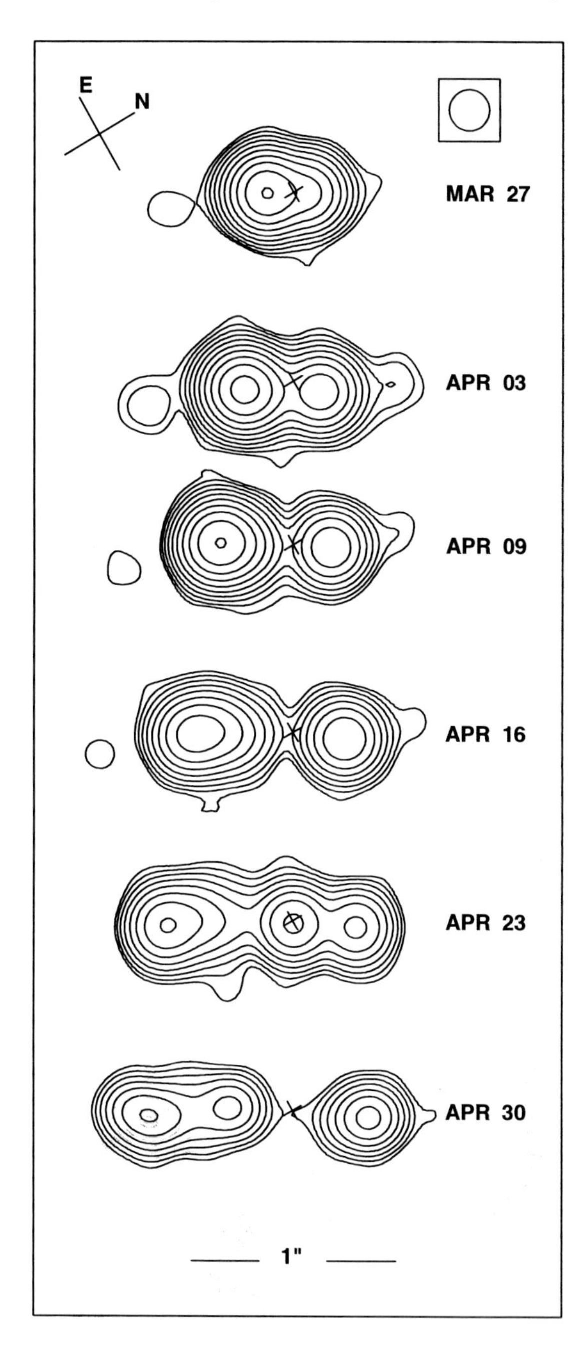

FIGURE 13. Contour map of the pair of radio condensations moving away from the hard X-ray source GRS 1915+105. These Very Large Array maps were made at λ3.6-cm for the 1994 epochs on the right side of the figure. Contours are 1,2,4,8,32,64,128,256 and 512 times 0.2 mJy/beam for all epochs except for March 27 where the contour levels are in units of 0.6 mJy/beam. The half power beam width of the observations, 0.2 arc sec, is shown in the top left corner. The position of the stationary core (where the X ray binary is located) is indicated with a small cross. The maps have been rotated 60° clockwise for easier display and comparison. Note in the first four epochs the presence of a fainter pair of condensations moving ahead the bright ones and in the last epoch the presence of a new southern component. Note also that on April 23 the core is detectable, and then turns off in the following epoch.

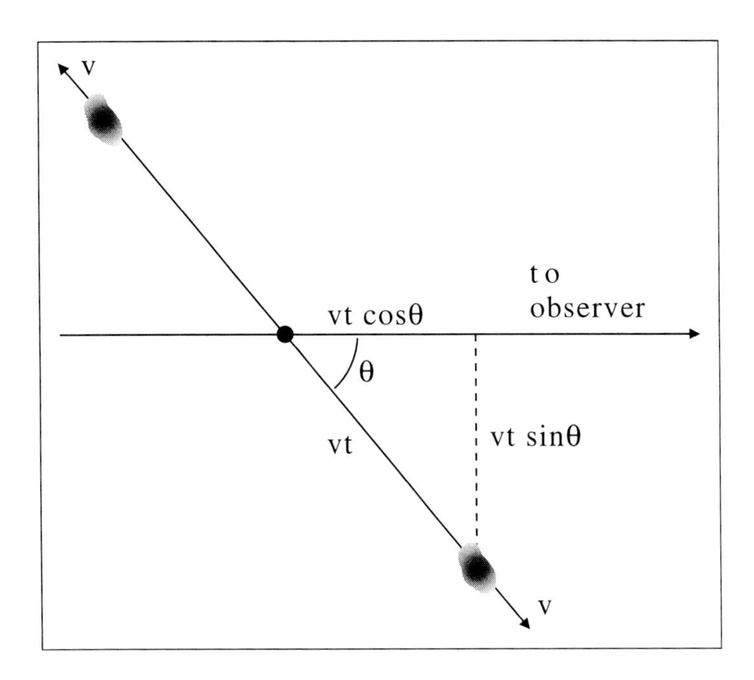

FIGURE 14. Diagram of a bipolar ejection.

while observing relativistically moving objects, first discussed in the extragalactic context by Martin Rees.

Consider an object that ejects to the surrounding medium twin, oppositely directed clouds (see Figure 14), what astronomers use to call a bipolar ejection. The relativistic illusions appear in the observation of ejecta moving close to the speed of light. The ejecta is moving so fast that it nearly catches up with its own electromagnetic radiation. After a time t from the moment of ejection, the condensations, with a true velocity v, have moved a distance vt. As seen in projection by the observer, the displacement seems to be $vt \ sin \ \theta$, where θ is the angle between the line-of-sight and the axis of ejection. However, since the approaching condensation is now closer to the observer by a distance $vt \ cos \ \theta$, the time t' in which the observer sees the condensation move from the origin to its present position is smaller than t and is given by $t' = t - (vt \ cos \ \theta/c)$. The apparent velocity of the approaching condensation in the plane of the sky is then $v_a = v \ sin \ \theta/(1-(v \ cos \ \theta/c))$, that can exceed c. By a similar reasoning, the apparent velocity of the receding condensation in the plane of the sky is then $v_r = v \ sin \ \theta/(1+(v \ cos \ \theta/c))$, always less than c. With the two equations for the velocities in the plane of the sky for the approaching and receding condensations, it is possible to solve for the two unknowns of the bipolar ejection, v and θ. For GRS 1915+105, v = 0.92c and $\theta = 70°$.

Additionally, for an object moving at relativistic speeds, the emitted radiation "focuses" in the direction of motion (the relativistic beaming), an effect that makes the approaching condensation look brighter than the receding condensation. In remote objects, such as quasars, the approaching ejecta can be observed only when their apparent brightness is very highly enhanced due to beaming to the observer. This Doppler-favoritism for the approaching jet implies the opposite effect for the receding jet, rendering it undetectable in practice. In the case of GRS 1915+105 this favoritism is not as extreme and this circumstance allowed for the first time the detection of relativistic receding ejecta (the cloud in the right side of Figure 13). The relations between the parameters of the approaching and receding clouds are given by

$$\frac{\mu_a}{\mu_r} = \frac{\nu_a}{\nu_r} = \frac{L_a}{L_r} = \frac{(1+\beta\ cos\ \theta)}{(1-\beta\ cos\ \theta)}, \quad and \tag{10.7}$$

$$\frac{S_a}{S_r} = \left(\frac{1+\beta\ cos\ \theta}{1-\beta\ cos\ \theta}\right)^{\kappa-\alpha}, \tag{10.8}$$

where μ is the proper motion of the clouds, ν is the observed frequency of a line emitted by the moving clouds, L is the observed length of the condensation, S is the observed flux density, κ is a constant equal to 2 or 3 according if the ejecta is continuous or formed by discrete clouds, α is the spectral index of the observed radiation, defined by $S_\nu \propto \nu^\alpha$ and measured to be -0.8 for the GRS 1915+105 ejecta, and $\beta = v/c$. The subscripts a and r refer to the approaching and receding components. The relative brightness of the approaching and receding condensations in GRS 1915+105 are in good agreement with the expectations of special relativity.

It is also interesting to point out that relativistic bipolar ejections can be used to determine the distance to the source, if in addition to the proper motions one could detect a line from one of the moving clouds. If one measures the proper motions of the approaching and receding ejecta, μ_a and μ_r, two independent equations are obtained:

$$\mu_{a,r} = \frac{v\ sin\ \theta}{(1 \mp (v/c)\ cos\ \theta)\ D}, \tag{10.9}$$

where D is the distance to the source.

Measuring the Doppler factor $\delta_{a,r} = \nu_{a,b}/\nu_{rest}$ of spectral lines arising in either the approaching or receding jets, a third equation can be chosen from:

$$\delta_{a,r} = \gamma^{-1}(1 \mp (v/c)\ cos\ \theta)^{-1}, \tag{10.10}$$

where $\gamma = [1-(v/c)^2]^{-1/2}$. One can then resolve the system of 3 equations and find the three unknowns: v, θ, and D, the distance to the source. Unfortunately, despite intensive efforts, no line emission has been detected in the ejecta from GRS 1915+105 to check if the distance obtained by this independent method agrees with that obtained from HI absorption.

A very exciting recent development in the study of the superluminal source GRS 1915+105 has been the detection of fast quasi-periodic oscillations (QPOs) of the X-ray power with a maximum stable frequency of 67 Hz (Morgan *et al.* 1997). One possible interpretation is that this frequency corresponds to phenomena taking place at the last stable circular orbit around the black hole, at 3 times the Schwarzschild radius for a non-rotating black hole. Gas located at distances below the last stable circular orbit will fall almost radially into the black hole. The period of rotation in the last stable circular orbit, as seen by an external observer, is

$$P_{last} = 2\pi \left(\frac{6}{c^2}\right)^{3/2} GM_{BH}, \tag{10.11}$$

so for the period of the QPOs observed in GRS 1915+105 (0.015 seconds), a blach hole mass of $M_{BH} = 33\ M_\odot$ is implied.

In a Kerr (rotating) black hole this frequency depends on the black hole's mass and spin, as well as on the rotation direction of the accretion disk. This offers the prospect of inferring the spin of black holes with masses independently determined. In any case, it appears firmly established that the fast QPOs detected in the X rays originate very close to the horizon of the black hole.

Since the detection of the superluminal motions in GRS 1915+105, a second superluminal source has been found, GRO J1655-40 (Tingay *et al.* 1995; Hjellming & Rupen 1995). At present, there are eight sources of relativistic jets known in the Galaxy. In four of them (SS 433, GRS 1915+105, GRO J1655-40, and Cygnus X-3) the velocity of the ejecta on the plane of the sky has been measured. These galactic sources, that mimic on a smaller scale the remarkable relativistic phenomena observed in distant quasars are called microquasars. Their discovery sets on a firmer basis the study of relativistic ejections seen elsewhere in the universe, and opens new perspectives for understanding the physics of strong-field relativistic gravity near the event horizon of black holes.

11. Challenges for the future

I will end this chapter by listing a series of challenges to consider and solve in the future. Despite great observational and theoretical advance, there is a long way to go.

• More powerful models for the acceleration and collimation of jets are needed. At present there is a consensus that a magnetized accretion disk provides the acceleration and possibly the collimation of the relativistic jets observed in active galactic nuclei and X ray binaries, and also of the "classic" ($v \leq c$) jets observed in association with very young stars (Rodríguez 1997). The available models, in particular the classic contribution of Blandford & Payne (1982), suggest that a magnetohydrodynamically accelerated and focused flow can be established but cannot predict many of the detailed features (i. e. why are the terminal jet velocities of the same order of the escape velocity from the gravitational well of the central object? What is the mass ratio between accreted and ejected gas?)

• The microquasars as laboratories of gravity in the strong-field regime. As mentioned before, the fast quasiperiodic variations observed in the X rays from GRS 1915+105 are probably originating close to the horizon of the black hole and perhaps could be used in the future to test the physics of accretion disks and the metric of gravity in the strong-field regime.

• The use of relativistic two-sided ejection to determine distances. This is an observational problem. The main difficulty resides in the detection of lines that are strongly Doppler broadened since they arise from plasma clouds that not only move but also expand at relativistic speeds. However, distance determinations are a key problem in astronomy and the method should be pursued first in galactic and later in extragalactic sources.

• Where are the intermediate mass black holes? Black holes in X ray binaries are believed to have stellar masses, while those in active galactic nuclei have masses in excess of $10^6\ M_{\odot}$. Some of the models for the formation of supermassive black holes work up from smaller black holes, and there is no a priori reason for not having intermediate mass black holes.

• An unequivocal signature for a black hole. Even with all the evidence in favor of black holes, many skeptical astronomers are not fully convinced of their existence and would like to see an unambiguous indicator of its existence. It is possible that in the coming century the study of gravitational waves from merging compact objects could provide this unequivocal signature.

I thank I. F. Mirabel for useful discussion and R. Remillard for providing the figure summarizing mass determinations for compact objects in binary systems. L. F. R. acknowledges the continued support of CONACyT, México and DGAPA, UNAM.

REFERENCES

BAAN, W. A., WOOD, P. A. D., & HASCHICK, A. D. 1982 Broad hydroxyl emission in IC 4553 *The Astrophysical Journal (Letters)* **260**. L49–L52.

BEGELMAN, M. C., & REES, M. J. 1995 *Gravity's Fatal Attraction: Black Holes in the Universe.* Scientific American Library.

BLANDFORD, R. D. & PAYNE, D. G. 1982 Hydromagnetic flows from accretion discs and the production of radio jets *Monthly Notices of the Royal Astroomical Society* **199**, 883–903.

BOLTON, C. T. 1972 Dimensions of the Binary System HDE 226868: Cygnus X-1 *Nature: Physical Sciences* **240**, 124-127.

CASTRO-TIRADO, A. J., BRANDT, S., LUND, N., LAPSHOV, I., SUNYAEV, R. A., SHLYAPNIKOV, A. A., GUZIY, S., & PAVLENKO, E. P. 1994 Discovery and observations by watch of the X-ray transient GRS 1915+105 *The Astrophysical Journal (Supplements)* **92**, 469–472.

CHANDRASEKHAR, S. 1931 The Maximum Mass of Ideal White Dwarfs *The Astrophysical Journal* **74**, 81–82.

CLAUSSEN, M. J., HEILIGMAN, G. M., & LO, K. Y. 1984 Water-vapor maser emission from galactic nuclei *Nature* **310**, 298–300.

COHEN, M. H., CANNON, W., PURCELL, G. H., SHAFFER, D. B., BRODERICK, J. J., KELLERMANN, K. I., & JAUNCEY, D. L. 1971 The small-scale structure of radio galaxies and quasi-stellar radio sources at 3.8 centimeters *The Astrophysical Journal* **170**, 207–217.

DOS SANTOS, P. M., & LEPINE, J. R. D. 1979 Detection of strong H_2O emission from the galaxy NGC4945 *Nature* **278**, 34–36.

ECKART, A., & GENZEL, R. 1996 Observations of stellar proper motions near the Galactic Centre *Nature* **383**, 415–417.

ECKART, A., & GENZEL, R. 1997 Stellar proper motions in the central 0.1 pc of the Galaxy *Monthly Notices of the Royal Astronomical Society* **284**, 576–598.

FABIAN, A. C., & REES, M. J. 1979 SS 433 - A double jet in action *Monthly Notices of the Royal Astronomical Society* **187**, 13P–16P.

FORD, H. C., HARMS, R. J., TSVETANOV, Z. I., HARTIG, G. F., DRESSEL, L. L., KRISS, G. A., BOHLIN, R., DAVIDSEN, A. F., MARGON, B., KOCHHAR, A. K. 1994 Narrowband HST images of M87: Evidence for a disk of ionized gas around a massive black hole *The Astrophysical Journal (Letters)* **435**, L27–L30.

FRANK, J., KING, A., & RAINE, D. 1992 *Accretion Power in Astrophysics.* Cambridge University Press.

GENZEL, R. & TOWNES, C. H. 1987 Physical conditions, dynamics, and mass distribution in the center of the Galaxy *Annual Reviews of Astronomy & Astrophysics* **25**, 377–423.

GIACCONI, R., GURSKY, H., PAOLINI, F. R., & ROSSI, B. B. 1962 Evidence for X-Rays from Sources outside the Solar System *Physical Review Letters* **9**, 439–443.

HARMS, R. J., FORD, H. C., TSVETANOV, Z. I., HARTIG, G. F., DRESSEL, L. L., KRISS, G. A., BOHLIN, R., DAVIDSEN, A. F., MARGON, B., KOCHHAR, A. K. 1994 HST FOS spectroscopy of M87: Evidence for a disk of ionized gas around a massive black hole *The Astrophysical Journal (Letters* **435**, L35–L38.

HERRNSTEIN, J. R., MORAN, J. M., GREENHILL, L. J., DIAMOND, P. J., MIYOSHI, M., NAKAI, N., & INOUE, M. 1997 Discovery of a Subparsec Jet 4000 Schwarzschild Radii Away from the Central Engine of NGC 4258 *The Astrophysical Journal (Letters* **482**, L17–L20.

HEWISH, A., BELL, S. J., PILKINGTON, J. D. H., SCOTT, P. F., & COLLINS, R. A. 1968 Observation of a Rapidly Pulsating Radio Source *Nature* **217**, 709–713.

HIRATA, K., KAJITA, T., KOSHIBA, M., NAKAHATA, M. & OYAMA, Y. 1987 Observation of a neutrino burst from the supernova SN1987 A *Physical Review Letters* **58**, 1490–1493.

HJELLMING, R. M. & JOHNSTON, K. J. 1981 An analysis of the proper motions of the SS 433 radio jets *The Astrophysical Journal (Letters)* **246**, L141–L145.

HJELLMING, R. M. & RUPEN, M. P. 1995 Episodic Ejection of Relativistic Jets by the X-ray Transient GRO J1655-40 *Nature* **375**, 464–466.

LINDQVIST, M., HABING, H., & WINNBERG, A. 1992 OH/IR stars close to the Galactic Centre. II - Their spatial and kinematic properties and the mass distribution within 5-100 pc from the galactic centre *Astronomy & Astrophysics* **259**, 118–127.

MAMMANO, A., CIATTI, F., & VITTONE, A. 1980 The unique spectrum of SS 433, a star inside a supernova remnant *Astronomy & Astrophysics* **85**, 14–19.

MARGON, B., STONE, R. P. S., KLEMOLA, A., FORD, H. C., KATZ, J. I., KWITTER, K. B., & ULRICH, R. K. 1979 The bizarre spectrum of SS 433 *The Astrophysical Journal (Letters)* **230**, L41–L45.

MARGON, B. 1984 Observations of SS 433 *Annual Reviews of Astronomy & Astrophysics* **22**, 507–536.

MCGINN, M. T., SELLGREN, K., BECKLIN, E. E., & HALL, D. N. B. 1989 Stellar kinematics in the Galactic center *The Astrophysical Journal* **338**, 824–840.

MILGROM, M. 1979 On the interpretation of the large variations in the line positions in SS 433 *Astronomy & Astrophysics* **76**, L3–L6.

MIRABEL, I. F., RODRÍGUEZ, L. F., CORDIER, B., PAUL, J., & LEBRUN, F. 1992 A Double-sided Radio Jet from The Compact Galactic Centre Annihilator 1E1740.7-2942 *Nature* **358**, 215–217.

MIRABEL, I. F. & RODRÍGUEZ, L. F. 1994 A Superluminal Source in the Galaxy *Nature* **371**, 46–48.

MIYOSHI, M., MORAN, J. M., HERRNSTEIN, J. R., GREENHILL, L. J., NAKAI, N., DIAMOND, P. J., & INOUE, M. 1995 Evidence for a black-hole from high rotation velocities in a sub-parsec region of NGC 4258 *Nature* **373**, 127–129.

MORGAN, E. H., REMILLARD, R. A., & GREINER, J. 1997 RXTE Observations of QPOs in the Black Hole Candidate GRS 1915+105 *The Astrophysical Journal* **482**, 993–1010.

NAKAI, N., INOUE, M., & MIYOSHI, M. 1993 Extremely-high-velocity H_2O maser emission in the galaxy NGC4258 *Nature* **361**, 45–47.

OROSZ, J. A. & BAILYN, C. D. 1997 Optical Observations of GRO J1655-40 in Quiescence. I. A Precise Mass for the Black Hole Primary *The Astrophysical Journal* **477**, 876–896.

PAGE, D. 1994 Geminga: A Cooling Superfluid Neutron Star *The Astrophysical Journal* **428**, 250–260.

REBOLO, R., MARTIN, E. L., BASRI, G., MARCY, G. W., & ZAPATERO-OSORIO, M. R. 1996 Brown Dwarfs in the Pleiades Cluster Confirmed by the Lithium Test *The Astrophysical Journal (Letters)* **469**, L53–L56.

REES, M. 1984 Black Hole Models for Active Galactic Nuclei *Annual Reviews of Astronomy & Astrophysics* **22**, 471–506.

REID, M. J. & MORAN, J. M 1988 Astronomical masers In *Galactic and extragalactic radio astronomy (2nd edition)* (ed. G. L. Verschuur & K. I. Kellermann) pp. 255-294. Spinger-Verlag.

RIEKE, G. H., & RIEKE, M. J. 1988 Stellar velocities and the mass distribution in the Galactic center *The Astrophysical Journal (Letters)* **330**, L33–L37.

RODRÍGUEZ, L. F. & CHAISSON, E. J. 1979 The temperature and dynamics of the ionized gas in the nucleus of our Galaxy *The Astrophysical Journal* **228**, 734–739.

RODRÍGUEZ, L. F., MIRABEL, I. F., & MARTÍ, J. 1992 The Radio Counterpart of the Hard X-Ray Source GRS 1758-258 *The Astrophysical Journal (Letters)* **401**, L15–L18.

RODRÍGUEZ, L. F., GERARD, E., MIRABEL, I. F., GÓMEZ, Y., & VELÁZQUEZ, A. 1995 Radio Monitoring of GRS 1915+105 *The Astrophysical Journal (Supplements)* **101**, 173–179.

RODRÍGUEZ, L. F. 1997 Thermal Radio Jets In *Herbig-Haro Flows and the Birth of Low Mass Stars* (ed. B. Reipurth & C. Bertout) pp. 83–92. Dordrecht-Kluwer.

RYBICKI, G. B. & LIGHTMAN, A. P. 1979 *Radiative Processes in Astrophysics.* Wiley-Interscience.

SALPETER, E. E. 1955 The Luminosity Function and Stellar Evolution *The Astrophysical Journal* **121**, 161–167.

SARGENT, W. L. W., YOUNG, P. J., LYNDS, C. R., BOKSENBERG, A., SHORTRIDGE, K., HARTWICK, F. D. A. 1978 Dynamical evidence for a central mass concentration in the galaxy M87 *The Astrophysical Journal* **221**, 731–744.

SHAKURA, N. I. & SUNYAEV, R. A. 1973 Black holes in binary systems. Observational appearance *Astronomy & Astrophysics* **24**, 337–355.

SHAPIRO, I. I., ASH, M. E., INGALLS, R. P., SMITH, W. B., CAMPBELL, D. B., DYCE, R. B., JURGENS, R. F. & PETTENGILL, G. H. 1971 Fourth test of general relativity: New radar result *Physical Review Letters* **26**, 1132–1135.

STEPHENSON, C. B. & SANDULEAK, N. 1977 New Hα emission stars in the Milky Way *The Astrophysical Journal (Supplements)* **33**, 459–469.

SVOBODA, R., BRATTON, C. B., CASPER, D., CIOCIO, A. & CLAUS, R. 1987 Neutrinos from SN1987A in the IMB detector. In *ESO Workshop on the SN1987 A* Proceedings (A88-35301 14-90). pp. 229–235. ESO.

TAYLOR, J. H. & WEISBERG, J. M. 1982 A new test of general relativity - Gravitational radiation and the binary pulsar PSR 1913+16 *The Astrophysical Journal* **253**, 908–920.

THORNE, K. S. 1994 *Black Holes & Time Warps: Einstein's Outrageous Legacy.* Norton.

TINGAY, S. J., JAUNCEY, D. L., PRESTON, R. A., REYNOLDS, J. E., MEIER, D. L., MURPHY, D. W., TZIOUMIS, A. K., MCKAY, D. J., KESTEVEN, M. J., LOVELL, J. E. J., CAMPBELL-WILSON, D. ELLINGSEN, S. P., GOUGH, R., HUNSTEAD, R. W., JONES, D. L., MCCULLOCH, P. M., MIGENES, V., QUICK, J., SINCLAIR, M. W., & SMITS, D. 1995 Relativistic Motion in a Nearby Bright X-ray Source *Nature* **374**, 141–143.

WEBSTER, B. L. & MURDIN, P 1972 Cygnus X-1: A Spectroscopic Binary with a Heavy Companion? *Nature* **235**, 37–38.

WHITNEY, A. R., SHAPIRO, I. I., ROGERS, A. A. E., ROBERTSON, D. S., KNIGHT, C. A., CLARK, T. A., GOLDSTEIN, R. M., MARINDINO, G. E., & VANDENBERG, N. R. 1971 Quasars Revisited: Rapid Time Variations Observed Via Very-Long-Baseline Interferometry *Science* **173**, 225–230.

WOLLMAN, E. R., GEBALLE, T. R., LACY, J. H., TOWNES, C. H., & RANK, D. M. 1976 Spectral and spatial resolution of the 12.8 micron Ne II emission from the galactic center. *The Astrophysical Journal (Letters* **205**, L5–L9.

Nucleosynthesis Basics and Applications to Supernovae

By F.-K. THIELEMANN[1,5]
T. RAUSCHER[1],
C. FREIBURGHAUS[1,5], K. NOMOTO[2,5],
M. HASHIMOTO[3],
B. PFEIFFER[4] AND K.-L. KRATZ[4]

[1]Departement für Physik und Astronomie, Universität Basel, CH-4056 Basel, Switzerland

[2]Department of Astronomy and Research Center for the Early Universe, University of Tokyo, Tokyo 113, Japan

[3]Department of Physics, Faculty of Science, Kyushu University, Fukuoka 810, Japan

[4]Institut für Kernchemie, Universität Mainz, D-55128 Mainz, Germany

[5]Institute for Theoretical Physics, University of California, Santa Barbara, CA 93106-4030

This review concentrates on nucleosynthesis processes in general and their applications to massive stars and supernovae. A brief initial introduction is given to the physics in astrophysical plasmas which governs composition changes. We present the basic equations for thermonuclear reaction rates and nuclear reaction networks. The required nuclear physics input for reaction rates is discussed, i.e. cross sections for nuclear reactions, photodisintegrations, electron and positron captures, neutrino captures, inelastic neutrino scattering, and beta-decay half-lives. We examine especially the present state of uncertainties in predicting thermonuclear reaction rates, while the status of experiments is discussed by others in this volume (see M. Wiescher). It follows a brief review of hydrostatic burning stages in stellar evolution before discussing the fate of massive stars, i.e. the nucleosynthesis in type II supernova explosions (SNe II). Except for SNe Ia, which are explained by exploding white dwarfs in binary stellar systems (which will not be discussed here), all other supernova types seem to be linked to the gravitational collapse of massive stars ($M>8M_\odot$) at the end of their hydrostatic evolution. SN1987A, the first type II supernova for which the progenitor star was known, is used as an example for nucleosynthesis calculations. Finally, we discuss the production of heavy elements in the r-process up to Th and U and its possible connection to supernovae.

---

1. Thermonuclear Rates and Reaction Networks

In this section we want to outline the essential features of thermonuclear reaction rates and nuclear reaction networks. This serves the purpose to define a unified terminology to be used throughout the review, more detailed discussions can be found in Fowler, Caughlan, & Zimmerman (1967,1975), Clayton (1983), Rolfs & Rodney (1988), Thielemann, Nomoto, & Hashimoto (1994), and Arnett (1996).

1.1. *Thermonuclear Reaction Rates*

The nuclear cross section for a reaction between target j and projectile k is defined by

$$\sigma = \frac{\text{number of reactions target}^{-1}\text{sec}^{-1}}{\text{flux of incoming projectiles}} = \frac{r/n_j}{n_k v}. \tag{1.1}$$

The second equality holds for the case that the relative velocity between targets with the number density n_j and projectiles with number density n_k is constant and has the value v. Then r, the number of reactions per cm^3 and sec, can be expressed as $r = \sigma v n_j n_k$.

More generally, when targets and projectiles follow specific distributions, r is given by

$$r_{j,k} = \int \sigma |\vec{v}_j - \vec{v}_k| d^3 n_j d^3 n_k. \tag{1.2}$$

The evaluation of this integral depends on the type of particles and distributions which are involved. For nuclei j and k in an astrophysical plasma, obeying a Maxwell-Boltzmann distribution,

$$d^3 n_j = n_j (\frac{m_j}{2\pi kT})^{3/2} \exp(-\frac{m_j v_j^2}{2kT}) d^3 v_j, \tag{1.3}$$

Eq.(1.2) simplifies to $r_{j,k} =< \sigma v > n_j n_k$. The thermonuclear reaction rates have the form (Fowler, Caughlan, & Zimmerman 1967, Clayton 1983)

$$r_{j,k} = < \sigma v >_{j,k} n_j n_k \tag{1.4a}$$

$$< j,k >: = < \sigma v >_{j,k} = (\frac{8}{\mu\pi})^{1/2} (kT)^{-3/2} \int_0^\infty E\sigma(E) \exp(-E/kT) dE. \tag{1.4b}$$

Here μ denotes the reduced mass of the target-projectile system. In astrophysical plasmas with high densities and/or low temperatures, effects of electron screening become highly important. This means that the reacting nuclei, due to the background of electrons and nuclei, feel a different Coulomb repulsion than in the case of bare nuclei. Under most conditions (with non-vanishing temperatures) the generalized reaction rate integral can be separated into the traditional expression without screening [Eq.(1.4)] and a screening factor (see e.g. Salpeter & van Horn 1969, Itoh, Totsuji, & Ichimaru 1977, Hansen, Torrie, & Veillefosse 1977, Alastuey & Jancovici 1978, Itoh et al. 1979, Ichimaru, Tanaka, Iyetomi 1984, Ichimaru & Utsumi 1983, 1984, Thielemann & Truran 1987, Fushiki & Lamb 1987, Itoh et al. 1990, Schramm & Koonin 1990, Ichimaru 1993, Chabrier & Schatzman 1994, Kitamura & Ichimaru 1995, Brown & Sawyer 1997)

$$< j,k >^* = f_{scr}(Z_j, Z_k, \rho, T, Y_i) < j,k > . \tag{1.5}$$

This screening factor is dependent on the charge of the involved particles, the density, temperature, and the composition of the plasma. Here Y_i denotes the abundance of nucleus i defined by $Y_i = n_i/(\rho N_A)$, where n_i is the number density of nuclei per unit volume and N_A Avogadro's number. At high densities and low temperatures screening factors can enhance reactions by many orders of magnitude and lead to *pycnonuclear ignition*. In the extreme case of very low temperatures, where reactions are only possible via ground state oscillations of the nuclei in a Coulomb lattice, Eq.(1.5) breaks down, because it was derived under the assumption of a Boltzmann distribution (for recent references see Fushiki & Lamb 1987, Itoh et al. 1990, Schramm & Koonin 1990, Ichimaru 1993, Chabrier & Schatzman 1994, Ichimaru 1996).

When in Eq.(1.2) particle k is a photon, the relative velocity is always c and quantities in the integral are not dependent on $d^3 n_j$. Thus it simplifies to $r_j = \lambda_{j,\gamma} n_j$ and $\lambda_{j,\gamma}$ results from an integration of the photodisintegration cross section over a Planck distribution for photons of temperature T

$$d^3 n_\gamma = \frac{1}{\pi^2 (c\hbar)^3} \frac{E_\gamma^2}{\exp(E_\gamma/kT) - 1} dE_\gamma \tag{1.6a}$$

$$r_j = \lambda_{j,\gamma}(T) n_j = \frac{\int d^3 n_j}{\pi^2 (c\hbar)^3} \int_0^\infty \frac{c\sigma(E_\gamma) E_\gamma^2}{\exp(E_\gamma/kT) - 1} dE_\gamma. \tag{1.6b}$$

There is, however, no direct need to evaluate photodisintegration cross sections, because, due to detailed balance, they can be expressed by the capture cross sections for the inverse reaction $l + m \rightarrow j + \gamma$ (Fowler et al. 1967)

$$\lambda_{j,\gamma}(T) = (\frac{G_l G_m}{G_j})(\frac{A_l A_m}{A_j})^{3/2}(\frac{m_u kT}{2\pi\hbar^2})^{3/2} < l, m > \exp(-Q_{lm}/kT). \tag{1.7}$$

This expression depends on the reaction Q-value Q_{lm}, the temperature T, the inverse reaction rate $< l, m >$, the partition functions $G(T) = \sum_i (2J_i + 1)\exp(-E_i/kT)$ and the mass numbers A of the participating nuclei in a thermal bath of temperature T.

A procedure similar to Eq.(1.6) is used for electron captures by nuclei. Because the electron is about 2000 times less massive than a nucleon, the velocity of the nucleus j is negligible in the center of mass system in comparison to the electron velocity ($|\vec{v}_j - \vec{v}_e| \approx |\vec{v}_e|$). The electron capture cross section has to be integrated over a Boltzmann, partially degenerate, or Fermi distribution of electrons, dependent on the astrophysical conditions. The electron capture rates are a function of T and $n_e = Y_e \rho N_A$, the electron number density (Fuller, Fowler, & Newman 1980, 1982, 1985). In a neutral, completely ionized plasma, the electron abundance is equal to the total proton abundance in nuclei $Y_e = \sum_i Z_i Y_i$ and

$$r_j - \lambda_{j,e}(T, \rho Y_e) n_j. \tag{1.8}$$

The same authors generalized this treatment for the capture of positrons, which are in a thermal equilibrium with photons, electrons, and nuclei. At high densities ($\rho > 10^{12}$gcm^{-3}) the size of the neutrino scattering cross section on nuclei and electrons ensures that enough scattering events occur to thermalize a neutrino distribution. Then also the inverse process to electron capture (neutrino capture) can occur and the neutrino capture rate can be expresses similar to Eqs.(1.6) or (1.8), integrating over the neutrino distribution (e.g. Fuller & Meyer 1995). Also inelastic neutrino scattering on nuclei can be expressed in this form. The latter can cause particle emission, like in photodisintegrations (e.g. Woosley et al. 1990, Kolbe et al. 1992, 1993, 1995, Qian et al. 1996). It is also possible that a thermal equilibrium among neutrinos was established at a different location than at the point where the reaction occurs. In such a case the neutrino distribution can be characterized by a chemical potential and a temperature which is not necessarily equal to the local temperature. Finally, for normal decays, like beta or alpha decays with half-life $\tau_{1/2}$, we obtain an equation similar to Eqs.(1.6) or (1.8) with a decay constant $\lambda_j = \ln 2/\tau_{1/2}$ and

$$r_j = \lambda_j n_j. \tag{1.9}$$

1.2. *Nuclear Reaction Networks*

The time derivative of the number densities of each of the species in an astrophysical plasma (at constant density) is governed by the different expressions for r, the number of reactions per cm^3 and sec, as discussed above for the different reaction mechanisms which can change nuclear abundances

$$(\frac{\partial n_i}{\partial t})_{\rho=const} = \sum_j N_j^i r_j + \sum_{j,k} N_{j,k}^i r_{j,k} + \sum_{j,k,l} N_{j,k,l}^i r_{j,k,l}. \tag{1.10}$$

The reactions listed on the right hand side of the equation belong to the three categories of reactions: (1) decays, photodisintegrations, electron and positron captures and neutrino induced reactions ($r_j = \lambda_j n_j$), (2) two-particle reactions ($r_{j,k} =< j,k > n_j n_k$), and (3) three-particle reactions ($r_{j,k,l} =< j,k,l > n_j n_k n_l$) like the triple-alpha process, which can be interpreted as successive captures with an intermediate unstable target (see e.g. Nomoto, Thielemann, & Miyaji 1985, Görres, Wiescher, & Thielemann 1995). The individual N^i's are given by: $N^i_j = N_i$, $N^i_{j,k} = N_i/\prod_{m=1}^{n_m} |N_{j_m}|!$, and $N^i_{j,k,l} = N_i/\prod_{m=1}^{n_m} |N_{j_m}|!$. The $N_i's$ can be positive or negative numbers and specify how many particles of species i are created or destroyed in a reaction. The denominators, including factorials, run over the n_m different species destroyed in the reaction and avoid double counting of the number of reactions when identical particles react with each other (for example in the $^{12}C+^{12}C$ or the triple-alpha reaction; for details see Fowler et al. 1967). In order to exclude changes in the number densities $\dot{n}_i$, which are only due to expansion or contraction of the gas, the nuclear abundances $Y_i = n_i/(\rho N_A)$ were introduced. For a nucleus with atomic weight A_i, $A_i Y_i$ represents the mass fraction of this nucleus, therefore $\sum A_i Y_i = 1$. In terms of nuclear abundances Y_i, a reaction network is described by the following set of differential equations

$$\dot{Y}_i = \sum_j N^i_j \lambda_j Y_j + \sum_{j,k} N^i_{j,k} \rho N_A < j,k > Y_j Y_k + \sum_{j,k,l} N^i_{j,k,l} \rho^2 N_A^2 < j,k,l > Y_j Y_k Y_l. \quad (1.11)$$

Eq.(1.11) derives directly from Eq.(1.10) when the definition for the $Y_i's$ is introduced. This set of differential equations is solved with a fully implicit treatment. Then the stiff set of differential equations can be rewritten (see e.g. Press et al. 1986, §15.6) as difference equations of the form $\Delta Y_i/\Delta t = f_i(Y_j(t+\Delta t))$, where $Y_i(t+\Delta t) = Y_i(t) + \Delta Y_i$. In this treatment, all quantities on the right hand side are evaluated at time $t + \Delta t$. This results in a set of non-linear equations for the new abundances $Y_i(t + \Delta t)$, which can be solved using a multi-dimensional Newton-Raphson iteration procedure. The total energy generation per gram, due to nuclear reactions in a time step Δt which changed the abundances by ΔY_i, is expressed in terms of the mass excess $M_{ex,i}c^2$ of the participating nuclei (Audi & Wapstra 1995)

$$\Delta\epsilon = -\sum_i \Delta Y_i N_A M_{ex,i} c^2 \quad (1.12a)$$

$$\dot{\epsilon} = -\sum_i \dot{Y}_i N_A M_{ex,i} c^2. \quad (1.12b)$$

As noted above, the important ingredients to nucleosynthesis calculations are decay half-lives, electron and positron capture rates, photodisintegrations, neutrino induced reaction rates, and strong interaction cross sections. Beta-decay half-lives for unstable nuclei have been predicted by Takahashi, Yamada, & Kondo (1973), Klapdor, Metzinger, & Oda (1984), Takahashi & Yokoi (1987, also including temperature effects) and more recently with improved quasi particle RPA calculations (Staudt et al. 1989, 1990, Möller & Randrup 1990, Hirsch et al. 1992, Pfeiffer & Kratz 1996, Möller, Nix, & Kratz 1997, Borzov 1996, 1997). Electron and positron capture calculations have been performed by Fuller, Fowler, & Newman (1980, 1982, 1985) for a large variety of nuclei with mass numbers between A=20 and A=60. For revisions see also Takahara et al. (1989) and for heavier nuclei Aufderheide et al. (1994), Sutaria, Sheikh, & Ray (1997). Rates for inelastic neutrino scattering have been presented by Woosley et al. (1990) and Kolbe et al. (1992, 1993, 1995). Photodisintegration rates can be calculated via detailed balance from the reverse capture rates. Experimental nuclear rates for light nuclei have been discussed in detail in the reviews by Rolfs, Trautvetter, & Rodney (1987), Filippone (1987), the book by Rolfs & Rodney (1988), the recent review on 40 years after B^2FH by Wallerstein et al. (1997), and the NuPECC report on nuclear and particle astrophysics (Baraffe et al. 1997). The most recent experimental charged particle rate compilations are the ones by Caughlan & Fowler (1988) and Arnould et al. (1997). Experimental neutron capture cross sections are summarized by Bao & Käppeler (1987, 1997), Beer, Voss, & Winters (1992), and Wisshak et al. (1997). Rates for unstable (light) nuclei are given by Malaney & Fowler (1988, 1989), Wiescher et al. (1986, 1987, 1988ab, 1989ab, 1990), Thomas et al. (1993,1994), van Wormer et al. (1994), Rauscher et al. (1994), and Schatz et al. (1997). For additional information see the article by M. Wiescher (this volume). For the vast number of medium and heavy nuclei which exhibit a high density of excited states at capture energies, Hauser-Feshbach (statistical model) calculations are applicable. The most recent compilations were provided by Holmes et al. (1975), Woosley et al. (1978), and Thielemann, Arnould, & Truran (1987, for a detailed discussion of the methods involved and neutron capture cross sections for heavy unstable nuclei see also section 3.4 and the appendix in Cowan, Thielemann, Truran 1991). Improvements in level densities (Rauscher, Thielemann, & Kratz 1997), alpha potentials, and the consistent treatment of isospin mixing will lead to the next generation of theoretical rate predictions (Rauscher et al. 1998). Some of it will be discussed in the following section.

2. Theoretical Cross Section Predictions

Explosive nuclear burning in astrophysical environments produces unstable nuclei, which again can be targets for subsequent reactions. In addition, it involves a very large number of stable nuclei, which are not fully explored by experiments. Thus, it is necessary to be able to predict reaction cross sections and thermonuclear rates with the aid of theoretical models. Explosive burning in supernovae involves in general intermediate mass and heavy nuclei. Due to a large nucleon number they have intrinsically a high density of excited states. A high level density in the compound nucleus at the appropriate excitation energy allows to make use of the statistical model approach for compound nuclear reactions [e.g. Hauser & Feshbach (1952), Mahaux & Weidenmüller (1979), Gadioli & Hodgson (1992)] which averages over resonances. Here, we want to present recent results obtained within this approach and outline in a clear way, where in the nuclear chart and for which environment temperatures its application is valid. It is often colloquially termed that the statistical model is only applicable for intermediate and heavy nuclei. However, the only necessary condition for its application is a large number of resonances at the appropriate bombarding energies, so that the cross section can be described by an average over resonances. This can in specific cases be valid for light nuclei and on the other hand not be valid for intermediate mass nuclei near magic numbers.

In astrophysical applications usually different aspects are emphasized than in pure nuclear physics investigations. Many of the latter in this long and well established field were focused on specific reactions, where all or most "ingredients", like optical potentials for particle and alpha transmission coefficients, level densities, resonance energies and widths of giant resonances to be implementated in predicting E1 and M1 gamma-transitions, were deduced from experiments. This of course, as long as the statistical model prerequisites are met, will produce highly accurate cross sections. For the majority of nuclei in astrophysical applications such information is not available. The real challenge is thus not the well established statistical model, but rather to provide all these necessary ingredients in as reliable a way as possible, also for nuclei where none of such informations are available. In addition, these approaches should be on a similar level as e.g. mass models, where the investigation of hundreds or thousands of nuclei is possible with managable computational effort, which is not always the case for fully microscopic calculations.

The statistical model approach has been employed in calculations of thermonuclear reaction rates for astrophysical purposes by many researchers [Truran et al. (1966), Michaud & Fowler (1970, 1972), Truran (1972)], who in the beginning only made use of ground state properties. Later, the importance of excited states of the target was pointed out by Arnould (1972). The compilations by Holmes et al. (1976), Woosley et al. (1978), Thielemann et al. (1987), and Cowan, Thielemann & Truran (1991) are presently the ones utilized in large scale applications in all subfields of nuclear astrophysics, when experimental information is unavailable. Existing global optical potentials, mass models to predict Q-values, deformations etc., but also the ingredients to describe giant resonance properties have been quite successful in the past [see e.g. the review by Cowan et al. (1991)].

Besides possibly necessary improvements in global alpha potentials (see Mohr et al. 1997), the major remaining uncertainty in all existing calculations stems from the prediction of nuclear level densities, which in earlier calculations gave uncertainties even beyond a factor of 10 at the neutron separation energy for Gilbert & Cameron (1965), about a factor of 8 for Woosley et al. (1978), and a factor of 5 even in the most recent calculations [e.g. Thielemann et al. (1987); see Fig.3.16 in Cowan et al. (1991)]. In nuclear reactions the transitions to lower lying states dominate due to the strong energy dependence. Because the deviations are usually not as high yet at low excitation energies, the typical cross section uncertainties amounted to a smaller factor of 2–3.

The implementation of a novel treatment of level density descriptions [Iljinov et al. (1992), Ignatyuk et al. (1978)] where the level density parameter is energy dependent and shell effects vanish at high excitation energies, improves the level density accuracy. This is still a phenomenological approach, making use of a back-shifted Fermi-gas model rather than a combinatorial approach based on microscopic single-particle levels. But it is the first one leading to a reduction of the average cross section uncertainty to a factor of about 1.4, i.e. an average deviation of about 40% from experiments, when only employing global predictions for all input parameters and no specific experimental knowledge.

2.1. *Thermonuclear Rates from Statistical Model Calculations*

A high level density in the compound nucleus permits to use averaged transmission coefficients T, which do not reflect a resonance behavior, but rather describe absorption via an imaginary part in the (optical) nucleon-nucleus potential as described in Mahaux & Weidenmüller (1979). This leads to the well known expression

$$\sigma_i^{\mu\nu}(j,o;E_{ij}) = \frac{\pi\hbar^2/(2\mu_{ij}E_{ij})}{(2J_i^\mu+1)(2J_j+1)} \times \sum_{J,\pi}(2J+1)\frac{T_j^\mu(E,J,\pi,E_i^\mu,J_i^\mu,\pi_i^\mu)T_o^\nu(E,J,\pi,E_m^\nu,J_m^\nu,\pi_m^\nu)}{T_{tot}(E,J,\pi)} \quad (2.13)$$

for the reaction $i^\mu(j,o)m^\nu$ from the target state i^μ to the exited state m^ν of the final nucleus, with a center of mass energy E_{ij} and reduced mass μ_{ij}. J denotes the spin, E the corresponding excitation energy in the compound nucleus, and π the parity of excited states. When these properties are used without subscripts they describe the compound nucleus, subscripts refer to states of the participating nuclei in the reaction $i^\mu(j,o)m^\nu$ and superscripts indicate the specific excited states. Experiments measure

$\sum_\nu \sigma_i^{0\nu}(j,o;E_{ij})$, summed over all excited states of the final nucleus, with the target in the ground state. Target states μ in an astrophysical plasma are thermally populated and the astrophysical cross section $\sigma_i^*(j,o)$ is given by

$$\sigma_i^*(j,o;E_{ij}) = \frac{\sum_\mu (2J_i^\mu+1)\exp(-E_i^\mu/kT)\sum_\nu \sigma_i^{\mu\nu}(j,o;E_{ij})}{\sum_\mu (2J_i^\mu+1)\exp(-E_i^\mu/kT)} \quad . \tag{2.14}$$

The summation over ν replaces $T_o^\nu(E,J,\pi)$ in Eq.(2.13) by the total transmission coefficient

$$T_o(E,J,\pi) = \sum_{\nu=0}^{\nu_m} T_o^\nu(E,J,\pi,E_m^\nu,J_m^\nu,\pi_m^\nu) + \int_{E_m^{\nu_m}}^{E-S_{m,o}} \sum_{J_m,\pi_m} T_o(E,J,\pi,E_m,J_m,\pi_m)\rho(E_m,J_m,\pi_m)dE_m \quad . \tag{2.15}$$

Here $S_{m,o}$ is the channel separation energy, and the summation over excited states above the highest experimentally known state ν_m is changed to an integration over the level density ρ. The summation over target states μ in Eq.(2.14) has to be generalized accordingly.

In addition to the ingredients required for Eq.(2.13), like the transmission coefficients for particles and photons, width fluctuation corrections $W(j,o,J,\pi)$ have to be employed. They define the correlation factors with which all partial channels for an incoming particle j and outgoing particle o, passing through the excited state (E,J,π), have to be multiplied. This takes into account that the decay of the state is not fully statistical, but some memory of the way of formation is retained and influences the available decay choices. The major effect is elastic scattering, the incoming particle can be immediately re-emitted before the nucleus equilibrates. Once the particle is absorbed and not re-emitted in the very first (pre-compound) step, the equilibration is very likely. This corresponds to enhancing the elastic channel by a factor W_j. In order to conserve the total cross section, the individual transmission coefficients in the outgoing channels have to be renormalized to T_j'. The total cross section is proportional to T_j and, when summing over the elastic channel (W_jT_j') and all outgoing channels $(T_{tot}' - T_j')$, one obtains the condition $T_j = T_j'(W_jT_j'/T_{tot}') + T_j'(T_{tot}' - T_j')/T_{tot}'$. We can (almost) solve for T_j'

$$T_j' = \frac{T_j}{1+T_j'(W_j-1)/T_{tot}'} \quad . \tag{2.16}$$

This requires an iterative solution for T' (starting in the first iteration with T_j and T_{tot}), which converges fast. The enhancement factor W_j has to be known in order to apply Eq.(2.16). A general expression in closed form was derived by Verbaatschot et al. (1986), but is computationally expensive to use. A fit to results from Monte Carlo calculations by Tepel et al. (1974) gave

$$W_j = 1 + \frac{2}{1+T_j^{1/2}} \quad . \tag{2.17}$$

For a general discussion of approximation methods see Gadioli & Hodgson (1992) and Ezhov & Plujko (1993). Eqs.(2.16) and (2.17) redefine the transmission coefficients of Eq.(2.13) in such a manner that the total width is redistributed by enhancing the elastic channel and weak channels over the dominant one. Cross sections near threshold energies of new channel openings, where very different channel strengths exist, can only be described correctly when taking width fluctuation corrections into account. Of the thermonuclear rates presently available in the literature, only those by Thielemann et al. (1987) and Cowan et al. (1991) included this effect, but their level density treatment still contained large uncertainties. The width fluctuation corrections of Tepel et al. (1974) are only an approximation to the correct treatment. However, Thomas et al. (1986) showed that they are quite adequate.

The important ingredients of statistical model calculations as indicated in Eqs.(2.13) through (2.15) are the particle and gamma-transmission coefficients T and the level density of excited states ρ. Therefore, the reliability of such calculations is determined by the accuracy with which these components can be evaluated (often for unstable nuclei). In the following we want to discuss the methods utilized to estimate these quantities and recent improvements.

2.2. *Transmission Coefficients*

The transition from an excited state in the compound nucleus (E, J, π) to the state $(E_i^\mu, J_i^\mu, \pi_i^\mu)$ in nucleus i via the emission of a particle j is given by a summation over all quantum mechanically allowed partial waves

$$T_j^\mu(E, J, \pi, E_i^\mu, J_i^\mu, \pi_i^\mu) = \sum_{l=|J-s|}^{J+s} \sum_{s=|J_i^\mu - J_j|}^{J_i^\mu + J_j} T_{j_{ls}}(E_{ij}^\mu). \tag{2.18}$$

Here the angular momentum $\vec{l}$ and the channel spin $\vec{s} = \vec{J}_j + \vec{J}_i^\mu$ couple to $\vec{J} = \vec{l} + \vec{s}$. The transition energy in channel j is $E_{ij}^\mu = E - S_j - E_i^\mu$.

The individual particle transmission coefficients T_l are calculated by solving the Schrödinger equation with an optical potential for the particle-nucleus interaction. All early studies of thermonuclear reaction rates by Truran et al. (1966), Michaud & Fowler (1972), Arnould (1972), Truran (1972), Holmes et al. (1976), and Woosley et al. (1978) employed optical square well potentials and made use of the black nucleus approximation. Thielemann et al. (1987) employed the optical potential for neutrons and protons given by Jeukenne, Lejeune, & Mahaux (1977), based on microscopic infinite nuclear matter calculations for a given density, applied with a local density approximation. It includes corrections of the imaginary part by Fantoni et al. (1981) and Mahaux (1982). The resulting s-wave neutron strength function $< \Gamma^o/D > |_{1\mathrm{eV}} = (1/2\pi) T_{n(l=0)}(1\mathrm{eV})$ is shown and discussed in Thielemann et al. (1983) and Cowan et al. (1991), where several phenomenological optical potentials of the Woods-Saxon type and the equivalent square well potential used in earlier astrophysical applications are compared. The purely theoretical approach gives the best fit. It is also expected to have the most reliable extrapolation properties for unstable nuclei. We show here in Fig. 1 the ratio of the s-wave strength functions for the Jeukenne, Lejeune, & Mahaux potential over the black nucleus, equivalent square well approach for different energies. A general overview on different approaches can be found in Varner et al. (1991).

Deformed nuclei were treated in a very simplified way by using an effective spherical potential of equal volume, based on averaging the deformed potential over all possible angles between the incoming particle and the orientation of the deformed nucleus. In most earlier compilations alpha particles were also treated by square well optical potentials. Thielemann et al. (1987) employed a phenomenological Woods-Saxon potential by Mann (1978), based on extensive data from McFadden & Satchler (1966). For future use, for alpha particles and heavier projectiles, it is clear that the best results can probably be obtained with folding potentials [e.g. Satchler & Love (1979), Chaudhuri et al. (1985), Oberhummer et al. (1996), and Mohr et al. (1997)].

The gamma-transmission coefficients have to include the dominant gamma-transitions

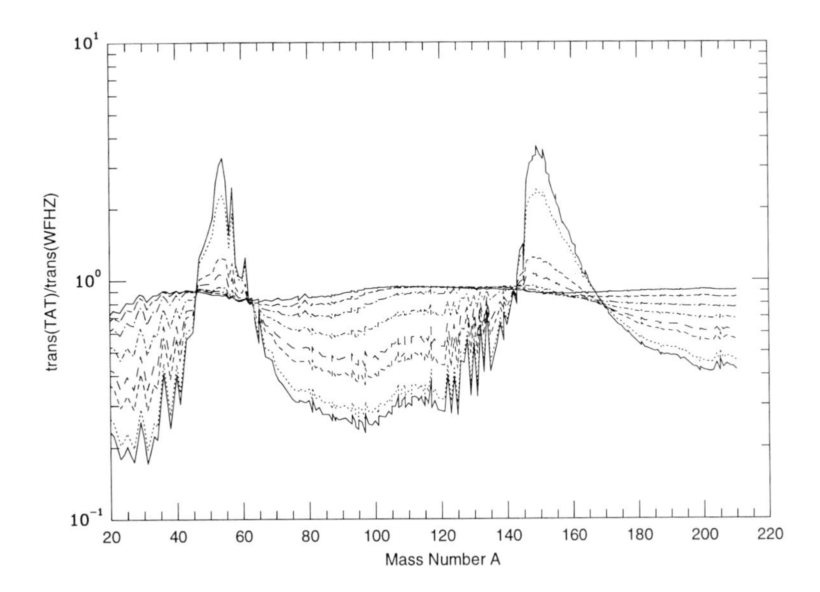

FIGURE 1. Ratios of transmission functions obtained with the Jeukenne et al. (1977) potential and the equivalent square well description of Woosley et al. (1978). Shown are the ratios for s-wave neutrons. Different line styles denote different energies: 0.01 MeV (solid), 0.1 MeV (dotted), 1.0 MeV (short dashes), 2.0 MeV (long dashes), 5.0 MeV (dot – short dash), 10.0 MeV (dot – long dash), 15.0 MeV (short dash – long dash), 20.0 MeV (solid).

(E1 and M1) in the calculation of the total photon width. The smaller, and therefore less important, M1 transitions have usually been treated with the simple single particle approach $T \propto E^3$ of Blatt & Weisskopf (1952), as also discussed in Holmes et al. (1976). The E1 transitions are usually calculated on the basis of the Lorentzian representation of the Giant Dipole Resonance (GDR). Within this model, the E1 transmission coefficient for the transition emitting a photon of energy E_γ in a nucleus ${}^A_N Z$ is given by

$$T_{E1}(E_\gamma) = \frac{8}{3}\frac{NZ}{A}\frac{e^2}{\hbar c}\frac{1+\chi}{mc^2}\sum_{i=1}^{2}\frac{i}{3}\frac{\Gamma_{G,i}E_\gamma^4}{(E_\gamma^2 - E_{G,i}^2)^2 + \Gamma_{G,i}^2 E_\gamma^2} \quad . \tag{2.19}$$

Here $\chi(= 0.2)$ accounts for the neutron-proton exchange contribution as discussed in Lipparini & Stringari (1989), and the summation over i includes two terms which correspond to the split of the GDR in statically deformed nuclei, with oscillations along (i=1) and perpendicular (i=2) to the axis of rotational symmetry. Many microscopic and macroscopic models have been devoted to the calculation of the GDR energies (E_G) and widths (Γ_G). Analytical fits as a function of A and Z were also used, e.g. in Holmes et al. (1976) and Woosley et al. (1978). Thielemann et al. (1987) employed the (hydrodynamic) droplet model approach by Myers et al. (1977) for E_G, which gives an excellent fit to the GDR energies and can also predict the split of the resonance for deformed nuclei, when making use of the deformation, calculated within the droplet model. In that case, the two resonance energies are related to the mean value calculated by the relations $E_{G,1} + 2E_{G,2} = 3E_G$, $E_{G,2}/E_{G,1} = 0.911\eta + 0.089$ of Danos (1958). η is the ratio of the diameter along the nuclear symmetry axis to the diameter perpendicular to it, and can be obtained from the experimentally known deformation or mass model predictions.

Cowan et al. (1991) also give a detailed description of the microscopic-macroscopic approach utilized to calculate Γ_G, based on dissipation and the coupling to quadrupole surface vibrations. This is the method applied to predict the gamma-transmission coefficients for the cross section determinations shown in the following.

2.3. *Level Densities*

While the method as such is well seasoned, considerable effort has been put into the improvement of the input for statistical Hauser-Feshbach models. However, the nuclear level density has given rise to the largest uncertainties in cross section determinations of Holmes et al. (1976), Thielemann et al. (1987), Thielemann et al. (1988), and Cowan et al. (1991). For large scale astrophysical applications it is also necessary to not only find reliable methods for level density predictions, but also computationally feasible ones.

Such a model is the non-interacting Fermi-gas model by Bethe (1936). Most statistical model calculations use the back-shifted Fermi-gas description of Gilbert & Cameron (1965). More sophisticated Monte Carlo shell model calculations, e.g. by Dean et al. (1995), as well as combinatorial approaches [see e.g. Paar (1997)], have shown excellent agreement with this phenomenological approach and justified the application of the Fermi-gas description at and above the neutron separation energy. Rauscher, Thielemann, & Kratz (1997) applied an energy-dependent level density parameter a together with microscopic corrections from nuclear mass models, which leads to improved fits in the mass range $20 \leq A \leq 245$.

The back-shifted Fermi-gas description of Gilbert & Cameron (1965) assumes an even distribution of odd and even parities [however, see e.g. Pichon (1994) for doubts on the validity of this assumption at energies of astrophysical interest]

$$\rho(U, J, \pi) = \frac{1}{2}\mathcal{F}(U, J)\rho(U), \tag{2.20}$$

with

$$\rho(U) = \frac{1}{\sqrt{2\pi}\sigma}\frac{\sqrt{\pi}}{12a^{1/4}}\frac{\exp(2\sqrt{aU})}{U^{5/4}}, \qquad \mathcal{F}(U, J) = \frac{2J+1}{2\sigma^2}\exp\left(\frac{-J(J+1)}{2\sigma^2}\right) \tag{2.21}$$

$$\sigma^2 = \frac{\Theta_{\mathrm{rigid}}}{\hbar^2}\sqrt{\frac{U}{a}}, \qquad \Theta_{\mathrm{rigid}} = \frac{2}{5}m_{\mathrm{u}}AR^2, \qquad U = E - \delta \quad .$$

The spin dependence $\mathcal{F}$ is determined by the spin cut-off parameter σ. Thus, the level density is dependent on only two parameters: the level density parameter a and the backshift δ, which determines the energy of the first excited state.

Within this framework, the quality of level density predictions depends on the reliability of systematic estimates of a and δ. The first compilation for a large number of nuclei was provided by Gilbert & Cameron (1965). They found that the backshift δ is well reproduced by experimental pairing corrections (Cameron & Elkin 1965). They also were the first to identify an empirical correlation with experimental shell corrections $S(Z, N)$

$$\frac{a}{A} = c_0 + c_1 S(Z, N), \tag{2.22}$$

where $S(Z, N)$ is negative near closed shells. The back-shifted Fermi-gas approach diverges for $U = 0$ (i.e. $E = \delta$, if δ is a positive backshift). In order to obtain the correct behavior at very low excitation energies, the Fermi-gas description can be combined with the constant temperature formula [Gilbert & Cameron (1965), Gadioli & Hodgson (1992) and references therein]

$$\rho(U) \propto \frac{\exp(U/T)}{T} \quad . \tag{2.23}$$

The two formulations are matched by a tangential fit determining T. There have been a number of compilations for a and δ, or T, based on experimental level densities, as e.g. the ones by von Egidy et al. (1986,1988). An improved approach has to consider the energy dependence of the shell effects, which are known to vanish at high excitation energies, see e.g. Iljinov et al. (1992). Although, for astrophysical purposes only energies close to the particle separation thresholds have to be considered, an energy dependence can lead to a considerable improvement of the global fit. This is especially true for strongly bound nuclei close to magic numbers.

An excitation-energy dependent description was initially proposed by Ignatyuk et al. (1975) and Ignatyuk et al. (1979) for the level density parameter a

$$a(U,Z,N) = \tilde{a}(A)\left[1 + C(Z,N)\frac{f(U)}{U}\right] \tag{2.24a}$$

$$\tilde{a}(A) = \alpha A + \beta A^{2/3} \tag{2.24b}$$

$$f(U) = 1 - \exp(-\gamma U). \tag{2.24c}$$

The values of the free parameters α, β and γ are determined by fitting to experimental level density data available over the whole nuclear chart.

The shape of the function $f(U)$ permits the two extremes: (i) for small excitation energies the original form of Eq.(2.22) $a/A = \alpha + \alpha\gamma C(Z,N)$ is retained with $S(Z,N)$ being replaced by $C(Z,N)$, (ii) for high excitation energies a/A approaches the continuum value α, obtained for infinite nuclear matter. In both cases we neglected β, which is realistic as discussed below. Previous attempts to find a global description of level densities used shell corrections S derived from comparison of liquid-drop masses with experiment ($S \equiv M_{\rm exp} - M_{\rm LD}$) or the "empirical" shell corrections $S(Z,N)$ given by Gilbert & Cameron (1965). A problem connected with the use of liquid-drop masses arises from the fact that there are different liquid-drop model parametrizations available in the literature which produce quite different values for S, as shown in Mengoni & Nakayama (1994).

However, in addition, the meaning of the correction parameter inserted into the level density formula Eq.(2.24) has to be reconsidered. The fact that nuclei approach a spherical shape at high excitation energies (temperatures) has to be included. Therefore, the correction parameter C should describe properties of a nucleus differing from the *spherical* macroscopic energy and contain those terms which are finite for low and vanishing at higher excitation energies. The latter requirement is mimicked by the form of Eq.(2.24). Therefore, the parameter $C(Z,N)$ should rather be identified with the so-called "microscopic" correction $E_{\rm mic}$ than with the shell correction. The mass of a nucleus with deformation ϵ can then be written in two ways

$$M(\epsilon) = E_{\rm mic}(\epsilon) + E_{\rm mac}(\text{spherical}) \tag{2.25a}$$

$$M(\epsilon) = E_{\rm mac}(\epsilon) + E_{\rm s+p}(\epsilon), \tag{2.25b}$$

with $E_{\rm s+p}$ being the shell-plus-pairing correction. This confusion about the term "microscopic correction", being sometimes used in an ambiguous way, is also pointed out in Möller et al. (1995). The above mentioned ambiguity follows from the inclusion of deformation-dependent effects into the macroscopic part of the mass formula.

Another important ingredient is the pairing gap Δ, related to the backshift δ. Instead of assuming constant pairing as in Reisdorf (1981) or an often applied fixed dependence on the mass number via e.g. $\pm 12/\sqrt{A}$, the pairing gap Δ can be determined from differences in the binding energies (or mass differences, respectively) of neighboring nuclei. Thus, similar to Ring & Schuck (1980), Wang et al. (1992) obtained for the neutron pairing gap Δ_n

$$\Delta_n(Z,N) = \frac{1}{2}\left[M(Z,N-1) + M(Z,N+1) - 2M(Z,N)\right], \tag{2.26}$$

where $M(Z,N)$ is the ground state mass excess of the nucleus (Z,N). Similarly, the proton pairing gap Δ_p can be calculated.

2.4. *Results*

In our study we utilized the microscopic corrections of the recent mass formula by Möller et al. (1995), calculated with the Finite Range Droplet Model FRDM (using a folded Yukawa shell model with Lipkin-Nogami pairing) in order to determine the parameter $C(Z,N)=E_{\rm mic}$. The backshift δ was calculated by setting $\delta(Z,N)=1/2\{\Delta_n(Z,N)+\Delta_p(Z,N)\}$ and using Eq.(2.26). The parameters α, β, and γ were obtained from a fit to experimental data for s-wave neutron resonance spacings of 272 nuclei at the neutron separation energy. The data were taken from the compilation by Iljinov et al. (1992). Similar investigations were recently undertaken by Mengoni & Nakajima (1994), who made, however, use of a slightly different description of the energy dependence of a and of different pairing gaps.

As a quantitative overall estimate of the agreement between calculations and experiments, one usually quotes the ratio

$$g \equiv \left\langle \frac{\rho_{\rm calc}}{\rho_{\rm exp}} \right\rangle = \exp\left[\frac{1}{n}\sum_{i=1}^{n}\left(\ln\frac{\rho^i_{\rm calc}}{\rho^i_{\rm exp}}\right)^2\right]^{1/2}, \tag{2.27}$$

with n being the number of nuclei for which level densities ρ are experimentally known. As best fit we obtain an averaged ratio $g = 1.48$ with the parameter values $\alpha = 0.1337$, $\beta = -0.06571$, $\gamma = 0.04884$. This corresponds to $a/A = \alpha = 0.134$ for infinite nuclear matter, which is approached for high excitation energies. The ratios of experimental to predicted level densities (i.e. theoretical to experimental level spacings D) for the nuclei considered are shown in Fig. 2. As can be seen, for the majority of nuclei the absolute deviation is less than a factor of 2. This is a satisfactory improvement over theoretical level densities used in previous astrophysical cross section calculations, where deviations of a factor 3–4, or even in excess of a factor of 10 were found [for details see Cowan et al. (1991)]. Such a direct comparison as in Fig. 2 was rarely shown in earlier work. In most cases the level density parameter a, entering exponentially into the level density, was displayed.

Although we quoted the value of the parameter β above, it is small in comparison to α and can be set to zero without considerable increase in the obtained deviation. Therefore, actually only two parameters are needed for the level density description. Rauscher, Thielemann, & Kratz (1997) also tested the sensitivity with respect to the employed mass formula. This phenomenological approach, still in the framework of the back-shifted Fermi gas model, but with an energy dependent level density parameter, based on microscopic corrections of nuclear mass models, gives better results than a recent BCS approach by Goriely (1996), which tried to implement level spacings from the ETFSI model (Extended Thomas-Fermi with Strutinski Integral, Aboussir et al. 1995) in a consistent combinatorial fashion.

With these improvements, the uncertainty in the level density is now comparable to

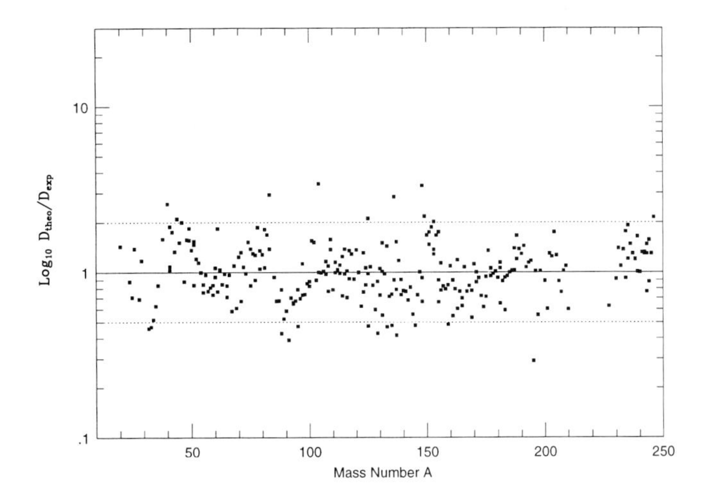

FIGURE 2. Ratio of predicted to experimental Iljinov et al. (1992) level densities at the neutron separation energy. The deviation is less than a factor of 2 (dotted lines) for the majority of the considered nuclei.

uncertainties in optical potentials and gamma transmission coefficients which enter the determinations of capture cross sections. The remaining uncertainty of extrapolations is the one due to the reliability of the nuclear structure model applied far from stability which provides the microscopic corrections and pairing gaps. We will discuss this in more detail in section 5 (see also the contribution by M. Wiescher, this volume).

2.5. *Applicability of the Statistical Model*

Having a reliable level density description also permits to analyze when and where the statistical model approach is valid. Generally speaking, in order to apply the model correctly, a sufficiently large number of levels in the compound nucleus is needed in the relevant energy range, which can act as doorway states to the formation of the compound nucleus. In the following this is discussed for neutron-, proton- and alpha-induced reactions with the aid of the level density approach presented above. This section is intended to be a guide to a meaningful and correct application of the statistical model.

The nuclear reaction rate per particle pair at a given stellar temperature T is determined by folding the reaction cross section with the Maxwell-Boltzmann (MB) velocity distribution of the projectiles, as displayed in Eq.(1.4). Two cases have to be considered, reactions between charged particles and reactions with neutrons.

2.5.1. *The Effective Energy Window*

The nuclear cross section for charged particles is strongly suppressed at low energies due to the Coulomb barrier. For particles having energies less than the height of the Coulomb barrier, the product of the penetration factor and the MB distribution function at a given temperature results in the so-called Gamow peak, in which most of the reactions will take place. Location and width of the Gamow peak depend on the charges of projectile and target, and on the temperature of the interacting plasma.

When introducing the astrophysical S factor $S(E) = \sigma(E)E\exp(2\pi\eta)$ (with η being the Sommerfeld parameter, describing the s-wave barrier penetration), one can easily see the two contributions of the velocity distribution and the penetrability in the integral

$$<\sigma v>= \left(\frac{8}{\pi\mu}\right)^{1/2} \frac{1}{(kT)^{3/2}} \int_0^\infty S(E) \exp\left[-\frac{E}{kT} - \frac{b}{E^{1/2}}\right] \quad , \tag{2.28}$$

where the quantity $b = 2\pi\eta E^{1/2} = (2\mu)^{1/2}\pi e^2 Z_j Z_k/\hbar$ arises from the barrier penetrability. Taking the first derivative of the integrand yields the location E_0 of the Gamov peak, and the effective width Δ of the energy window can be derived accordingly

$$E_0 = \left(\frac{bkT}{2}\right)^{2/3} = 1.22(Z_j^2 Z_k^2 A T_6^2)^{1/3}\,\mathrm{keV}, \tag{2.29a}$$

$$\Delta = \frac{16 E_0 kT}{3}^{1/2} = 0.749(Z_j^2 Z_k^2 A T_6^5)^{1/6}\,\mathrm{keV}, \tag{2.29b}$$

as shown in Fowler et al. (1967) and Rolfs & Rodney (1988), where the charges Z_j, Z_k, the reduced mass A of the involved nuclei in units of m_u, and the temperature T_6 given in 10^6 K, enter.

In the case of neutron-induced reactions the effective energy window has to be derived in a slightly different way. For s-wave neutrons ($l = 0$) the energy window is simply given by the location and width of the peak of the MB distribution function. For higher partial waves the penetrability of the centrifugal barrier shifts the effective energy E_0 to higher energies. For neutrons with energies less than the height of the centrifugal barrier this was approximated by Wagoner (1969)

$$E_0 \approx 0.172 T_9 \left(l + \frac{1}{2}\right) \quad \mathrm{MeV}, \tag{2.30a}$$

$$\Delta \approx 0.194 T_9 \left(l + \frac{1}{2}\right)^{1/2} \quad \mathrm{MeV}. \tag{2.30b}$$

The energy E_0 will always be comparatively close to the neutron separation energy.

2.5.2. *A Criterion for the Application of the Statistical Model*

Using the above effective energy windows for charged and neutral particle reactions, a criterion for the applicability can be derived from the level density. For a reliable application of the statistical model a sufficient number of nuclear levels has to be within the energy window, thus contributing to the reaction rate. For narrow, isolated resonances, the cross sections (and also the reaction rates) can be represented by a sum over individual Breit-Wigner terms. The main question is whether the density of resonances (i.e. level density) is high enough so that the integral over the sum of Breit-Wigner resonances may be approximated by an integral over the statistcial model expressions of Eq.(2.13), which assume that at any bombarding energy a resonance of any spin and parity is available [see Wagoner (1969)].

Numerical test calculations have been performed by Rauscher et al. (1997) in order to find the average number of levels per energy window which is sufficient to allow this substitution in the specific case of folding over a MB distribution. To achieve 20% accuracy, about 10 levels in total are needed in the effective energy window in the worst case (non-overlapping, narrow resonances). This relates to a number of s-wave levels smaller than 3. Application of the statistical model for a level density which is not sufficiently large, results in general in an overestimation of the actual cross section, unless a strong s-wave resonance is located right in the energy window [see the discussion in van Wormer et al. (1994)]. Therefore, we will assume in the following a conservative

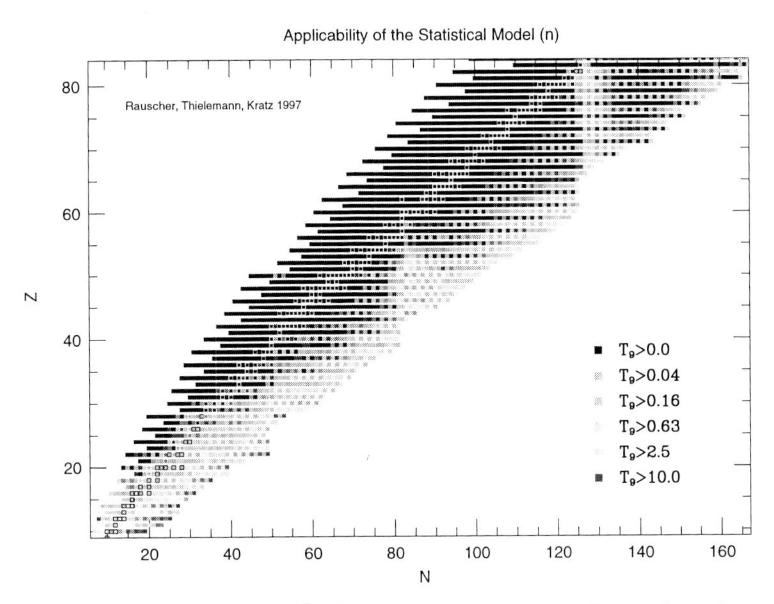

FIGURE 3. Stellar temperatures (in 10^9 K) for which the statistical model can be used. Plotted is the compound nucleus of the neutron-induced reaction n+Target. Stable nuclei are marked.

limit of 10 contributing resonances in the effective energy window for charged and neutral particle-induced reactions.

To obtain the necessary number of levels within the energy window of width Δ can require a sufficiently high excitation energy, as the level density increases with energy. This combines with the thermal distribution of projectiles to a minimum temperature for the application of the statistical model. Those temperatures are plotted in a logarithmic grey scale in Figs. 3–5. For neutron-induced reactions Fig. 3 applies, Fig. 4 describes proton-induced reactions, and Fig. 5 alpha-induced reactions. Plotted is always the minimum stellar temperature T_9 (in 10^9 K) at the location of the compound nucleus in the nuclear chart. It should be noted that the derived temperatures will not change considerably, even when changing the required level number within a factor of about two, because of the exponential dependence of the level density on the excitation energy.

This permits to read directly from the plot whether the statistical model cross section can be "trusted" for a specific astrophysical application at a specified temperature or whether single resonances or other processes (e.g. direct reactions) have also to be considered. These plots can give hints on when it is safe to use the statistical model approach and which nuclei have to be treated with special attention for specific temperatures. Thus, information on which nuclei might be of special interest for an experimental investigation may also be extracted.

3. Nucleosynthesis Processes in Stellar Evolution and Explosions

Nucleosynthesis calculations can in general be classified into two categories: (1) nucleosynthesis during hydrostatic burning stages of stellar evolution on long timescales and (2) nucleosynthesis in explosive events (with different initial fuel compositions, specific to the event). In the following we want to discuss shortly reactions of importance for both conditions and the major burning products.

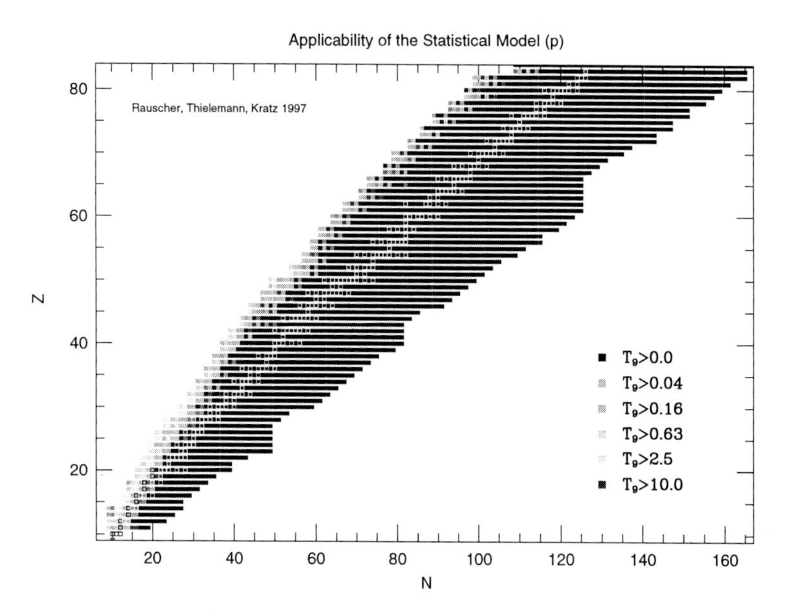

FIGURE 4. Stellar temperatures (in 10^9) for which the statistical model can be used. Plotted is the compound nucleus of the proton-induced reaction p+Target. Stable nuclei are marked.

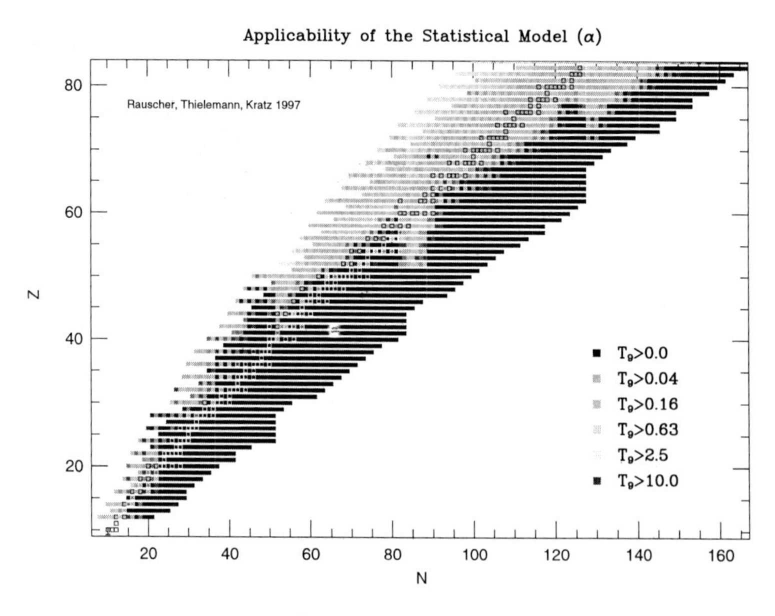

FIGURE 5. Stellar temperatures (in 10^9) for which the statistical model can be used. Plotted is the compound nucleus of the alpha-induced reaction alpha+Target. Stable nuclei are marked.

3.1. *Hydrostatic Burning Stages in Presupernova Evolution*

The main hydrostatic burning stages and most important reactions are:
H-burning: there are two alternative reaction sequences, the different pp-chains which convert ^{1}H into ^{4}He, initiated by $^1\mathrm{H}(p,e^+)^2\mathrm{H}$, and the CNO cycle which converts ^{1}H into ^{4}He by a sequence of (p,γ) and (p,α) reactions on C, N, and O isotopes and subsequent beta-decays. The CNO isotopes are all transformed into ^{14}N, due to the fact that the reaction $^{14}\mathrm{N}(p,\gamma)^{15}\mathrm{O}$ is the slowest reaction in the cycle.
He-burning: the triple-alpha reaction $^4\mathrm{He}(2\alpha,\gamma)^{12}\mathrm{C}$ and $^{12}\mathrm{C}(\alpha,\gamma)^{16}\mathrm{O}$ are the main reactions.
C-burning: $^{12}\mathrm{C}(^{12}\mathrm{C},\alpha)^{20}\mathrm{Ne}$ and $^{12}\mathrm{C}(^{12}\mathrm{C},p)^{23}\mathrm{Na}$. Most of the ^{23}Na nuclei will react with the free protons via $^{23}\mathrm{Na}(p,\alpha)^{20}\mathrm{Ne}$.
Ne-Burning: $^{20}\mathrm{Ne}(\gamma,\alpha)^{16}\mathrm{O}$, $^{20}\mathrm{Ne}(\alpha,\gamma)$ ^{24}Mg and $^{24}\mathrm{Mg}(\alpha,\gamma)^{28}\mathrm{Si}$. It is important that photodisintegrations start to play a role when $30kT{\approx}Q$ (as a rule of thumb), with Q being the Q-value of a capture reaction. For those conditions sufficient photons with energies $>Q$ exist in the high energy tail of the Planck distribution. As $^{16}\mathrm{O}(\alpha,\gamma)^{20}\mathrm{Ne}$ has an exceptionally small Q-value of the order 4 MeV, this relation holds true for $T{>}1.5\times10^9$K, which is the temperature for (hydrostatic) Ne-burning.
O-burning: $^{16}\mathrm{O}(^{16}\mathrm{O},\alpha)^{28}\mathrm{Si}$, $^{16}\mathrm{O}(^{16}\mathrm{O},p)^{31}\mathrm{P}$, and $^{16}\mathrm{O}(^{16}\mathrm{O},n)^{31}\mathrm{S}(\beta^+)^{31}\mathrm{P}$. Similar to carbon burning, most of the ^{31}P is destroyed by a (p,α) reaction to ^{28}Si.
Si-burning: Si-burning is initiated like Ne-burning by photodisintegration reactions which then provide the particles for capture reactions. It ends in an equilibrium abundance distribution around Fe (thermodynamic equilibrium). As this includes all kinds of Q-values (on the average 8-10 MeV for capture reactions along the valley of stability), this translates to temperatures in excess of $3{\times}10^9$K, being larger than the temperatures for the onset of Ne-burning. In such an equilibrium (also denoted nuclear statistical equilibrium, NSE) the abundance of each nucleus is only governed by the temperature T, density ρ, its nuclear binding energy B_i and partition function $G_i = \sum_j (2J^i_j + 1)\exp(-E^i_j/kT)$

$$Y_i = (\rho N_A)^{A_i-1}\frac{G_i}{2^{A_i}}A_i^{3/2}(\frac{2\pi\hbar^2}{m_u kT})^{\frac{3}{2}(A_i-1)}\exp(B_i/kT)Y_p^{Z_i}Y_n^{N_i}, \qquad (3.31)$$

while fulfilling mass conservation $\sum_i A_iY_i = 1$ and charge conservation $\sum_i Z_iY_i = Y_e$ (the total number of protons equals the net number of electrons, which is usually changed only by weak interactions on longer timescales). This equation is derived from the relation between chemical potentials (for Maxwell-Boltzmann distributions) in a thermal equilibrium ($\mu_i = Z_i\mu_p + N_i\mu_n$), where the subscripts n and p stand for neutrons and protons. Intermediate quasi-equilibrium stages (QSE), where clusters of neighboring nuclei are in relative equilibrium via neutron and proton reactions, but different clusters have total abundances which are offset from their NSE values, are important during the onset of Si-burning before a full NSE is reached and during the freeze-out from high temperatures, which will be discussed in section 3.2.
s-process: the slow neutron capture process leads to the build-up of heavy elements during core and shell He-burning, where through a series of neutron captures and beta-decays, starting on existing heavy nuclei around Fe, nuclei up to Pb and Bi can be synthesized. The neutrons are provided by a side branch of He-burning, $^{14}\mathrm{N}(\alpha,\gamma)^{18}\mathrm{F}(\beta^+)^{18}\mathrm{O}(\alpha,\gamma)$ $^{22}\mathrm{Ne}(\alpha,n)^{25}\mathrm{Mg}$. An alternative stronger neutron source in He-shell flashes is the reaction $^{13}\mathrm{C}(\alpha,n)^{16}\mathrm{O}$, which requires admixture of hydrogen and the production of ^{13}C via proton capture on ^{12}C and a subsequent beta-decay.

An extensive overview over the major and minor reaction sequences in all burning stages from helium to silicon burning in massive stars is given in Arnett & Thielemann (1985), Thielemann & Arnett (1985), Woosley & Weaver (1995), Hix & Thielemann (1996) and Nomoto et al. (1997) (see also Arnett 1996 and for the status of experimental rate uncertainties M. Wiescher, this volume). For less massive stars which burn at higher densities, i.e. experience higher electron Fermi energies, electron captures are already important in O-burning and lead to a smaller Y_e or larger neutron excess $\eta = \sum_i (N_i - Z_i) Y_i = 1 - 2Y_e$. For a general overview of the s-process see Käppeler, Beer, & Wisshak (1989), Käppeler et al. (1994), Wisshak et al. (1997), and Gallino & Busso (1997).

Most reactions in hydrostatic burning stages proceed through stable nuclei. This is simply explained by the long timescales involved. For a $25M_\odot$ star, which is relatively massive and therefore experiences quite short burning phases, this still amounts to: H-burning 7×10^6 years, He-burning 5×10^5 y, C-burning 600 y, Ne-burning 1 y, O-burning 180 days, Si-burning 1 d. Because all these burning stages are long compared to beta-decay half-lives, with a few exceptions of long-lived unstable nuclei, nuclei can decay back to stability before undergoing the next reaction. Examples of such exceptions are the s-process branchings with a competition between neutron captures and beta-decays of similar timescales (see e.g. Gallino & Busso 1997).

3.2. *Explosive Burning*

Many of the hydrostatic burning processes discussed in section 3.1 can occur also under explosive conditions at much higher temperatures and on shorter timescales. The major reactions remain still the same in many cases, but often the beta-decay half-lives of unstable products are longer than the timescales of the explosive processes under investigation. This requires in general the additional knowledge of nuclear cross sections for unstable nuclei.

Extensive calculations of explosive carbon, neon, oxygen, and silicon burning, appropriate for supernova explosions, have already been performed in the late 60s and early 70s with the accuracies possible in those days and detailed discussions about the expected abundance patterns (for a general review see Trimble 1975; Truran 1985). More recent overviews in the context of stellar models are given by Trimble (1991) and Arnett (1995). Besides minor additions of ^{22}Ne after He-burning (or nuclei which originate from it in later burning stages, see section 3.1), the fuels for explosive nucleosynthesis consist mainly of alpha-particle nuclei like ^{12}C, ^{16}O, ^{20}Ne, ^{24}Mg, or ^{28}Si. Because the timescale of explosive processing is very short (a fraction of a second to several seconds), only few beta-decays can occur during explosive nucleosynthesis events, resulting in heavier nuclei, again with N$\approx$Z. However, a spread of nuclei around a line of N=Z is involved and many reaction rates for unstable nuclei have to be known. Dependent on the temperature, explosive burning produces intermediate to heavy nuclei. We will discuss the individual burning processes below. For the processes discussed in this section, nuclei within a few mass units from stability are encountered, where nuclear masses and decay half-lives are known experimentally.

Two processes differ from the above scenario, where either a large supply of neutrons or protons is available, the r-process and the rp-process, denoting rapid neutron or proton capture (the latter also termed explosive hydrogen burning). In such cases, nuclei close to the neutron and proton drip lines can be prodruced and beta-decay timescales can be short in comparison to the process timescales. In this section we will only discuss the possible connection between explosive Si-burning and the r-process.

Burning timescales in stellar evolution are dictated by the energy loss timescales of stellar environments. Processes like hydrogen and helium burning, where the stellar energy loss is dominated by the photon luminosity, choose temperatures with energy generation rates equal to the radiation losses. For the later burning stages neutrino losses play the dominant role among cooling processes and the burning timescales are determined by temperatures where neutrino losses are equal to the energy generation rate (see the long series of investigations by Itoh and collaborators, e.g. Itoh et al. 1993, 1994, 1996ab). Explosive events are determined by hydrodynamics, causing different temperatures and timescales for the burning of available fuel. We can generalize the question by defining a burning timescale according to Eq.(1.11) for the destruction of the major fuel nuclei i

$$\tau_i = |\frac{Y_i}{\dot{Y}_i}|. \tag{3.32}$$

These timescales for the fuels $i \in$ H, He, C, Ne, O, Si are determinated by the major distruction reaction. They are in all cases temperature dependent. Dependent on whether this is (i) a decay or photodisintegration, (ii) a two-particle or (iii) a three-particle fusion reaction, they are (i) either not density dependent or have an inverse (ii) linear or (iii) quadratic density dependence. Thus, in the burning stages which involve a fusion process, the density dependence is linear, with the exception of He-burning, where it is quadratic. Ne- and Si-burning, which are dominated by (γ,α) distructions of ^{20}Ne and ^{28}Si, have timescales only determined by the burning temperatures. The temperature dependences are typically exponential, due to the functional form of the corresponding $N_A < \sigma v >$ expressions. We have plotted these burning timescales as a function of temperature (see Figs. 6 and 7), assuming a fuel mass fraction of 1. The curves for (also) density dependent burning processes are labeled with a typical density. He-burning has a quadratic density dependence, C- and O-burning depend linearly on density. If we take typical explosive burning timescales to be of the order of seconds (e.g. in supenovae), we see that one requires temperatures to burn essential parts of the fuel in excess of 4×10^9K (Si-burning), 3.3×10^9K (O-burning), 2.1×10^9K (Ne-burning), and 1.9×10^9K (C-burning). Beyond 10^9K He-burning is determined by an almost constant burning timescale. We see that essential destruction on a time scale of 1s is only possible for densities $\rho > 10^5$g cm^{-3}. This is usually not encountered in He-shells of massive stars. In a similar way explosive H-burning is not of relevance for massive stars, but important for explosive burning in accreted H-envelopes in binary stellar evolution (these issues are discussed by M. Wiescher, this volume).

3.2.1. *Explosive He-Burning*

Explosive He-burning is chararcterized by the same reactions as hydrostatic He-burning, producing ^{12}C and ^{16}O. Fig. 6 indicated that even for temperatures beyond 10^9K high densities ($>10^5$g cm^{-3}) are required to burn essential amounts of He. During the passage of a 10^{51}erg supernova shockfront through the He-burning zones of a 25M$_\odot$ star, maximum temperatures of only (6-9)$\times10^8$K are attained and the amount of He burned is negligible. However, neutron sources like ^{22}Ne$(\alpha,n)^{25}$Mg [or ^{13}C$(\alpha,n)^{16}$O], which sustain an s-process neutron flux in hydrostatic burning, release a large neutron flux under explosive conditions. This leads to partial destruction of ^{22}Ne and the build-up of ^{25,26}Mg via ^{22}Ne$(\alpha,n)^{25}$Mg$(n,\gamma)^{26}$Mg. Similarly, ^{18}O and ^{13}C are destroyed by alpha-induced reactions. This releases neutrons with $Y_n \approx 2\times10^{-9}$ at a density of $\approx 8.3\times10^3$g cm^{-3}, corresponding to $n_n \approx 10^{19}$cm^{-3} for about 0.2s, and causes neutron processing (Truran, Cowan, & Cameron 1978, Thielemann, Arnould, & Hillebrandt 1979, Thielemann, Metzinger, & Klapdor 1983, Cowan, Cameron, & Truran 1983). This is, however,

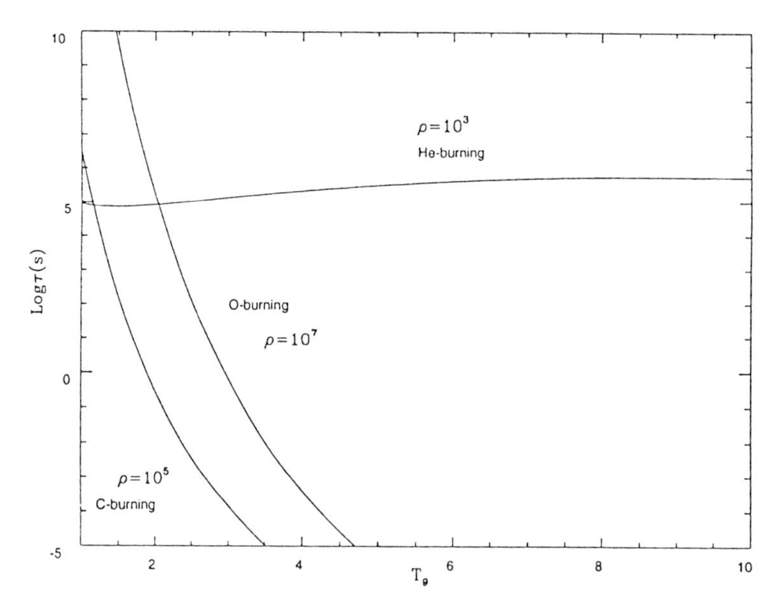

FIGURE 6. Burning timescales for fuel destruction of He-, C-, and O-burning as a function of temperature. A 100% fuel mass fraction was assumed. The factor $N^i_{i,i} = N_i/N_i!$ cancels for the destruction of identical particles by fusion reactions, as N_i=2. For He-burning the destruction of three identical particles has to be considered, which changes the leading factor $N^i_{i,i,i}$ to 1/2. The density-dependent burning timescales are labeled with the chosen typical density. They scale linearly for C- and O-burning and quadraticly for He-burning. Notice that the almost constant He-burning timescale beyond T_9=1 has the effect that efficient destruction on explosive timescales can only be attained for high densities.

not an r-process (Blake et al. 1981). More detailed calculations for such mass zones have recently been performed by Howard, Meyer, & Clayton (1992).

3.2.2. *Explosive C- and Ne-Burning*

The main burning products of explosive neon burning are ^{16}O, ^{24}Mg, and ^{28}Si, synthesized via the reaction sequences $^{20}\mathrm{Ne}(\gamma,\alpha)^{16}\mathrm{O}$ and $^{20}\mathrm{Ne}(\alpha,\gamma)^{24}\mathrm{Mg}(\alpha,\gamma)^{28}\mathrm{Si}$, similar to the hydrostatic case. The mass zones in supernovae which undergo explosive neon burning must have peak temperatures in excess of 2.1×10^9K. They undergo a combined version of explosive neon and carbon burning (see Figs. 6 and 7). Mass zones which experience temperatures in excess of 1.9×10^9K will undergo explosive carbon burning, as long as carbon fuel is available. This is often not the case in type II supernovae originating from massive stars. Besides the major abundances, mentioned above, explosive neon burning supplies also substantial amounts of ^{27}Al, ^{29}Si, ^{32}S, ^{30}Si, and ^{31}P. Explosive carbon burning contributes in addition the nuclei ^{20}Ne, ^{23}Na, ^{24}Mg, ^{25}Mg, and ^{26}Mg. Many nuclei in the mass range $20<A<30$ can be reproduced in solar proportions. This was confirmed for realistic stellar conditions by Morgan (1980). As photodisintegrations become important in explosive Ne-burning, also heavier pre-existing nuclei in such burning shells, from previous s- or r-processing (originating from prior stellar evolution or earlier stellar generations), can undergo e.g. (γ,n) or (γ,α) reactions. These can produce rare proton-rich stable isotopes of heavy elements. The relation to the so-called p-process is discussed e.g. in Woosley & Howard (1978), Rayet, Prantzos, & Arnould (1990), Howard, Meyer, & Woosley (1991), and Rayet et al. (1995).

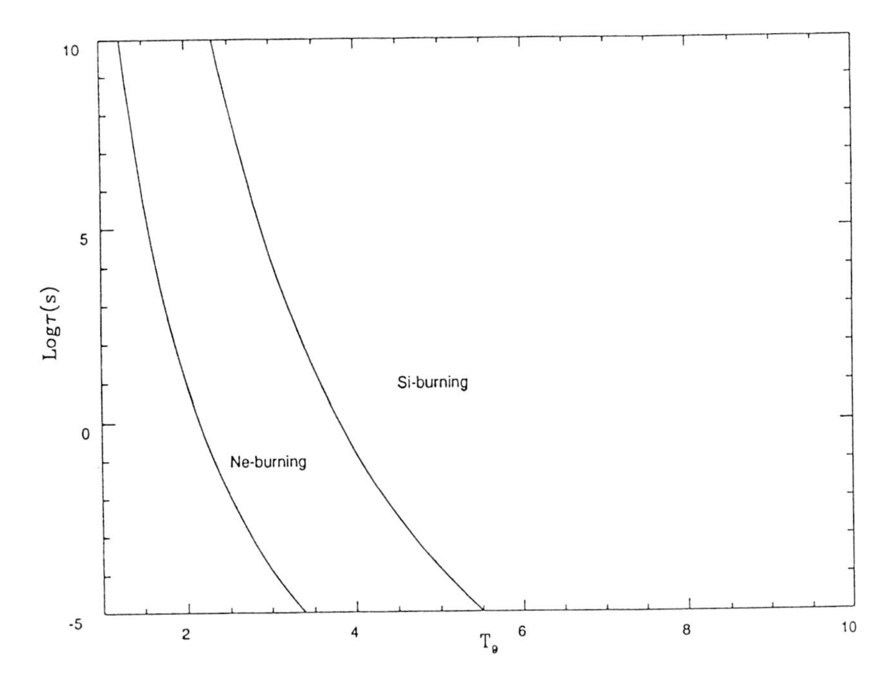

FIGURE 7. Burning timescales for fuel destruction of Ne- and Si-burning as a function of temperature. These are burning phases initiated by photodisintegrations and therefore not density-dependent.

3.2.3. *Explosive O-Burning*

Temperatures in excess of roughly 3.3×10^9K lead to a quasi-equilibrium (QSE) in the lower QSE-cluster which extends over the range 28<A<45 in mass number, while the path to heavier nuclei is blocked by small Q-values and reaction cross sections for reactions out of closed shell nuclei with Z or N=20 (see already Woosley, Arnett, & Clayton 1973 or Hix & Thielemann 1997). A full NSE with dominant abundances in the Fe-group cannot be attained. The main burning products are ^{28}Si, ^{32}S, ^{36}Ar, ^{40}Ca, ^{38}Ar, and ^{34}S, while ^{33}S, ^{39}K, ^{35}Cl, ^{42}Ca, and ^{37}Ar have mass fractions of less than 10^{-2}. In zones with temperatures close to 4×10^9K there exists some contamination by the Fe-group nuclei ^{54}Fe, ^{56}Ni, ^{52}Fe, ^{58}Ni, ^{55}Co, and ^{57}Ni.

The abundance distribution within the QSE-cluster is determined by alpha, neutron, and proton abundances. Because electron captures during explosive processing are negligible, the original neutron excess stays unaltered and fixes the neutron to proton ratio. Under those conditions the resulting composition is dependent only on the alpha to neutron ratio at freeze-out. In an extensive study Woosley, Arnett, and Clayton (1973) noted that with a neutron excess η of 2×10^{-3} the solar ratios of ^{39}K/^{35}Cl ^{40}Ca/^{36}Ar, ^{36}Ar/^{32}S, ^{37}Cl/^{35}Cl, ^{38}Ar/ ^{34}S, ^{42}Ca/^{38}Ar, ^{41}K/^{39}K, and ^{37}Cl/^{33}S are attained within a factor of 2 for freeze-out temperatures in the range $(3.1 - 3.9) \times 10^9$K. This is the typical neutron excess resulting from solar CNO-abundances, which are first transformed into ^{14}N in H-burning and then into ^{22}Ne in He-burning via $^{14}\mathrm{N}(\alpha,\gamma)^{18}\mathrm{F}(\beta^+)^{18}\mathrm{O}(\alpha,\gamma)^{22}\mathrm{Ne}$. Similar results were obtained earlier by Truran and Arnett (1970), while for lower values of the neutron excess (as expected for stars of lower metallicity) essentially only the alpha nuclei ^{28}Si, ^{32}S, ^{36}Ar, ^{40}Ca are produced in sufficient amounts (Truran and Arnett 1971). This affects element abundances and causes an odd-even staggering in Z.

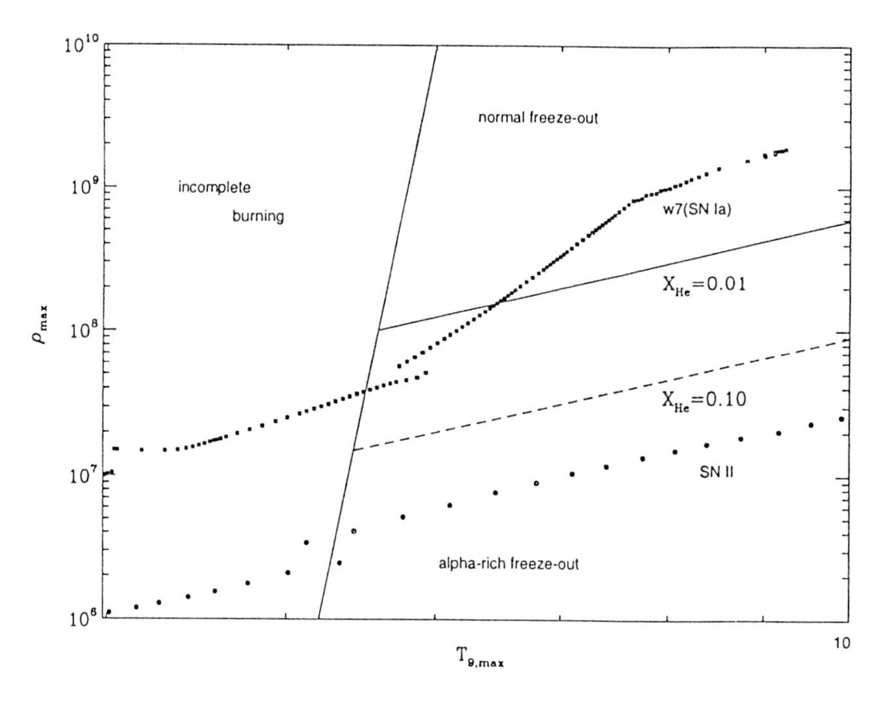

FIGURE 8. Division of the $\rho_{max} - T_{max}$-plane for adiabatic expansions from ρ_{max} and T_{max} with an adiabatic index of 4/3 and a hydrodynamic timescale equal to the free fall timescale. Conditions separate into incomplete and complete Si-burning with normal and alpha-rich freeze-out. Contour lines of constant ^{4}He mass fractions in complete burning are given for levels of 1 and 10%. They coincide with lines of constant radiation entropy per gram of matter. For comparision also the maximum $\rho - T$-conditions of individual mass zones in type Ia and type II supernovae are indicated.

3.2.4. *Explosive Si-Burning*

Zones which experience temperatures in excess of 4.0–5.0$\times 10^9$K undergo explosive Si-burning. For $T > 5\times 10^9$K essentially all Coulomb barriers can be overcome and a nuclear statistical equilibrium is established. Such temperatures lead to complete Si-exhaustion and produce Fe-group nuclei. The doubly-magic nucleus ^{56}Ni, with the largest binding energy per nucleon for $N=Z$, is formed with a dominant abundance in the Fe-group in case Y_e is larger than 0.49. Explosive Si-burning can be devided into three different regimes: (i) incomplete Si-burning and complete Si-burning with either (ii) a normal or (iii) an alpha-rich freeze-out. Which of the three regimes is encountered depends on the peak temperatures and densities attained during the passage of supernova shock front (see Fig. 20 in Woosley, Arnett, and Clayton 1973, and for applications to supernova calculations Fig. 4 in Thielemann et al. 1996b and Fig. 5 in Thielemann, Hashimoto, & Nomoto 1990 – combined here as Fig. 8). One recognizes that the mass zones of SNe Ia and SNe II experience different regions of complete Si-burning.

At high temperatures in complete Si-burning or also during a normal freeze-out, the abundances are in a full NSE and given by Eq.(3.31). An alpha-rich freeze-out is caused by the inability of the triple-alpha reaction ^{4}He$(2\alpha,\gamma)^{12}$C, transforming ^{4}He into ^{12}C, and the ^{4}He$(\alpha n,\gamma)^9$Be reaction, to keep light nuclei like n, p, and ^{4}He, and intermediate mass nuclei beyond A=12 in an NSE during declining temperatures, when the densities are small. The latter enter quadratically for these rates, causing during the fast expansion and cooling in explosive events a large alpha abundance after charged particle freeze-out, which shifts the QSE groups to heavier nuclei, tranforming e.g. ^{56}Ni, ^{57}Ni, and ^{58}Ni into ^{60}Zn, ^{61}Zn, and ^{62}Zn. This also leads to a slow supply of carbon nuclei still during freeze-out, leaving traces of alpha nuclei, ^{32}S, ^{36}Ar, ^{40}Ca, ^{44}Ti, ^{48}Cr, and ^{52}Fe, which did not fully make their way up to ^{56}Ni. Figs. 9ab show this effect, typical for SNe II, as a function of remaining alpha-particle mass fraction after freeze-out. It is clearly seen that the major NSE nuclei ^{56}Ni, ^{57}Ni, and ^{58}Ni get depleted when the remaining alpha fraction increases, while all other species mentioned above increase.

Incomplete Si-burning is characterized by peak temperatures of $4-5\times10^9$K. Temperatures are not high enough for an efficient bridging of the bottle neck above the proton magic number Z=20 by nuclear reactions. Besides the dominant fuel nuclei ^{28}Si and ^{32}S we find the alpha-nuclei ^{36}Ar and ^{40}Ca being most abundant. Partial leakage through the bottle neck above Z=20 produces ^{56}Ni and ^{54}Fe as dominant abundances in the Fe-group. Smaller amounts of ^{52}Fe, ^{58}Ni, ^{55}Co, and ^{57}Ni are encountered. All explosive burning phases discussed above will be applied in more detail to SNe II nucleosynthesis in section 4.

3.2.5. *The r-Process*

The operation of an r-process is characterized by the fact that 10 to 100 neutrons per seed nucleus (in the Fe-peak or somewhat beyond) have to be available to form all heavier r-process nuclei by neutron capture. For a composition of Fe-group nuclei and free neutrons that translates into a neutron excess of $\eta=0.4-0.7$ or Y_e=0.15-0.3. Such a high neutron excess can only by obtained through capture of energetic electrons (on protons or nuclei) which have to overcome large negative Q-values. This can be achieved by degenerate electrons with large Fermi energies and requires a compression to densities of $10^{11}-10^{12}$g cm^{-3}, with a beta equilibrium between electron captures and β^--decays (Cameron 1989) as found in neutron star matter (see also Meyer 1989).

Another option is an extremely alpha-rich freeze-out in complete Si-burning with moderate neutron excesses η and Y_e's (0.16 or 0.42, respectively). After the freeze-out of charged particle reactions in matter which expands from high temperatures but relatively low densities, 70, 80, 90 or 95% of all matter can be locked into ^{4}He with N=Z. Figure 9 showed the onset of such an extremely alpha-rich freeze-out by indicating contour lines for He mass fractions of 1 and 10%. These contour lines correspond to T_9^3/ρ=const, which is proportional to the entropy per gram of matter of a radiation dominated gas. Thus, the radiation entropy per gram of baryons can be used as a measure of the ratio between the remaining He mass-fraction and heavy nuclei. Similarly, the ratio of neutrons to Fe-group (or heavier) nuclei (i.e. the neutron to seed ratio) is a function of entropy and permits for high entropies, with large remaining He and neutron abundances and small heavy seed abundances, neutron captures which proceed to form the heaviest r-process nuclei (Woosley & Hoffman 1992, Meyer et al. 1992, Takahashi et al. 1994, Woosley et al. 1994b, Hoffman et al. 1996, 1997).

A different situation surfaces for maximum temperatures below freeze-out conditions for charged particle reactions with Fe-group nuclei. Then reactions among light nuclei which release neutrons, like (α,n) reactions on ^{13}C and ^{22}Ne, can sustain a neutron flux. The constraint of having 10-100 neutrons per heavy nucleus, in order to attain r-process conditions, can only be met by small abundances of Fe-group nuclei. Such conditions were expected when a shock front passes the He-burning shell and enhances the ^{22}Ne(α,n) reaction by orders of magnitude. However, Blake et al. (1981) and Cowan, Cameron, & Truran (1983) could show that this neutron source is not strong enough for an r-process in realistic stellar models (see also subsection 3.2.1). Recent research, based on additional neutron release via inelastic neutrino scattering (Epstein, Colgate, and Haxton 1988),

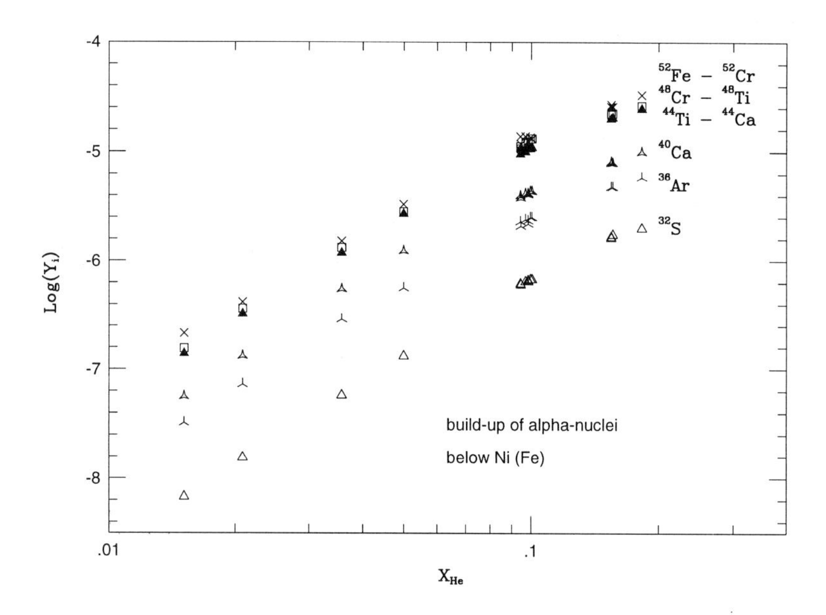

FIGURE 9. Display of the composition up to Cr and from Mn to Ni (after decay) as a function of remaining alpha mass-fraction X_{He} from an alpha-rich freeze-out. Lighter nuclei, being produced by alpha-captures from a remaining alpha reservoir, have larger abundances for more pronounced alpha-rich freeze-outs. Nuclei beyond Fe and Ni behave similarly, because of shifts in the dominant Fe-group nuclei caused by alpha-captures. Therefore, the dominant Fe-group nuclei like Fe and Ni show the opposite effect.

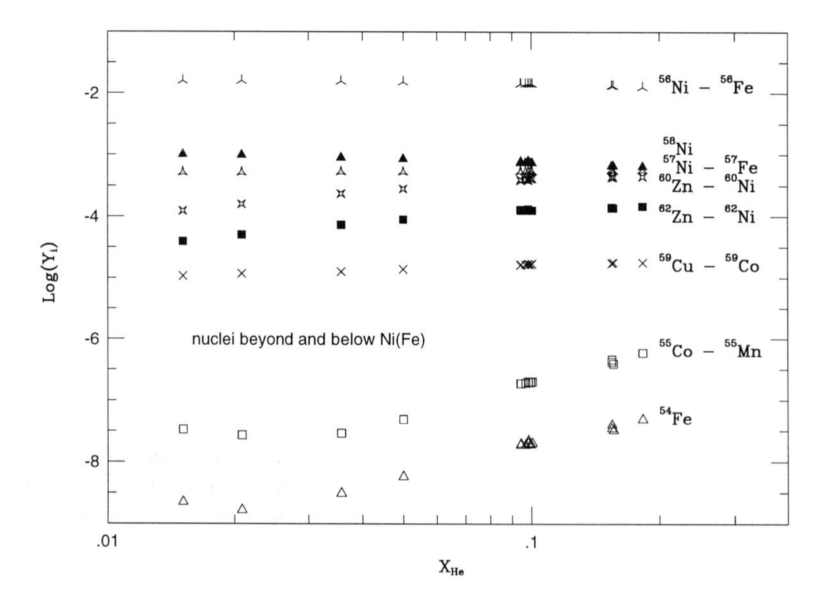

can also not produce neutron densities which are required for such a process to operate (see also Woosley et al. 1990 and Meyer 1995).

r-process calculations independent of a specific astrophysical site, and just based on the goal to find the required neutron number densities and temperatures which can reproduce the solar abundance pattern of heavy elements, have been performed for a number of years. The latest ones are e.g. Kratz et al. (1988), Thielemann et al. (1993), Kratz et al. (1993), Thielemann et al. (1994a), Chen et al. (1995), Bouquelle et al. (1996), Pfeiffer, Kratz, & Thielemann (1997), Kratz, Pfeiffer, & Thielemann (1997) and Freiburghaus et al. (1997b). They, together with applications to the astrophysical sites listed above, will be discussed in section 5.

3.3. *Nucleosynthesis in Supernovae*

In the following section 4 we will apply these explosive burning processes to nucleosynthesis calculations in supernova explosions from massive stars (SNe II) for nuclei with A<70. The discussion of the explosive production of heavier nuclei in supernovae will be given in section 5. There exist many original and review articles about the mechanisms of SNe II (e.g. Bruenn 1989ab, Cooperstein & Baron 1990, Wilson & Mayle 1988, Mayle & Wilson 1990, Bethe 1990, Bruenn & Haxton 1991; Wilson & Mayle 1993, Herant et al. 1994, Janka & Müller 1995,1996, Burrows 1996; Mezzacappa et al. 1997), so that we do not intend to repeat such a discussion here. We rather want to concentrate on the accompanying nucleosynthesis processes.

One of the major free parameters in stellar evolution, and thus for the pre-supernova models, is the still uncertain $^{12}C(\alpha,\gamma)^{16}O$ reaction (see Filippone, Humblet, & Langanke 1989, Caughlan et al. 1985, Caughlan and Fowler 1988, Barker & Kajino 1991, Buchmann et al. 1993, Zhao et al. 1993ab, Azuma et al. 1994, Langanke and Barnes 1996, Buchmann et al. 1996,1997). The permitted uncertainty range still spans almost over a factor of 3. However, also the treatment of convection in stellar evolution is not a settled one, especially the issue of overshooting and semiconvection. This has an influence on the possible growth of the He-burning core, which causes mixing in of fresh He at higher temperatures, and consequently also enhances the O/C ratio. Stellar evolution calculations by Langer and Henkel (1995) show that the total amount of ^{16}O can also vary by almost a factor of 3, for the extreme choices of the semi-convection parameter.

Thus, only the combination of these two uncertain parameters can be determined by comparison with abundance observations from supernova explosions. The calculations, presented in this review, are based on stellar models which made use of the Schwarzschild criterion of convection (Nomoto & Hashimoto 1988; Hashimoto et al. 1995) and employed the $^{12}C(\alpha,\gamma)^{16}O$-rate by Caughlan et al. (1985), which is one choice within the permitted uncertainty window.

4. Type II Supernova Explosions

All stars with main sequence masses $M>8M_\odot$ (e.g. Nomoto & Hashimoto 1988, Hashimoto, Iwamoto & Nomoto 1993, Weaver & Woosley 1993) produce a collapsing core after the end of their hydrostatic evolution, which proceeds to nuclear densities (for a review see e.g. Bethe 1990). The total energy released, 2-3$\times10^{53}$erg, equals the gravitational binding energy of a neutron star. Because neutrinos are the particles with the longest mean free path, they are able to carry away that energy in the fastest fashion. This was proven by the neutrino emission of supernova 1987A, detected in the Kamiokande, IMB and Baksan experiments (see Burrows 1990 for an overview).

The most promising mechanism for supernova explosions is based on neutrino heating beyond the hot proto-neutron star via the dominant processes $\nu_e + n \rightarrow p + e^-$ and $\bar{\nu}_e + p \rightarrow n + e^+$ with a (hopefully) about 1% efficiency in energy deposition (see also M. Guidry, this volume). The neutrino heating efficiency depends on the neutrino luminosity, which in turn is affected by neutrino opacities (e.g. Bruenn 1985, Sawyer 1989, Schinder 1990, Horowitz & Wehrberger 1992, Mezzacappa & Bruenn 1993, Keil & Janka 1995, Reddy & Prakash 1997, Reddy, Prakash, & Lattimer 1997). The explosion via neutrino heating is delayed after core collapse for a timescale of seconds or less. The exact delay time t_{de} depends on the question whether neutrinos diffuse out from the core (>0.5s), weak convection occurs due to composition gradients, or convective turnover due to entropy gradients shortens this escape time substantially (e.g. Burrows & Fryxell 1992, Janka & Müller 1993, Wilson & Mayle 1993, Herant et al. 1994, Bruenn, Mezzacappa, & Dineva 1995, Janka & Müller 1995,1996, Burrows 1996; Mezzacappa et al. 1997). The behavior of t_{de} as a function of stellar mass is still an open question and quantitative results of self-consistent calculations should still be taken with care, suggesting instead to make use of the fact that typical kinetic energies of 10^{51} erg are observed and light curve as well as explosive nucleosynthesis calculations can be performed by introducing a shock of appropriate energy in the pre-collapse stellar model (see e.g. Woosley & Weaver 1986, Shigeyama, Nomoto & Hashimoto 1988, Thielemann, Hashimoto, & Nomoto 1990, Aufderheide, Baron, & Thielemann 1991, Weaver and Woosley 1993, Woosley & Weaver 1995, Thielemann, Nomoto, & Hashimoto 1996, Nomoto et al. 1997). Due to these remaining open questions, present explosive nucleosynthesis calculations for SNe II are still based on such induced supernova explosions by either depositing thermal energy or invoking a piston with a given kinetik energy of the order 10^{51} erg, in order to process and eject matter outside the collapsed Fe-core of a massive star.

These are not self-consistent calculations, which would also precisely determine a mass cut between the central neutron star and the ejected envelope. Although self-consistent calculations show promising results in recent years, on the one hand one expects changes from 2D to more realistic 3D calculations, on the other hand issues like the mass cut are not consistently solved yet, and some models would eject very unwanted nucleosythesis products. Induced calculations, with the constraint of requiring ejected ^{56}Ni-masses from the innermost explosive Si-burning layers in agreement with supernova light curves, being powered by the decay chain ^{56}Ni-^{56}Co-^{56}Fe, are preferable at this point and can also serve as guidance to the solution of the whole supernova problem. Such mass cuts, based on ^{56}Ni in the ejecta, are always the "final" cuts, not necessarily the position of the high entropy bubble where neutrino heating causes the explosion. Massive stars will have some fallback, caused by reverse shocks reflected at density jumps in the outer layers. Recent observations of massive type II supernovae with very small amounts of ^{56}Ni are an indication for just this effect (Schmidt 1997, Turatto et al. 1997, Sollerman et al. 1998). Thus, when we will use the expression mass cut in the following, it will always relate to the final cut after fallback.

The composition of the innermost ejected layers is crucial and reflects aspects of the total energy in the shock and the temperatures attained due to it (responsable for ^{56}Ni), the neutronization of matter in form of Y_e, affecting the Fe-group composition in general and especially the ^{57}Ni/^{56}Ni ratio, and finally the entropy of the material which determines the degree of the alpha-rich freeze-out and with it the amount of some intermediate-mass alpha-elements like radioactive ^{44}Ti. Comparison with abundances from specific supernova observations or supernova remnants can teach a lot about these details and the supernova mechanism as a function of progenitor mass. The amount of detected ^{16}O and ^{12}C or products from carbon and explosive oxygen burning can constrain our knowledge of the *effective* ^{12}C$(\alpha,\gamma)^{16}$O rate in He-burning. The ^{57}Ni/^{56}Ni ratio can give constraints on Y_e in the innermost ejected zones. This helps to estimate the necessary delay time between collapse and the neutrino-driven explosion. Provided that the stellar pre-collapse models are reliable, this allows additional insight into the exact working of the supernova explosion mechanism.

All these aspects can be explored when being guided by comparison to observations (e.g. SN1987A, a $20M_\odot$ during the main sequence stage – see e.g. Arnett et al. 1989, McCray 1993, Fransson & Kozma 1993, Suntzeff et al. 1992, 1997, Kozma & Fransson 1997; SN 1993J, a $14\pm1M_\odot$ star during main sequence – see e.g. Nomoto et al. 1993, Shigeyama et al. 1994, Woosley et al. 1994a, Houck & Fransson 1996; type Ib and Ic supernova light curves like SN 1994I, which due to the lack of a large H-envelope and their early X-ray and gamma-ray losses are steeper than those of SNe II, but are also core collapse events – see e.g. Shigeyama et al. 1990, Nomoto et al. 1994, Iwamoto et al. 1994; the $^{57}Ni/^{56}Ni$ ratio deduced from γ-rays of the $^{56,57}Co$ decay or from spectral features changing during the decay time – see e.g. Clayton et al. 1992, Kurfess et al. 1992, Kumagai et al. 1993, Fransson & Kozma 1993, Varani et al. (1990); or supernova remnants like G292.0+1.8, N132D, CAS A – Hughes and Singh 1994, Blair et al. 1994, Iyudin et al. (1994), Dupraz et al. (1997), Hartmann et al. (1997); and comparison with abundances in low metallicity stars, which reflect the average SNe II composition (Wheeler, Sneden, & Truran 1989, Lambert 1989, Pagel 1991, Zhao & Magain 1990, Gratton & Sneden 1991, Edvardsson et al. 1993, Nissen et al. 1994, McWilliam et al. 1995, Schuster et al. 1996, Ryan, Norris, & Beers 1996, Norris, Ryan, & Beers 1996, Barbuy er al. 1997, Beers, Ryan, & Norris 1997, McWilliam 1997).

We concentrate here on the composition of the ejecta from such core collapse supernovae as an extension to earlier work (Hashimoto, Nomoto & Shigeyama 1989; Thielemann, Hashimoto & Nomoto 1990; Thielemann, Nomoto & Hashimoto 1993, 1994, 1996; Hashimoto et al. 1993, 1995; and Nomoto et al. 1997).

4.1. *Basic Nucleosynthesis Features*

The calculations were performed by depositing a total thermal energy of the order $E = 10^{51}$erg + the gravitational binding energy of the ejected envelope into several mass zones of the stellar Fe-core. A first overview of results from the explosion calculations can be seen in Table 3 of Thielemann et al. (1996a) for element abundances and Table 1 of Nomoto et al. (1997) for isotopic abundances in the supernova ejecta as a function of progenitor star mass. They can be characterized by the following behavior: the amount of ejected mass from the unaltered (essentially only hydrostatically processed) C-core and from explosive Ne/C-burning (C, O, Ne, Mg) varies strongly over the progenitor mass range, while the amount of mass from explosive O- and Si-burning (S, Ar, and Ca) is almost the same for all massive stars. Si has some contribution from hydrostatic burning and varies by a factor of 2-3. The amount of Fe-group nuclei ejected depends directly on the explosion mechanism. The values listed for the $20M_\odot$ star have been chosen to reproduce the $0.07M_\odot$ of ^{56}Ni deduced from light curve observations of SN 1987A. The choice for the other progenitor masses is also based on supernova light curve observations, but their uncertain nature should be underlined and a clearer picture is only emerging now with the observation of varying amounts of ^{56}Ni for varying progenitor star masses (see Blanton, Schmidt, & Kirshner 1995, Schmidt 1997, Turatto et al. 1997, Sollerman, Cumming, & Lundquist 1998).

Thus, we have essentially three types of elements, which test different aspects of supernovae, when comparing with individual observations. (i) The first set (C, O, Ne, Mg) tests the stellar progenitor models, (ii) the second (Si, S, Ar, Ca) the progenitor models and the explosion energy in the shock wave, while (iii) the Fe-group (beyond Ti) probes clearly in addition the actual supernova mechanism. Only when all three aspects of the predicted abundance yields can be verified with individual observational checks, it will be reasonably secure to utilize these results in chemical evolution calculations of galaxies (see e.g. Tsujimoto et al. 1995; Timmes et al. 1995; Pagel & Tautvaisiene 1995, 1997; Tsujimoto et al. 1997). In general we should keep in mind, that as long as the explosion mechanism is not completely and quantitatively understood yet, one has to assume a position of the mass cut which causes (not predicts!) a specific amount of ^{56}Ni ejecta. Dependent on that position, which is a function of the delay time between collapse and final explosion, the ejected mass zones will have a different neutron excess or $Y_e = < Z/A >$ of the nuclear composition, determining the ratio ^{57}Ni/^{57}Ni. The nature and amount of the energy deposition affects the entropy in the innermost ejected layers, and with it the degree of the alpha-rich freeze-out and amount of ^{44}Ti ejecta. We will discuss this in more detail in the following subsections.

4.2. *Ni(Fe)-Ejecta and the Mass Cut*

Figs. 10ab (both presenting a 13M$_\odot$ star) make clear how strongly a Y_e change can affect the resulting composition. Fig. 10a makes use of a constant Y_e=0.4989 in the inner ejcta, experiencing incomplete and complete Si-burning. Figure 10b makes use of the original Y_e, resulting from the pre-collapse burning phases. Here Y_e drops to 0.4915 for mass zones below $M(r)$=1.5M$_\odot$. Huge changes in the Fe-group composition can be noticed. The change in Y_e from 0.4989 to 0.4915 causes a tremendous change in the isotopic composition of the Fe-group for the affected mass regions (<1.5M$_\odot$). In the latter case the abundances of ^{58}Ni and ^{56}Ni become comparable. All neutron-rich isotopes increase (^{57}Ni, ^{58}Ni, ^{59}Cu, ^{61}Zn, and ^{62}Zn), the even-mass isotopes (^{58}Ni and ^{62}Zn) show the strongest effect. In Fig. 10 one can also recognize the increase of ^{40}Ca, ^{44}Ti, ^{48}Cr, and ^{52}Fe with an increasing remaining He mass fraction. These are direct consequences of a so-called alpha-rich freeze-out with increasing entropy.

While these calculations were performed by depositing energy at a specific radius inside the Fe-core and letting the shock wave propagate outward, this should involve the outer structure of the star after collapse and a the time t_{de}, when the successful shock wave is initiated. Instead they were taken at the onset of core collapse, which would corresponds to a prompt explosion without delay. In case of a delayed explosion, accretion onto the proto-neutron star will occur until finally after a delay period t_{de} a shock wave is formed, leading to the ejection of the outer layers. Aufderheide et al. (1991) performed a calculation with a model at t_{de}=0.29s after core collapse for a 20M$_\odot$ star, when the prompt shock had failed, and found an accretion caused increase of the mass cut by roughly ΔM_{acc}=0.02M$_\odot$. A delayed explosion could set in after a delay of up to 1s, with the exact time being somewhat uncertain and dependent on the details of neutrino transport (Wilson & Mayle 1993, Herant et al. 1994, Bruenn, Mezzacappa, & Dineva 1995, Janka & Müller 1996, Burrows 1996).

The outer boundary of explosive Si-burning with complete Si-exhaustion is given by T=5×10^{9}K and is also the outer boundary of ^{56}Ni production. From pure energetics it can be shown that this corresponds approximately to a radius r_5=3700 km for $E_{SN}\approx 10^{51}$ erg, independent of the progenitor models (Woosley 1988, Thielemann, Hashimoto, & Nomoto 1990). Therefore, the mass cut would be at

$$M_{cut} = M(r_5) - M_{ej}(^{56}\mathrm{Ni}). \tag{4.33}$$

In case of a delayed explosion, we have to ask the question from which radius $r_{0,5}(t=0)$ matter fell in, which is located at radius $r_5(t=t_{de})$=3700km when the shock wave emerges

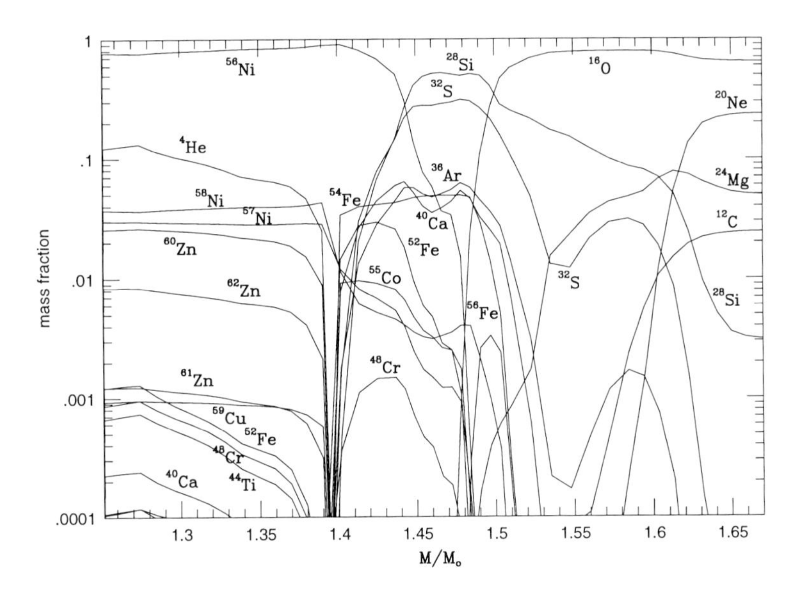

FIGURE 10. Isotopic composition of the ejecta for a core collapse supernova from a $13M_{\odot}$ star ($3.3M_{\odot}$ He-core). Only the dominant abundances of intermediate mass nuclei are plotted, while the Fe-group composition is presented in full detail. The exact mass cut in $M(r)$ between neutron star and ejecta depends on the details of the delayed explosion mechanism. Figures 10a and 10b show how strongly a Y_e-change can affect the resulting composition. Figure 10a makes use of a constant Y_e=0.4989 in the inner ejcta, Figure 10b makes use of the original Y_e, resulting from the pre-collapse burning phases, which drops to 0.4915 at the position for matter resulting from core O-burning, which experienced high densities and electron captures.

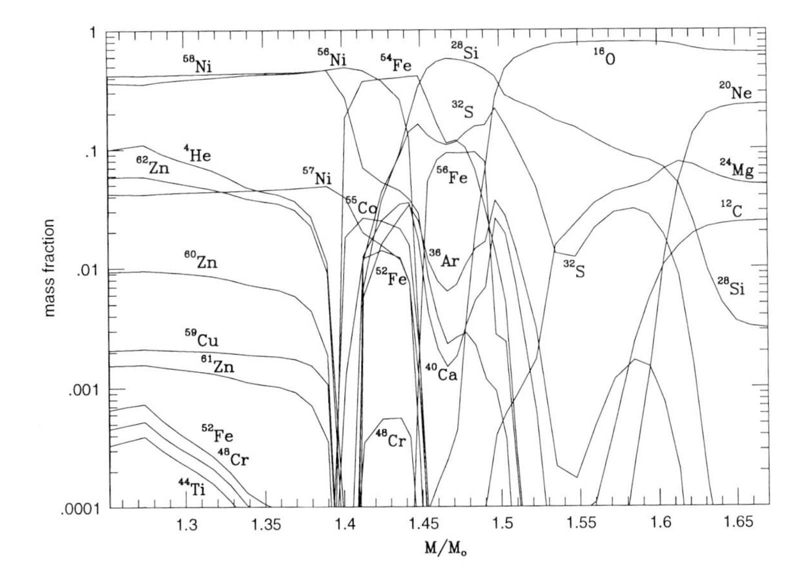

at time t_{de}. This effect of accretion as a function of delay time t_{de} has been studied in detail (Thielemann et al. 1996a). Here we want to present only the quantitative results.

In Figs. 11ab we display the Y_e-distributions of a 13 and a 20M$_\odot$ star and the position of the outer boundary of explosive Si-burning with complete Si exhaustion, M_{ex-Si}, as a function of the delay time t_{de}. We consider for each star delay times of 0, 0.3, 0.5, 1, and 2s, resulting in $r_{0,5}$=3700, 4042, 4412, 5410, and 7348km. Inside this boundary ^{56}Ni is produced as the dominant nucleus and the mass cuts would have to be positioned at $M_{cut}=M(r_{ex-Si}) - M_{ej}(^{56}\mathrm{Ni})=M(r_{0,5}(0)) - M_{ej}(^{56}\mathrm{Ni})$. When Ni-ejecta of 0.15 and 0.07M$_\odot$ are used for 13 and 20M$_\odot$ stars, mass cuts M_{cut} of 1.27 and 1.61M$_\odot$ result for a vanishing delay time. For $t_{de,i}$=0.3, 0.5, 1, and 2s the accreted masses $\Delta M_{acc,i}$ of 0.02, 0.03, 0.07-0.08, and 0.14-0.16M$_\odot$ have to be added to M_{cut}. It is recognizable that especially for the 13M$_\odot$ star the Y_e's encountered for these different delay times vary strongly, and differences of the Fe-group composition can be expected. Assuming that the stellar models are correct, all delay times less than 1s for the 13M$_\odot$ star are not compatible with the chemical evolution of our galaxy, as will be discussed below. On the other extreme, the Y_e in the innermost ejecta of a 25M$_\odot$ star are not affected at all by the available choices. A more detailed discussion for the 20M$_\odot$ star will follow.

The neutron star boundary would have to be moved outward, accordingly, by adding $\Delta M_{acc,i}$ mentioned in the previous paragraph. Whether a neutron star or black hole is formed depends on the permitted maximum neutron star mass, which is somewhat uncertain and related to the still limited understanding of the nuclear equation of state beyond nuclear densities (e.g. Glendenning 1991, Weber & Glendenning 1991, Brown & Bethe 1994, Prakash et al. 1997). A proto-neutron star with a baryonic mass M_b will release a binding energy E_{bin} in form of black body radiation in neutrinos during its contraction to neutron star densities. The gravitational mass is then given by

$$M_g = M_b - E_{bin}/c^2. \tag{4.34}$$

For reasonable uncertainties in the equation of state, Lattimer and Yahil (1989) obtained a relatively tight relation between gravitational mass and binding energy. Applying their expression results in a gravitational mass of the formed neutron star M_g. An error of roughly $\pm$15% for the difference $M_b - M_g$ applies. ΔM_{acc}, due to the uncertainty of the accretion period or delay time, and the choice of $M_{ej}(^{56}\mathrm{Ni})$ which determines M_{cut}, dominate the error in M_g of 1.16+(0-0.11)+(0.15-$M_{ej}(^{56}\mathrm{Ni})$) for the example of the 13M$_\odot$ star and 1.45+(0-0.12)+(0.07-$M_{ej}(^{56}\mathrm{Ni})$) for the 20M$_\odot$ and possible delay periods between 0 and 2s. The first bracket includes uncertainties in t_{de}, the second one in the actually ejected ^{56}Ni mass. A delay time of about 1s is expected to be an upper bound for the delayed explosions. This is close to a pure neutrino diffusion time scale without any convective turnover.

The results indicate a clear spread of neutron star masses. This spread would be preserved in real supernova events, unless a possible conspiracy in the combination of proto-neutron star masses, delay times, and explosion energetics (i.e. the explosion mechanism in general) leads to a smaller range in neutron star masses. A certain spread is also found in neutron star masses from observations (e.g. Nagase 1989, Page and Baron 1990, van Paradijs 1991, van Kerkwijk, van Paradijs, & Zuiderwijk 1995, van Paradijs & McClintock 1995, Thorsett 1996) but it is not clear to which extent it is just due to large observational errors. It is possible that the range predicted here already includes the uncertain upper mass limit of neutron stars due to the nuclear equation of state (Baym 1991, Weber and Glendenning 1991, Prakash et al. 1997). If it does, we would expect

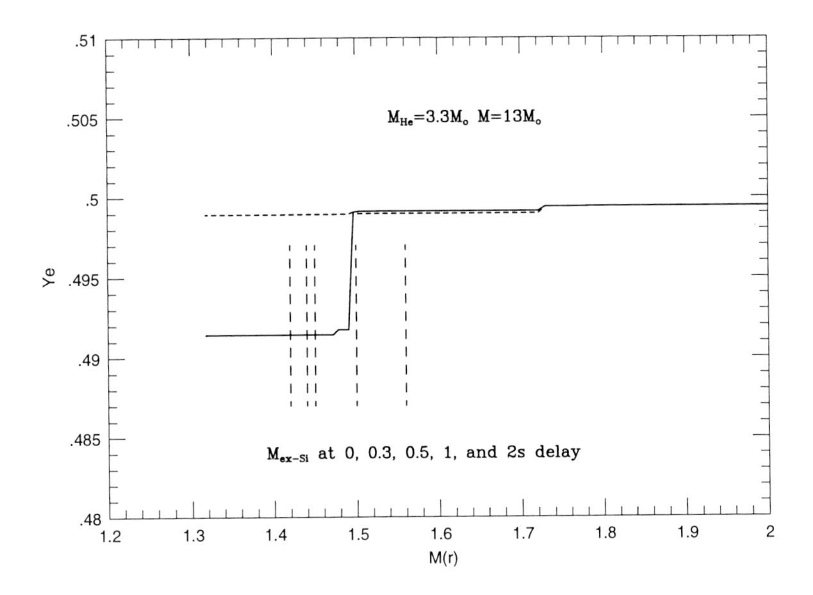

FIGURE 11. Figures 11ab present the Y_e-distributions of a 13 and 20M$_\odot$ star and the position of the outer boundary of explosive Si-burning with complete Si-exhaustion, M_{ex-Si}, as a function of the delay/accretion period t_{de}. For each star delay times of 0, 0.3, 0.5, 1, and 2s are considered, resulting in $r_{0,5}$=3700, 4042, 4412, 5410, and 7348km. ^{56}Ni is produced inside this boundary $r_{0,5}$ as the dominant nucleus. For a given amount of Ni-ejecta, mass cuts would have to be positioned at $M_{cut}=M(r_{ex-Si}) - M_{ej}(^{56}\mathrm{Ni})=M(r_{0,5}(0)) - M_{ej}(^{56}\mathrm{Ni})$. The delay times t_{de} and the required $M_{ej}(^{56}\mathrm{Ni})$ determine Y_e in the ejected material (solid=original, dashed=experienced for sufficiently large t_{de}, when low Y_e-matter is accreted onto the neutron star. The steep drop in Y_e corresponds to the edge of core O-burning.

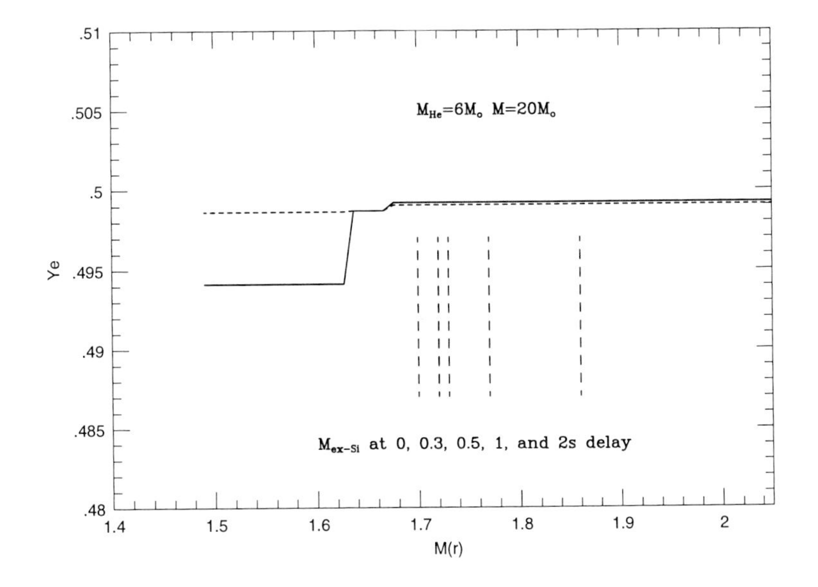

for these cases the formation of a central black hole during the delay period. Thus, no supernova explosion would occur and no yields be ejected. Different maximum stable masses between the initially hot and a cold neutron star (see e.g. Brown & Bethe 1994, Prakash et al. 1997 and the discussion of kaon condensates) could result in a supernova explosion *and* afterwards the formations of a central black hole. Timmes, Woosley, & Weaver (1996) had the stamina to predict a neutron star initial mass function based on ideas similar to the ones presented above. This is probably a somewhat bold undertaking, given the fact that we do not understand the supernova explosion mechanism fully, yet, and the ΔM_{acc} can vary widely. But there is one aspect which is worth mentioning. Due to the temperature dependence of the $^{12}C(\alpha,\gamma)^{16}O$ rate, stars above a critical mass limit will leave after core He-burning less than a critical amount of ^{12}C ($\approx$0.1M$_\odot$), which leads to radiative rather than convective core C-burning and finally the formation of large Fe-cores, which probably form black holes rather than supernovae. This would also agree with observations based on the O/Fe ratio in early galactic evolution, which requires an upper mass limit of about 25-50M$_\odot$ in order to avoid too high a production of oxygen (Maeder et al. 1992, Tsujimoto et al. 1997).

4.3. *Observational Constraints*

There exist a number of quantitative comparisons for SN1987A (a 20M$_\odot$ star during its main sequence evolution) between nucleosynthesis predictions and observations [see e.g. Table 2 in Danziger et al. (1990), section IVb in Thielemann et al. (1990) or McCray (1993), Fransson, Houck, & Kozma (1996), Chugai (1994)], which show reasonable agreement for C, O, Si, Cl, Ar, Co, and Ni (or Fe) between observation and theory. We want to concentrate here on a crucial aspect, the O abundance.

The amount of ^{16}O is closely linked to the "effective" $^{12}C(\alpha,\gamma)^{16}O$ rate during core He-burning. This effective rate is determined by three factors: (1) the actual nuclear rate, (2) the amount of overshooting, mixing fresh He-fuel into the core at late phases of He-burning, when the temperatures are relatively high and favor alpha-captures on ^{12}C, and (3) the stellar mass or He-core size, which determine the central temperature during He-burning.

We have discussed above the nuclear rate and its uncertainties and used in Thielemann et al. (1996a) the rate by Caughlan et al. (1985) based on an astrophysical S-factor of S_{tot}(0.3MeV)=0.24MeV barn. The S-factor is composed of an E1 component in the range 0.08±0.020 and an E2-component with a much larger uncertainty of 0.066-0.064+0.104 MeV barn, thus ranging in total from 0.062 to 0.270 Mev barn (the references where quoted in section 3). As the rate by Caughlan et al. (1985) seems to be close to the upper limit, it is crucial to check the observations for individual stellar models, in order to normalize the O-production correctly. The model calculations for a 20 M$_\odot$ star predict 1.48M$_\odot$ of ejected ^{16}O. This is within observational constraints by Franson, Houck & Kozma (1993) who found about 1.5M$_\odot$ and Chugai (1994) who determined 1.2-1.5M$_\odot$. It should, however, be clear that these observations test only the combined effect of nuclear rate and convection treatment (here Schwarzschild without overshooting). Similar results were found by Werner et al. (1995) when analyzing spectra of young white dwarfs with models of d'Antona and Mazzitelli (1992).

The O-determinations for SN 1993J from Houck & Fransson (1996) result in $\approx$0.5M$_\odot$. Thielemann et al. (1996a) predicted 0.423M$_\odot$ for a 15M$_\odot$ main sequence star, which agrees fairly well, SN 1993J was determined to be a 14±1M$_\odot$ star. This leads to the conclusion that the Caughlan et al. (1985) rate, used in conjunction with the Schwarzschild criterion for convection and no overshooting, gives a very good agreement with observations for individual supernovae. Only a comment about this combined usage can be made. Statements about the $^{12}C(\alpha,\gamma)^{16}O$ rate alone, in addition based on hydrostatic rather than explosive yields (Weaver & Woosley 1993) should be taken with some caution, and progress should preferably be made by improving nuclear cross sections and stellar convection treatment independently.

Recently other diagnostics became available for abundance determinations in supernova remnants. In that case the progenitor mass is not known, but the relative abundance ratios between different elements can be tested for consistency with abundance predictions for a variety of progenitor masses. Hughes and Singh (1994) made use of X-ray spectra of the supernova remnant G292.0+1.8 and found remarkable agreement for all element ratios from O through Ar with our $25M_\odot$ calculations (15% rms deviation). This tests implicitely the effective $^{12}C(\alpha,\gamma)^{16}O$-rate, as it is also reflected in the ratios between C-burning products like Ne and Mg and explosive O-burning products like Ar and S. Comparisons with other model predictions (Woosley & Weaver 1995) led to larger deviations. UV and optical observations of supernova remnant N132D by Blair, Raymond & Long (1994) give very good agreement with our element predictions for a $20M_\odot$ star, with slight deviations for Mg. Thus, we have direct observations of supernovae and supernova remnants ranging from 15 over 20 to $25M_\odot$, which agree well with our model predictions and indicate that their application for other purposes should be quite reliable. This has recently also been demonstrated for galactic chemical evolution calculations (Tsujimoto et al. 1995, 1997, Timmes, Woosley & Weaver 1995, Pagel & Tautvaisiene 1995,1997).

The formation of the nuclei $^{58,61,62}Ni$, which are produced in form of the neutron-rich species ^{58}Ni and $^{61,62}Zn$, is strongly dependent on Y_e and varies therefore with the position of the mass cut between ejected matter and the remaining neutron star (see the discussion in Thielemann et al. 1990 and Kumagai et al. 1991, 1993). Especially for the Ni-abundances the position of the mass cut is crucial. The $^{57}Ni/^{56}Ni$ ratio is correlated with the abundances of stable Ni isotopes, predominantly ^{58}Ni, i.e. with $^{58}Ni/^{56}Ni$. Light curve observations of SN1987A (Elias et al. 1991, Bouchet et al. 1991, Suntzeff et al. 1992) could be interpreted with a high 57/56 ratio of 4 times solar, but this would also have required too large stable Ni abundances not substantiated from observations (Witteborn et al. 1989, Wooden et al. 1993, 1997). In order to meet the stable Ni constraints of $3\text{-}5\times10^{-3}M_\odot$ (Danziger et al. 1990, Witteborn et al. 1989, and Wooden et al. 1993) only an upper limit of 1.4-1.7 times solar is permitted for the 57/56 ratio from our results, given in detail in Thielemann et al. (1996). This also agrees well with the observations by Varani et al. (1990) and γ-ray line observations by GRO (Kurfess et al. 1992, Clayton et al. 1992). The apparent 57/56 discrepancy was solved by correct light curve and spectra modeling with a non-equilibrium treatment of the involved ionization stages at late times (Fransson & Kozma 1993). This gives a consistent picture for observations of stable Ni, light curve observations which are sensitive to ^{56}Co and ^{57}Co decay, and the γ-ray lines emitted from both decays.

This corresponds to a Y_e at the mass cut of 0.4987 within the little nitch in Figure 11b. A mass cut at deeper layers, where Y_e decreases to 0.494, would imply 57/56 ratios larger than 2.5 times solar. A mass cut further out, implying a Y_e of 0.4989 results in a 57/56 ratio of the order of 1 times solar. This means that in order to meet the Y_e-constraint with an ejection of $0.075M_\odot$ of ^{56}Ni, we have a required delay time of 0.3-0.5s. Keeping all uncertainties of the model in mind, this can be taken as a support that SN 1987A did not explode via a prompt explosion, and did not experience a delayed explosion with a long delay time $t_{de}>0.5$s. The latter would correspond more to a pure neutrino diffusion case, while this result supports the understanding that larger neutrino luminosities are required than in the purely diffusive case (Herant et al. 1994; Burrows 1996; Janka & Müller 1996; Mezzacappa et al. 1997).

^{44}Ti is produced as a result of a strong alpha-rich freeze-out from explosive Si-burning, as discussed in section 3.2.4. Fig. 9 displays nicely that ^{44}Ti provides a measure of the entropy in the explosively processed matter. Exactly such conditions prevail in the innermost ejecta as can be seen in Figs. 10ab. Thus, we have another important observational constraint besides ^{56}Ni and ^{57}Ni, witnessing temperature, entropy and Y_e close to the mass cut. The predictions for ^{44}Ti ejecta range from 2×10^{-5} to 1.7×10^{-4} $M_\odot$ for stars ranging from 13 to 40 $M_\odot$ (Woosley & Weaver 1995, Thielemann, Nomoto, & Hashimoto 1996, Nomoto et al. 1997). Observational limits for supernova remnants have been described in Timmes et al. (1996) and recent GRO, COMPTEL gamma-ray observations of CAS A (Iyudin et al. 1994, Dupraz et al. 1997, Hartmann et al. 1997) yield $(1.27 \pm 0.34 \times 10^{-4})M_\odot$ with the new half life determinations between 59 and 62 y of Norman (1997), Görres et al. (1997), and Ahmad et al. (1997). This is a nice confirmation of nucleosynthesis predictions. Recent light curve calculations, based on the radioactive decay energies of ^{56}Ni, ^{57}Ni, and ^{44}Ti (Kozma & Fransson 1997), when compared with late time light curve observations of SN 1987A (Suntzeff et al. 1997), also come to the conclusion of $\approx 10^{-4}M_\odot$ ejecta of ^{44}Ti in good agreement with the predictions by Thielemann et al. (1996a), who obtained yields typically somewhat larger than Woosley & Weaver (1995), probably because energy deposition provides a somewhat larger entropy for the inner layers than induced explosions with the aid of a piston.

Unfortunately, we do not yet have similar observational and computational results for other supernovae. This would be a strong test for the explosion mechanism as a function of progenitor mass. It is important to explore the whole progenitor mass range with multidimensional explosion calculations in order to find out what Y_e and entropy self-consistent calculations would predict for the inner ejecta. Taken at face value, our 13$M_\odot$ model would ask for a delay time >1s, in order to avoid pollution of the galaxy with an unwanted Fe-group composition.

A further test for the correct behavior of the ejecta composition as a function of progenitor mass is the comparison with abundances in low metallicity stars. These reflect the average SNe II composition, integrated over an initial mass function of progenitor stars. First individual tests were done in Thielemann et al. (1990, 1996a). Applications to full chemical evolution calculations of the galaxy were performed by e.g. Tsujimoto et al. (1995), Timmes et al. (1995), Pagel & Tautvaisiene (1995, 1997), and Tsujimoto et al. (1997) and prove to be a clear testing ground for supernova models. A verification of SNe II ejecta in such a way permits a correct application in chemical evolution calculations together with SNe Ia and planetary nebula ejecta (stars of initial mass $M < 8M_\odot$ which form white dwarfs and eject their H- and He-burned envelopes).

5. The r-Process

The rapid neutron-capture process (r-process) leads to the production of highly unstable nuclei near the neutron drip-line and functions via neutron captures, (γ,n)-photodisintegrations, β^--decays and beta-delayed processes. Neutrino-induced reactions may also play a possible role. The r-process abundances witness the interplay between nuclear structure far from beta-stability and the appropriate astrophysical environment. Observations of heavy elements in low metallicity stars with abundances of Fe/H being 1/1000 to 1/100 of solar give information about stellar surface abundances, which are the abundances of the interstellar gas from which stars formed early in galactic evolution. Such observations show on the one hand an apparently completely solar r-process abundance pattern, at least for $A>130$, indicating that during such early times in galactic evolution only r-process sources and no s-process sources contributed to the production of heavy elements (Sneden et al. 1996, Cowan et al. 1997). That is consistent with the picture discussed in section 3.1 of the s-process origin in low and intermediate mass stars, which set in only at evolution times $>10^8$y.

On the other hand, it is also recognized that the r-process abundances come in with a delay of $>10^7$y after Fe and O (Mathews, Bazan, & Cowan 1992), which excludes the higher mass SNe II as r-process sources, because such massive stars beyond 10-12$M_\odot$ have shorter evolution times. The r-process has generally been associated with the inner ejecta of type II supernovae [see e.g. the reviews by Cowan et al. (1991) and Meyer (1994)], but also the decompression of neutron star matter was suggested by Lattimer et al. (1977), Meyer (1989), and Eichler et al. (1989) and is consistent with the above mentioned low metallicity observations. Both these environments provide or can possibly provide high neutron densities and high temperatures. Models trying to explain the whole r-process composition by low neutron density ($<10^{20}\,\mathrm{cm}^{-3}$) and temperature ($<10^9\,\mathrm{K}$) environments, like e.g. explosive He-burning in massive stars Thielemann et al. (1979), were clearly invalidated by Blake et al. (1981). The high entropy wind of the hot neutron star following type II supernova explosions has been suggested as a promising site for r-process nucleosynthesis by Woosley & Hoffman (1992), Woosley et al. (1994b), and Takahashi et al. (1994).

Actual r-process calculations usually followed two different approaches. Some studies, focusing mostly on nuclear physics issues far from stability, made use of a model-independent approach for the r-process as a function of neutron number densities n_n and temperatures T, extending for a duration time τ [see e.g. Kratz et al. (1988), Kratz et al. (1993), Thielemann et al. (1994a), Chen et al. (1995), Bouquelle et al. (1996), Pfeiffer et al. (1997), and Kratz et al. (1997)]. Other studies usually stayed closer to a specific astrophysical environment and followed the expansion of matter on expansion timescales τ with an initial entropy S, passing through declining temperatures and densities until the freeze-out of all reactions [see e.g. Woosley and Hoffman (1992), Meyer et al. (1992), Howard et al. (1993), Hoffman et al. (1996), Qian & Woosley (1996), Hoffman et al. (1997), Meyer & Brown (1997ab), and Surman et al. (1997)]. Here we compare the similarities and differences between the two approaches and whether there actually exists a one-to-one relation. Special emphasis is given to constraints, resulting from a comparison with solar r-process abundances in either approach, on nuclear properties far from stability. In addition, investigations are presented to test whether some features can also provide clear constraints on the permitted astrophysical conditions. This relates mostly to the $A<110$ mass range, where the high entropy scenario in supernovae faces problems.

5.1. *Model-Independent Studies*

The sequence of neutron captures, (γ,n)-photodisintegrations and beta-decays (and possibly additional reactions like beta-delayed neutron emission, fission etc.) have in principle to be followed with a detailed reaction network, given by a system of (several thousand) coupled differential equations with a dimension equal to the number of isotopes. This can be done efficiently, as shown in Cowan et al. (1991), however, approximations are also applicable for neutron densities and temperatures well in excess of $n_n>10^{20}\,\mathrm{cm}^{-3}$ and $T>10^9\,\mathrm{K}$, which cause reaction timescales as short as $\approx 10^{-4}\,\mathrm{s}$ [see Cameron et al. (1983), Bouquelle et al. (1996), and Goriely & Arnould (1996)]. As the beta-decay half-lives are longer, roughly of the order of $10^{-1}\,\mathrm{s}$ to a few times $10^{-3}\,\mathrm{s}$, an equilibrium can set in for neutron captures and photodisintegrations. Such conditions allow to make use of the "waiting point approximation", sometimes also called the "canonical r-process", which is equivalent to an $(n,\gamma)-(\gamma,n)$-equilibrium $[n_n \langle\sigma v\rangle_{n,\gamma}^{Z,A} Y_{(Z,A)} = \rho N_A \langle\sigma v\rangle_{n,\gamma}^{Z,A} Y_n Y_{(Z,A)} = \lambda_{\gamma,n}^{Z,A+1} Y_{(Z,A+1)}$, see Eq.(1.11)] for all nuclei in an isotopic chain with charge number Z. As the photodisintegration rate $\lambda_{\gamma,n}^{Z,A+1}$ is related to the capture rate $\langle\sigma v\rangle_{n,\gamma}^{Z,A}$ by detailed balance and proportional to the capture rate times $\exp(-Q/kT)$, as shown in Eq.(1.7), the maximum abundance in each isotopic chain (where $Y_{(Z,A)} \approx Y_{(Z,A+1)}$) is located at the same neutron separation energy S_n, being the neutron-capture Q-value of nucleus (Z,A). This permits to express the location of the "r-process path", i.e. the contour lines of neutron separation energies corresponding to the maximum in all isotopic chains, in terms of the neutron number density n_n and the temperature T in an astrophysical environment, when smaller effects like ratios of partition functions are neglected, as reviewed in Cowan et al. (1991).

The nuclei in such r-process paths, which are responsible for the solar r-process abundances, are highly neutron-rich, unstable, and located $15-35$ units away from β-stability with neutron separation energies of the order $S_n = 2-4$ MeV. These are predominantly nuclei not accessible in laboratory experiments to date. The exceptions in the $A = 80$ and 130 peaks were shown in Kratz et al. (1988) and Kratz et al. (1993) and continuous efforts are underway to extend experimental information in these regions of the closed shells N=50 and 82 with radioactve ion beam facilities. The dependence on nuclear masses or mass model predictions enters via S_n. The beta-decay properties along contour lines of constant S_n towards heavy nuclei [see e.g. Fig. 4 in Thielemann et al. (1994a) or Fig. 12 below for the region around the N=82 shell closure] are responsible for the resulting abundance pattern. The build-up of heavy nuclei is governed within the waiting point approximation only by effective decay rates λ_β^Z of isotopic chains. Then the environment properties n_n and T (defining the S_n of the path), and the duration time τ, predict the abundances. In case the duration time τ is larger than the longest half-lives encountered in such a path, also a steady flow of beta-decays will follow, making the abundance ratios independent of τ ($\lambda_\beta^Z Y_{(Z)} = const.$ for all Z's, where $Y_{(Z)}$ is the total abundance of an isotopic chain and λ_β^Z its effective decay rate).

One has to recognize a number of idealizations in this picture. It assumes a constant $S_n(n_n,T)$ over a duration time τ. Then the nuclei will still be existent in form of highly unstable isotopes, which have to decay back to beta-stability. In reality n_n and T will be time-dependent. As long as both are high enough to ensure the waiting point approximation, this is not a problem, because the system will immediately adjust to the new equilibrium and only the new $S_n(n_n,T)$ is important. The prominent question is whether the decrease from equilibrium conditions in n_n and T (neutron freeze-out), which initially ensure the waiting point approximation, down to conditions where the competition of neutron captures and beta-decays has to be taken into account explicitely, will affect the abundances strongly. In our earlier investigations we considered a sudden drop in n_n and T, leading to a sudden "freeze-out" of this abundance pattern, and only beta-decays and also beta-delayed properties [neutron emission and fission] have to be taken into account for the final decay back to stability [see e.g. the effect displayed in Fig. 9 of Kratz et al. (1993)].

When following this strategy, the analysis of the solar-system isotopic r-process abundance pattern showed that a minimum of three components with different S_n's, characterizing different r-process paths, was necessary for correctly reproducing the three peaks at $A \simeq 80$, 130, and 195 and the abundances in between [Thielemann et al. (1993a), Kratz et al. (1993)]. The "low-A wings" of the peaks (when making use of experimental beta-decay properties at the magic neutron numbers $N = 50$ and 82), as well as

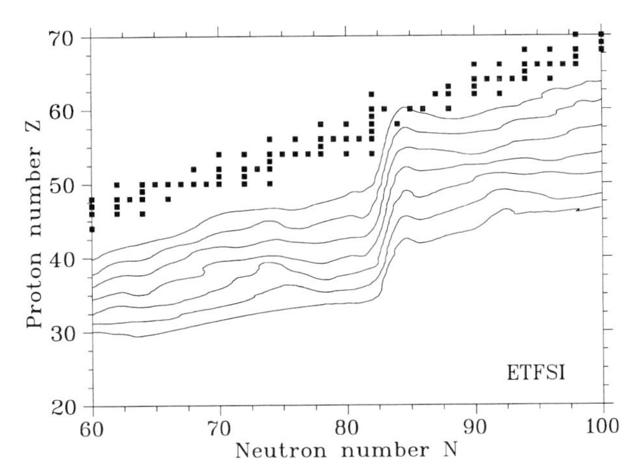

FIGURE 12. Contour plots of constant neutron separation energies S_n=1,2,3,4,5,6, and 7 MeV in the $80 \leq A \leq 140$ mass region for the ETFSI mass model Aboussir et al. (1995). The saddle point behavior before the shell closure at $N = 82$, also existing when using the FRDM masses by Möller et al. (1995), causes a deep trough before the peak at $A = 130$ (see upper part of Fig. 13), as the step from the abundance maximum of an isotopic chain Z to $Z+1$ can also cause a large jump in N or equivalently A, leading to a large number of unpopulated mass numbers A.

the abundance pattern down to the next peak, could be reproduced, even with the assumption of a steady flow of beta-decays. This indicates that the astrophysical duration timescales τ are large in comparison to most of the beta-decay half-lives encountered and only comparable to the longest half-lives in the peaks (where the path comes closest to stability, see e.g. a 2 MeV contour line in Fig. 12), which control the leaking out to larger A's. A continuous superposition of components with varying n_n, T or $S_n(n_n, T)$ (rather than only three), as expected in an astrophysical environment, with equidistant steps in S_n between 2 and 4 MeV and τ between 1 and 2.5 s led to a slight, but not dramatic, change/improvement of the abundance curve in Kratz et al. (1994).

When the calculations of Kratz et al. (1993) were supplemented by use of the most modern mass formula data [Finite Range Droplet Model FRDM by Möller et al. (1995) and Extended Thomas-Fermi model with Strutinski Integral ETFSI by Aboussir et al. (1995), instead of using a somewhat dated but still very successful droplet model by Hilf, von Groote, & Takahashi (1976), we could show that abundance troughs appeared before (and after) the 130 and 195 abundances peaks, due to the behavior of the S_n contour lines of these mass models [Thielemann et al. (1994a), Chen et al. (1995)]. The location in N of an r-process path with a given S_n does not behave smoothly as a function of Z. Fig. 12 indicates a sudden jump to the position of the magic neutron number, where the contour lines show a saddle point behavior for the FRDM as well as ETFSI mass models. The population gap of nuclei as a function of A leads after decay to the abundance trough of Fig. 13. The upper part of Fig. 13 shows the abundance curve obtained with ETFSI [Aboussir et al. (1995)] nuclear masses and beta-decay properties from a quasi-particle random-phase approximation [QRPA, Möller et al. (1997)]. When using FRDM masses by Möller et al. (1995) instead of the ETFSI predictions, a similar picture is obtained as shown in Thielemann et al. (1994a) and Bouquelle et al. (1996).

Additional tests were performed in order to see how this pattern could be avoided with different nuclear structure properties far from stability. The problem could be resolved in Chen et al. (1995), if for very neutron-rich nuclei the shell gap at the magic

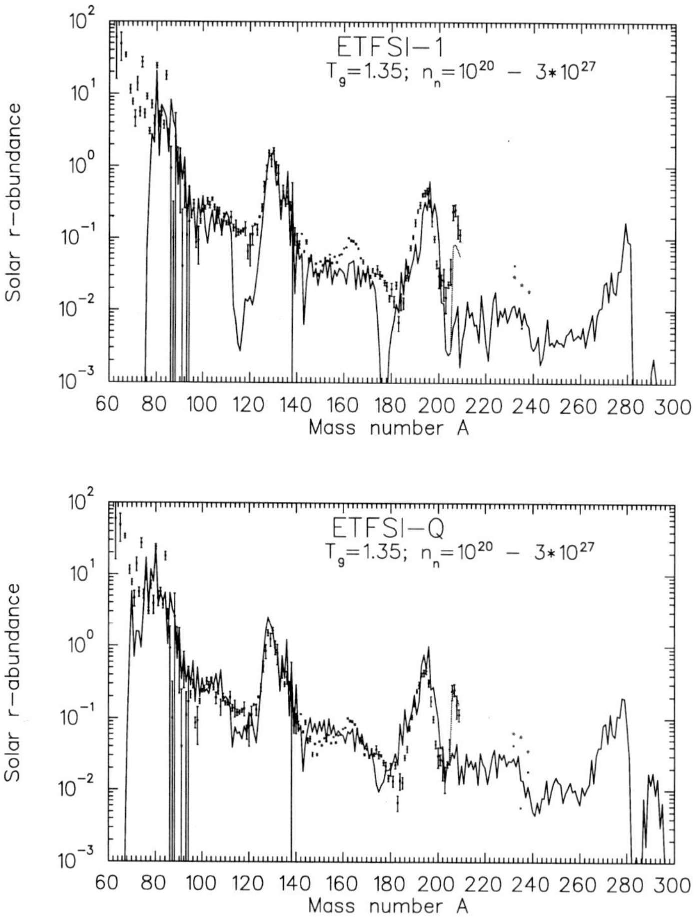

FIGURE 13. Fits to solar r-process abundances by Käppeler et al. (1989), obtained with 17 equidistant $S_n(n_n, T)$ components from 1 to 4 MeV. In the upper part, the result is presented for ETFSI masses with half-lives $\tau_{1/2}$ and beta-delayed neutron emission P_n values from QRPA calculations. In the lower part, the ETFSI-Q mass model by Pearson et al. (1996) was applied, which introduced a phenomenological quenching of shell effects, comparable to HFB calculations with the Skyrme force SkP of Dobaczewski, Nazarewicz, & Werner (1995). The quenching of the $N = 82$ shell gap leads to a filling of the abundance troughs and to a better overall reproduction of the heavy mass region. These results by Pfeiffer et al. (1997) are also the first which show a good fit to the r-process Pb and Bi contributions after following the decay chains of unstable heavier nuclei. For ^{232}Th, ^{235,238}U the solar and r-process production abundances are shown, allowing apparently for a increasing amount of decay with decreasing decay half-lives (in the sequence 232, 238, 235).

neutron number $N = 82$ is less pronounced, i.e. quenched, than predicted by the global macroscopic-microscopic mass models. In light nuclei, the quenching of shells in neutron-rich isotopes is well established and a long-studied effect [see Orr (1991), Campi et al. (1975), Fukunishi, Otsuka, & Sebe (1992), and Sorlin et al. (1993)]. The Hartree-Fock-Bogoliubov calculations by Werner et al. (1994), Dobaczewski et al. (1994), and Dobaczewski, Nazarewicz, & Werner (1995) with a specific Skyrme force had exactly the expected effect on the r-process path and the resulting abundance curve, as shown in Chen et al. (1995). This effect was recently also confirmed by Pearson et al. (1996), when the ETFSI mass formula was phenomenologically quenched in a similar way as the HFB results and led to a very good agreement with solar r-abundances in a more systematic study by Pfeiffer et al. (1997) shown in the lower part of Fig. 13. An experimental investigation of shell quenching along the $N = 50$ and 82 shell towards more neutron-rich nuclei (and approaching the r-process path for $N = 126$) is a highly desirable goal in order to test the nuclear structure responsible for the solar abundances of heavy nuclei.

There are two aspects which have to be considered when trying to relate these simplified, model-independent results to astrophysics: (a) what kind of environments can produce the required conditions, and (b) do the nuclear structure conclusions drawn from the sudden freeze-out approximation stay valid for actual freeze-out timescales encountered in a specific environment? The second question cannot be answered in general, but only case by case. The question whether we understand fully all astrophysical sites leading to an r-process is not a settled one. There are strong indications that it is associated with type II supernovae. But galactic evolution timescales indicate that these can probably only be the low mass SNe II with longer evolution timescales Cowan et al. (1991), Mathews et al. (1992), while neutron star mergers or still other sites are not necessarily excluded Lattimer et al. (1977), Meyer (1989), Eichler et al. (1989).

5.2. *Parameter Studies for High Entropies*

5.2.1. *The Model and Nuclear Input*

Recent r-process studies by Woosley et al. (1994b), Takahashi et al. (1994), Qian & Woosley (1996), and Hoffman et al. (1997) have concentrated on the hot, neutron-rich environment in the innermost ejecta of type-II supernovae, also called the neutrino wind. These are the layers heated by neutrino emission and evaporating from the hot proto-neutron star after core collapse. These calculations obtain neutron separation energies of the r-process path S_n of $2 - 4\,\mathrm{MeV}$, in agreement with the conclusions of section 5.1. Whether the entropies required for these conditions can really be attained in supernova explosions has still to be verified. In relation to the questions discussed in section 5.1, it also has to be investigated whether a sudden freeze-out is a good approximation to these astrophysical conditions. In order to test this, and how explosion entropies can be translated into n_n and T (or S_n) of the model independent approach, we performed a parameter study based on the entropy S and the total proton to nucleon ratio Y_e (which measures the neutron-richness of the initial composition), in combination with an expansion timescale (for the radius of a blob of matter) of typically 0.05 s as in Takahashi et al. (1994), and varied nuclear properties (i.e. mass models) like in section 5.1.

Thus, a hot blob of matter with entropy S, (i) initially consisting of neutrons, protons and some alpha-particles in NSE ratios given by Y_e, expands adiabatically and cools, (ii) the nucleons and alphas combine to heavier nuclei (typically Fe-group) with some neutrons and alphas remaining, (iii) for high entropies an alpha-rich freeze-out from charged-particle reactions occurs for declining temperatures, leading to nuclei in the mass range $A \approx 80 - 100$, and (iv) finally these remaining nuclei with total abundance Y_{seed} can capture the remaining neutrons Y_n and undergo an r-process. We chose a parameterized model for the expansion, essentially to introduce an expansion timescale, which makes these calculations independent of any specific supernova environment. But we will have to test later whether the expansion timescale employed is relevant to the supernova problem. The calculations were performed for a grid of entropies S and electron abundances Y_e ($S = 3, 10, 20, 30, \ldots 390 k_B$/baryon and $Y_e = 0.29, 0.31, \ldots 0.49$). Neutron capture rates were calculated with the new version of the statistical model code

SMOKER by Rauscher et al. (1997), discussed in section 2. The β^--rates came from experimental data or QRPA calculation by Möller et al. (1997).

Different mass zones have different initial entropies, which leads therefore to a superposition of different contributions in the total ejecta. For each pair of parameters Y_e and S, the calculations were initially started with a full charged particle nuclear network up to Pd. After the α-rich freeze-out, an r-process network containing only neutron induced reactions and beta-decay properties followed the further evolution. The dynamical r-process calculations were performed in the way as described in Cowan, Thielemann, & Truran (1991) and Rauscher et al. (1994). The amount of subsequent r-processing depends on the available number of neutrons per heavy nucleus Y_n/Y_{seed} ($Y_{seed} = \sum_{A>4} Y_{(Z,A)}$). In Fig. 14a the Y_n/Y_{seed}-ratio is plotted in the (S, Y_e)-plane. A simple scaling with Y_e is clearly visible. Fig. 14b also shows that low Y_e-values would be one mean to avoid the very high entropies required to obtain large Y_n/Y_{seed}-ratios.

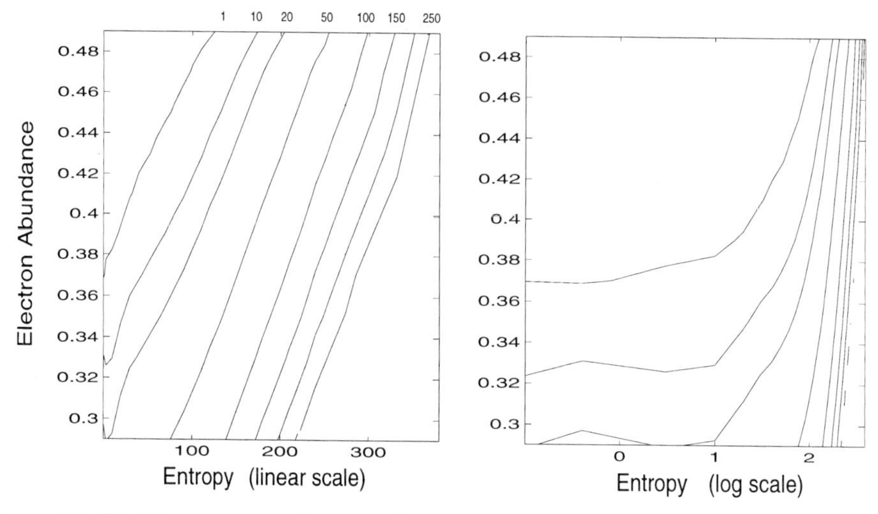

FIGURE 14. Y_n/Y_{seed} contour plots as a function of intial entropy S and Y_e for an expansion time scale of 0.05s, as expected from type II supernova conditions.

5.2.2. *Superpositions of Entropies*

The remaining question is, what kind of superposition of entropies the astrophysical environment provides. The calculations of Witti et al. (1994) showed that the amount of mass ejected per entropy interval was relatively constant at late phases (when the higher entropy matter was expelled) and declining slightly at early phases (lower entropies) as a function of time (i.e. with increasing entropy). We did not perform complete hydro calculations, but rather followed these findings within a parametrized way, which allows to optimize for the best possible fit to the solar abundance distribution via a weighting function $g(S_i)$ with $g(S_i) = x_1 e^{-x_2 S_i}$, where i is the index of the components. We restrict ourselves here to two different Y_e-sequences with Y_e=0.45 and 0.49 shown in Fig. 15.

Entropies from about 200 to about 350 give Y_n/Y_{seed}-ratios growing from approximately 30 to 150. The α-rich freeze-out always produces seed nuclei in the range $90 < A < 120$. This material can then be "r-processed", leading to a fully neutron dominated process as discussed in section 5.1 and the components have a very similar abundance pattern in the mass range $A = 110 - 200$. Thus, it is possible for this entropy range to establish a one-to-one correspondence for abundances obtained in r-process conditions between entropy and expansion timescale (S, τ) in one type of calculation and a neutron separation energy of the r-path and timescale $(S_n(n_n, T), \tau)$ in the calculations discussed in section 5.1. The neutron separation energy S_n of the r-path is the one obtained during neutron capture freeze-out in the entropy based calculations. This correspondence can, however, only be established for entropies producing nuclei with $A > 110$.

Matter for $A < 110$ is a result of lower entropies with a neutron-poor and alpha-rich freeze-out, where the abundance of heavy nuclei is dominated by nuclei with alpha separation energies of ≈ 6 MeV and a $Z/A = Y_{e,heavy}$ of the dominating heavy nucleus after charged particle freeze-out, resulting from $Y_{e,global} = \sum_i Z_i Y_i \approx 0.5 X_\alpha + Y_{e,heavy} X_{heavy}$ with mass fractions $X_i = A_i Y_i$ [for more details see Freiburghaus et al. (1997ab)]. None of the entropies produces an abundance peak at charged particle freeze-out with $A < 80$, leaving a sufficient amount of neutrons for an r-processing which would reproduce the typical neutron-induced abundance features in the range $A = 80 - 110$. A different choice of $Y_{e,global}$ (shown here for 0.45 and 0.49) can influence that pattern somewhat in avoiding very large spikes for $A \approx 90$ and $N = 50$ isotopes, but the overall features stay.

Beyond $A = 110$ different mass models (in Fig. 15 only ETFSI is shown) give fits of similar quality as those displayed in section 5.1. The discrepancies below the $A = 130$ r-process peak, in form of a pronounced trough, occur again for the FRDM and ETFSI mass model. Thus, our results and conclusions from 5.1 can be translated also to "realistic" astrophysical applications for this mass region. The nuclear structure properties leading to agreement and deficiencies apply in the same way, due to the nature of a fast freeze-out, which preserves the abundances as they result from an initial $(n, \gamma) - (\gamma, n)$-equilibrium at high temperatures, even when neutron captures and photodisintegrations are followed independently. Figure 16 shows the neutron number densities as a function of time. Low entropies ($S = 3 - 150$) that contribute to the mass range between $90 < A < 140$ lead to an r-process with a fast (almost sudden) freeze out on short timescales of $\tau \approx 0.04$ s. Thus it is not surprising that the trough before $A = 130$ (due to shell structure far from stability and its effect on abundance patterns in $(n, \gamma) - (\gamma, n)$-equilibrium) survives.

There is possibly one difference to the conclusions given with Figure 13. As can be seen from Figure 16, the calculations experiencing the highest entropies have the longest neutron freeze-out timescales. On the other hand, they are responsible for the heaviest nuclei with the largest neutron capture cross sections. Our results show that the trough before the $A = 195$ peak, resulting in case of the ETFSI mass model and a waiting point approach, does not survive [see Thielemann et al. (1994a), Chen et al. (1995), Bouquelle et al. (1996), and Pfeiffer et al. (1997)]. This r-process abundance region is changed by ongoing (non-equilibrium) captures during the freeze-out and does not directly witness nuclear properties far from stability at the $N = 126$ shell closure. In Fig. 15 we actually observe a filling of the minimum before the $A = 195$ peak and even the ETFSI masses, that produced the largest trough in the waiting point calculations, seem to give a good fit.

There have been suggestions that neutrino-induced spallation of nuclei in the $A = 130$ peak, caused by the strong neutrino wind from the hot neutron star, could fill the abundance trough Qian et al. (1996). We refer to a more detailed discussion of this effect in Thielemann et al. (1997) and Freiburghaus (1997ab), including the requirements on neutrino luminosities and distances of matter from the neutron star at the time of the neutron freeze-out. We come to the conclusion that as an alternative interpretation the nuclear structure effects (shell quenching far from stability) outlined in detail in section 5.1 are still preferred, especially as they are already observed experimentally for lighter nuclei.

What can we learn from these entropy based studies and the fact that the r-process

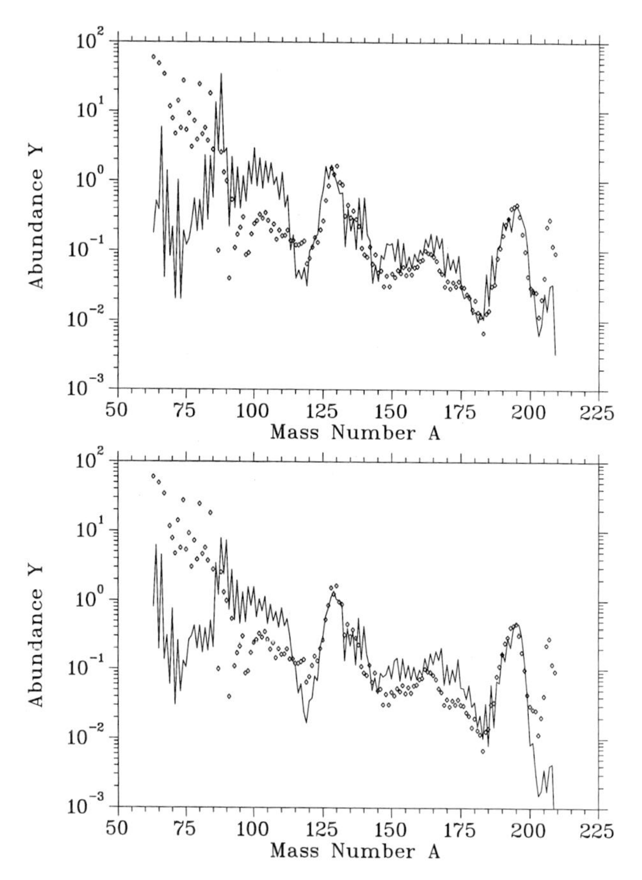

FIGURE 15. Similar to Fig 13 with the ETFSI mass formula, making use of a superposition of entropies $g(S)$ to attain an overall good fit to solar r-process abundances from the high entropy neutrino wind in type II supernovae. These calculations were performed with $Y_e = 0.45$, and 0.49, but similar results are obtained in the range 0.30 – 0.49, only requiring a scaling of entropy. The trough below $A = 130$ behaves similar to Fig. 13. This shows that a time dependent freeze-out (with a full treatment of neutron captures and photodisintegrations) resulting from a more realistic astrophysical scenario, can cause the same abundance deficiencies due to specific nuclear structure features as obtained in an instantaneous freeze-out from $(n, \gamma) - (\gamma, n)$-equilibrium. The trough before the $A = 195$ peak existing e.g. for the ETFSI mass formula in the waiting point approximation and an instantaneous freeze-out is filled due to non-equilibrium freeze-out neutron captures. The strong deficiencies in the abundance pattern below $A = 110$ are due to the alpha-rich freeze-out and thus related to the astrophysical scenario rather than to nuclear structure.

abundances below A=110 cannot be reproduced correctly? There are several possible conclusions: (a) the high entropy wind is not the correct r-process site (on the one hand due to the inherent deficiencies in the abundance pattern below $A = 110$ and the problems to obtain the high entropies in SNe II explosions required for producing the massive r-process nuclei up to $A \simeq 195$ and beyond, or (b) the high entropy wind overcomes the problems to attain the high entropies and produces only the masses beyond $A = 110$, avoiding or diluting the ejection of the lower entropy matter. In the latter case another site is responsible for the lower mass region. An extension of Y_e to smaller values, as low as 0.3, could also solve the problem, and constraints on ν_e and $\bar{\nu}_e$ fluxes and mean energies in the supernova environment have been explored by Qian & Woosley (1996)

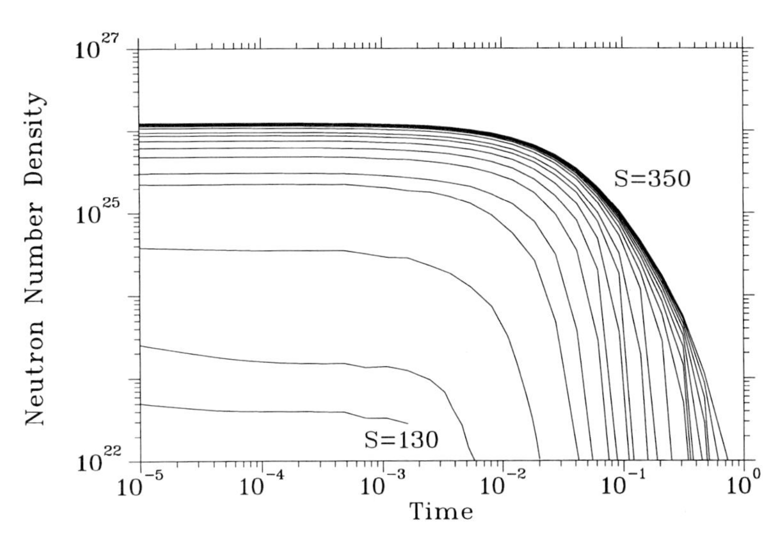

FIGURE 16. $n_n(t)$ in s, displayed for different entropies S in units of k_B per baryon.

to achieve this goal. But in addition to a lower Y_e, also lower entropies are required as they might come from cold high density matter in beta-equilibrium (see section 3.2.5 and Cameron 1989, Meyer 1989, and Hillebrandt, Takahashi, & Kodama 1976). This can be deduced from Figure 14b. It shows the Y_n/Y_{seed}-ratio plotted in the (S, Y_e)-plane with a logarithmic entropy axis and extends down to entropies as low as $S = 10^{-2}$, where a normal and not an alpha-rich freeze-out is encountered.

Whether such an interpretation ($A < 130$ from low Y_e and S conditions, $A > 130$ from high S conditions) is the solution might eventually be answered by observations. There seems to exist meteoritic evidence, discussed by Wasserburg et al. (1996), that the last r-process contributions to the solar system for $A > 130$ and $A < 130$ came at different times, i.e. from different types of events deduced from the extinct radioactivities ^{107}Pd, ^{129}I and ^{182}Hf in meteoritic matter. It is highly desirable to have an independent verification of this from observations of low metallicity stars, which apparently show a completely solar r-process composition for nuclei with $A > 130$ [see Sneden et al. (1996), and Cowan et al. (1997)], possibly stemming from the first events in our galaxy which produce r-process nuclei (Mathews et al. 1992). It is also necessary to explore the abundances of nuclei with $A < 130$ in such observations, in order to test whether the solar pattern will also be found there or is absent due to different evolution time scales of two independent stellar sources for these different mass ranges of r-process nuclei.

The results presented in this study benefitted from discussions with W.D. Arnett, R. Azuma, W. Benz, A. Burrows, S. Bruenn, J.J. Cowan, S. Goriely, C. Fransson, R. Hoffman, J. Hughes, H.-T. Janka, E. Kolbe, S. Kumagai, K. Langanke, J. Lattimer, A. Mezzacappa, P. Möller, M. Pearson, M. Prakash, Y.-Z. Qian, A. Ray, S. Reddy, R. Sawyer, P. Vogel, and M. Wiescher. This work was supported by the Swiss Nationalfonds (grant 20-47252.96), the grant-in-Aid for Scientific Research (05242102, 06233101) and COE research (07CE2002) of the Ministry of Education, Science, and Culture in Japan, the German BMBF (grant 06Mz864) and DFG (grant Kr80615), the US NSF (grant PHY 9407194) and the Austrian Academy of Sciences. Some of us (FKT, CF, and KN) thank the ITP at the Univ. of California, Santa Barbara, for hospitality and inspiration during the supernova program.

REFERENCES

Aboussir, Y., Pearson, J. M., Dutta, A. K., & Tondeur, F. 1995, At. Data Nucl. Data Tables, 61, 127

Ahmad, I., Bonino, G., Castagnoli, G.C., Fischer, S.M., Kutschera, W., Paul, M. 1997, Phys. Rev. Lett., submitted

Alastuey, A., Jancovici, B. 1978, Ap. J. 226, 1034

Arnett, W.D. 1995, Ann. Rev. Astron. Astrophys. 33, 115

Arnett, W.D. 1996, *Nucleosynthesis and Supernovae*, Princeton Univ. Press

Arnett, W.D., Bahcall, J.N., Kirshner, R.P., Woosley, S.E. 1989, Ann. Rev. Astron. Astrophys. 27, 629

Arnett, W.D., Thielemann, F.-K. 1985, Ap. J. 295, 589

Arnould, M., 1972, A & A 19, 92

Arnould, M. 1997, At. Data Nucl. Data Tables, in press

Audi, G., Wapstra, A.H. 1995, Nucl. Phys. A595, 409

Aufderheide, M., Baron, E., Thielemann, F.-K. 1991, Ap. J. 370, 630

Aufderheide, M., Fushiki, I., Woosley, S.E., Hartmann, D. 1994, Ap. J. Suppl. 91, 389

Azuma, R. et al. 1994, Phys. Rev. C50, 1194

Bao, Z. Y., Käppeler, F. 1987, At. Data Nucl. Data Tables, 36, 411

Bao, Z. Y., Käppeler, F. 1997, private communication

Baraffe, I., Bloemen, H., Borzov, I.N., Busso, M., Cooper, S., Dobaczewski, J., Durrer, R., von Feilitzsch, F., Gallino, R., Isern, J., Janka, H.-T., Lorenz, E., Pethick, C., Rolfs, C., Thielemann, F.-K., Vervier, J., Wiescher, M. 1997, *NuPECC Report on Nuclear and Particle Astrophysics*, http://quasar.physik.unibas.ch/nupecc

Barbuy, B. et al. 1997, A & A 317, L63

Barker, F.C., Kajino, T. 1991, Aust. J. Phys. 44, 369

Baym, G. 1991, in *Neutron Stars, Theory and Observations*, eds., J. Ventura and D. Pines, Kluwer Acad. Publ., Dordrecht, p.37

Beer, H., Voss, F., Winters, R.R. 1992, Ap. J. Suppl. 80, 403

Beers, T.C., Ryan, S.G., Norris, J.E. 1997, Nucl. Phys. A621, 37c

Bethe, A.H. 1936, Phys. Rev. 50, 332

Bethe, H.A. 1990, Rev. Mod. Phys. 62, 801

Blair, W.P., Raymond, J.C., Long, K.S. 1994, Ap. J. 423, 334

Blake, J.B., Woosley, S.E., Weaver, T.A., Schramm, D.N. 1981, Ap. J. 248, 315

Blanton, E.L., Schmidt, B.P., Kirshner, R.P. 1995, A. J. 110, 2868

Blatt, J.M., Weisskopf V.F. 1952, *Theoretical Nuclear Physics* Wiley, New York

Borzov, I.N. 1996, Z. Phys. A335, 125; 1997, Nucl. Phys. A621, 307c

Bouchet, P., Danziger, I.J. Lucy, L. 1991, in *SN1987A and Other Supernovae*, ed. I.J. Danziger (ESO, Garching), p. 217

Bouquelle, V., Cerf, N., Arnould, M., Tachibana, M., Goriely, S. 1996, A & A, 305, 1005

Brown, G.E., Bethe, H.A. 1994, Ap. J. 423, 659

Brown, L.S., Sawyer, R.F. 1997, Rev. Mod. Phys. 69, 411

Bruenn, S.W. 1985, Ap. J. Suppl. 58, 77

Bruenn, S.W. 1989, Ap. J. 340, 955

Bruenn, S.W. 1989, Ap. J. 341, 385

Bruenn, S.W., Haxton, W.C. 1991, Ap. J. 376, 678

Bruenn, S.W., Mezzacappa, A., Dineva, T. 1995, Phys. Rep. 256, 69

Buchmann, L. et al. 1993, Phys. Rev. Lett. 70, 726

Buchmann, L. 1996, Ap. J. 468, L127; 1997, Ap. J. 479, L153

Buchmann, L., Azuma, R.E., Barnes, C.A., Humblet, J., Langanke, K. 1996, Phys. Rev. C54, 393; 1997, Nucl. Phys. A621, 153

Burrows, A. 1990, Ann. Rev. Nucl. Part. Sci. 40, 181

Burrows, A., Fryxell, B. 1992, Science 258, 430

Burrows, A., Hayes, J., Fryxell, B. 1995, Ap. J. 450, 830

Burrows, A., 1996, Nucl. Phys. A606, 151

Cameron, A.G.W. 1989, in *Cosmic Abundances of Matter*, ed. C.J. Waddington, AIP Conf. Proc. 183, 349

Cameron, A.G.W., Elkin, R.M. 1965, Can. J. Phys. 43, 1288

Cameron, A.G.W., Cowan, J., Truran, J.W. 1983, Astopys. Space Sci., 91, 235

Campi, X., Flocard, H., Kerman, A.K., Koonin, S. 1975, Nucl. Phys. A, A251, 193

Caughlan, G.R., Fowler, W.A. 1988, At. Data Nucl. Data Tables, 40, 283

Caughlan, G.R., Fowler, W.A., Harris, M.J, Zimmerman, G.E. 1985, At. Data Nucl. Data Tables 32, 197

Chabrier, G., Schatzman, E. 1994, *The Equation of State in Astrophysics*, IAU Coll. 147, Cambridge Univ. Press

Chaudhuri, A.K., Basu, D.N., Sinha, B. 1985, Nucl. Phys. A439, 415

Chen, B., Dobaczewski, J., Kratz, K.-L., Langanke, K., Pfeiffer, B., Thielemann, F.-K., Vogel, P. 1995, Phys. Lett., B355, 37

Chugai, N.N. 1994, Ap. J. 428, L17

Clayton, D.D. 1968, 1983, *Principles of Stellar Evolution and Nucleosynthesis*, Univ. of Chicago Press

Clayton, D.D., Leising, M.D., The, L.-S., Johnson, W.N., Kurfess, J.D. 1992, Ap. J. 399, L14

Cooperstein, J., Baron, E. 1989, in *Supernovae*, ed. A. Petschek, Springer-Verlag, New York, p. 213

Cowan, J.J., Cameron, A. G.W., Truran, J.W. 1983, Ap. J., 265, 429

Cowan, J.J., McWilliam, A., Sneden, C., Burris, D.L. 1997, Ap. J., 480, 246

Cowan, J. J., Thielemann, F.-K., Truran, J. W. 1991, Phy. Rep., 208, 267

Danos, M. 1958, Nucl. Phys. 5, 23

d'Antona, Mazzitelli, I. 1991, in IAU Symposium 145 *Evolution of Stars: The Photospheric Abundance Connection*, eds. G. Michaud, A. Tutukov, Kluwer Acad. Publ., Dordrecht, p.399

Danziger, I.J., Bouchet, P., Gouiffes, C., Lucy, L. 1990, in *Supernovae*, ed S.E. Woosley, Springer-Verlag, New York, p.69

Dean, D.J., Koonin, S.E., Langanke, K., Radha, P.B., Alhassid, Y. 1995, Phys. Rev. Lett. 74, 2909

Dobaczewski, J., Hamamoto, I., Nazarewicz, W., Sheikh, J. A. 1994, Phys. Rev. Lett., 72, 981

Dobaczewski, J., Nazarewicz, W., Werner, T. R. 1995, Phys. Scr., T56, 15

Dupraz, C., Wessolowski, U., Oberlack, U., Georgii, R., Lichti, R., Iyudin, A., Schönfelder, V., Strong, A.W., Bloemen, H., Morns, D., Ryan, J., Winkler, C., Knödlseder, J., von Ballmoos, P. 1997, A & A 324, 683

Edvardsson, B., Andersen, J., Gustafsson, B. Lambert, D.L., Nissen, P.E., Tomkin, J. 1993, A & A 275, 101

Eichler, D., Livio, M., Piran, T., Schramm, D. N. 1989, Nature, 340, 126

Elias, J.H., Depoy, D.L., Gregory, B., Suntzeff, N.B. 1991, in *SN1987A and Other Supernovae*, ed. I.J. Danziger (ESO, Garching), p. 293

Epstein, R.I., Colgate, S.A., Haxton, W.C. 1988, Phys. Rev. Lett. 61, 2038

Ezhov, S.N., Plujko, V. A. 1993, Z. Phys. 346, 275

Fantoni, S., Friman, B.L., Pandharipande, V.R. 1981, Phys. Rev. Lett. 48, 1089

Filippone, B.W. 1987, Ann. Rev. Nucl. Part. Sci. 36, 717

Filippone, B.W., Humblet, J., Langanke, K. 1989, Phys. Rev. C40, 515

Fowler, W. A., Caughlan, G. E., Zimmermann, B. A. 1967, Ann. Rev. Astron. Astrophys. 5, 525

Fowler, W.A., Caughlan, G.E., Zimmerman, B.A. 1975, Ann. Rev. Astron. Astrophys. 13, 69

Fransson, C., Kozma, C. 1993, Ap. J. 408, L25

Fransson, C., Houck, J., Kozma, C. 1996, in IAU Coll. 145 *Supernovae and SN Remnants*, ed. R. McCray, Cambridge Univ. Press, p.211

Freiburghaus, C., Kolbe, E., Rauscher, T., Thielemann, F.-K., Kratz, K.-L., Pfeiffer, B. 1997, Nucl. Phys. A621, 405c

Freiburghaus, C., Rembges, F., Rauscher, T., Kolbe, E., Thielemann, F.-K., Kratz, K.-L., Pfeiffer, B. 1997, Ap. J., submitted

Fukunishi, N., Otsuka, T., & Sebe, T. 1992, Phys. Lett., B296, 279

Fuller, G.M., Fowler, W.A., Newman, M. 1980, Ap. J. Suppl. 42, 447

Fuller, G.M., Fowler, W.A., Newman, M. 1982, Ap. J. Suppl. 48, 279

Fuller, G.M., Fowler, W.A., Newman, M. 1985, Ap. J. 293, 1

Fuller, G., Meyer, B. S. 1995, ApJ, 453, 792

Fushiki, I., Lamb, D.Q. 1987, Ap. J. 317, 368

Gadioli, E., Hodgson, P. E., 1992, *Pre-Equilibrium Nuclear Reactions* , Clarendon Press, Oxford

Gallino, R., Busso, M. 1997, in *Presolar Grains*, eds. T. Bernatowicz, E. Zinner, AIP, New York, in press

Gilbert, A., Cameron, A.G.W. 1965, Can. J. Phys. 43, 1446

Glendenning, N.K. 1991, Nucl. Phys. B24B, 110

Görres, J. et al. 1997, Phys. Rev. Lett., submitted

Görres, J., Wiescher, M., Thielemann, F.-K. 1995, Phys. Rev. C51, 392

Goriely, S. 1996, Nucl. Phys. A605, 28

Goriely, S., Arnould, M. 1996, A & A, 312, 327

Gratton, R.G., Sneden, C. 1991, A & A 241, 501

Hansen, J.P., Torrie, G.M., Veillefosse, P. 1977, Phys. Rev. A16, 2153

Hartmann, D.H. et al. 1997, Nucl. Phys. A621, 83c

Hashimoto, M., Iwamoto, K., Nomoto, K., 1993, Ap.J. 414, L105

Hashimoto, M., Nomoto, K., Shigeyama, T. 1989, A & A 210, L5

Hashimoto, M., Nomoto, K., Tsujimoto, T., Thielemann, F.-K. 1993, in *Nuclei in the Cosmos II*, ed. F. Käppeler, IOP Bristol, p.587

Hashimoto, M., Nomoto, K., Thielemann, F.-K. 1995, in *Nuclei in the Cosmos III*, eds. M. Busso, R. Gallino, C. Raiteri, IOP, Bristol, p.419

Hauser, W., Feshbach, H. 1952, Phys. Rev. A 87, 366

Herant, M., Benz, W., Hix, W. R., Fryer, C.L., Colgate, S. A. 1994, Ap. J. 435, 339

Hilf, E. R., von Groote, H., & Takahashi, K. 1976, in Proceedings of the 3rd International Conference on Nuclei far from Stability, Geneva, CERN-Rep 76-13, p.142

Hillebrandt, W., Takahashi, K., and Kodama, T. 1976, A & A 52, 63

Hirsch, M., Staudt, A., Muto, K., Klapdor-Kleingrothaus, H.V. 1992, Nucl. Phys. A535, 62

Hix, W.R., Thielemann F.-K. 1996, Ap. J. 460, 869

Hix, W.R., Thielemann F.-K. 1997, Ap. J., submitted

Hoffman, R. D., Woosley, S. E., Fuller, G. M., Meyer, B. S. 1996, Ap. J., 460, 478

Hoffman, R. D., Woosley, S. E., Qian, Y.-Z. 1997, ApJ, 482, 951

Holmes, J.A., Woosley, S.E., Fowler, W.A., Zimmerman, B.A. 1976, At. Data Nucl. Data Tables 18, 306

Horowitz, C.J., Wehrberger, K., Phys. Lett. B226, 236

Houck, J.C., Fransson, C. 1996, Ap. J. 456, 811

Howard, W. M., Goriely, S., Rayet, M., & Arnould, M. 1993, Ap. J., 417, 713

Howard, W.M., Meyer, B.S., Clayton, D.D. 1992, Meteoritics 27, 404

Howard, W.M., Meyer, B.S., Woosley, S.E. 1991, Ap. J. Lett. 373, L5

Hughes, J.P., Singh, K.P. 1994, Ap.J. 422, 126

Ichimaru, S. 1993, Rev. Mod. Phys. 65, 255

Ichimaru, S. 1996, Publ. Astr. Soc. Japan 48, 613

Ichimaru, S., & Utsumi, K. 1983, Ap. J. 269, L51

Ichimaru, S., Tanaka, S., Iyetomi, H. 1984, Phys. Rev. A29, 2033

Ichimaru, S., Utsumi, K. 1984, Ap. J. 286, 363

Ignatyuk, A.V., Smirenkin, G.N., Tishin, A.S. 1975, Yad. Phys. 21, 485

Ignatyuk, A.V., Istekov, K.K., Smirenkin, G.N. 1979, Sov. J. Nucl. Phys. 29, 450

Iljinov, A.S. et al. 1992, Nucl. Phys. A543, 517

Itoh, N., Hayashi, H., Kohyama, Y. 1993; 1994, Ap. J. 418, 405; 436. 418

Itoh, N., Hayashi, H., Nishikawa, A., Kohyama, Y. 1996, Ap. J. Suppl. 102, 411

Itoh, N., Kuwashima, F., Munakata, H. 1990, Ap. J. 362, 620

Itoh, N., Nishikawa, A., Kohyama, Y. 1996, Ap. J. 470, 1015

Itoh, N., Totsuji, H., Ichimaru, S. 1977, Ap. J. 218, 477; Ap. J. 220, 742

Itoh, N., Totsuji, H., Ichimaru, S., DeWitt, H.E. 1979, Ap. J. 234, 1079; Ap. J. 239, 415

Itoh, M., Kumagai, S., Shigeyama, T., Nomoto, K., Nishimura, J. 1987 Nature 330, 233

Iwamoto, K., Nomoto, K., Höflich, P., Yamaoka, H., Kumagai, S., Shigeyama, T. 1994, Ap. J. 437, L115

Iyudin, A.F., Diehl, R., Bloemen, H., Hermsen, W., Lichti, G.G., Morns, D., Ryan, J., Schönfelder, V., Steinle, H., Varendorff, M., de Vries, C., Winkler, C. 1994, A& A 284, L1

Janka, H.-T., Müller, E. 1993, in *Frontiers of Neutrino Astrophysics*, eds. Y. Suzuki, K. Nakamura, Universal Academy Press, Tokyo, p.203

Janka, H.-T., Müller, E. 1995, Phys. Rep. 256, 135

Janka, H.-T., Müller, E. 1996, A&A, 306, 167

Jeukenne, J.P., Lejeune, A., Mahaux, C. 1977, Phys. Rev. C16, 80

Käppeler, F., Beer, H., Wisshak, K. 1989, Rep. Prog. Phys. 52, 945

Käppeler, F. et al. 1994, Ap. J. 437, 396

Keil, W., Janka, H.-T. 1995, A & A 296, 145

Kitamura, H., Ichimaru, S. 1995, Ap. J. 438, 300

Klapdor, H.V., Metzinger, J., Oda, T. 1984, At Data Nucl. Data Tables 31, 81

Kolbe, E., Krewald, S., Langanke, K., Thielemann, F.-K. 1992, Nucl. Phys. A540, 599

Kolbe, E., Langanke, K., Krewald, S., Thielemann, F.-K. 1993, Phys. Rep. 227, 37

Kolbe, E., Langanke, K., Thielemann, F.-K., Vogel, P. 1995, Phys. Rev. C52, 3437

Kozma, C., Fransson, C. 1997, Ap. J., in press

Kratz, K.-L. 1995, in AIP Conf. Proc. 327 of Nuclei in the Cosmos III, ed. M. Busso, R. Gallino, C. M. Raiteri, AIP Press, p.113

Kratz, K.-L., Bitouzet, J.-P., Thielemann, F.-K., Möller, P., Pfeiffer, B. 1993, Ap. J. 402, 216

Kratz, K.-L., Pfeiffer, B., Thielemann, F.-K., 1997, Nucl. Phys. A, in press

Kratz, K.-L., Thielemann, F.-K., Hillebrandt, W., Möller, P., Harms, V., Truran, J. W. 1988, J. Phys. G, 14, 331

Kratz, K.-L., Thielemann, F.-K., Möller, P., Pfeiffer, B. 1994, in Proc. 8th Int. Symp. on Capture

Gamma-Ray Spectroscopy, ed. J. Kern, IOP Bristol, p.724

Kumagai, S., Shigeyama, T., Hashimoto, M., Nomoto, K. 1991, A & A 243, L13

Kumagai, S., Nomoto, K., Shigeyama, T., Hashimoto, M., Itoh, M. 1993, A & A 273, 153

Kurfess, J. D. et al. 1992, Ap. J. 399, L137

Lambert, D.L. 1989, in *Cosmic Abundances of Matter*, ed. C.J. Waddington, AIP Conf. Proc. 183, p.168

Langanke, K., Barnes, C.A. 1996, Adv. Nucl. Phys. 22, 173

Langanke, K., Vogel, P., Kolbe, E. 1996, Phys. Rev. Lett. 76, 2629

Langer, N., Henkel, C 1995, in *Nuclei in the Cosmos III*, eds. M. Busso, R. Gallino, C.M. Raiteri, AIP Press, p.413

Lattimer, J. M., Mackie, F., Ravenhall, D. G., Schramm, D. N. 1977, Ap. J., 213, 225

Lattimer, J.M., Yahil, A. 1989, Ap. J. 340, 426

Lipparini, E., Stringari, S. 1989, Phys. Rep. 175, 103

Maeder, A. 1992, A & A 264, 105

Mahaux, C., 1982, Phys. Rev. C82, 1848

Mahaux, C., Weidenmüller, H.A. 1979, Ann. Rev. Part. Nucl. Sci. 29, 1

Malaney, R.A., Fowler, W.A. 1988, Ap. J. 333, 14

Malaney, R.A., Fowler, W.A. 1989, Ap. J. 345, L5

Mann, F.M. 1978, Hanford Engineering (HEDL-TME 78-83)

Mathews, G.J., Bazan, G., Cowan, J.J. 1992, Ap. J. 391, 719

Mayle, R. W., Wilson, J. R. 1990, in *Supernovae*, ed. S. E. Woosley, p.333

McCray, R. 1993, Ann. Rev. Astron. Astrophys. 31, 175

McFadden, L. Satchler, G.R., 1966, Nucl. Phys. 84, 177

McWilliam, A. 1997, Ann. Rev. Astron. Astrophys. 35, 503

McWilliam, A., Preston, G.W., Sneden, C., Searle, C. 1995, A. J. 109, 2757

Mengoni, A., Nakajima, Y. 1994, J. Nucl. Sci. Techn. 31, 151

Meyer, B. S. 1989, Ap. J., 343, 254

Meyer, B. S. 1994, Ann. Rev. Astron. Astrophys, 32, 153

Meyer, B.S., 1995, Ap. J. 449, L55

Meyer, B. S., Brown, J. S. 1997, Nucl. Phys. A621, 409c

Meyer, B. S., Brown, J. S. 1997, Ap. J. Suppl. 112, 119

Meyer, B.S., Mathews, G.J., Howard, W.M., Woosley, S.E., Hoffman, R.D. 1992, Ap. J 399, 656

Mezzacappa, A., Bruenn, S.W. 1993, Ap. J. 405, 637

Mezzacappa, A., Calder, A.C., Bruenn, S.W., Blondin, J.M., Guidry, M.W., Strayer, M.R., Umar, A.S., 1997, Ap. J., in press

Michaud, G. Fowler, W.A. 1970, Phys. Rev. C2, 2041

Michaud, G. Fowler, W.A. 1972, Ap. J. 173, 157

Möller, P., Nix, J.R., Kratz, K.-L. 1997, At. Data Nucl. Data Tables 66, 131

Möller, P., Nix, J.R., Myers, W.D., Swiatecki, W.J. 1995, At. Data Nucl. Data Tables 59, 185

Möller, P., Randrup, J. 1990, Nucl. Phys. A514, 1

Mohr, P., Rauscher, T., Oberhummer, H., Maté, Z., Fülöp, Zs., Somorjai, E., Jaeger, M., Staudt, G. 1997, Phys. Rev. C55, 1523

Morgan, J.A. 1980, Ap. J. 238, 674

Myers, W.D., Swiatecki, W.J., Kodama, T., El-Jaick, L.J., Hilf, E.R. 1977, Phys. Rev. C15, 2032

Nagase, F. 1989, Publ. Astron. Soc. Japan 41, 1

Nissen, P.E., Gustafsson, B., Edvardsson, B., Gilmore, G. 1994, A & A 285, 440

Nomoto, K., Hashimoto, M. 1988, Phys. Rep. 163, 13

Nomoto, K., Hashimoto, M., Tsujimoto, T., Thielemann, F.-K., Kishimoto, N., Kubo, Y., Nakasato, N. 1997, Nucl. Phys., A161, 79c

Nomoto, K., Suzuki, T., Shigeyama, T., Kumigai, S., Yamaoka, H., Saio, H. 1993, Nature 364, 507

Nomoto, K., Thielemann, F.-K., Miyaji, S. 1985, A & A 149, 239

Nomoto, K., Yamaoka, H., Pols, O. R., van den Heuvel, E. P. J., Iwamoto, K., Kumagai, S., Shigeyama, T. 1994, Nature 371, 227

Norman, E.B. et al. 1997, Phys. Rev. C, submitted

Oberhummer, H., Herndl, H., Rauscher, T., Beer, H. 1996, Surveys in Geophys. 17, 665

Orr, N. A. 1991, Phys. Lett. B258, 29

Paar, V. et al. 1997, in *Proc. Int. Conf. Gamma-Ray Spectroscopy and Related Topics*, ed. G. Molnar, Springer Hungarica, in press

Page, D., Baron, E. 1990, Ap. J. 354, L17

Pagel, B.E.J. 1991, Phys. Scr. T36, 7

Pagel, B.E.J., Trautvaisiene, G., 1995, MNRAS 276, 505

Pagel, B.E.J., Trautvaisiene, G., 1997, MNRAS 288, 108

Pearson, J. M., Nayak, R. C., Goriely, S. 1996, Phys. Lett. B387, 455

Pfeiffer, B., Kratz, K.-L. 1996, KCh Mainz Report, unpublished

Pfeiffer, B., Kratz, K.-L., Thielemann, F.-K. 1997, Z. f. Phys. A357, 235

Pichon, B. 1994, Nucl. Phys. A568, 553

Prakash, M. et al. 1997, Phys. Rep. 280, 1

Press, W.H., Flannery, B.P., Teukolsky, S.A., Vetterling, W.T. 1986, *Numerical Recipes*, Cambridge University Press

Qian, Y.-Z., Vogel, P., Wasserburg, G.J. 1997, Ap. J., submitted

Qian, Y.-Z., Woosley, S. E. 1996, Ap. J., 471, 331

Qian, Y.-Z., Woosley, S.E., Haxton, W.C., Langanke, K., Vogel, P. 1996, Phys. Rev. C55, 1532

Rauscher, T., Applegate, J.H., Cowan, J.J., Thielemann, F.-K., Wiescher, M. 1994, Ap. J., 429, 499

Rauscher, T., Thielemann, F.-K., Kratz, K.-L. 1997, Phys. Rev. C56, 1613

Rauscher, T., Thielemann, F.-K. 1998, At. Data Nucl. Data Tables, to be published

Rayet, M., Prantzos, N., Arnould, M. 1990, A & A 227, 211

Rayet, M., Arnould, M., Hashimoto, M., Prantzos, N., Nomoto, K. 1995, A & A 298, 517

Reddy, S., Prakash, M. 1997, Ap. J. 423, 689

Reddy, S., Prakash, M., Lattimer, J.M. 1997, Nucl. Phys. A, submitted

Reisdorf 1981, Z. Phys. A300, 227

Ring, P., Schuck, P. 1980, *The Nuclear Many-Body Problem*, Springer-Verlag, Berlin

Rolfs, C., Rodney, W.S. 1988, *Cauldrons in the Cosmos*, University of Chicago Press, Chicago

Rolfs, C., Trautvetter, H.P., Rodney, W.S. 1987, Rep. Prog. Phys. 50, 233

Ryan, S.G., Norris, J.E., Beers, T.C. 1996, Ap. J. 471, 254

Salpeter, E.E., van Horn, H.M. 1969, Ap. J. 155, 183

Satchler, G.R., Love, W.G. 1979, Phys. Rep. 55, 183

Sawyer, R.F. 1989, Phys. Rev. C40, 865

Schatz, H., Aprahamian, A,. Görres, J., Wiescher, M., Rauscher, T., Rembges, J.F., Thielemann, F.-K., Kratz, K.-L., Pfeiffer, B., Möller, P., Herndl, B., Brown, B.A. 1997c, Phys. Rep., in press

Schinder, P.J. 1990, Ap. J. Suppl. 74, 249

Schmidt, B.P 1997, in *SN1987A: Ten Years After*, eds. M.M. Phillips, N.B. Suntzeff, Astron.

Soc. Pac., in press

Schramm, S., Koonin, S.E. 1990, Ap. J. 365, 296

Shigeyama, T., Nomoto, K., Hashimoto, M. 1988, A & A 196, 141

Shigeyama, T., Nomoto, K., Tsujimoto. T.,Hashimoto, M. 1990, Ap. J. 361, L23

Shigeyama, T., Suzuki, T., Kumagai, S., Nomoto, K., Saio, H., Yamaoka. H., 1994, Ap. J. 420, 341

Sneden, C., McWilliam, A., Preston, G. W., Cowan, J. J., Burris, D. I., Armosky, B. J. 1996, Ap. J. 467, 819

Sollermann, J., Cumming, R.J., Lundquist, P. 1998, Ap. J. in press

Sorlin, O. et al. 1993, Phys. Rev. C47, 2941

Staudt, A., Bender, E., Muto, K., Klapdor, H.V. 1989, Z. Phys. A334, 47

Staudt, A., Bender, E., Muto, K., Klapdor, H.V. 1990, At. Data Nucl. Data Tables 44, 79

Suntzeff, N.B., Phillips, M.M., Elias, J.H., Walker, A.R., Depoy, D.L. 1992, Ap. J. 384, L33

Suntzeff, N.B. et al. 1997, in *SN1987A: Ten Years After*, eds. M.M. Phillips, N.B. Suntzeff, Astron. Soc. Pac., in press

Surman, R., Engel, J., Bennett, J. R., Meyer, B. S. 1997, Phys. Rev. C, in press

Sutaria, F.K., Sheikh, J.H., Ray, A. 1997, Nucl. Phys. A621, 375c

Takahashi, K., Witti, J., Janka, H.-T. 1994, A&A, 286, 857

Takahashi, K., Yamada, M., Kondo, Z. 1973, At. Data Nucl. Data Tables 12, 101

Takahashi, K., Yokoi, K. 1988, At. Data Nucl. Data Tables 36, 375

Takahara, M., Nino, M., Oda, T., Muto, K., Wolters, A.A., Claudemans, P.W.M., Sato, K. 1989, Nucl. Phys. A504, 167

Tepel, J.W., Hoffmann, H.M., Weidenmüller, H.A. 1974, Phys. Lett. B49, 1

Thielemann, F.-K., Arnett, W.D. 1985, Ap. J. 295, 604

Thielemann, F.-K., Arnould, M., Hillebrandt, W. 1979, A & A, 74, 175

Thielemann, F.-K., Arnould, M., Truran, J. W. 1987, in *Advances in Nuclear Astrophysics*, ed. E. Vangioni-Flam, Gif sur Yvette, Editions Frontière, p.525

Thielemann, F.-K., Arnould, M., Truran, J.W. 1988, *Capture Gamma-Ray Spectroscopy*, eds. K. Abrahams, P. van Assche, IOP, Bristol, p. 730.

Thielemann, F.-K., Bitouzet, J.-P., Kratz, K.-L., Möller, P., Cowan, J. J., Truran, J. W. 1993, Phys. Rep., 227, 269

Thielemann, F.-K., et al. 1997, in *Capture Gamma-Ray Spectroscopy*, ed. G. Molnar, Springer Hungarica, in press

Thielemann, F.-K., Hashimoto, M., Nomoto, K. 1990, Ap. J. 349, 222

Thielemann, F.-K., Kratz, K.-L., Pfeiffer, B., Rauscher, T., van Wormer, L., & Wiescher, M. C. 1994, Nucl. Phys. A570, 329c

Thielemann, F.-K., Metzinger, J., Klapdor, H.V. 1983, Z. Phys. A 309, 301

Thielemann, F.-K. Nomoto, K., Hashimoto, M., 1993 in *Origin and Evolution of the Elements*, ed. N. Prantzos, E. Vangioni-Flam, M. Cassé, Cambridge Univ. Press, p.297

Thielemann, F.-K., Nomoto, K., Hashimoto, M. 1994, in *Supernovae, Les Houches, Session LIV*, eds. S. Bludman, R. Mochkovitch, J. Zinn-Justin, Elsevier, Amsterdam, p. 629

Thielemann, F.-K., Nomoto, K., & Hashimoto, M. 1996, Ap. J. 460, 408

Thielemann, F.-K. Nomoto, K., Iwamoto, K., Brachwitz, F. 1996, in *Thermonuclear Supernovae*, eds. R. Canal, P. Ruiz-Lapuente, p. 485

Thielemann, F.-K., Truran, J.W. 1987, in *Advances in Nuclear Astrophysics*, eds. E. Vangioni-Flam et al., Editions Frontières, Gif sur Yvette, p. 541

Thomas, D., Schramm, D.N., Olive, K.A., Fields, B.D. 1993, Ap. J. 406, 569

Thomas, D., Schramm, D.N., Olive, K.A., 1994, Ap. J. 430, 291

Thomas, J., Zirnbauer, M.R., Langanke, K. 1986, Phys. Rev. C33, 2197

Thorsett, S.E. 1996, Phys. Rev. Lett. 77, 1432 Phys. Rev. C33, 2197

Timmes, F., Woosley, S.E., Weaver, T.A. 1995, Ap. J. Suppl. 98, 617

Timmes, F., Woosley, S.E., Weaver, T.A. 1996, Ap. J. 457, 834

Trimble, V. 1975, Rev. Mod. Phys. 47, 877

Trimble, V. 1991, Astron. Astrophys. Rev. 3, 1

Truran, J.W. 1972, Astrophys. Space Sci. 18, 308

Truran, J.W. 1985, Ann. Rev. Nucl. Part. Sci. 34, 53

Truran, J.W. , Arnett, W.D. 1970, Ap. J. 160, 181

Truran, J.W. , Arnett, W.D. 1971, Astrophys. Space Sci. 11, 430

Truran, J.W., Cameron, A.G.W., Gilbert, A. 1966, Can. J. Phys. 44, 563

Truran, J.W., Cowan, J.J., Cameron, A.G.W. 1978, Ap. J. 222, L63

Tsujimoto, T., Nomoto, K., Yoshii, Y., Hashimoto, M., Yanagida, S., Thielemann, F.-K. 1995, MNRAS 277, 945

Tsujimoto, T. Yoshii, Y., Nomoto, K., Matteucci, F., Thielemann, F.-K. 1997, Ap. J. 483, 228

Turatto, M. et al. 1997, in *SN1987A: Ten Years After*, eds. M.M. Phillips, N.B. Suntzeff, Astron. Soc. Pac., in press

van Kerwijk, M.A., van Paradijs, J.A., Zuiderwijk, E.S. 1995, A& A 303, 497

van Paradijs, J.A. 1991, in *Neutron Stars, Theory and Observations*, eds., J. Ventura and D. Pines, Kluwer Acad. Publ., Dordrecht, p.289

van Paradijs, J.A., McClintock, J.E. 1995, in *X-Ray Binaries*, eds. W.H.G. Lewin, J. v. Paradijs, E.J.P. van den Heuvel, Cambridge Univ. Press, p. 111

van Wormer, L., Görres, J., Iliadis, C., Wiescher, M., Thielemann, F.-K. 1994, Ap. J. 432, 267

Varani, G.-F., Meikle, W.P.S., Spyromilio, J., Allen, D.A. 1990, MNRAS 245, 570

Varner, R.L. Thompson, W.J., McAbee, T.L. Ludwig, E.J., Clegg, T.B. 1991, Phys. Rep. 201, 57

Verbaatschot, J.J.M., Weidenmüller, H.A., Zirnbauer, M.R. 1984, Phys. Rep. 129, 367

von Egidy, T., Behkami, A.N., Schmidt, H.H. 1986, Nucl. Phys. A376, 405

von Egidy, T., Schmidt, H.H., Behkami, A.N. 1988, Nucl. Phys. A481, 189

Wagoner, R.V. 1969, Ap. J. Suppl. 18, 247

Wallerstein, G. et al. 1997, Rev. Mod. Phys. 69, 995

Wang, R.-P., Thielemann, F.-K., Feng, D.H., Wu, C.-L. 1992, Phys. Lett. B284, 196

Wasserburg, G., Busso, M., Gallino, R. 1996, Ap. J. 466, L109

Weaver, T.A., Woosley, S.E. 1993, Phys. Rep. 227, 65

Weber, F., Glendenning, N.K. 1991, Phys. Lett. B265, 1

Werner K., Dreizler S., Heber U., Rauch T. 1995, *Nuclei in the Cosmos III*, eds. M. Busso, R. Gallino, C.M. Raiteri, AIP Press, p.45

Werner, T. R., Sheikh, J. A., Nazarewicz, W., Strayer, M. R., Umar, A. S., Misu, M. 1994, Phys. Lett. B333, 303

Wheeler, J.C., Sneden, C., Truran, J.W. 1989, Ann. Rev. Astron. Astrophys. 27, 279

Wiescher, M., Görres, J., Thielemann, F.-K., Ritter, H. 1986, A & A 160, 56

Wiescher, M., Harms, V., Görres, J., Thielemann, F.-K., Rybarcyk, L.J. 1987, Ap. J. 316, 162

Wiescher, M., J. Görres 1988, Z. Phys. A329, 121

Wiescher, M., Görres, J., Thielemann, F.-K. 1988, Ap. J. 326, 384

Wiescher, M., Görres, J., Thielemann, F.-K. 1990, Ap. J. 363, 340

Wiescher, M., Görres, J., Graaf, S, Buchmann, L., Thielemann, F.-K. 1989, Ap. J. 343, 352

Wiescher, M., Steininger, R., Käppeler, F. 1989, Ap. J. 344, 464

Wilson, J.R., Mayle, R.W. 1988, Phys. Rep. 163, 63

Wilson, J.R., Mayle, R.W. 1993, Phys. Rep. 227, 97

Wisshak, K., Voss, F., Käppeler, F., Kerzakov, I. 1997, Nucl. Phys. A621, 270c

Witteborn, F.C., Bregman, J.D., Wooden, D.H., Pinto, P.A., Rank, D.M., Woosley, S.E., Cohen, M. 1989, Ap. J. 338, L9

Witti, J., Janka, H.-T., Takahashi, K. 1994, A&A, 286, 841

Wooden, D.H., Rank, D.M., Bregman, J.D., Witteborn, E.C., Cohen, M., Pinto, P.A., Axelrod, T.S. 1993, Ap. J.Suppl. 88, 477

Wooden, D.H. 1997, in *Presolar Grains*, eds. T. Bernatowicz, E. Zinner, AIP, New York, in press

Woosley, S.E. 1988, Ap. J. 330, 218

Woosley, S.E., Arnett, W.D., Clayton, D.D. 1973, Ap. J. Suppl. 26, 231

Woosley, S.E., Fowler, W.A., Holmes, J.A., Zimmerman, B.A. 1978, At. Data Nucl. Data Tables 22, 371

Woosley, S.E., Hartmann, D., Hoffman, R.B., Haxton, W.C. 1990, Ap. J. 356, 272

Woosley, S. E., Hoffman, R. D. 1992, Ap. J. 395, 202

Woosley, S.E., Howard, W.M. 1978, Ap. J. Suppl. 36, 285

Woosley, S.E., Weaver, T.A. 1986, Ann. Rev. Astron. Astrophys. 24, 205

Woosley, S.E., Eastman, R.G., Weaver, T.A., Pinto P.P. 1994, Ap. J. 429, 300

Woosley, S.E., Weaver, T.A. 1994, in *Les Houches, Session LIV, Supernovae*, eds. S.R. Bludman, R. Mochkovitch, J. Zinn-Justin, Elsevier Science Publ., p. 63

Woosley, S.E., Weaver, T.A. 1995, Ap. J. Suppl. 101, 181

Woosley, S. E., Wilson, J. R., Mathews, G. J., Hoffman, R. D., Meyer, B. S. 1994, Ap. J. 433, 229

Zhao, G., Magain, P. 1990, A & A 238, 242

Zhao, Z., France, R.H., Lai, K.S., Rugari, S.L., Gai, M., Wilds, E.L. 1993, Phys. Rev. Lett. 70, 2066

Zhao, Z., France, R.H., Lai, K.S., Rugari, S.L., Gai, M., Wilds, E.L., Kryger, R.A., Winger, J.A., Beard, K.B. 1993, Phys. Rev. C48, 429

Signatures of Nucleosynthesis in Explosive Stellar Processes

By MICHAEL WIESCHER

Department of Physics, University of Notre Dame, Notre Dame, IN 46556, USA

This paper presents a discussion of the characteristic observables of stellar explosions and compares the observed signatures such as light curve and abundance distribution with the respective values predicted in nucleosynthesis model calculations. Both the predicted energy generation as well as the abundance distribution in the ejecta depends critically on the precise knowledge of the reaction rates and decay processes involved in the nucleosynthesis reaction sequences. The important reactions and their influence on the production of the observed abundances will be discussed. The nucleosynthesis scenarios presented here are all based on explosive events at high temperature and density conditions. Many of the nuclear reactions involve unstable isotopes and are not well understood yet. To reduce the experimental uncertainties several radioactive beam experiments will be discussed which will help to come to a better understanding of the correlated nucleosynthesis.

1. Introduction

Historically, the field of nuclear astrophysics has been concerned with the interpretation of the observed elemental and isotopic abundance distribution (Anders & Grevesse 1989) and with the formulation and description of the originating nucleosynthesis processes (Burbidge *et al.* 1957; Wagoner 1973; Fowler 1984). Each of these nucleosynthesis processes can be characterized by a specific signature in luminosity and/or in the resulting abundance distribution. Improved observational instrumentation and spectroscopic detection methods allow detailed observation and analysis of the luminosity curve as well as of the elemental abundances in the ejecta of explosive stellar events (Smith *et al.* 1990; Jorissen *et al.* 1992; Kraft 1994; Sneden *et al.* 1996; Hauschildt *et al.* 1995; Höflich *et al.* 1997). A direct comparison between the predicted and observed abundances will yield information about the temperature, density and hydrodynamical conditions in the stellar explosion. Satellite-based gamma-ray telescopes like GRO/COMPTEL and GRO/OSSE allow the search for long-lived galactic radioactivity produced in stellar burning scenario (Prantzos & Diehl 1995) as well as in novae and supernovae explosions (Iyudin *et al* 1995; Diehl *et al* 1995). This offers an effective tool to monitor the production of specific radionuclides during specific nucleosynthesis scenarios. Again, direct comparison between the predicted and observed abundance of the radionuclide can reveal valuable information about the macroscopic conditions of the respective nucleosynthesis process.

After a short discussion of stellar reaction and decay rates necessary for the interpretation of nucleosynthesis, I will present a general overview of static stellar nucleosynthesis processes and a short summary of the remaining experimental problems and uncertainties. In the following sections I will concentrate on a selected set of explosive nucleosynthesis scenarios in which nuclear reaction and decay data for radioactive isotopes are of particular interest. I will therefore discuss some specific examples which require radioactive beam facilities.

2. Reaction Rates and Decay Rates

Energy generation and nucleosynthesis in stellar burning processes can be simulated by large scale nuclear reaction network calculations for the stellar temperature and density conditions characteristic of a particular stellar scenario. The simulation of nucleosynthesis follows the time evolution of the isotopic abundances $Y_i = X_i/A_i$ (mass fraction divided by mass number) and determines the reaction flux which defines the actual reaction path for nucleosynthesis. This path depends strongly on the macroscopic temperature and density conditions in the stellar event, which might change on a short time scale. Therefore it is important to compare and correlate the macroscopic time scale with the nuclear time scale of the process.

The reaction network is defined by a set of differential equations for the various isotopic abundances. The time derivative of the abundance for each isotope is expressed in terms of the reaction rates of the different production and depletion reactions,

$$\frac{dY_i}{dt} = \sum_j N_j^i \lambda_j Y_j + \sum_{j,k} N_{j,k}^i \rho N_A < j,k > Y_j Y_k + \sum_{j,k,l} N_{j,k,l}^i \rho^2 N_A^2 < j,k,l > Y_j Y_k Y_l. \tag{2.1}$$

The reactions represented by the three terms on the right side of the equation are categorized into decay and photodisintegration processes, λ_j, two particle capture processes, $\rho N_A < j,k >$, and three particle interactions, $\rho^2 N_A^2 < j,k,l >$. The particle induced reaction rate depends on the density ρ, and N_A represents Avogadro's number. The individual N^i's are given by $N_j^i = N_i$, $N_{j,k}^i = N_i/(N_j!N_k!)$, and $N_{j,k,l}^i = N_i/(N_j!N_k!N_l!)$. Each N^i represents a positive or negative number specifying how many particles of species i are created or destroyed in the reaction. The denominators, including factorials, avoid double counting of the number of reactions when identical particles react with each other. For a detailed discussion see (Fowler *et al.* 1967, 1975).

The time integrated net reaction flow between two isotopes i and j is defined by

$$F_{i,j} = \int \left[\frac{dY_i}{dt}_{(i \to j)} - \frac{dY_j}{dt}_{j \to i)} \right] dt \tag{2.2}$$

The maximum flux $F_{i,j}$ defines the main reaction path along which nucleosynthesis will take place.

The observed luminosity of a stellar event is strongly correlated to the energy production. The energy production ϵ in the nucleosynthesis process can be directly determined from the net-reaction flow and the corresponding Q-values $Q_{i,j}$ of the contributing reactions along the reaction path,

$$\epsilon = \sum_{i,j} F_{1,j} \cdot Q_{i,j} \tag{2.3}$$

The time evolution of the isotopes, the time-integrated reaction flux, and the energy production depend critically on the different thermonuclear reaction rates involved in the process. An accurate knowledge of the reaction rates is therefore essential for a reliable interpretation of the nucleosynthesis process.

General nuclear reaction network calculations include decay processes, two particle interactions, photodisintegration processes and three particle interactions. In the following I will concentrate on the rates for β-decay and two-particle interactions only. A detailed discussion of the rates for photodisintegration and three particle interactions is beyond the scope of this paper and can be found in the literature (Fowler *et al.* 1967, 1975; Cowan *et al.* 1991; Görres *at al.* 1995: Schatz *et al.* 1997*a*).

The decay rates of β-unstable particles, λ_i are typically determined from the experimental lifetimes τ_i or half lives $T^i_{1/2}$,

$$\lambda_i = \frac{1}{\tau_i} = \frac{ln2}{T^i_{1/2}}. \tag{2.4}$$

If no experimental half-life is known the decay rate can be calculated in terms of the β-strength function $S_\beta(E)$ by

$$\lambda_i = \frac{g_v^2 m_e^5 c^4}{2\pi^3 \hbar^7} \int_0^{Q_\beta} S_\beta(E) f_0(Z, Q_\beta - E) dE. \tag{2.5}$$

In this equation g_v represents the vector coupling constant, and $f_0(Z, Q_\beta - E)$ the Fermi function. The β-strength function

$$S_\beta(E) = \sum_{J,\pi} B(E, J, \pi) \rho(E, J, \pi) \tag{2.6}$$

describes the energy dependence of the transition probability and is expressed in terms of the reduced transition probability $B(E_i, J, \pi)$ to a final state at excitation energy E_i with spin and parity J^π, weighted with the level density for each spin and parity (Fuller *et al.* 1980).

For interactions between two particles j, k (e.g. proton or alpha capture processes on heavier mass particles) the stellar reaction rate $N_A < j, k >$, can be derived by the convolution of the energy dependent reaction cross section $\sigma(E)$ with the Maxwell-Boltzmann distribution of the interacting particles in the stellar gas. The reaction rate can be expressed in terms of the particle energy E, and the stellar temperature T by,

$$N_A < j, k > = \left[\frac{8}{\mu\pi}\right]^{1/2} (kT)^{-3/2} \int_0^\infty E\sigma_{j,k}(E) \cdot exp(-E/kT) dE. \tag{2.7}$$

Here μ denotes the reduced mass of the target projectile system. The reaction cross section depends critically on the Coulomb barrier between the two interaction particles and on the nuclear structure of the compound nucleus. For charged particle interactions the cross section decreases rapidly with decreasing energy, therefore only particles at the high energy tail of the Maxwell Boltzmann distribution have sufficient energy to tunnel through the Coulomb-barrier. The effective energy range of nuclear burning is called the Gamow window (Fowler *et al.* 1967, 1975).

In the case of high-level density the cross section for a reaction between two particles i and j, $\sigma_{i,j}$ is determined by many overlapping resonance contributions and appears non-resonant. The cross section can be well approximated by the statistical Hauser Feshbach model (Sargood 1982; Rauscher *et al.* 1997) which describes the cross section in terms of the transmission functions T_{in}, T_{out} of the reaction entrance and exit channel potential conditions

$$\sigma_{j,k} = \frac{\pi\hbar^2}{2\mu_{j,k} E_{j,k}} \cdot \frac{1}{(2j_j + 1)(2j_k + 1} \cdot \sum_J (2J + 1) \frac{T_{in} \cdot T_{out}}{T_{tot}} \tag{2.8}$$

with $\mu_{j,k}$ as the reduced mass of the projectile-target system and $E_{j,k}$ as the center of mass energy. T_{tot} is the total transmission coefficient summed over all open channels.

In the case of low-level density in the compound system, the reaction cross section is determined by a sum of Breit-Wigner terms in the cross section which correspond to unbound states in the compound nucleus, and by nonresonant terms in the cross section which correspond to transitions to bound states in the final nucleus,

These low level density conditions are predominantly observed in light nuclei or in nuclei near closed shells. The resonant term of the reaction rate reaction rate $N_A < j,k >_{\mathrm{r}}$ [in $\mathrm{cm}^3\mathrm{s}^{-1}$mole] is derived by the analytical intergration over the Breit-Wigner cross section of a single resonance. The rate is typically expressed in terms of the resonance energy E_{r} [in MeV] and the resonance strength $\omega\gamma$ [in MeV] as a function of temperature T_9 [in 10^9K] in the notation of (Fowler *et al.* 1975).

$$N_A < j,k >_{\mathrm{r}} = 1.54 \cdot 10^{11} (\mu_{j,k} T_9)^{-3/2} \omega\gamma \cdot \exp\left(\frac{-11.605 E_{\mathrm{r}}}{T_9}\right) \quad . \tag{2.9}$$

The resonance strength $\omega\gamma$ depends on the spin J_T of the target, the spin J of the resonance state, the partial width Γ_{in} for the entrance channel, the partial width Γ_{out} for the exit channel, and the total width Γ_{tot},

$$\omega\gamma = \frac{(2J+1)}{2(2J_T+1)} \frac{\Gamma_{in} \cdot \Gamma_{out}}{\Gamma_{\mathrm{tot}}} \quad . \tag{2.10}$$

The nonresonant reaction cross sections are mainly determined by direct interaction processes or by tail contributions of broad resonances. The nonresonant reaction rate $N_A < j,k >_{\mathrm{nr}}$ [in $\mathrm{cm}^3\mathrm{s}^{-1}$mole] is determined by the astrophysical S–factor $S(E_0)$ in the energy range of the Gamow window (Fowler *et al.* 1967, 1975, Rolfs & Rodney 1988),

$$N_A < j,k >_{\mathrm{nr}} = 7.83 \cdot 10^9 \left(\frac{Z_j Z_k}{\mu_{j,k} T_9^2}\right)^{1/3} S(E_0) \cdot \exp\left(-4.29 \left[\frac{\mu_{j,k} Z_j^2 Z_k^2}{T_9}\right]^{1/3}\right) \tag{2.11}$$

where Z_j, Z_k denote the charge number of the interacting nuclei j and k, respectively. The total S–factor [in MeV-barn] corresponds directly to the total nonresonant reaction cross section σ_{nr},

$$S(E_0) = \sigma_{nr} \cdot E_0 \cdot \exp(2\pi\eta). \tag{2.12}$$

The exponential term approximates the influence of the Coulomb barrier on the reaction cross section with η as the Sommerfeld parameter (Fowler *et al.* 1967, 1975). Experimental determination of a stellar reaction rate for resonant processes therefore requires the measurement of the resonance parameters like resonance energy E_r, and resonance strength $\omega\gamma$. For the determination of the nonresonant contributions the measurement of the cross section $\sigma_{nr}(E)$ over a wide energy range is necessary to determine the astrophysical S–factor in the stellar energy range. Figure 1 shows the effective energy range of a reaction for the nonresonant (upper part) as well as for the resonant case (lower part).

In the nonresonant part, the range is determined by the energy and width of the Gamow peak, while the range of the resonant part is determined by the energy and width of the resonance.

The total reaction rate consists of the sum of all resonant and nonresonant reaction contributions,

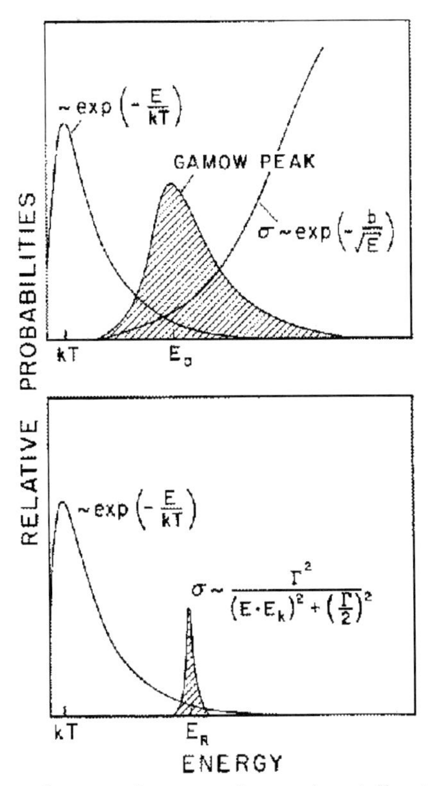

FIGURE 1. Effective energy range for a nuclear reaction under stellar temperature conditions for the nonresonant case (Gamow window) and for the case of a single resonance (resonance width).

$$N_A < j,k >= N_A < j,k >_{\rm r} + N_A < j,k >_{\rm nr} \,. \tag{2.13}$$

From the total reaction rate $N_A < j,k >$ the lifetime $\tau_k(j)$ of the isotope j against interaction with the isotope k can be calculated as a function of temperature and density in the stellar environment,

$$\tau_k(j) = \frac{A_k}{\rho \cdot X_k \cdot N_A < j,k >} \tag{2.14}$$

with X_k, A_k as the mass fraction and the atomic number of the isotope k, respectively. Unlike β-decay processes, the life-time for charged particle interactions depends sensitively on the stellar temperature and density conditions and are therefore subject to rapid changes during explosive nucleosynthesis.

3. Overview of Stellar Nucleosynthesis Processes

Static stellar burning is characterized by fairly "low" temperature conditions in the burning zone, T$\approx 2 \cdot 10^7$ to $5 \cdot 10^7$ K in massive main sequence stars, and T$\approx 5 \cdot 10^7$ to $3 \cdot 10^8$ K for red giant stars. This corresponds to a Gamow energy range from 40 to 120 keV for hydrogen burning and from 80 to 450 keV for helium burning, respectively. Typical reaction cross sections at these energies are in the sub-nanobarn range and therefore extremely difficult to measure. For this reason most of the presently available reaction rates are based on the extrapolation of high energy measurements down to the stellar energy range. This introduces significant uncertainties into the rates and the subsequent interpretation of stellar nucleosynthesis.

In the following sections various stellar burning conditions will be outlined where additional experimental data are required to improve our current understanding of the associated nucleosynthesis.

3.1. *Stellar hydrogen burning through the pp–chains and the CNO cycles*

The pp–chains and the CNO cycles are the dominant energy sources in main sequence stars. While the pp-chains determine the energy generation in stars with $M \leq 1.5\ M_\odot$, in more massive stars the CNO reactions dominate. The latter are also important for the understanding of energy generation and nucleosynthesis in the outer hydrogen burning shells during the later stages of stellar evolution.

Figure 2 shows the reaction sequences of the three pp-chains by which hydrogen is converted to helium. All pp-chains are initiated by the slow ^{1}H(p,e$^+\nu$)^{2}H reaction which determines the energy generation rate. The pp-chain reactions are responsible for the production of the solar neutrinos and detailed experimental studies are required for the investigation of the discrepancy between predicted and observed solar neutrinos, the so-called Solar Neutrino Problem (Bahcall & Pinsonneault 1992). Most of the reactions in

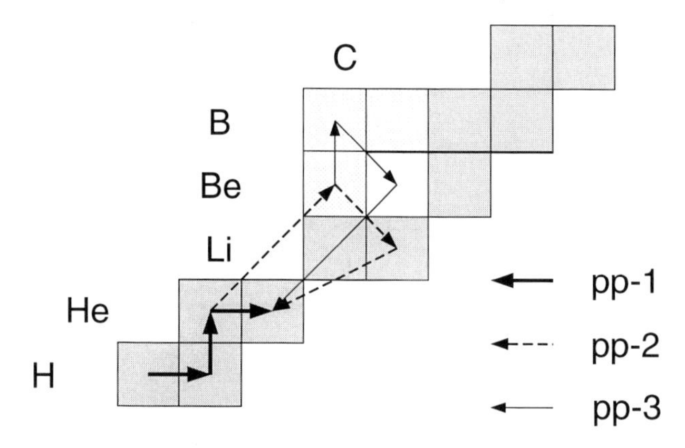

FIGURE 2. The pp-chain reactions which determine the energy generation by hydrogen fusion in low mass ($M \leq 1.5\ M_\odot$) main sequence stars.

the pp-chains have been extensively studied over a wide energy range. A detailed discussion of the various reaction cross sections is given by Rolfs & Rodney (1988). However, over the last few years considerable effort has been made to extend the measurements down to low energies using the underground accelerator facility LUNA (Fiorentini *et al.* 1995). Several reaction cross sections of the pp–chain have been measured down or close to the Gamow window. Of particular importance is the measurement of the pp-1 reaction ^{3}He(^{3}He,2p)^{4}He in the Gamow range (Arpesella *et al.* 1996). It has been speculated that a possible low energy resonance may enhance the reaction rate dramatically and such reduce the high energy neutrino flux ($E_\nu \geq 0.8$ MeV) which is originated in the pp-2 and pp-3 chains. The experimental results indicate a purely nonresonant direct reaction mechanism and rule out such a resonance. The cross section of the ^{7}Be(p,γ)^{8}B reaction (pp-3 chain) is one of the few remaining nuclear physics related uncertainties in the prediction of the solar neutrino flux. Aside from uncertainties in the extrapolation of high energy data to stellar energies there exists also unacceptable discrepancies in the various theoretical calculations and the measurements of the absolute cross section values (Bahcall & Pinsonneault 1992; Castellani *et al.* 1997). Therefore, the study of ^{7}Be(p,γ)^{8}B is extremely important for the interpretation of the results of recent and future solar neutrino searches like HOMESTAKE, GALLEX, SAGE, GNO, (SUPER)–KAMIOKANDE, BOREXINO, SNO, and HELLAZ. Presently, several groups have focused on this reaction using a large variety of experimental techniques, ranging from low energy proton capture studies using a radioactive ^{7}Be target, to studies in inverse kinematics with a radioactive ^{7}Be beam, and to studies of Coulomb dissociation using a high energy radioactive ^{8}B beam. The various experimental results will hopefully resolve the persistent uncertainty in this reaction rate in the near future. Figure 3 shows the four classical CNO cycles which allow catalytic hydrogen fusion to helium. A detailed discussion of the various reaction processes is given by Rolfs & Rodney (1988). Most of the CNO

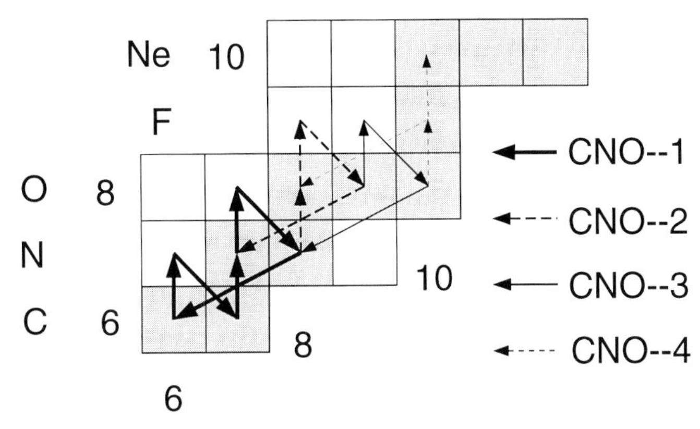

FIGURE 3. The classical CNO-cycles which determine the energy generation and the nucleosynthesis by catalytic hydrogen burning in massive (M$\geq$1.5 M$_\odot$) main sequence stars.

reactions have been well studied down to E$\approx$100 keV. However, this still requires further extrapolation of the reaction cross sections down to the Gamow range (Rolfs & Barnes 1990). Presently the major uncertainties in the interpretation of the CNO cycles are the reaction rates of ^{14}N(p,γ)^{15}O and ^{17}O(p,α)^{14}N. The first reaction is the slowest reaction in the first CNO cycle at stellar temperatures. It therefore determines the energy generation in hydrogen burning phase of the star. The reaction cross section has been extensively measured at higher energies and a reaction rate has been derived (Schröder *et al.* 1987). The main uncertainty in the stellar reaction rate are possible contributions from a -504 keV subthreshold s-wave resonance. Detailed low energy measurements are necessary to determine possible contributions from the high energy tail of this level. The second reaction rate is dominated by a low energy resonance at 70 keV. The reaction rate determines determines the amount of ^{17}O produced by the first and second dredge-up in intermediate mass stars and during hot–bottom burning in AGB stars. Recent experimental efforts to study the low energy resonant contributions led to contradictory results (Berheide *et al.* 1992; Blackmon *et al.* 1995), Further experimental efforts are clearly necessary.

3.2. *Stellar helium burning and neutron sources for the s–process*

Figure 4 shows the dominant reaction sequences in red giant stars. The energy generation is dominated by the energy release from the triple-α-process which fuses three α particles to form ^{12}C (Rolfs & Rodney 1988). The reaction ^{12}C$(\alpha,\gamma)^{16}$O competes with the triple-α-process in the consumption of helium and is therefore (together with ^{16}O$(\alpha,\gamma)^{20}$Ne)

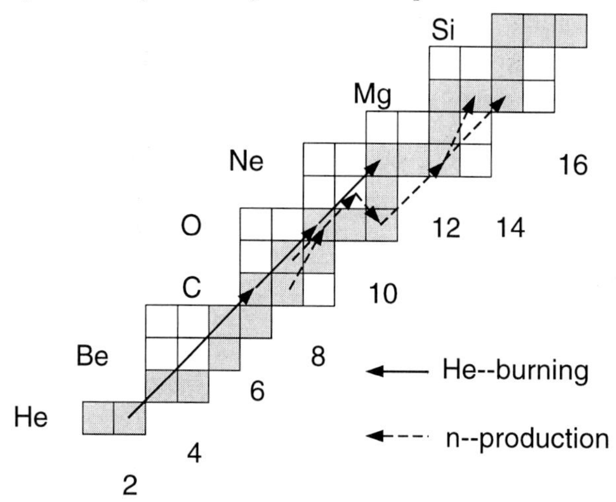

FIGURE 4. Reaction sequences in stellar helium burning in massive red giant stars. The ^{13}C(α,n) neutron source operates in low mass AGB stars after hydrogen is mixed into the carbon rich intra-shell regions during dredge-up (Busso *et al.* 1995).

the key reaction for understanding stellar helium burning because their rates determine the final ^{12}C/^{16}O abundance ratio. This is crucial for the interpretation of later burning stages in massive stars (carbon-burning, oxygen-burning) as well as for supernovae nucleosynthesis (Weaver & Woosley 1993). The $^{12}C(\alpha,\gamma)^{16}O$ reaction is extremely difficult to study directly because the reaction cross section at stellar temperatures is dominated by interference effects between higher energy resonances and two subthreshold states in the compound nucleus ^{16}O. A recent measurement of the β–delayed α–decay of ^{16}N permitted an improved determination of the low energy E1 component of the reaction cross section (Azuma *et al.* 1994). The remaining major uncertainty is the possible E2–contribution from a 2^+ subthreshold state and the direct capture to the ground state in ^{16}O as well as possible interference between the three contributions. The E2–component has been predicted to be as large as 50% of the total cross section (see Langanke & Barnes (1996)). A summary of the present experimental status is given by Buchmann *et al.* (1996).

The study of α–capture reactions on light nuclei from ^{14}N up to Mg is directly related to the production of neutrons for the weak s–process component in the helium burning core of massive stars. Figure 4 shows the neutron production sequence triggered by α capture on ^{14}N isotopes produced in the preceeding hydrogen burning phase. Specifically important for a reliable prediction of the neutron production are low energy measurements of the reactions $^{14}N(\alpha,\gamma)^{18}F$, $^{18}O(\alpha,\gamma)^{22}Ne$, $^{22}Ne(\alpha,\gamma)^{26}Mg$, and $^{22}Ne(\alpha,n)^{25}Mg$. Although these reactions have been extensively investigated in recent years, the rapidly decreasing low energy cross sections does not yet allow a reliable determination of all reaction rate contributions. This causes significant uncertainties in the reaction rates at stellar temperatures, making it difficult to predict the neutron production in massive star helium burning (Käppeler *et al.* 1994).

3.3. *Convective burning processes in AGB stars*

Recent observations of surface abundance distributions of asymptotic giant branch (AGB) stars and the results of the analysis of abundances in meteoritic inclusions have triggered a strong interest in convective nucleosynthesis in the various hydrogen, helium and carbon burning shells of AGB stars. Deep convective and semiconvective mixing between the various burning zones (Lattanzio 1995) initiates complex nucleosynthesis conditions like hot bottom burning (Blöcker & Schönberner 1995, Cannon *et al.* 1995, Lattanzio *et al.* 1997). The processed material is mixed to the surface of the star in the various dredge-up phases and modifies the surface composition with the products of core and shell hydrogen and helium burning (Busso *et al.* 1995). AGB stars are also presently considered to be one of the sources for the long-lived isotope ^{26}Al, which has been observed by satellite based γ-observatories (Prantzos & Diehl 1995). The proposed production mechanism is hot-bottom hydrogen burning at temperatures of up to 10^8 K, which initiates the hot CNO cycle, but also reactions in the NeNa-cycle, the MgAl-, and the SiP-cycle. AGB stars are also considered to be the likely source for graphite and SiC grain inclusions in meteorites (Zinner *et al.* 1991). The observed abundance ratios of oxygen, aluminum, silicon, and titanium isotopes is therefore directly connected to the nucleosynthesis in AGB stars (Nittler *et al.* 1997).

4. Explosive stellar processes and radioactive beam experiments

Radioactive beam experiments are necessary for the measurement of reactions on radioactive nuclei which can take place at the high temperature conditions typical in explosive stellar events.

Capture measurements with radioactive beams in inverse kinematics are among the major goals for radioactive beam facilities. The previous measurements on first generation facilities were handicapped by insufficient detector systems with relatively low detection efficiencies or limited background reduction capabilities. These conditions can be considerably improved by the utilization of recoil-mass-separators. These guarantee a $\leq 10^{-12}$ rejection of the primary beam and further background reduction by particle identification methods while maintaining a high detection efficiency.

Fragment mass separators can be used for the production of neutron–rich or neutron–deficient radioactive isotopes between the region of stability and the neutron or proton drip–line, to perform mass or half–life measurements. These nuclear structure results are extremely important for determining the reaction path of the r– and rp–process and the associated time scales for explosive events (Cowan *et al.* 1991; Schatz *et al.* 1997*a*).

In the following sections I will discuss a few examples of explosive burning which require experimental data that can only be obtained at radioactive beam facilities.

5. Nucleosynthesis of ^{15}N in Novae

Novae have been interpreted as thermonuclear runaways on the surface of an accreting white dwarf in close binary star systems (Truran 1982, 1990; Starrfield 1989). Accretion processes or mass exchange between the two binary stars can take place when the second companion star fills its Roche-Lobe, the gravitational equipotential surface enclosing both stars. The matter of the extended star accretes through the Roche-Lobe onto the surface of the second companion which represents a deeper gravitational potential. The accreted material forms a thin, but high-density, electron-degenerate envelope at the surface of the white dwarf. Dredge-up of white dwarf material (^{4}He, ^{12}C, ^{16}O in the case of an CO-white dwarf, ^{16}O, ^{20}Ne, and ^{24}Mg in the case of an ONeMg-white dwarf) into the envelope leads to an enrichment of the accreted material in heavier isotopes (Glasner *et al.* 1996).

After a "critical" mass has been accreted, thermonuclear ignition takes place at the bottom of the accreted envelope. Ignition depends critically on the mass of the white dwarf and the accretion rate which determines the pressure conditions at the bottom of the envelope. The ignition occurs presumably via the pp-chains, causing a rapid increase in temperature at constant pressure and density at degenerate conditions. This "thermonuclear runaway" is further enhanced by the subsequent ignition of the hot CNO cycles on the high abundances of ^{12}C and ^{16}O until the degeneracy is lifted after the Fermi temperature T_F has been reached. The Fermi temperature depends on the density ρ [in g/cm^3] and the composition of the accreted material along with μ_e, the electron mean molecular weight (Starrfield 1989; Hansen & Kawaler 1994),

$$T_F = 3.03 \cdot 10^5 \left(\frac{\rho}{\mu_e}\right)^{2/3} . \tag{5.15}$$

The higher the density of the material, the higher the peak temperature which can be reached in the thermal runaway before degeneracy is lifted. However, if the shell temperature rises too rapidly the peak temperature can exceed the Fermi temperature before the electron gas is sufficiently degenerate to initiate expansion. Due to the rapid temperature increase at the bottom of the envelope a convective zone develops which gradually grows to the surface as the temperature continues to increase. This allows rapid energy transport to the surface within the convective time-scale of $t_{conv} \approx 10^2$s. Within that short time-scale, an appreciable fraction of the long-lived β^+ emitters which are produced by the hot CNO cycles is carried to the surface. The release of decay energy further increases the luminosity to values above 10^5 $L_\odot$. The large amount of energy deposited in the outer layers on the short convective time scale, coupled with high luminosity often exceeding the Eddington limit L_{edd} (Shore *et al.* 1994), causes rapid expansion of the outer layers and the ejection of the outer shells. Typical novae are characterized by thermal runaways with densities of approximately $\rho \approx 10^3$ g/cm^3 and typical peak temperatures between $1{\cdot}10^8$ K and $4{\cdot}10^8$ K (Starrfield 1989; Starrfield *et al.* 1997).

5.1. *CNO Nucleosynthesis in Ne Novae*

A large fraction of the observed nova ejecta indicate enhancements in neon (Truran & Livio 1986; Weiss & Truran 1990; Livio & Truran 1994) as compared to the solar abundances (Anders & Grevesse 1989). This has been interpreted as the result of a thermonuclear runaway on the surface of an accreting ONeMg white dwarf after dredge up of white dwarf material into the accreted envelope (Weiss & Truran 1990). In this case the white dwarf material is enriched in ^{16}O, ^{20}Ne, and ^{24}Mg (Law & Ritter 1983). The observation of heavier elements up to sulfur in the ejecta has been interpreted as the results of nucleosynthesis at high temperature conditions in the thermonuclear runaway (Politano *et al.* 1995; Coc *et al.* 1995; Jose *et al.* 1997; Starrfield *et al.* 1997). The nucleosynthesis process is characterized by the NeNa- and the MgAl cycle, or at higher temperatures by the rp-process (Van Wormer *et al.* 1994; Herndl *et al.* 1995). These reaction sequences are triggered by the initially high ^{20}Ne and ^{24}Mg abundances in the dredged-up material.

The abundance distribution of Ne-novae ejecta is also characterized by a high nitrogen abundance (Williams 1982). This can be interpreted as the result of hot CNO nucleosynthesis triggered on the high ^{16}O abundances in the dredged-up white dwarf material via $^{16}O(p,\gamma)^{17}F(p,\gamma)^{18}Ne(\beta^+\nu)^{18}F(p,\alpha)^{15}O$. The fairly long-lived radioactive isotope ^{15}O ($T_{1/2}$=107 s) subsequently decays in the expansion phase after being mixed to the surface of the envelope by rapid convection processes. The ejected outer shell therefore contains a large fraction of freshly produced ^{15}N. A quantitative interpretation of the involved nucleosynthesis processes, however, requires an exact knowledge of the reaction rates involved. The reaction sequence includes capture reactions on the radioactive fluorine isotopes ^{17}F and ^{18}F. An experimental determination of the reaction rates requires detailed measurements of the cross sections for the various resonant and nonresonant reaction contributions. Because both ^{17}F and ^{18}F are rather short-lived these experiments can only be performed by producing radioactive fluorine beams.

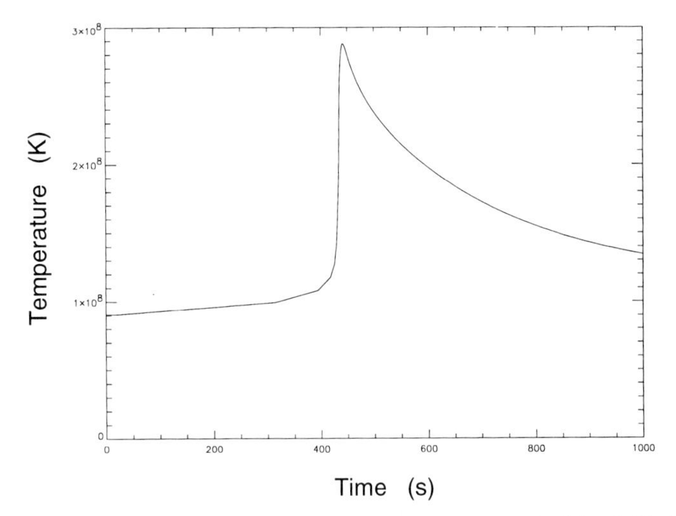

FIGURE 5. Temperature curve for the thermonuclear runaway in a 1.25 $M_\odot$ O-Ne-Mg white dwarf model. The peak temperature reaches $\approx 3 \cdot 10^8$ K (for details see Starrfield *et al.* (1997)).

The impact of the various reaction rates on the production of particular isotopes during the thermonuclear runaway can be studied by simulating the nucleosynthesis in a reaction network code. The source for the production of ^{15}O as radioactive progenitor of the stable ^{15}N is the reaction sequence ^{16}O(p,γ)^{17}F(p,γ)^{18}Ne ($\beta^+\nu$)^{18}F(p,α)^{15}O. This sequence is part of the hot CNO cycles (Audouze *et al.* 1973) triggered on the abundance of ^{16}O which has been dredged-up into the accreted hydrogen layer. The reaction rate for ^{16}O(p,γ)^{17}F is dominated by the direct capture to the ground state and the first excited state in ^{17}F and has been studied extensively (Rolfs, 1973; Caughlan & Fowler 1988). The proton capture rate for the ^{17}F(p,γ)^{18}Ne reaction has been estimated from the nuclear structure parameters of the proton unbound states in the compound nucleus ^{18}Ne (Wiescher *et al.* 1988*a*; Garcia *et al.* 1991; Hahn *et al.* 1996). The reaction rate is dominated by the resonant contribution of the 3^+ state at 4.56 MeV which has been observed only in a ^{16}O(^{3}He,n)^{18}Ne two-neutron transfer experiment (Garcia *et al.* 1991). The resonance strength is therefore highly uncertain and radioactive beam measurements with ^{17}F similar to these described above are necessary to confirm the present estimates. Based on the estimated resonance strength the reaction rate of ^{17}F(p,γ)^{18}Ne is calculated to be approximately one order of magnitude smaller than the rate of the progenitor reaction ^{16}O(p,γ)^{17}F. Comparison with the β-decay rate of ^{17}F shows that for temperatures above $T \approx 2 \cdot 10^8$K the proton capture becomes faster than the competing β-decay, leading to the production of ^{18}Ne. The time-scale for its production therefore depends sensitively on the small rate of the ^{17}F(p,γ)^{18}Ne reaction which is predicted to be close to the β-decay rate of ^{17}F at typical nova temperatures Because ^{19}Na is proton unbound no further proton capture can occur (for a detailed discussion see Görres *et al.* (1995)), and ^{18}Ne ($T_{1/2}$=1.67 s) β^+-decays to ^{18}F. Further nucleosynthesis depends on the branching ratio between ^{18}F(p,γ)^{19}Ne, which may trigger a break-out flow from the CNO cycles, and ^{18}F(p,α)^{15}O, which would cause the production of ^{15}N. A detailed

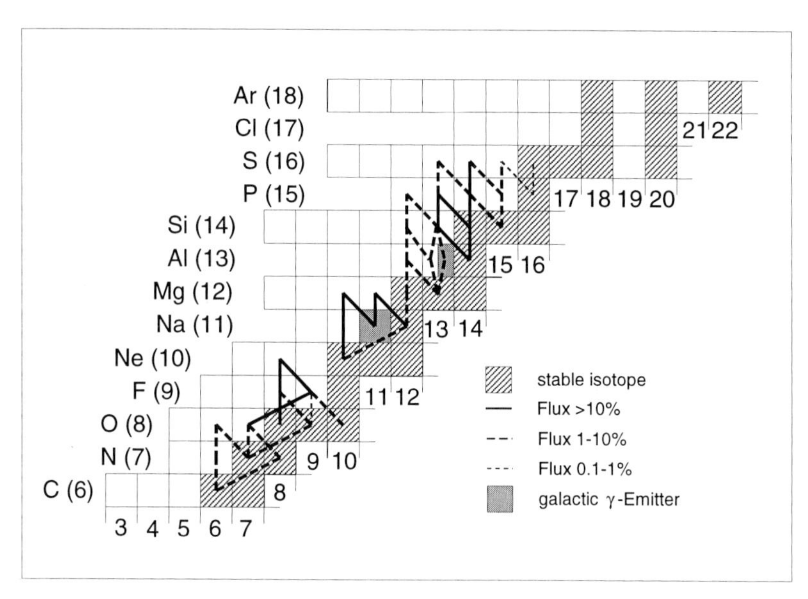

FIGURE 6. Reaction flux during explosive hydrogen burning in novae. The net flow is integrated over the period of high temperature conditions in the burning zone

study of the ^{19}F(^{3}He,t)^{19}Ne(α)^{15}O and ^{19}F(^{3}He,t)^{19}Ne(p)^{18}F reactions indicates the the α-channel of the proton unbound states in ^{19}Ne dominates (by more than three orders of magnitude) (Utku *et al.* 1997). A recent measurement of the ^{18}F(p,α)^{15}O reaction in inverse kinematics at the Louvain-la-Neuve radioactive beam facility has focused on the direct study of the lower energy resonances (Coszach *et al.* 1995; Graulich *et al.* 1997). Two resonances have been observed at E_{cm}=0.33 MeV and 0.638 MeV, and the resonance strengths of these levels have been measured directly. The reaction rate of ^{18}F(p,α)^{15}O is dominated by the contributions of these two resonances as well as by the low energy tail of the rather broad ($\Gamma \approx$40 keV) resonance at 638 keV (Coszach *et al.* 1995; Rehm *et al.* 1996; Utku *et al.* 1997). The reaction rate is large, which translates into a rather short effective lifetime for ^{18}F at nova conditions, since all the freshly produced ^{18}F is immediately processed further to the long-lived ^{15}O isotope.

Figure 5 shows the temperature profile for the thermonuclear runaway at the surface of a 1.25 $M_\odot$ accreting white dwarf (Starrfieldt *et al.* 1997). This model suggests that peak temperatures of up to T=3·10^8 K at initial densities of $\rho \approx$5·10^3 g/cm^3 can be reached. Figure 6 shows the reaction flux calculated for the hydrogen burning zone at the bottom of the accreted layer. Due to dredge-up of ^{16}O from the upper layers of the white dwarf, the reaction path is mainly characterized by the hot CNO cycles. Because of the ^{20}Ne and ^{24}Mg content in the white dwarf material, the NeNa- and the MgAl-cycles also operate. The nucleosynthesis for ^{16}O, ^{18}Ne, ^{18}F, and ^{15}O has been calculated for these conditions in a self consistent network. The change of the isotopic abundances with time are shown in figure 7. Correlated with the increase in temperature the initial abundance of ^{16}O is rapidly depleted and converted into ^{18}Ne, which has a rather long lifetime of τ_β(^{18}Ne)=2.41 s. ^{18}Ne decays, feeding ^{18}F which is rapidly depleted by the fast (p,α) reaction to ^{15}O. The overall equilibrium abundance of ^{18}F remains small during the entire

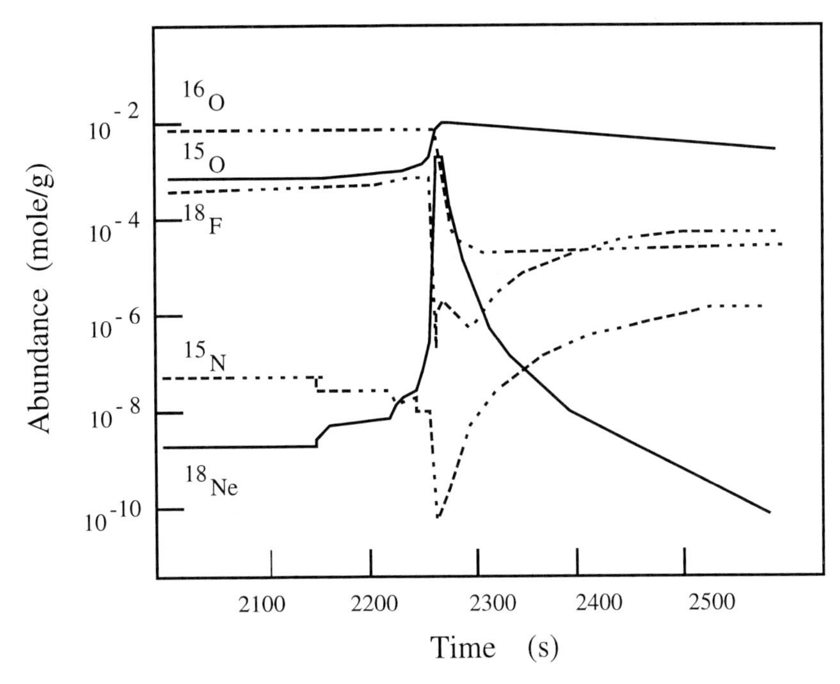

FIGURE 7. Nucleosynthesis in the hot CNO cycles in nova burning. The initial amount of ^{16}O is rapidly converted to ^{15}O which subsequently decays to ^{15}N. The time-scale has an off-set time of t_{off}=1800 s.

process due to its fast depletion rate. The initial amount of ^{16}O is completely converted to ^{15}O which subsequently decays, forming ^{15}N slowly on a time-scale comparable to that of the fast convective mixing of ^{15}O to the surface of the ejected envelope.

5.2. ^{22}*Na Production in Ne-Novae*

Gamma-ray emission from long-lived radionuclides is considered to be one of the dominant signatures for explosive nucleosynthesis (Leising & Clayton 1987; Prantzos & Diehl 1995). Characteristic lines which can be observed with today's gamma-ray telescopes include the ^{26}Al γ-line at 1.809 MeV ($T_{1/2}$ $\approx$7.16$\cdot$10^5 y), the ^{44}Ti γ-line at 1.157 MeV (for a discussion of the halflife see next section), and the ^{22}Na line at 1.274 MeV ($T_{1/2}$ = 2.603 y). While the present observation of galactic ^{26}Al and ^{44}Ti γ-radioactivity is mainly associated with known supernova remnants (Dupraz *et al.* 1997), novae have been discussed as a possible source for the production of ^{22}Na and ^{26}Al (Higdon & Fowler 1987, 1989). Various nova model simulations (Politano *et al.* 1995; Coc *et al.* 1995; Wanajo *et al.* 1997; Starrfield *et al.* 1997) have indeed predicted an appreciable production of ^{22}Na in Ne-novae mainly due to nucleosynthesis in the NeNa-cycle (Marion & Fowler 1957). Figure 6 shows the reaction flow in the Ne-Na region integrated over the thermonuclear runaway of a Ne-nova. The flow indicates that the reaction sequence ^{20}Ne(p,γ)^{21}Na(p,γ)^{22}Mg($\beta^+\nu$) leads to the production of ^{22}Na. While the production rate is dominated by the lowest rate in the reaction chain - which depends critically on the temperature - the depletion is determined by the ^{22}Na(p,γ)^{23}Mg reaction (Seuthe *et al.* 1989; Schmidt *et al.* 1995; Stegmüller *et al.* 1996). Once ejected, ^{22}Na could be observable by the characteristic 1.275 MeV γ-ray following its β^+-decay (2.6 yr half-life) (Weiss & Truran 1990; Starrfield *et al.* 1993; Politano *et al.* 1995). Considerable efforts were undertaken in the search for a galactic 1.274 MeV γ-line, but so far only upper limits could be derived. The most stringent limit is based on recent COMPTEL observations in the direction of relatively close neon novae like Nova Cyg 1992 or Nova Her 1991. The observational upper limits indicate that neon novae produce substantially less ^{22}Na than suggested by model calculations (Iyudin *et al.* 1995).

Figure 6 indicates that ^{22}Na is mainly produced by the β^+-decay of ^{22}Mg (Coc *et al.* 1995). Like other even Z, $T = 1$ nuclei, ^{22}Mg becomes strongly enriched. Its β-decay is rather slow ($T_{1/2}$=3.86 s), the proton capture is determined by a very weak resonance (Wiescher *et al.* 1988*b*), and the net flow is reduced by strong inverse photodisintegration of the weakly bound ^{23}Al. It has been pointed out recently (Herndl *et al.* 1995) that despite the strong photodisintegration of ^{23}Al a considerable net proton capture flow through ^{22}Mg could be established via proton capture on the small ^{23}Al equilibrium abundance (for details on 2p-capture reactions see Görres *et al.* (1995) and Schatz *et al.* (1997*a*)). Then the net proton capture rate on ^{22}Mg depends only on the proton separation energy of ^{23}Al, which fixes the equilibrium ^{23}Al abundance, and the ^{23}Al(p,γ)^{24}Si rate. This mechanism could reduce the effective lifetime of ^{22}Mg and thus the amount of ^{22}Na produced in neon novae, which would be in better agreement with observations. However, a conclusive determination of the effectiveness of this mechanism has been hampered by the large uncertainties in the ^{23}Al(p,γ)^{24}Si reaction rate, since no experimental information on ^{24}Si was available except for the ground state mass.

5.3. *Experimental study of the ^{23}Al(p,γ)^{24}Si rate*

The most recent calculation of the ^{23}Al(p,γ)^{24}Si reaction rate (Herndl *et al.* 1995) is based on shell model calculations of the compound nucleus ^{24}Si and on experimental information from the mirror nucleus ^{24}Ne. The results suggest that the reaction rate is dominated by a single resonance corresponding to the second excited state in ^{24}Si. However, a reliable calculation of the reaction rate is strongly handicapped by the large uncertainty in the resonance energy (typically 100-150 keV for sd-shell nuclei) which goes through the exponential dependence into an uncertainty of 4-7 orders of magnitude in the reaction rate at 0.2–0.4 GK.

In a recent experiment at the Indiana cyclotron facility the ^{28}Si(α,^{8}He)^{24}Si reaction has been studied to determine with high accuracy the excitation energies in ^{24}Si (Schatz *et al.* 1997*b*). The reaction products were detected in the focal plane of the high resolution K600 spectrometer. Fig. 8 shows the ^{8}He position spectrum after 70 hours of beam time. The transitions to the ^{24}Si ground state are clearly observable, as well as transitions to the first and second excited states in ^{24}Si . Also shown in Fig. 8 is the simultaneously obtained ^{6}He spectrum. Well known states in ^{26}Si, ^{27}Si, and ^{28}Si are populated via the ^{28}Si(α,^{6}He)^{26}Si, ^{29}Si(α,^{6}He)^{27}Si, and ^{30}Si(α,^{6}He)^{28}Si reactions. The peaks in the ^{6}He spectrum are well distributed around the ^{8}He peaks and provide an excellent focal plane calibration.

The resulting excitation energies of the first two excited states in ^{24}Si relative to the ^{24}Si ground state are E_x=1.879 $\pm$ 0.011 MeV and E_x=3.441 $\pm$ 0.010 MeV, respectively. Our energies for the first and second excited states in ^{24}Si are 71 keV and 179 keV below the shell model predictions. The dominant resonance in the ^{23}Al(p,γ)^{24}Si reaction rate corresponds to the second excited state in ^{24}Si. Based on the new results the resonance energy is determined to be 141 keV$\pm$30 keV which is well below the predicted shell model value of $\approx$320 keV. The error in the new resonance energy is dominated by the 25 keV uncertainty in the ^{23}Al mass. The reaction rate can be calculated from the resonance strength which is based on the the shell model parameters for the second excited state (Herndl *et al.* 1995). The reaction rate carries large uncertainties due to the experimental

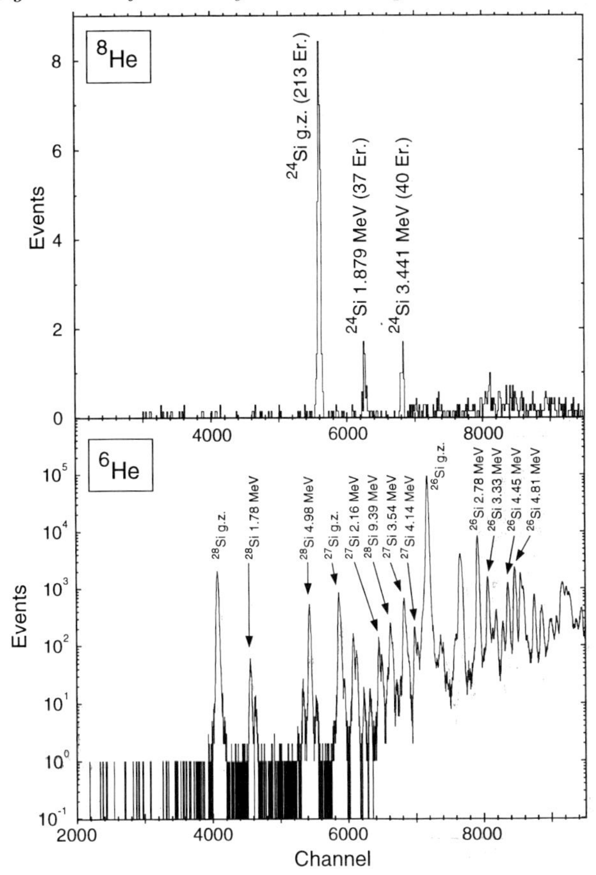

FIGURE 8. The simultaneously obtained ^{6}He and ^{8}He position spectra from the Si target. In the ^{8}He spectrum the ground and first two excited states of ^{24}Si can be identified. The ^{6}He spectrum shows states in ^{26}Si, ^{27}Si and ^{28}Si. The low-lying states indicated were used for the focal plane calibration.

uncertainties in the resonance energy and the theoretical uncertainties in the resonance strength. Nevertheless the impact of this reaction on the ^{22}Na production in novae can be estimated by calculating the temperatures and densities that are required to process more than 50% of the total reaction flow via net proton capture on ^{22}Mg. This would lead to a significant reduction in the ^{22}Na production.

The maximum possible reaction flow through net proton capture on ^{22}Mg is given by the ^{22}Mg(p,γ)^{23}Al reaction with the subsequent ^{23}Al(p,γ)^{24}Si reaction. Figure 9 indicates the corresponding loss of ^{22}Mg towards ^{24}Si via ^{22}Mg(2p,γ)^{24}Si for different temperature and density conditions. It therefore represents the amount of ^{22}Mg which does not contribute to the formation of ^{22}Na in the nova burning zone. It can be seen that only at very high temperature and density conditions an appreciable fraction of ^{22}Mg is converted to ^{24}Si. At the lower temperature and density conditions typical for novae (see figure 5) most of the ^{22}Mg produced decays to ^{22}Na.

Figure 10 shows the change of the isotopic Na and Mg abundances as a function of temperature for the thermonuclear runaway on the surface of a 1.25 $M_\odot$ white dwarf. The corresponding temperature conditions in the burning layer have been shown in figure 5. At the initially low temperature conditions ^{22}Na is produced by the reaction sequence

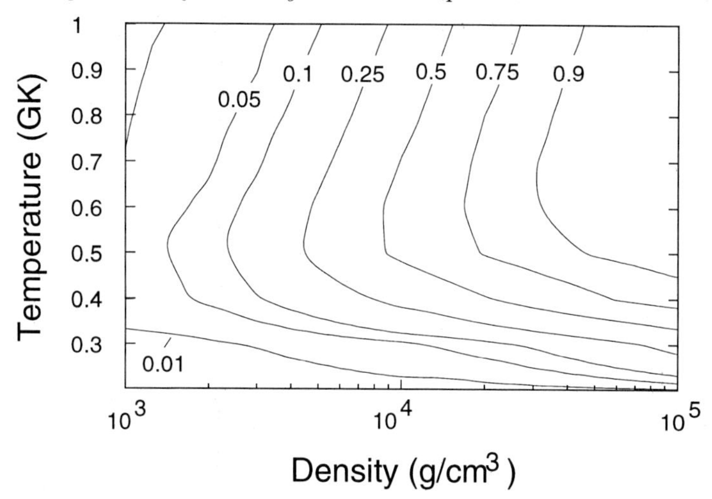

FIGURE 9. Fraction of ^{22}Mg processed by proton capture towards ^{23}Al and subsequently to ^{24}Si as a function of density and temperature.

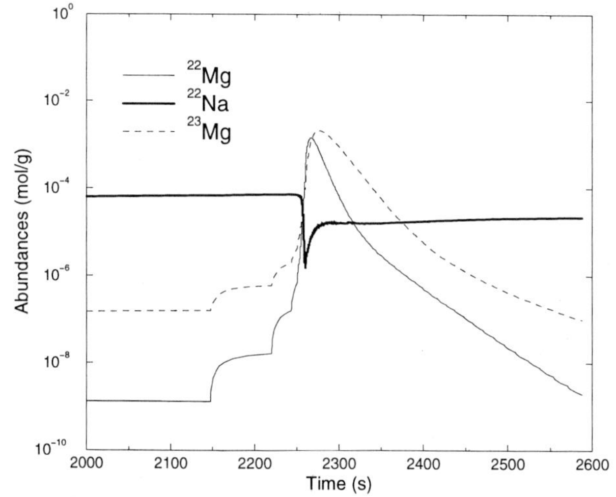

FIGURE 10. Nucleosynthesis in the hot NeNa cycles in Ne-nova burning. The initial amount of ^{20}Ne is converted to ^{22}Mg which subsequently decays to ^{22}Na. The ^{22}Na is then converted to ^{23}Mg by proton capture.

^{20}Ne(p,γ)^{21}Na($\beta^+\nu$)^{21}Ne(p,γ)^{22}Na and the production rate is determined by the slow ^{20}Ne(p,γ)^{21}Na process (Rolfs *et al.* 1975). Only at rapidly increasing temperature during the thermonuclear runaway a slight decrease in the initial ^{20}Ne abundance can be observed. Due to this temperature increase, however, the newly produced ^{22}Na is depleted rapidly by the ^{22}Na(p,γ)^{23}Mg reaction (Stegmüller *et al.* 1996). Large amounts of ^{22}Mg are formed by ^{20}Ne(p,γ)^{21}Na(p,γ)^{22}Mg. ^{22}Mg decays to ^{22}Na which is immediately processed further via ^{22}Na(p,γ) to ^{23}Mg which becomes largely enriched. After the freeze out, ^{22}Mg and ^{23}Mg β-decay while the abundance of ^{22}Na essentially remains at a constant value. Only at the later phase a slight increase over the equilibrium value can be observed due to the decay of ^{22}Mg. There are no indications of any appreciable mass loss towards ^{23}Al and ^{24}Si via the two-proton capture sequence on ^{22}Mg. The discrepancy between the predicted ^{22}Na and the observed ^{22}Na γ-intensity can not be explained in terms of nucleosynthesis, but must be due to fast mixing processes between the different mass layers of the accreted envelope. Improved 2-dimensional nova model calculations that include fast convection and mixing processes (Glasner *et al.* 1996) will help to investigate whether the necessary conditions can be reached to bring the predicted ^{22}Na abundance into agreement with observations.

6. ^{44}Ti in Supernovae

The luminosity of the later phase in supernovae explosions is determined by the decay of radioactive nuclei (Arnett 1996). The dominant radioisotope produced in supernovae explosions is ^{56}Ni ($T_{1/2}$=6.10 d) which decays into the longer-lived ^{56}Co ($T_{1/2}$=77.12 d). Its decay dictates the initial characteristic light curve of supernovae. As an example, figure 11 shows the tail of the light curve of the Type II supernovae 1987A where the development after the luminosity peak is clearly characterized by the radioactive decay of ^{56}Co. However, other rather long-lived radioactive species like neutron-deficient ^{57}Co ($T_{1/2}$=271.78 d), neutron-rich ^{60}Co ($T_{1/2}$=5.272 a), and ^{44}Ti ($T_{1/2}$ ≈44-66 y) are produced during the supernova explosion. The slow decay of these isotopes contributes to

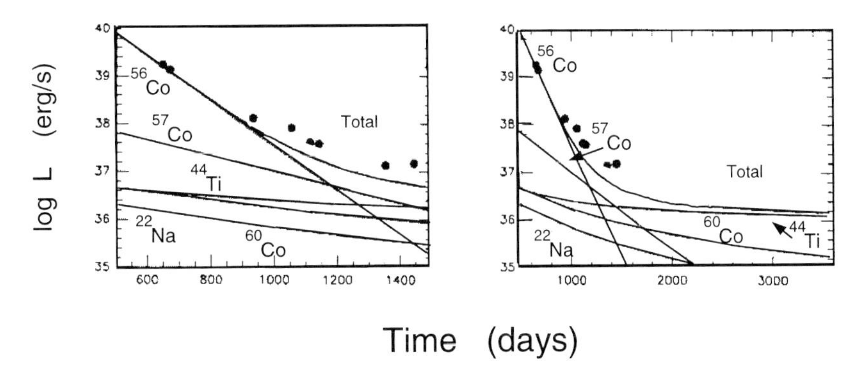

FIGURE 11. The late phase of the light curve of supernova 1987A, powered by the decay of the radionuclides which are produced in the explosive Si-burning; for details see Timmes *et al.* (1996).

the radioactive heating mainly in the later phase of the light curve. This can be seen in the right hand part of the figure which extrapolates the various contributions of the radioactive heating process of SN87A to the present time. The dots represent the observed bolometric luminosity up to 2000 days after the event. The lines in the figure indicate the contributions of the various radioactive isotopes produced in the α-rich freeze out in the deep layers of the supernova (Woosley & Weaver 1995; Thielemann *et al.* 1996; Nomoto *et al.* 1997). The solid line indicates the calculated total luminosity (Timmes *et al.* 1996). Clearly, the light curve should be dominated by the decay of ^{44}Ti.

Particular isotopes can also can be identified by their characteristic γ-emission which can be observed with satellite-based gamma-Ray observatories like COMPTEL and OSSE, which are sensitive to γ energies below 8 MeV (Schönfelder 1993). As ^{44}Ti is predicted to be strongly produced in the α-rich freeze out during a supernovae explosion, its characteristic γ-decay line at 1.157 MeV should also be observable with gamma-Ray telescopes. The measurement of the intensity can serve as an important test for the reliability of supernova model parameters.

Because of its relatively short lifetime, the observation of γ-emission from ^{44}Ti is also an ideal probe for galactic supernovae activity over the last few centuries. A systematic search for galactic γ-ray emission from ^{44}Ti with the COMPTEL observatory however showed no strong indication for γ-activity from ^{44}Ti (Dupraz *et al.* 1997). For most of the classical known supernova remnants only upper limits for the ^{44}Ti γ-flux could be established. Out of the six known recorded supernova events in the last millenium characteristic ^{44}Ti γ-activity has only been detected at the location of the supernova remnant Cassiopeia A (Cas A) (Iyudin *et al.* 1994) with a flux of $F_\gamma=(3.4\pm0.9)\cdot10^{-5}$ γcm^{-2}s^{-1} (Dupraz *et al.* 1997). This finding is in 1σ agreement with the result of the latest OSSE observation, $F_\gamma=(1.7\pm1.4)\cdot10^{-5}$ γcm^{-2}s^{-1} (The *et al.* 1996). These results for the characteristic γ flux allow one to determine the amount of ^{44}Ti M_4 [in 10^{-4} M$_\odot$], initially produced in the supernova explosion if the distance D_{kpc} [in kpc] and the age of the supernova remnant t [in years] are known,

$$M_4 = F_\gamma \cdot \frac{D_{kpc}^2}{0.72/\tau \cdot f_\gamma \cdot e^{-t/\tau}} \tag{6.16}$$

where f_γ is the 99.9% branching ratio for the 1.157 MeV γ-ray emission per ^{44}Ti decay. The lifetime τ of ^{44}Ti is an important parameter to reliably determine the amount of ^{44}Ti produced in the supernova event. The large uncertainty in the lifetime of ^{44}Ti therefore translates directly in a large uncertainty for the determination of the initially produced amount of ^{44}Ti in supernova Cas A. Figure 12 shows the ^{44}Ti yield [in 10^{-4}M$_\odot$]

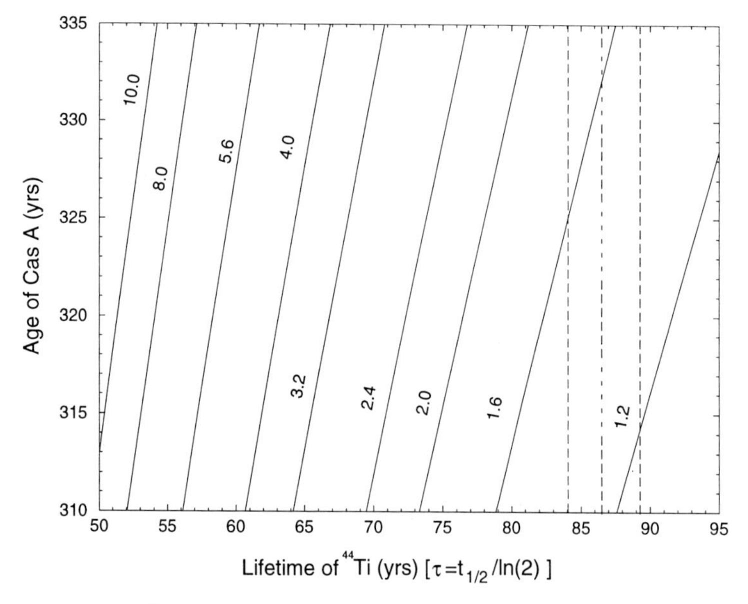

FIGURE 12. Amount of ejected ^{44}Ti from supernova Cas A calculated within the uncertainties of the age of the supernova remnant and the uncertainties for the lifetime of ^{44}Ti. The calculation is based on a distance of 2.8 kPc.

as a function of age and lifetime calculated for a γ-flux of F_γ=3.4$\cdot10^{-5}$ γcm^{-2}s^{-1} and a distance of 2.8 kpc. The figure indicates that within the uncertainties for age of the supernova remnant, t=310-335 y (Hartmann *et al.* 1997) and the lifetime, τ=64-94 y a ^{44}Ti yield between 1.0$\cdot10^{-4}$ M$_\odot$ and 10.0$\cdot10^{-4}$ M$_\odot$ has been observed.

These results can be compared with the predictions from supernovae nucleosynthesis model calculations. In a core collapse model, which describes Supernovae type II and Ib, the production of ^{44}Ti is most efficient within the α-rich freeze out at nuclear statistical equilibrium for relatively low density conditions (Arnett 1996; Woosley & Hoffmann 1992; Thielemann *et al.* 1996). The amount of ^{44}Ti ejected by the supernova does also depend critically on the mass and metallicity of the progenitor star, the entropy conditions for the α-rich freeze out and the mass cut between the remnant and the ejecta. The mass cut is typically chosen on the basis of the observed ^{56}Ni/^{57}Ni ratio (Thielemann *et al.* 1996). The results indicate that most of the ^{44}Ti is ejected together with large quantities of the synthesized ^{56}Ni. Typical predictions for the ^{44}Ti yield range between $1.0 \cdot 10^{-5}$ $M_\odot$ and $2.3 \cdot 10^{-4}$ $M_\odot$ (Woosley & Weaver 1995; Woosley *et al.* 1995; Thielemann *et al.* 1996; Nomoto *et al.* 1997). This seems slightly lower than the observed ^{44}Ti yield but the predictions depend also strongly on the mass of the progenitor star (Timmes *et al.* 1995, Thielemann *et al.* 1996; Nomoto *et al.* 1997).

6.1. *The Lifetime Measurement of* ^{44}*Ti*

The main "nuclear" uncertainty in the calculation of ^{44}Ti production is the uncertainty in the half-life of ^{44}Ti. Various measurements based on different experimental techniques yield half-life values between $T_{1/2}$=44 y and 66 y (Wing *et al.* 1965; Moreland & Heymann 1965; Frekers *et al.* 1983; Alburger & Harbottle 1990). Figure 13 compares the various experimental results which show significant discrepancies. Recently, a novel method for

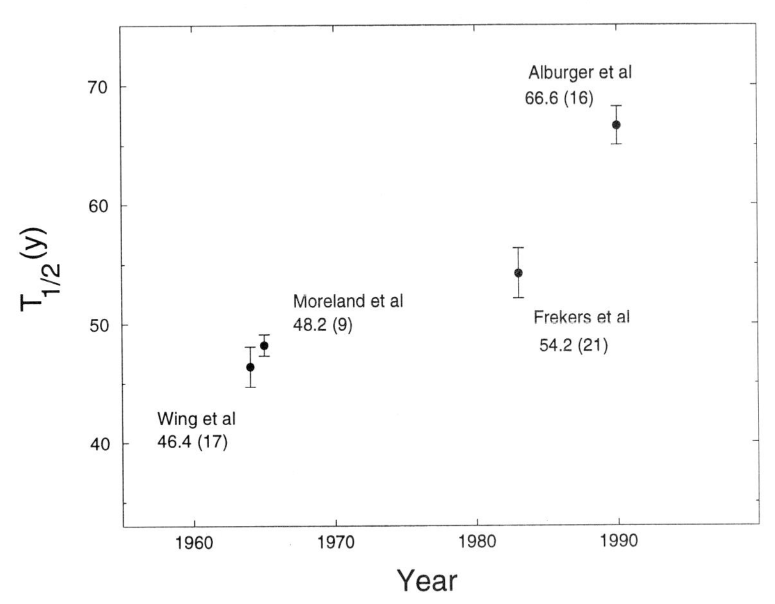

FIGURE 13. Experimental results of half life measurements on ^{44}Ti prior to 1995.

measuring this half-life has been employed. A mixed beam of radioactive ^{44}Ti, $N_{^{44}Ti}$, and ^{22}Na particles, $N_{^{22}Na}$ was implanted into an Al sample. The specific ^{44}Ti activity, $A_{^{44}Ti}$, was measured relative to the ^{22}Na activity, $A_{^{22}Na}$ with a Ge-detector. After the appropriate corrections for the energy dependence of the γ-ray efficiency of the detector, the lifetime of ^{44}Ti $\tau(^{44}Ti)$ was derived relative to the well known lifetime of ^{22}Na, $\tau(^{22}\mathrm{Na})=3.764\pm0.002$ y, by

$$\tau(^{44}Ti) = \left(\frac{A_{^{22}Na}}{A_{^{44}Ti}}\right) \cdot \left(\frac{N_{^{44}Ti}}{N_{^{22}Na}}\right) \cdot \tau(^{22}Na). \tag{6.17}$$

This lifetime depends only on the ratios of ^{44}Ti and ^{22}Na numbers and activities, therefore the systematic errors cancel out and a much higher accuracy than in an absolute measurement can be achieved.

The production of a suitable sample was carried out at the National Superconducting Cyclotron Laboratory at Michigan State University. With a primary $^{46}Ti^{12+}$ beam of E/A=70.6 MeV/u from the superconducting K-1200 cyclotron, a fragmented beam was produced on a ^{9}Be target (202mg/cm^{2}). Using the A1200 projectile fragment separator (Chen *et al.* 1993) ^{44}Ti particles can separated from most of the other fragments. The field conditions of the separator were optimized to accept both ^{22}Na and ^{44}Ti with good transmission to the focal plane. The accepted ^{22}Na and ^{44}Ti ions were implanted into a stack of Al foils located at the focal plane of the A1200. A 2 cm diameter collimator was positioned in front of the Al stack to define the area for the implantation. The beam intensity at the focal plane during implantation was $\approx 8 \cdot 10^{6}$ particles/s, far too large for on-line detection. To analyze the beam particles the primary ^{46}Ti beam was reduced by an attenuation factor of 10^{4} by introducing a collimator after the ECR ion source but before the K-1200 cyclotron.

For particle identification a ΔE (Si detector) - E (plastic detector) array replaced the Al stack. Different particle groups were identified at the implantation spot in terms of their energy loss ΔE and their time-of-flight (TOF) through the A1200 spectrograph. Figure 14 shows the (ΔE-TOF) histogram, in which both the ^{22}Na group and the ^{44}Ti group are well separated. In addition two another groups corresponding to the long-lived γ-emitters ^{46}Sc and ^{48}V could be identified. The analysis of the various particle groups in the histogram yields the ratio $N_{^{44}Ti}/N_{^{22}Na}$.

For the actual implantation the attenuation collimator in front of the cyclotron was removed and the ΔE-E detector system was replaced by the Al stack. The thicknesses of the Al foils were chosen such that all of the ^{44}Ti radioisotopes were completely stopped in the second foil. The ^{22}Na isotopes were stopped in the sixth foil because of their smaller energy losses.

The implantation was performed for approximately two days, during which the stack was removed every few hours and the detectors were inserted to monitor the ^{44}Ti to ^{22}Na ratio. The ratio remained constant within 1%. The total number of implanted ^{44}Ti was estimated at $N_{^{44}Ti} \approx 5 \cdot 10^{10}$. To determine the activity ratio $A_{^{22}Na}/A_{^{44}Ti}$, the characteristic 1157 keV line of the decay of ^{44}Ti and the 1274 keV line from the β^{+}-decay of ^{22}Na were measured independently. The ^{44}Ti implanted Al-foil and the ^{22}Na implanted Al-foil (without absorber or veto foil) were placed at different distances in front of a 31% HPGe γ-ray detector (resolution 1.9 keV at 1.33 MeV). The spectrum was recorded until sufficient counting statistics of $\leq$ 2% for the γ intensity of a single run was ensured. For each distance four individual measurements were performed. Figure 15 shows the corresponding γ-ray spectra taken with each of the samples. The spectrum measured for the sixth foil shows a strong line for the decay of the implanted ^{22}Na, no indication for transition from the other radioisotopes are observed. The spectrum measured for the second foil clearly shows the decay of ^{44}Ti. From the background corrected intensity in the respective γ-ray peaks, I_{1157} and I_{1274}, the ratio of the activities $A_{^{22}Na}/A_{^{44}Ti}$ could be calculated.

The ratio for the number of implanted radioisotopes $N_{^{44}Ti}/N_{^{22}Na}$ and the ratio of the activities $A_{^{22}Na}/A_{^{44}Ti}$ allows one to determine the lifetime of ^{44}Ti with high accuracy $\tau(^{44}Ti)$ = 86.7±2.7 y (2σ standard deviation). The uncertainty is determined by a 2% uncertainty in the ratio of the numbers of implanted radioisotopes and by a 1.3% uncertainty in the activity ratio. This result agrees well with the results of a recent independent measurement, $\tau(^{44}Ti)$ = 91.3±4.3 y (1σ standard deviation) Normann *et*

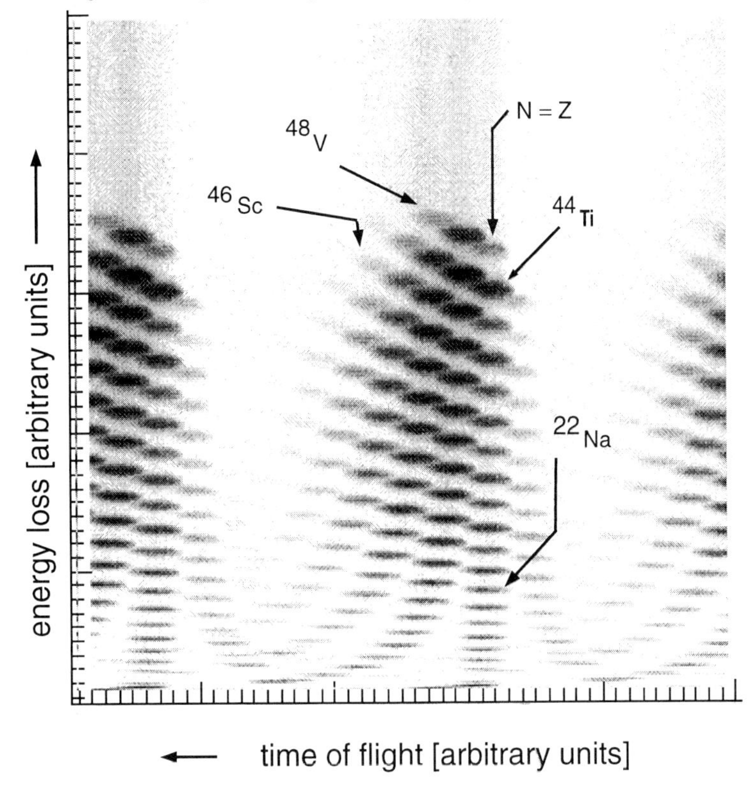

FIGURE 14. Time of flight versus energy loss particle identification at the focal plane of the A1200 fragment mass separator.

al. 1997). This experiment is based on a long-term (2 years) measurement of the γ-decay of a ^{44}Ti sample using Ge-detectors to determine the lifetime from the decay curve.

6.2. *Implications for the understanding of the ^{44}Ti production in supernovae*

The present result removes the previous uncertainties which had handicapped a reliable comparison between the observed ^{44}Ti γ flux from supernova Cas A with the COMPTEL observatory, F_γ=(3.4±0.9)$\cdot 10^{-5}$ γcm^{-2}s^{-1} (Dupraz *et al.* 1997) and the theoretical model predictions for the nucleosynthesis of ^{44}Ti. The observed amount of ^{44}Ti in Cas A, M_4, can be calculated using equation (6.16) if the distance D and the age of the supernova event t are known. The distance has been determined to $D = 3.4^{+0.3}_{-0.1}$ kpc (Reed *et al.* 1995). However, a standard distance of 2.8 kPc is often adopted (e.g. Timmes *et al.* 1995) which is used here for better comparison. It should however be noted that the measured distance increases the amount of ejected ^{44}Ti considerably. The date of the supernova event also carries a considerable uncertainty, it ranges from 1658±3 AD, which has been estimated from the expansion velocity of the remnant (van den Bergh & Kamper 1983) to 1680 AD, which is based on the analysis of observational results from that period (Ashworth 1980). A weak ≈ 6th mag star 3 Cas has been reported in 1680 AD by the first British Astronomer Royal Sir John Flamsteed (1725), but this star was not recorded by later observers. Identifying this event with the supernovae sets the age to t=312-313 years for the time range of the COMPTEL observations (Hartmann *et al.* 1997). This yields $M(^{44}\mathrm{Ti})=(1.27\pm0.34)\cdot 10^{-4}\ M_\odot$ for the the observed amount of ^{44}Ti. This range is also indicated by the dashed lines in figure 12. This value seems in good

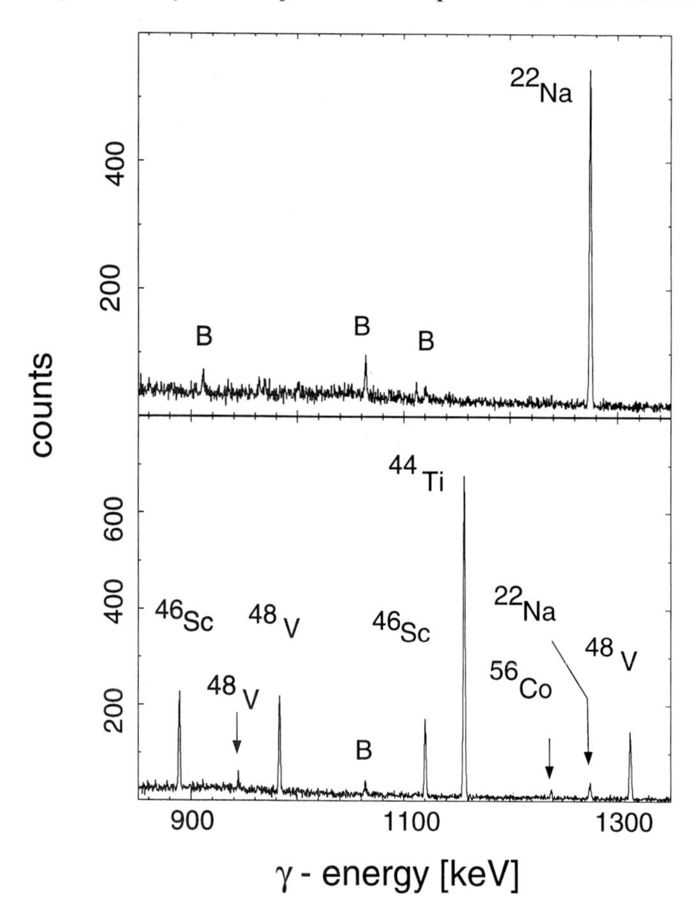

FIGURE 15. The γ spectra measured for the sixth foil implanted with ^{22}Na (top spectrum) and the second foil implanted with ^{44}Ti (bottom spectrum). Natural background lines are labeled with 'B'.

agreement with the predictions for the ^{44}Ti production in supernova type Ib explosions which range between $(0.6 - 1.3)\cdot 10^{-4}$ $M_\odot$ for a core mass of 10 to 20 $M_\odot$ (Woosley *et al.* 1995; Thielemann *et al.* 1996; Nomoto *et al.* 1997). A more massive presupernova star like a Wolf-Rayet star as suggested in the case of Cas A (Fesen & Becker 1991; Borkowski 1996) may increase the ^{44}Ti production to the observed value (Hartmann *et al.* 1997).

7. X-ray bursts as possible source for light p-nuclei?

X-ray bursts have been suggested as possible sites for hot temperature hydrogen burning via the rp- and αp-process (Wallace & Woosley 1981; Ayasli *et al.* 1982; Woosley & Weaver 1984; Taam 1985, Fujimoto *et al.* 1987; Taam *et al.* 1993). However, while x-ray bursts are frequently observed phenomena (Lewin *et al.* 1993), the nucleosynthesis and the correlated nuclear energy generation, are not completely embedded in the models (Schatz *et al.* 1997*a*). The standard models for type-I x-ray bursts are based on accretion processes in a close binary system similar to the nova scenario discussed previously. In the case of type-I x-ray bursts, accretion takes place from the filled Roche-Lobe of the extended companion star onto the surface of a neutron star. Typical predictions for the accretion rate vary from 10^{-10} to 10^{-9} $M_\odot$ y^{-1}. The accreted matter is continuously compressed by the freshly accreted material until it reaches sufficiently high pressure and temperature conditions which allow the thermonuclear ignition.

Nuclear burning is ignited at high density, $\rho \geq 10^6$ g/cm^3, in the accreted envelope, via the pp-chains, the hot CNO-cycles and the triple-α-process. The released energy triggers a thermonuclear runaway at the electron degenerate conditions at the bottom of the accreted layer near the surface of the neutron star. Peak temperatures up to $2 \cdot 10^9$ K can be reached before the degeneracy is completely lifted (Bildsten 1997). These temperatures are sufficiently high to trigger the rp- and the αp-process which cause rapid nucleosynthesis towards heavier mass regions. Also, the nuclear energy release fueling the thermal runaway is dominated by the rp-process reactions (Woosley & Weaver 1984; Taam 1985; Fujimoto *et al.* 1987). The time-scale for the thermal runaway and the

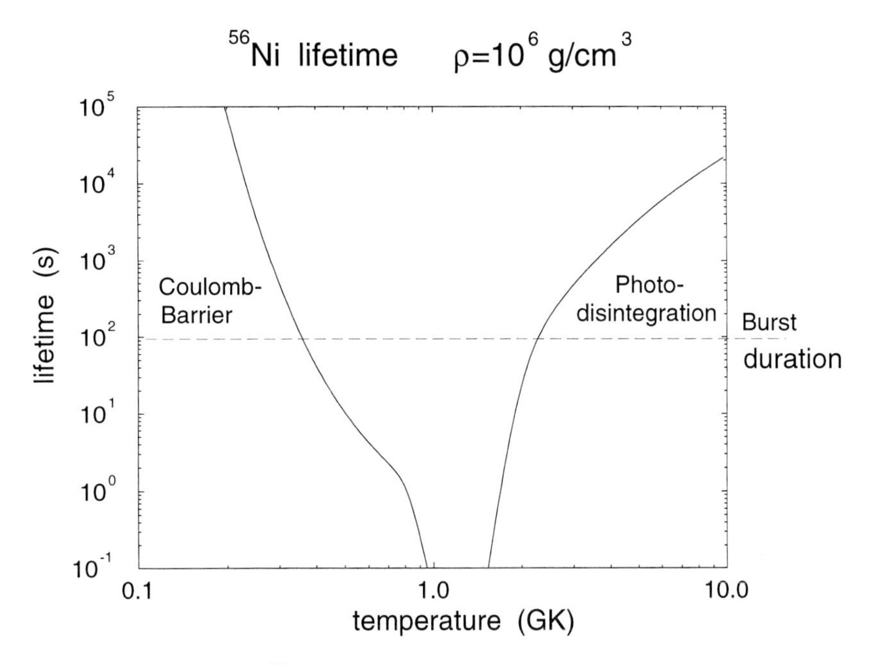

FIGURE 16. The effective lifetime of ^{56}Ni as a function of temperature calculated for a density of $\rho=10^6$ g/cm^3.

subsequent cooling phase varies between 10 s up to 100 s (Woosley & Weaver 1984) depending on the particular model parameters for the accretion process. Within this time-scale, the rp-process can proceed well up to ^{56}Ni (Wallace & Woosley 1981) or even further up to ^{96}Cd (Woosley & Weaver 1984; Wallace & Woosley 1984; Van Wormer *et al.* 1994; Schatz *et al.* 1997*a*).

7.1. *rp-process in X-ray bursts*

The rp-process was first suggested by Wallace and Woosley (1981) as the dominant nucleosynthesis process in explosive hydrogen burning at high temperature and density conditions. The process is characterized by a sequence of fast proton capture reactions and subsequent β-decay far away from the line of stability. The characteristics of the rp-process has been discussed extensively in the literature (Wallace & Woosley 1981; Van Wormer *et al.* 1994; Schatz *et al.* 1997*a*).

The fast proton capture reactions drive the path out towards the proton drip-line until a waiting point nucleus is reached where a (p,γ)-(γ,p) and/or a (p,γ)-(p)-equilibrium is achieved. This causes an enrichment in the waiting point isotopes until β-decay or, for nuclei with low Z, (α,p) reactions occur. Waiting point nuclei are typically even-even nuclei which are characterized by low or negative Q-values for subsequent proton capture.

The main waiting points are ^{55}Ni and ^{56}Ni, where the reaction flow is delayed by the ^{56}Ni(p,γ)–^{57}Cu(γ,p) and the ^{55}Ni(p,γ)–^{56}Cu(γ,p) equilibrium which causes considerable enrichment in ^{55}Ni and in ^{56}Ni. It has been shown (Van Wormer *et al.* 1994; Görres *et al.* 1995; Schatz *et al.* 1997*a*) that for high density conditions the effective lifetime for a waiting point nucleus can be considerably reduced by a sequential two-proton capture process. Subsequent proton capture on the equilibrium abundance of ^{57}Cu will lead to a fast leakage out of the equilibrium. For most of the waiting point nuclei the β-decay lifetime is rather short and the sequential proton capture rate is negligible. However, in the case of ^{56}Ni the effective lifetime is mainly determined by the proton capture rate on ^{57}Cu because of its rather slow decay rate. Figure 16 shows the effective lifetime for ^{56}Ni as a function of temperature for a constant density, $\rho=10^6$ g/cm^3. Towards higher temperatures the lifetime becomes rather large due to the inverse photodisintegration of ^{57}Cu while at lower temperatures the high Coulomb barrier prevents rapid proton capture on ^{56}Ni and its effective lifetime increases. In the temperature range between T$\approx 5\cdot 10^8$ K and T$\approx 2\cdot 10^9$ K at densities of $\rho=10^6$ g/cm^3 the effective lifetime of ^{56}Ni is $\tau \leq 100$ s. The effective lifetime remains low until at higher temperatures the inverse photodisintegration of ^{68}Zn reduces the effective ^{57}Cu(p,γ)^{58}Zn flow. Only within a relatively small temperature window (between $T \approx 0.8 - 1.5\cdot 10^9$ K) and for relatively high densities ($\rho \geq 10^5$ g/cm^3) fast (≤ 1000 s) rp-process nucleosynthesis beyond ^{56}Ni is possible.

7.2. *The rp-process in the thermal runaway of the x-ray burst*

The following section will focus on the rp-process characteristics calculated for temperature conditions during the thermonuclear runaway in a one mass zone x-ray burst model. For this model hydrostatic equilibrium is maintained. The hydrostatic evolution during the process in terms of pressure p and temperature T is determined by the radiative flux F which changes with the depth y of the accreted layer, the nuclear energy production ϵ, and the mass accretion rate $\dot{m}$,

$$\frac{\partial F}{\partial y} + \epsilon = C_p \left(\frac{\partial T}{\partial t} + \dot{m}\frac{\partial T}{\partial y} \right) - \frac{C_p T \dot{m}}{y} \nabla_{ad} \tag{7.18}$$

C_p is the specific heat at constant pressure and ∇_{ad} is the adiabatic change in the neutron star atmosphere. The pressure P in the burning zone is determined by the ion pressure P_{ion}, the pressure of the degenerate electron gas P_e, and the radiation pressure P_{rad}. The total pressure P is kept constant (Bildsten 1997). In cases where the thermonuclear runaway is faster than convective processes, the temperature conditions are only determined by the energy release from the thermonuclear reactions in the runaway, ϵ, and by the various cooling processes (Bildsten 1997; Fujimoto *et al.* 1981),

$$\frac{\partial F}{\partial y} + \epsilon = C_p \cdot \frac{\partial T}{\partial t}. \tag{7.19}$$

The cooling rate $\frac{\partial F}{\partial y}$ is dominated by the radiative energy losses ϵ_{rad} [in erg/g·s],

$$\frac{\partial F}{\partial y} \approx \epsilon_{rad} \propto \frac{T^4}{P^2} \tag{7.20}$$

and increases rapidly with increasing temperature T. When the cooling rate becomes equal to the nuclear energy generation, further temperature increase is prevented and the peak temperature of the burst is reached.

Figure 17 shows the energy production, the temperature and the density in the accreted envelope (burning-zone), and the hydrogen and helium abundances calculated as a function of time over the duration of the x-ray burst. The structure of the energy production curve is directly correlated with the waiting point concept of the rp-process (Schatz *et al.* 1997). A series of fast proton captures and β-decay processes causes a peak in the

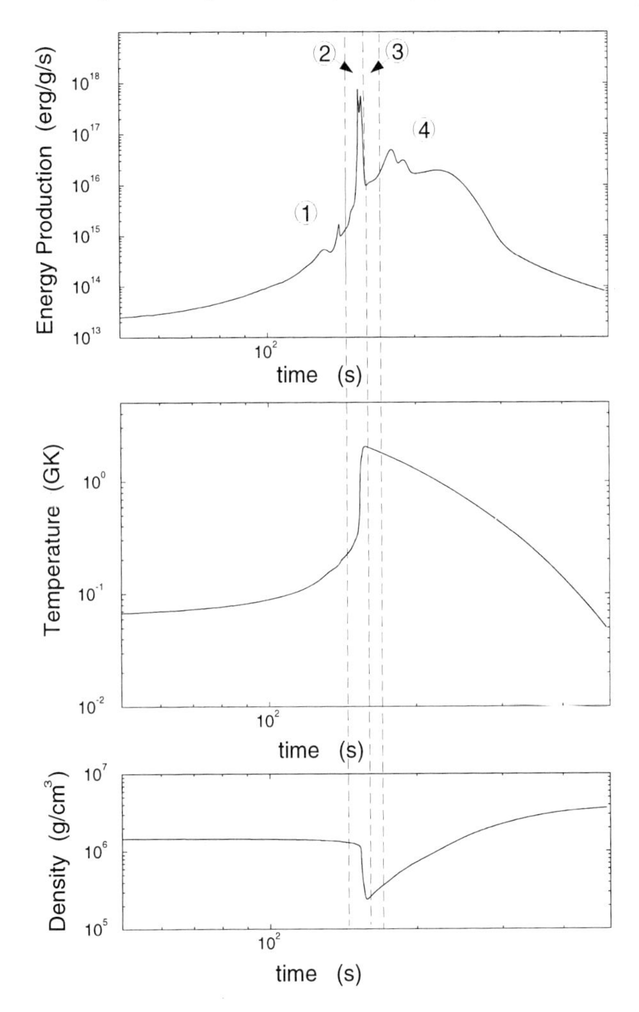

FIGURE 17. Shown is the energy production (luminosity), the temperature, and the density as a function of time for the thermonuclear runaway in a simplified x-ray-burst model. Indicated are different phases in the thermonuclear runaway. Phases 1, 2, 4 correspond to periods of rapid nucleosynthesis, phase 3 correspond to a dormant period at the maximum of the temperature curve. For details see text.

energy production. In contrast during periods where the material is stored in waiting point nuclei with slow β-decay, the process halts and the energy production drops. The lifetime of waiting point nuclei is also influenced by the rapid changes in temperature and density which modify the above discussed (p,γ)-(γ-p) equilibrium conditions quite rapidly. Initially the temperature increases dramatically due to the energy generation in the thermonuclear runaway and the density remains constant as long as the pressure is dominated by the pressure of the degenerate electron gas. With increasing temperature the ion pressure and the radiation pressure increase dramatically, this causes the gas to expand leading to a decrease in density before the peak temperature is reached. When the cooling rate becomes equal to the energy production rate of $\epsilon_{nucl} \approx 10^{18}$ erg/g·s the temperature reaches its peak value of $T \approx 2 \cdot 10^9$ K. At this point the pressure has dropped by approximately one order of magnitude to $\rho \approx 2.5 \cdot 10^5$ g/cm^3. With the decrease in temperature the density increases due to the condition of constant pressure.

The energy production in the burst is characterized by four periods, the ignition phase of the burst (1), the peak of the burst (2), the dormant phase of the burst (3), and the phase of the after-burst (4). These four different phases are characterized by different temperature and density conditions for the nucleosynthesis and energy generation.

Phase (1) is characterized by the hot CNO cycles triggered by proton capture reactions of the accreted hydrogen on the carbon, nitrogen, and oxygen isotopes which have not been destroyed by spallation processes in the outer atmosphere of the accreting neutron star (Bildsten *et al.* 1992). The energy production and the nucleosynthesis are shown in figure 18. The two peaks in the energy production are caused by the conversion of the

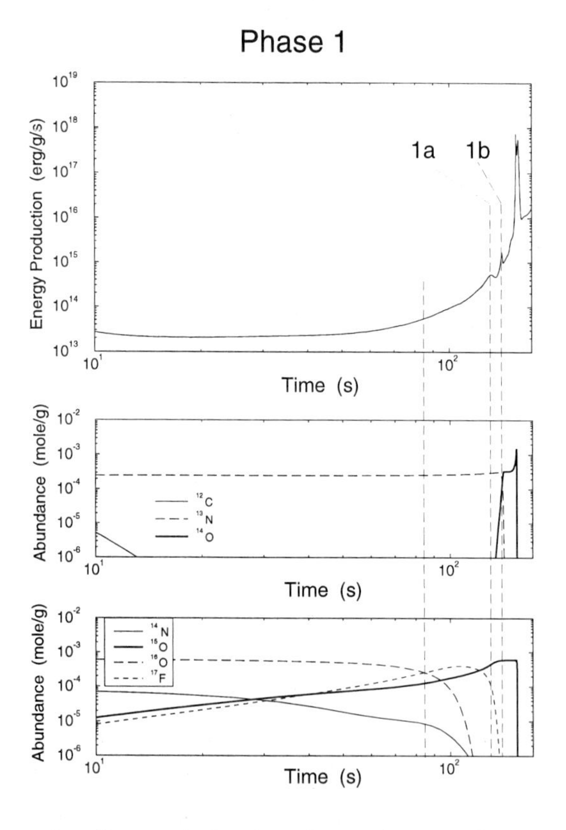

FIGURE 18. Energy production (luminosity) and nucleosynthesis in the early phase (1) of an x-ray burst. Peak 1a and 1b correspond to the burning of ^{17}F and ^{13}N.

initial abundance of ^{12}C into ^{14}O and of ^{16}O into ^{15}O by two subsequent proton capture reactions. Because of the slow decay of these two isotopes the CNO process is halted and the energy production drops. The main reaction flow is confined to the CNO cycles, but, as seen in figure 18 the temperature has already increased sufficiently to trigger phase (2) of the burst, the ignition of the triple-α-reaction.

Phase (2) is initiated at a temperature of $T \approx 2.4 \cdot 10^8$ K via the triple α-process. Parallel to that the ^{14}O and ^{15}O waiting point nuclei are rapidly depleted by α-capture processes. The fine structure of the energy burst is characterized by the different waiting points along the process path and is shown in figure 19 together with the abundances of the most important waiting point nuclei along the process path. During the

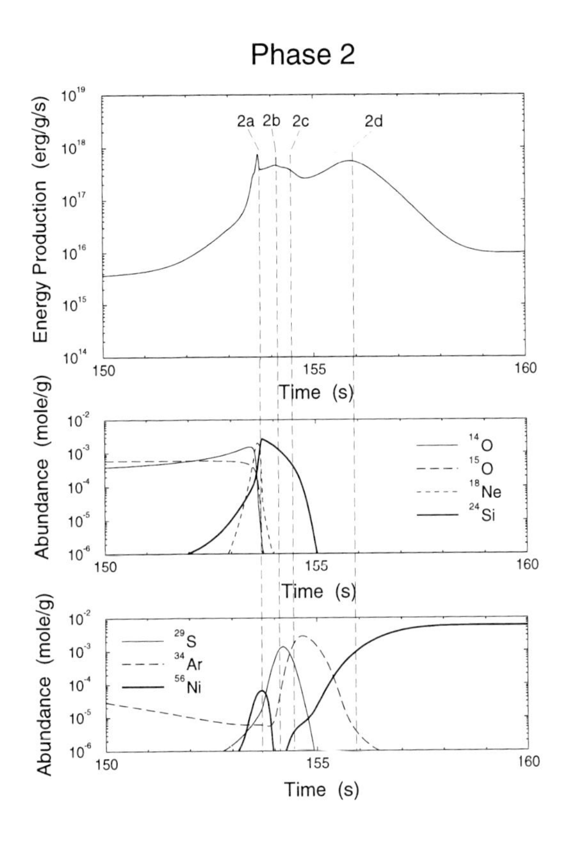

FIGURE 19. Peak conditions in the energy production of the x-ray burst. The structure of the energy generation curve is determined by the waiting point characteristics of the rp-process. For details see text.

first burst (2a) the waiting point isotopes ^{14}O, ^{15}O, and ^{18}Ne which have been produced by the hot CNO cycles are rapidly converted by the αp-process to ^{24}Si. Because the process has to wait for its β-decay, the energy production drops until it is reinitiated at higher temperatures by proton and α capture reactions leading to the production of the next waiting point isotopes ^{29}S (2b) and ^{34}Ar (2c). The rapid increase in temperature allows subsequent α-capture to bridge these waiting points and leads to a rapid conversion of the waiting point isotopes to ^{56}Ni (2d). At this time a continuous reaction flow converts the accreted He-abundance via the αp-process and the rp-process to ^{56}Ni. The reaction flow, integrated over the duration of the last phase (2d) of the burst is shown in figure 20. Because most of the reaction path is characterized by (α,p) reactions considerably more helium is burned than hydrogen. At this point, peak temperatures of $T \approx 1.5 \cdot 10^9$ K have been reached and further processing is halted by the ^{56}Ni(p,γ)-(γ,p) equilibrium (Schatz *et al.* 1997*a*). Therefore the energy production drops rapidly while most of the initial heavy isotope abundances as well as a large fraction of the initial helium remains stored

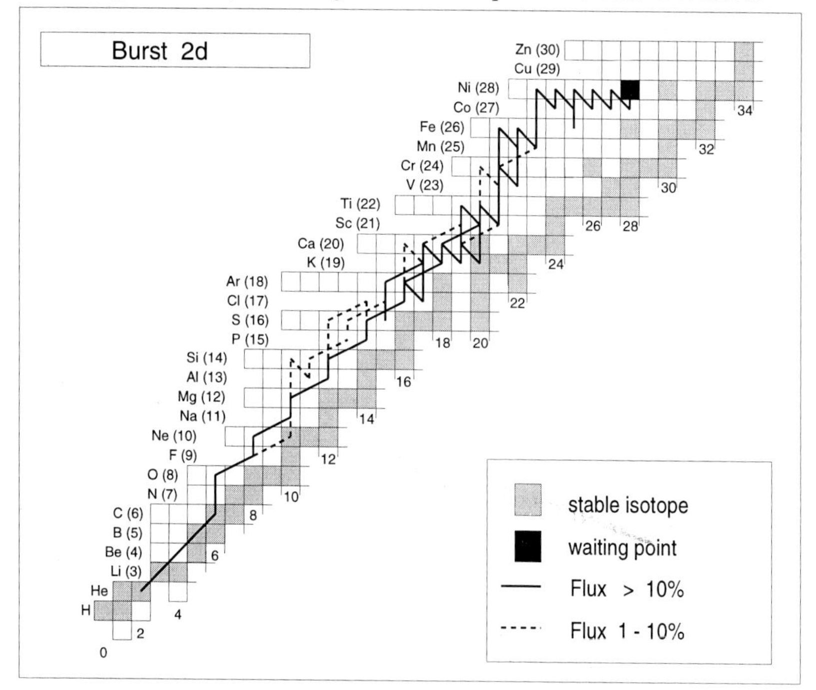

FIGURE 20. The αp- and rp-process reaction path in the thermonuclear runaway phase (2) of the x-ray burst.

in the waiting point nucleus ^{56}Ni. The drop in energy production causes a slow down in the temperature increase just before the peak temperature is reached.

At these peak temperature conditions the ^{56}Ni has a lifetime of approximately 100 s versus two-proton capture. However, with the decrease in temperature and the parallel increase in density the ^{56}Ni(p,γ)-(γ,p) falls out of equilibrium and the effective lifetime of ^{56}Ni decreases down to fractions of a second versus proton capture at temperatures $T \approx 1.5 \cdot 10^9$ - $1 \cdot 10^9$ K. In this temperature window, phase (4) of the energy burst is initiated which is characterized by nucleosynthesis via the rp-process beyond ^{56}Ni. Figure 21 shows the details of the burst structure in the cooling phase. Several peaks in the energy production are due to the depletion of ^{56}Ni and the further processing towards the waiting point ^{64}Ge and the subsequent nucleosynthesis towards ^{68}Se (see previous section). In the final phase the waiting point ^{68}Se is converted to heavier isotopes up to the mass 100 region. Figure 22 shows the reaction flow integrated over the duration of (4c). Notice that due to the lower temperature the αp-process is only dominant below sulfur, at higher masses the reaction path is characterized by the rp-process pattern leading up to ^{100}Sn. In the final phase most $\geq$90%) of the initial helium as well as most of the other isotopes has been converted to heavy isotopes with masses A$\geq$72. It is therefore of interest to consider the relative abundance distribution in this mass range during the final phase of the x-ray burst.

7.3. *x-ray burst nucleosynthesis of the light p-nuclei?*

The analysis of x-ray burst profiles (Lewin *et al.* 1984; Tawara *et al.* 1984*a*, 1984*b*; Haberl *et al.* 1987) indicates that for some of the observed events the luminosity exceeds the Eddington limit. The excess luminosity is transformed into kinetic energy causing radiation driven winds which lead to an expansion of the photosphere up to radii many

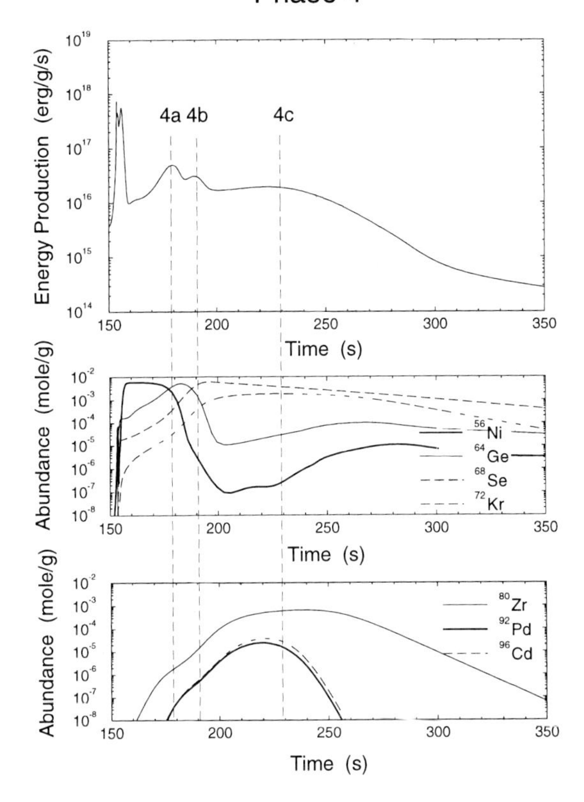

FIGURE 21. Energy production (luminosity) and nucleosynthesis in the late phase (4) of an x-ray burst. The peaks in the luminosity correspond to the destruction of waiting point nuclei.

times larger than the neutron star radius. A detailed discussion of the observational indications is given by Lewin, Paradijs & Taam (1993). Model calculations suggest that the peak luminosity of the burst remains close to the Eddington limit (Paczynski 1983; Paczynski & Anderson 1986; Joss & Melia 1987; Nobili *et al.* 1994) just balancing the gravitational forces, therefore the photosphere contracts again when the luminosity drops. These calculations are based on several simplified assumptions. First, the triple α reaction is considered to be the dominant nuclear energy source. Figure 23, however shows that full αp- and rp-process nucleosynthesis does increase the energy output by more than one order of magnitude. While the luminosity is nowhere near high enough to cause significant mass loss out of the gravitational potential of the neutron stars, small fractions of material in the outer photosphere may still escape.

Possible additional escape mechanisms are neutrino-driven winds and losses due to disk accretion. Approximately 10% to 30% of the energy originated in thermonuclear runaway is carried by fast neutrinos which are generated by the β-decay processes along the reaction path. Figure 24 compares the overall energy production with the neutrino losses during the process. Absorption of the neutrinos in the surface layer may initiate additional neutrino driven winds which may enhance the radiation-driven mass loss effects (Duncan *et al.* 1986; Chevalier 1993, 1996). While the question of possible mass loss effects in the x-ray burst requires much more sophisticated models than presently available it is nevertheless of interest to study the abundance distribution in the accreted

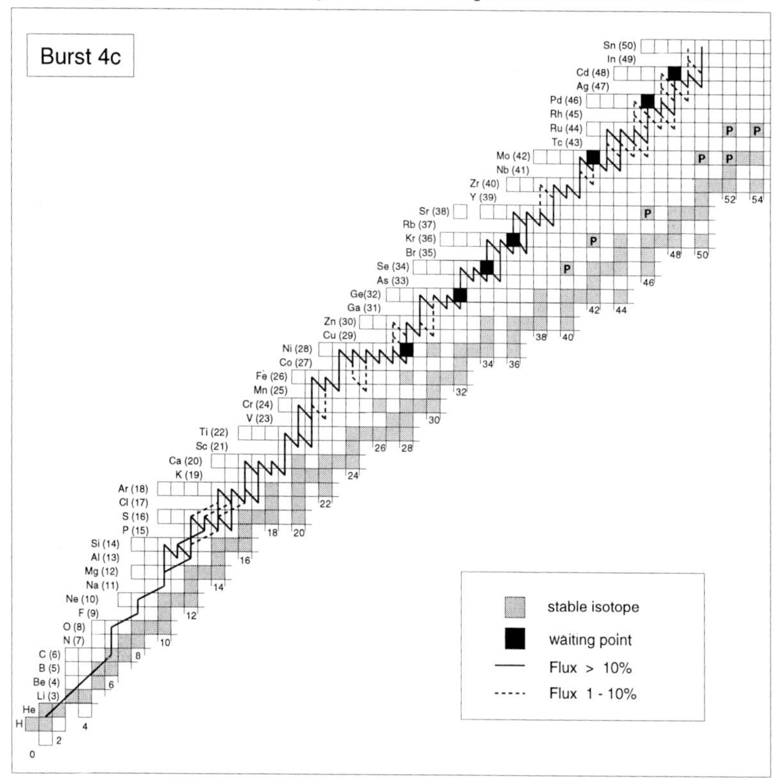

FIGURE 22. The reaction flux in the late phase (4d) of the x-ray burst. The overall flux is dominated by the rp-process. Only for very light Z nuclei ($Z \leq 14$) is the Coulomb barrier low enough for α capture reactions to compete with the proton capture and the β-decay. Also indicated are the light p-nuclei which are possibly produced in the x-ray burst. For discussion see section 6.4.

material after the thermal runaway. In the case of mass loss x-ray bursts may contribute to the observed abundance distribution. If mass loss is prevented by the gravitational potential of the neutron star the nuclei produced may become seed material for processes triggered by the formation of neutrons by electron capture on the remaining hydrogen in the high density zones ($\rho \geq 10^7$ g/cm^3) (Taam *et al.* 1996).

As discussed in the previous sections, the rp-process can proceed well into the A=80 to 100 mass region within the short time-scale of the event. Figure 26 shows in its upper part the abundance distribution in the material after the freeze out of the thermonuclear runaway. This means that the β-unstable isotopes along the process path have decayed back to the line of stability. While there is still an appreciable amount of hydrogen, the bulk of the material has been converted to nuclei with masses $A \geq 70$. Figure 26 shows in its lower part the comparison with the solar abundances. All isotopes above mass A=68 are enriched by more than four orders of magnitude compared to the solar abundances which have served as the initial distribution of the accreted material. Clearly peaking are the abundances of the light p-nuclei ^{74}Se, ^{78}Kr, ^{84}Sr, ^{92}Mo, ^{94}Mo, ^{96}Ru, ^{98}Ru (Lambert 1992). They are overproduced by up to seven orders of magnitudes compared to the solar abundances of these particular isotopes. This is in particular noticable because the relatively high observed abundances for these isotopes have not been explained yet in classical p-process scenarios (Woosley & Howard 1990; Rayet *et al.* 1990; Howard

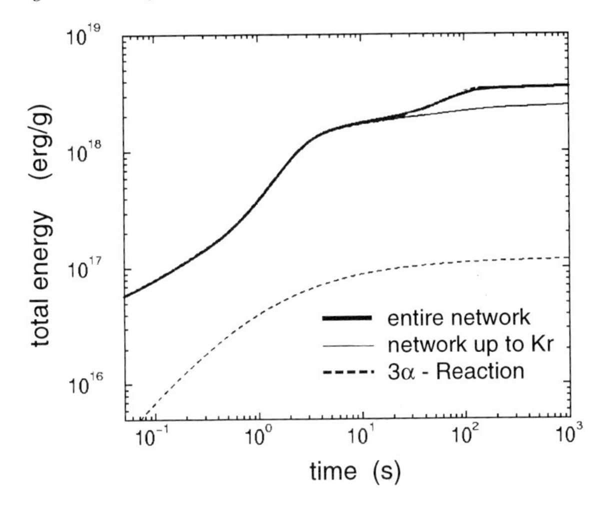

FIGURE 23. The total energy produced over the duration of the thermonuclear runaway. The triple alpha process contributes ≈10 of the energy.

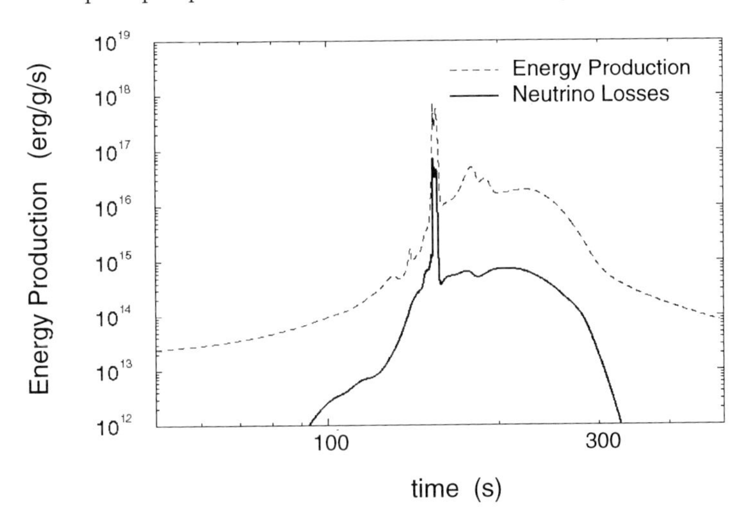

FIGURE 24. The neutrino losses in comparison with the overall energy production in the x-ray burst. Neutrinos carry only a small percentage of the produced energy but can cause .

et al. 1991; Rayet *et al.* 1995) or in alternative neutrino-induced processes in type II supernovae (Hoffmann *et al.* 1996). If a suffiently high mass loss out of the gravitational potential of the neutron star is possible, x-ray bursts may be a potential source for these isotopes.

8. Conclusion

Experimental Nuclear Astrophysics has proven itself to be a very powerful tool for understanding stellar processes. Besides the accumulation of observational data, it is the only experimental way to verify our interpretation of the dramatic events observed throughout our universe. One of its great successes is the formulation of nucleosynthesis processes responsible for the observed elemental and isotopic abundance distribution in our galaxy. The experimental data and the theoretical models correlated the long stable

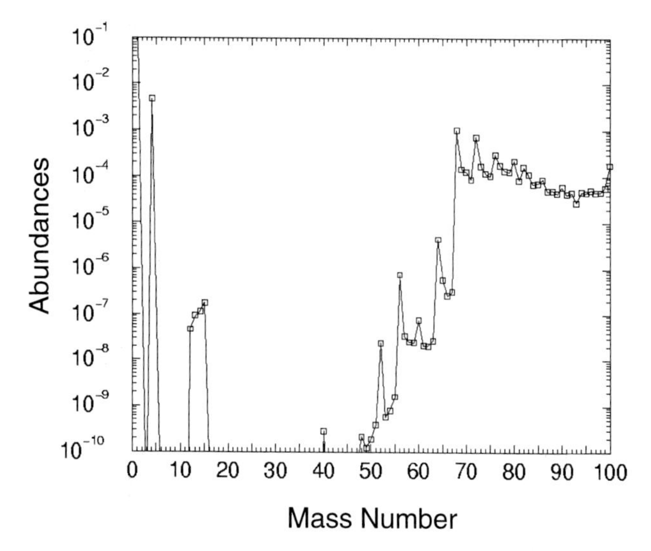

FIGURE 25. The abundance distribution after the thermonuclear runaway of a single x-ray burst.

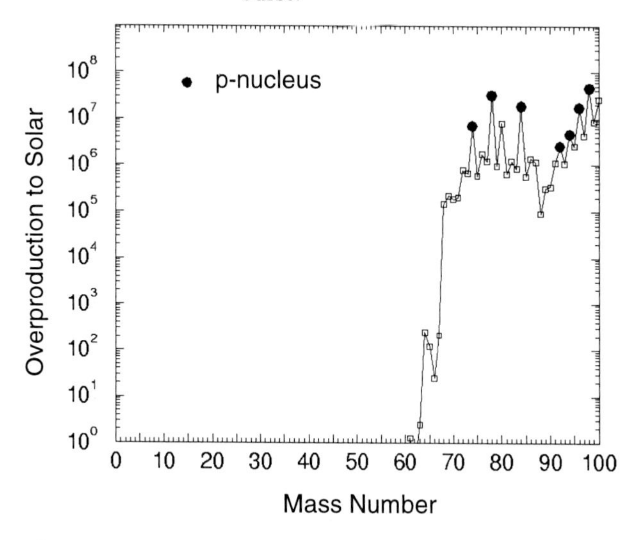

FIGURE 26. The ratio of the produced abundances and the initial solar abundances after the thermonuclear runaway of a single x-ray burst. The light p-nuclei are marked as black dots, their overabundance is about one order of magnitude larger than the of other isotopes. The abundances of the last two mass numbers are not reliable because they correspond to the end of the network.

phases during the stellar evolution, like the main-sequence, the red giant, and the AGB star phase, directly with the slow reactions on stable nuclei like the p+p reaction, the triple-α-process, and the $^{12}C+^{12}C$ fusion reaction. High intensity low energy accelerators coupled with sophisticated detector arrays are desperately needed to continue these studies down to the actual temperature regimes of quiet stellar burning. The use of radioactive ion beam techniques will allow us to continue and expand these studies from the static nucleosynthesis in stellar burning to the dynamic phases during cataclysmic stellar explosions. This will offer the key for our understanding of various stellar explosions which we observe as novae, x-ray bursts, and supernovae in our own as well as in far distant galaxies. The experimental goal of the field in the future is two-fold. First, the aim for more measurements towards lower energies to explore stellar nucleosynthesis at true stellar energies. Second, the aim for measurements towards the limits of stability to explore nucleosynthesis at the conditions of stellar explosions in which most of the heavy elements have been created.

Acknowledgments

I want to thank J. Görres, H. Schatz, and P. Tischhauser from the University of Notre Dame, and P. Leleux, from the Universite de Louvain-la-Neuve, for their help and input in the preparation of this manuscript. The here described work was supported by the NSF-Grant PHY94-02761, the DOE-Grant DE-FG02-95-ER40934, and the NATO-Grant CRG940213-97.

REFERENCES

Alburger, D. E. & Harbottle, G., 1990, *Phys.Rev.*C **41**, 2320

Anders, E. & Grevesse, M., 1989,*Geochim.Cosmochim.Acta* **53**, 197

Arnett, D. A., 1996, *Supernovae and Nucleosynthesis*, Princeton, Princeton University Press

Arpesella, C., *et al.* 1996, *Phys.Lett.* B **389**, 452

Ashworth, W. B., 1980, *J.Hist.Astr.* **11**, 1

Audouze, J., Truran, J. W. & Zimmerman, B. A., 1973, *Astrophys.J* **184**, 493

Ayasli, A. & Joos, P., 1982, *Astrophys.J.* **256**, 637

Azuma, R.E., *et al.* 1994, *Phys.Rev.* C **50**, 1194

Bahcall, J. N. & Pinsonneault, M. H., 1992 *Rev.Mod.Phys.* **64**, 885

Berheide, M., Rolfs, C., Schröder, U. & Trautvetter, H.P., 1992, *Z.Phys.* A **343**, 483

Bildsten, L., Salpeter, E. E. & Wassermann, I., 1992, *Astrophys.J.* **384**, 143

Bildsten, L., 1997, In *The many Faces of Neutron Stars*, (eds. A. Alpar, L. Buccheri, J. van Paradijs) Dordrecht

Blackmon, J.C., Champagne, A.E., Hofstee, M.A., Smith, M.S., Downing, R.G. & Lamaze, G.P. 1995, *Phys.Rev.Lett.* **74**, 2642

Blöcker, T. & Schönberner, D., 1995, in *Nuclei in the Cosmos III*, (eds. M. Busso, R. Gallino, C.M. Raiteri), AIP Conference Proceedings **327**, New York, American Institute of Physics, pp 399

Borkowski, K., *et al*, 1996, *Astrophys.J.*, in press

Buchmann, L., Azuma, R. E., Barnes, C.A., Humblet, J. & Langanke, K., 1996, *Phys.Rev.* C **54**, 393

Burbidge, E. M., Burbidge, G., Fowler, W. A. & Hoyle, F., 1957, *Rev.Mod.Phys.* **29**, 547

Busso, M., Lambert, D. L., Beglio, L., Gallino, R., Taiteri, C. M. & Smith, V. V., 1995, *Astrophys.J.* **446**, 775

Cannon, R. C., Frost, C. A., Lattanzio, J. C. & Wood, P. R., 1995, in *Nuclei in the Cosmos III*, (eds. M. Busso, R. Gallino, C.M. Raiteri), AIP Conference Proceedings **327**, New York, American Institute of Physics, pp 469

V. Castellani, V., Degl'Innocenti, S., Fiorentini, G., Lissia, M. & Ricci, B., 1997,*Phys. Rep.* in press

Caughlan, G. R. & Fowler, W. A., 1988, *At.Data Nucl.Data Tab* **40**, 284

Chen, B., *et al.* 1995, *Phys.Lett.* B **355**, 37

Chevalier, R. A., 1993, *Astrophys.J.* **411**, L33

Chevalier, R. A., 1996, *Astrophys.J.* **459**, 322

Coc, A., Mochkovitch, R., Oberto, Y., Thibaud, J.-P. & Vangioni-Flam, E., 1995, *Astr.& Astrophys.* **299**, 479 **78**, 85

Coszach, R., *et al.* 1995, *Phys.Lett.* B **353**, 184

Cowan, J. J., Thielemann, F. K. & Truran, J. W., 1991, *Phys.Rep.* **208**, 268

Decrock, P., *et al.* 1991 *Phys.Rev.Lett.* **67**, 808

Diehl, R., *et al*, 1995, *Astron.& Astrophys.* **298**, L25

Duncan, R., C., Shapiro, S. L. & Wasserman, I, 1986, *Astrophys.J.* **309**, 141

Dupraz, C., *et al.* 1997, *Astron.& Astrophys.* **324**, 683

Fesen, R.A. & Becker, R. H., 1991, *Astrophys.J.* **371**, 621

Flamsteed, J., 1725 in *Historia Coelestis Britannica, Vol.1*, ed. E. Halley, London

Fiorentini, G., Kavanagh, R. W. & Rolfs, C., 1995, *Z.Phys.* A **350**, 289

Fowler, W. A., Caughlan, G. R. & Zimmerman, B. A., 1967, *Ann. Rev. Astr. Astrophys.* **5**, 525

Fowler, W. A., Caughlan, G. R. & Zimmerman, B. A., 1975, *Ann. Rev. Astr. Astrophys.* **13**, 69 (1975)

Fowler, W. A., 1984, *Rev.Mod.Phys.* **56**, 159

Frekers, D., *et al* 1983, *Phys.Rev.*C **28**, 1765

Fujimoto, M. Y., Hanawa, T. & Miyaji, S., 1981, *Astrophys.J.* **247**, 267

Fujimoto, M., Sztajno, M., Lewin, W. & van Paradijs, J., 1987, *Astrophys.J.* **319**, 902

Fuller, G., Fowler, W. A. & Newman, M. J., 1980, *Astrophys.J.Suppl.* **42**, 447

Garcia, A., *et al.* 1991, *Phys.Rev.* C **43**, 2012

Glasner, S. A., Livne, E. & Truran, J. W., 1997, *Astrophys.J.* **475**, 754

Görres, J., Wiescher, M. & Thielemann, F. K., 1995, *Phys.Rev.* C **51**, 392

Graulich, J. S., *et al.* 1997, *Nucl.Phys.* A, submitted

Haberl, F., *et al* 1987, *Astrophys.J.* **314**, 266

Hahn, K. I., *et al.*, 1996, *Phys.Rev.* C **54**, 1999

Hansen, C. J. & Kawaler, S. D., 1994 *Stellar Interiors*, New York, Springer Verlag

Hartmann, D., *et al* 1997, *Nucl.Phys.* A **621**, 83c

Hauschildt, P., *et al* 1995, *Astrophys.J.* **447**, 829

Herndl, H., Görres, J., Wiescher, M., Brown, B. A. & Van Wormer, L., 1995, *Phys.Rev.* C **52**, 1078

Higdon, J. C. & Fowler, W. A., 1987, *Astrophys.J.* **317**, 710

Higdon, J. C. & Fowler, W. A., 1989, *Astrophys.J.* **339**, 956

Hoffman, R. D., Woosley, S. E., Fuller, G. M. & Meyer, B. S., 1996, *Astrophys.J.* **460**, 478

Höflich, P. *et al*, 1997, *Astron.& Astrophys.*, in press

Howard, W. M., Meyer, B. S. & Woosley, S. E., 1991, *Astrophys.J.* **373**, L5 **65**, 1256

Iyudin I., *et al.*, 1994, *Astr.& Astrophys.* **284**, L1

Iyudin I., *et al.*, 1995, *Astr.& Astrophys.* **300**, 422

Jorissen, A., Smith, V. V. $ Lambert, D. L. 1992, *Astron.& Astrophys.* **261**, 164

Jose, J., Hernanz, M. & Coc, A., 1996, *Astrophys.J.Lett.* **479**, L55

Joss, P. C. & Melia, F., 1987, *Astrophys.J.* **312**, 700

Käppeler, F., *et al.* 1994, *Astrophys.J.* **437**, 396

Kraft, R. P., 1994, *PASP* **106**, 553 **14**, 331

Lambert, D. L., 1992, *Astr.& Astrophys.Rev.* **3**, 201

Langanke, K. & Barnes, C.A. 1996, *Advances in Nuclear Physics* **22**, (eds. J. W. Negele, E. Vogt), New York, Plenum Press

Lattanzio, J. C., 1995, (in Nuclei in the Cosmos III), (eds. M. Busso, R. Gallino, C.M. Raiteri), AIP Conference Proceedings **327**, New York, American Institute of Physics, pp. 353

Lattanzio, J. C., Frost, C. A., Cannon, R. C. & Wood, P. R., 1997, *Nucl.Phys.* A **621**, 435c
Law, W. Y. & Ritter, H., 1983, *Astr.& Astrophys.* **63**, 265
Leising, M. D. & Clayton, D. D., 1987, *Astrophys.J.* **323**, 159
Lewin, W., van Paradijs, J. & Taam, R., 1993, *Space Sci.Rev.* **62**, 233
Lewin, W. H., Vacca, W. D. & Basinska, E., 1984, *Astrophys.J.* **277**, L57
Livio, M. & Truran, J. W., 1994, *Astrophys.J.* **425**, 797
Marion, J. B. & Fowler, W. A., 1957, *Astrophys.J.* **125**, 221
Moreland, P. E. & Heymann, D., 1965, *J.Inorg.Chem* **27**, 493
Nittler, L. R., Alexander, D., Gao, X., Walker, R.M. $ Zinner, E., 1997, *Nucl.Phys.* A **621**, 113c
Nobili, L., Turolla, R. & Lapidus, I., 1994, *Astrophys.J.* **433**, 276
Nomoto, K., *et al.*, *Nucl.Phys. A* **616**, 79c (1997)
Norman, E., *et al*, 1997, *Nucl.Phys.* A, **621** 92c
Paczynski, B., 1983, *Astrophys.J.* **267**, 315
Paczynski, B. & Anderson, N., 1986, *Astrophys.J.* **302**, 519
Politano, M., *et al.*, 1995, *Astrophys.J.* **448**, 807
Prantzos, N. & Diehl, R., 1995, *Phys.Rep* **267**, 1
Rauscher, T., Kratz, K.-L. & Thielemann, F.-K., 1997 *Phys.Rev.* C, **56**, 1613
Rayet, M., Prantzos, N. & Arnould, M., 1990, *Astr.& Astrophys.* **227**, 271
Rayet, M., Arnould, M., Hashimoto, M., Prantzos, N. & Nomoto, K., 1995, *Astr.& Astrophys.* **298**, 517
Reed, J. E., *et al* 1995, *Astrophys.J.* **440**, 706
Rehm, K. E., *et al.* 1996, *Phys.Rev.* C **53**, 1950
Rolfs, C., 1973, *Nucl.Phys* A **217**, 29
Rolfs, C., Rodney, W. S., Shapiro, M. H. & Winkler, H., 1975, *Nucl.Phys.* A **241**, 460
Rolfs, C. & Rodney, W.S., 1988, *Cauldrons in the Cosmos*, Chicago, Chicago University Press
Rolfs, C. & Barnes, C.A., 1990, *Ann.Rev.Nucl.Part.Sci.* **40**, 45
Sargood, D. G., 1982 *Phys.Rep.* **63**, 61
Schatz, H., *et al.* 1997*a*, *Phys.Rep.*, in press
Schatz, H., *et al.* 1997*b*, *Phys.Rev.Lett.* submitted
Schmidt, S., *et al.* 1995, *Nucl.Phys.* A **591**, 227
Schönfelder, V., *et al* 1993, *Astrophys.J.Suppl.***86**, 657
Schröder, U., *et al.* 1987, *Nucl.Phys.* A **467**, 240
Seuthe, S., *et al.* 1990, *Nucl.Phys.* A **514**, 471
Shore, S., *et al.* 1994 *Astrophys.J.* **421**, 344
Smith, V. V. & Lambert, D. L., 1990, *Astrophys.J.Suppl.* **72**, 387
Sneden, C., McWilliam, A., Preston, G. W., Cowan, J. J., Burris, D. L. & Armosky, B. J., 1996, *Astrophys.J.* **467**, 819
Starrfield, S., 1989. *Classical Novae*, (eds. M.F. Bode, A. Evans), New York, John Wiley & Sons
Starrfield, S., Truran, J. W., Wiescher, M. & Sparks, W.M., 1997, *M.N.R.A.S.*, in press
Stegmüller, F., *et al.* 1996, *Nucl.Phys.* A **601**, 168
Taam, R. M., 1985, *Ann.Rev.Nucl.Sci.* **35**, 1
Taam, R. M., Woosley, S. E., Weaver, T. & Lamb, D., 1993 *Astrophys.J.* **413**, 324
Taam, R. E. , Woosley, S. E. & Lamb, D. Q., 1996, *Astrophys.J.* **459**, 271
Tawara, Y., *et al* 1984*a*, *Astrophys.J.* **276**, L41
Tawara, Y., Hayakawa, S. & Kii, T., 1984*b*, *PASJ* **36**, 845
The, L. S., *et al* 1997, *Astr.& Astrophys.*, in press

Thielemann, F.-K., Nomoto, K. & Hashimoto, M., 1996, *Astrophys.J.* **460**, 408

Timmes, F. X., Woosley, S. E., Hartmann, D. H. & Hoffmann, R. D., 1995, *Astrophys.J.* **464**, 332

Truran, J. W., 1982, *Essays in Nuclear Astrophysics*, (eds. C.A. Barnes, D.D. Clayton, D.N. Schramm), Cambridge, Cambridge University Press

Truran, J. W. & Livio, M., 1986, *Astrophys.J* **308**, 721

Truran, J. W., 1990, *Physics of Classical Novae*, (eds. A. Cassatella, R. Vietti), Berlin, Springer Verlag

Van den Bergh, S. & Kamper, K. W., 1983, *Astrophys.J.* **268**, 129

Van Wormer, L., Görres, J., Iliadis, C., Wiescher, M. & Thielemann, F.-K., 1994, *Astrophys.J.* **432**, 326

Wagoner, R. V., 1973, *Astrophys.J.* **179**, 343

Wallace, R. & Woosley, S. E., 1981, *Astrophys.J.Suppl.* **45**, 389

Wallace, R. & Woosley, S. E., 1984, in *High Energy Transients in Astrophysics*, (ed. S.E Woosley), AIP Conference Proceedings **115**, New York, American Institute of Physics, pp. 319

Wanajo, S., Nomoto, K., Hashimoto, M., Kajino, T. & Kubono, S., 1997, *Nucl.Phys.* A, **616**, 91c - 96c

Weaver, T. A. & Woosley, S. E., 1993, *Phys.Rep.* **227**, 65

Weiss,A. & Truran, J.W., 1990, *Astr.& Astrophys.* **238**, 178

Wiescher, M., Görres, J. & Thielemann, F.-K., 1988*a* *Astrophys.J.* **326**, 384

Wiescher, M., *et al.* 1988*b* *Nucl.Phys.* A **484**, 90

Williams, R. E., 1982, *Astrophys.J.* **261**, L77

Wing, J., Walgren, M. A., Stevens, C. M. & Orlandi, K. A., 1965, *J.Inorg.Chem* **27**, 487 (1965)

Woosley S. E. & Weaver, T. A., 1984, in *High Energy Transients in Astrophysics*, (ed. S.E. Woosley), AIP Conference Proceedings **115**, New York, American Institute of Physics, pp. 273

Woosley, S. E. & Howard, E. M., 1990, *Astrophys.J.* **354**, L21

Woosley, S. E. & Hoffman, R. D., 1992, *Astrophys.J.* **395**, 202

Woosley, S. E., Langer, N. & Weaver, T. A., 1995, *Astrophys.J.*, **488**, 315

Zinner, E., Amari, S., Anders, A. & Lewis, R., 1991 *Nature* **349**, 51

Neutrino Transport and Large-Scale Convection in Core-Collapse Supernovae

By MIKE GUIDRY

Department of Physics and Astronomy, University of Tennessee, Knoxville, TN 37996-1200, USA

Theoretical and Computational Physics Section Oak Ridge National Laboratory, Oak Ridge, TN 37831-6373, USA

The mechanism for a core-collapse or type II supernova is a fundamental unresolved problem in astrophysics. Although there is general agreement on the outlines of the mechanism, a detailed model that includes microphysics self-consistently and leads to robust explosions having the observational characteristics of type II supernovae does not exist. Within the past five years supernova modeling has moved from earlier one-dimensional hydrodynamical simulations with approximate microphysics to multi-dimensional hydrodynamics on the one hand, and to much more detailed microphysics on the other. These simulations suggest that large-scale and rapid convective effects are common in the core during the first hundreds of milliseconds after core collapse, and may play a role in the mechanism. However, the most recent simulations indicate that the proper treatment of neutrinos is probably even more important than convective effects in producing successful explosions. In this series of lectures I will give a general overview of the core-collapse problem, and will discuss the role of convection and neutrino transport in the resolution of this problem.

---

1. Introduction

A type II supernova is one of the most spectacular events in nature, and is a likely source of the heavy elements that are produced in the rapid neutron capture or r-process. Considerable progress has been made over the past two decades in understanding the mechanisms responsible for such events. See the review by Bethe (1990) and the volume of Physics Reports edited by Brown (1988) for a good summary of the field at the beginning of this decade; a recent overview of the field may be found in Burrows (1997). This understanding was tested both qualitatively and quantitatively by the observation of Supernova 1987A in the nearby Magellanic Cloud—the brightest supernova observed from Earth since the time of Kepler. The observations of this event, and the continuing studies of its aftermath [McCray (1993)], provide compelling evidence that a type II supernova represents the death of a massive star in which a degenerate iron core of approximately 1.4 solar masses collapses catastrophically on millisecond timescales. This gravitational collapse is reversed as the inner core exceeds nuclear densities because of the stiff nuclear equation of state, and a pressure wave reflects from the center of the star and propagates outward, steepening into a shock wave as it passes into increasingly less dense material of the outer core.

However, the most realistic simulations of this event indicate that the shock wave loses energy rapidly in propagating through the outer core and that, if the core has a mass of more than about 1.1 solar masses, the shock stalls into an accretion shock within several hundred milliseconds of the bounce at a distance of several hundred kilometers from the center. Thus the "prompt shock" does not eject the outer layers of the star and fails to produce a supernova. This is the "supernova problem": there is good evidence that we understand the basics, but the details fail us.

For a decade there has been a growing concensus that the neutrinos, which are produced in prodigious quantities in the supernova and play a central role in the entire event, are somehow responsible for rescuing the nascent supernova from this embarassing state of affairs by reenergizing the shock, thereby allowing the explosion to proceed to a conclusion in accord with the observed properties of supernovae [Bethe & Wilson (1985]. In these lectures, I will review the current situation with respect to these ideas.

Let us begin our discussion of supernovae and the death of massive stars by introducing the basic principles governing the evolution of stars. Our discussion will be compact, summarizing only those aspects that are of direct relevance to the subsequent presentation. The reader is urged to consult standard references for a more thorough consideration of the issues addressed here: Hansen & Kawaler (1994), Kippenhahn & Wiegert (1994), Arnett (1996).

2. Hydrostatic Equilibrium

For much of their lives, stars are governed by an exquisite balance between the gravitational forces that would like to contract the star and energy production mechanisms that supply pressure to expand the star. For non-relativistic conditions, the requirement of hydrostatic equilibrium can be expressed as

$$\frac{dP}{dr} = -\frac{GM(r)}{r^2}\rho(r) = -g(r)\rho(r), \tag{2.1}$$

where $g(r)$ is the local gravitational acceleration, $P(r)$ is the pressure, and $\rho(r)$ is the density in a spherical shell at a radius r:

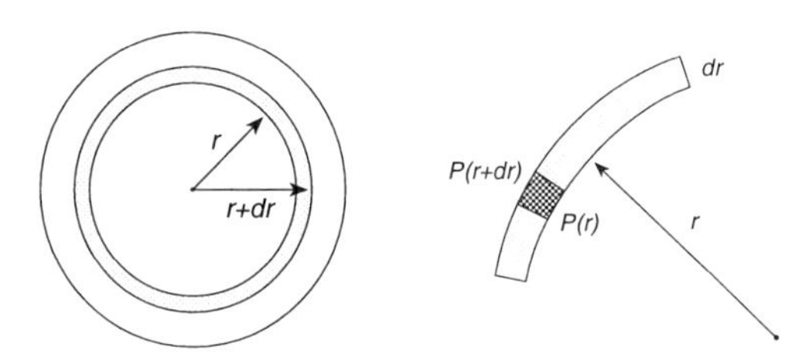

Furthermore, the mass $M(r)$ contained within a sphere of radius r must be

$$M(r) = \int_0^r 4\pi r^2 \rho(r) dr. \tag{2.2}$$

The solution of these equations requires additional information contained in the equation of state for the matter and radiation of the star.

3. Equation of State

An equation of state expresses a relationship among thermodynamical variables such as the pressure P, the temperature T, and the density ρ. For most ordinary stars, the equation of state is well approximated by that of an ideal gas:

$$P(r) = \frac{k}{\mu}\rho(r)T(r) \tag{3.3}$$

where $k = 1.38 \times 10^{-16}$ ergs K^{-1} (Boltzmann constant) and μ is the mean molecular weight of the particles in the gas. The total pressure is a sum of the contribution from the gas and that from radiation,

$$P(r) = P_{\text{gas}} + P_{\text{rad}} \tag{3.4}$$

where the radiation pressure is given in blackbody approximation by Stefan's Law,

$$P_{\rm rad} = \tfrac{1}{3}aT^4 \tag{3.5}$$

with $a = 7.565 \times 10^{-15}$ erg cm^{-3} K^{-4}. For low-mass stars, the pressure from the gas typically is much larger than that from radiation, but in stars of more than about 25 solar masses the radiation pressure exceeds the contribution from the gas pressure. The equation of state associated with photons follows from the Stefan–Boltzmann Law. For the pressure we have

$$P_{\rm rad} = \sigma T^4 = \frac{aT^4}{3}, \tag{3.6}$$

and the energy density of the radiation field is

$$u_{\rm rad} = aT^4 = 3P_{\rm rad}. \tag{3.7}$$

For a simple stellar model, it is often a good starting point to assume an ideal gas equation of state for the matter and a blackbody equation of state for the radiation. In that case we may write for the pressure P and specific energy ϵ,

$$P = \frac{\rho kT}{\mu} + \frac{aT^4}{3} \qquad \epsilon = \frac{u}{\rho} = \frac{3NkT}{2\mu} + \frac{aT^4}{\rho}, \tag{3.8}$$

where the first term in each case is the contribution of the ideal gas and the second term is that of the radiation.

Although most regions of normal stars are described well as an ideal gas, the late stages of stellar evolution may involve matter in states of extremely high density where the Pauli principle plays an important role. Such matter is termed *degenerate matter* and, as we now discuss, the equation of state is considerably different from that of an ideal gas.

4. Quantum Gases and Degenerate Matter.

Because the microscopic constituents of stellar matter are fermions, the Pauli principle has a significant effect on the equation of state if the density of the matter becomes very large. Such conditions are encountered in white dwarfs and the cores of massive stars for the electrons, and in neutron stars for the neutrons. We shall be particularly interested in the electrons in the cores of massive stars.

4.1. *Quantum Fermi Gases*

Let us consider the distinction between a classical gas and a quantum gas, and the corresponding implications for stellar structure. Identical fermions are described statistically by the *Fermi–Dirac distribution*

$$f_{\rm f}(\epsilon_p) = \frac{1}{\exp[(\epsilon_p - \mu)/kT] + 1} \qquad \text{(Fermi–Dirac)}, \tag{4.9}$$

where ϵ_p is given by

$$\epsilon_p^2 = p^2c^2 + m^2c^4 \tag{4.10}$$

$$\simeq mc^2 + \frac{p^2}{2m} \qquad \text{(nonrelativistic: } v/c \sim 0) \tag{4.11}$$

$$\simeq p^2c^2 \qquad \text{(ultrarelativistic: } v/c \sim 1) \tag{4.12}$$

and the *chemical potential* μ is introduced by including in the first law of thermodynamics the possibility of a change in the particle number N:

$$dU = TdS - PdV + \mu dN\,. \tag{4.13}$$

(Don't confuse this μ with the mean molecular weight introduced earlier.) We shall consider a gas to be a quantum gas if it is described by Eq. (4.9), and a classical gas if the condition

$$e^{(mc^2-\mu)/kT} >> 1 \tag{4.14}$$

is fulfilled. In that case, the states of lowest energy have $\epsilon_p \sim mc^2$ and (for fermions or bosons) the distribution function is well approximated by *Maxwell–Boltzmann statistics*,

$$f(\epsilon_p) = e^{-(\epsilon_p-\mu)/kT} \qquad \text{(Maxwell–Boltzmann)}, \tag{4.15}$$

where generally $f_p << 1$ for the classical gas. If we introduce a nonrelativistic quantum concentration variable n_{Q} through

$$n_{\text{Q}} = \left(\frac{2\pi mkT}{h^2}\right)^{3/2}. \tag{4.16}$$

it is not difficult to show that the classical condition (4.14) is equivalent to a requirement that the actual number density be small compared with that specified by Eq. (4.16); this constraint in turn is satisfied if on average the separation between particles in the gas is much larger than the de Broglie wavelength for the particles. Similar considerations apply for relativistic gases.

4.2. *Transition from Classical to Quantum Gas Behavior*

Thus, the classical approximation breaks down at high density and the gas then behaves as a quantum gas subject to the quantum statistics appropriate for the gas. Notice from (4.16) that with increasing gas density the least massive particles will be most prone to a deviation from classical behavior because the scale set by the quantum concentration is proportional to $m^{3/2}$. Thus photons, neutrinos, and electrons are most susceptible to such effects. The massless photons never behave as a classical gas, and the neutrinos (also either massless or nearly so) interact so weakly with matter that they leave the star unimpeded in most stages of stellar evolution.† In normal stellar environments, the electrons are most susceptible to a transition from classical to quantum gas behavior.

Presently in the Sun the average number density for electrons is $n \sim 6 \times 10^{29}$ m^{-3} and for $T = 6 \times 10^6$ K the nonrelativistic quantum concentration is $n_{\text{Q}}^{\text{NR}} \sim 3 \times 10^{31}$ m^{-3}. Thus, the electrons in the Sun are well approximated by a dilute classical gas and quantum corrections are small. However, the core of the Sun, as for all stars, will contract late in its life as its nuclear fuel is exhausted. It may be shown that $n/n_{\text{Q}} \approx R^{-3/2}$, where R is the radius of the star. Thus, we expect that as the core of a star contracts, at some point the electrons in its interior may begin to behave as a quantum rather than classical gas.

4.3. *The Degenerate Electron Gas*

From Eq. (4.16), the quantum gas condition $n >> n_{\text{Q}}$ is equivalent to a constraint

$$kT << h^2 n^{2/3}/2\pi m. \tag{4.17}$$

Thus, a quantum gas is a *cold gas*, but cold on a temperature scale set by the right side of Eq. (4.17); if the density is high enough, Eq. (4.17) indicates that the gas could be "cold" and still have a temperature of billions of degrees. The precise quantum

† An exception to this statement directly relevant to the present discussion occurs for the core of a type II supernova, where the densities and temperatures become high enough to trap neutrinos for significant periods relative to the dynamical timescales for the problem. We discuss such trapping more extensively below.

mechanical meaning of a cold electron gas is that the electrons are all concentrated in the lowest available quantum states; we say that such a gas is *degenerate*.

In the limit that the temperature may be neglected the Fermi–Dirac distribution becomes a step function in energy space,

$$f_{\rm f}(\epsilon_p) = \frac{1}{e^{(\epsilon_p - \mu)/kT} + 1} \underset{T\to 0}{\longrightarrow} \begin{cases} f(\epsilon_p) = 1 & \epsilon_p \leq \epsilon_{\rm f} \\ f(\epsilon_p) = 0 & \epsilon_p > \epsilon_{\rm f} \end{cases} \tag{4.18}$$

where the corresponding value of the chemical potential μ is denoted by $\epsilon_{\rm f}$ and is termed the *fermi energy*. Thus, the fermi energy gives the energy of the highest occupied state in the degenerate fermi gas. The corresponding value of the momentum is denoted by $p_{\rm f}$ and is termed the *fermi momentum*.

The number of electrons N in the degenerate gas is just the number of states with momentum less than the fermi momentum $p_{\rm f}$,

$$N = \int_0^{p_{\rm f}} g(p)\, dp = 4\pi V \frac{g_{\rm s}}{h^3} \int_0^{p_{\rm f}} p^2\, dp = \frac{8\pi V}{3h^3} p_{\rm f}^3, \tag{4.19}$$

where a statistical factor $g_{\rm s} = 2j + 1 = 2$ has been used for electrons. Introducing $n = N/V$ and solving for the fermi momentum, we find that $p_{\rm f}$ is determined by fundamental constants and the electron number density n

$$p_{\rm f} = \left(\frac{3n}{8\pi}\right)^{1/3} h. \tag{4.20}$$

(This result implies that the de Broglie wavelength $\lambda = h/p_{\rm f} \sim n^{-1/3}$ is approximately equal to the interparticle spacing.) We may construct the equation of state by evaluating the internal energy. Let us first consider the nonrelativistic and then the ultrarelativistic limits for the equation of state for degenerate electrons.

4.4. *Nonrelativistic Degenerate Electrons*

The nonrelativistic limit is defined by $p_{\rm f} << mc$, which implies that $n << (1/\lambda_{\rm c})^3 << (mc/h)^3$, where $\lambda_{\rm c} = 2.4 \times 10^{-12}$ m is the Compton wavelength for an electron. The internal energy is

$$U = \int_0^\infty \epsilon_p f(\epsilon_p) g(p)\, dp = N\left(mc^2 + \frac{3}{10m} p_{\rm f}^2\right), \tag{4.21}$$

where (4.10), (4.18), (4.19), and $g_{\rm s} = 2$ have been used. For a nonrelativistic gas the pressure is given by $\frac{2}{3}$ of the kinetic energy density; identifying the second term of Eq. (4.21) divided by the volume V as the kinetic energy density, we obtain

$$P = \frac{2}{3}\left(\frac{N}{V}\frac{3p_{\rm f}^2}{10m}\right) = n\frac{p_{\rm f}^2}{5m} = \frac{h^2}{5m}\left(\frac{3}{8\pi}\right)^{2/3} n^{5/3}, \tag{4.22}$$

where $n = N/V$ and Eq. (4.20) have been used. For sufficiently low density, the electrons are nonrelativistic and the electron pressure is given by

$$P_{\rm e} = \frac{h^2}{5m}\left(\frac{3}{8\pi}\right)^{2/3}\left(\frac{\rho}{m_{\rm p}\mu_{\rm e}}\right)^{5/3} = 1.0 \times 10^{13}\left(\frac{\rho}{\mu_{\rm e}}\right)^{5/3} \text{dyne/cm}^2, \tag{4.23}$$

where $\mu_{\rm e}$ is defined through $n \equiv n_{\rm e} = \rho/m_{\rm p}\mu_{\rm e}$.

4.5. *Ultrarelativistic Degenerate Electrons*

For ultrarelativistic electrons $n >> n_{\rm Q}^{\rm R}$ implies that $n >> (mc/h)^3$. Utilizing the ultra-relativistic limit $\epsilon_p = pc$ of Eq. (4.12), the internal energy is

$$U = \int_0^\infty \epsilon_p f(\epsilon_p) g(p)\, dp = \frac{8\pi V c}{h^3} \int_0^{p_{\rm f}} p^3\, dp = \frac{3}{4} N c p_{\rm f}. \tag{4.24}$$

For an ultrarelativistic gas the pressure is $\frac{1}{3}$ of the kinetic energy density. Identifying the kinetic energy density as Eq. (4.24) divided by the volume V for ultrarelativistic particles,

$$P = \frac{1}{3} \times (\frac{3}{4} c n p_{\rm f}) = \frac{hc}{4} \left(\frac{3}{8\pi} \right)^{1/3} n^{4/3}, \tag{4.25}$$

where $n = N/V$ and we have used Eq. (4.20). For densities $\rho \gtrsim 10^6$ g/cm^3 the electrons are highly relativistic and the corresponding degenerate equation of state takes the form implied by Eq. (4.25)

$$P_{\rm e} = \frac{hc}{4} \left(\frac{3}{8\pi} \right)^{1/3} \left(\frac{\rho}{m_{\rm p} \mu_{\rm e}} \right)^{4/3} = 1.24 \times 10^{15} \left(\frac{\rho}{\mu_{\rm e}} \right)^{4/3} \text{dyne/cm}^2. \tag{4.26}$$

This pressure is *independent of temperature*, for it derives from the exclusion principle and is present even at zero temperature. Imposing the condition of hydrostatic equilibrium, this electron degeneracy pressure leads to an equation limiting the masses that can be stabilized by this pressure (the *Chandrasekhar mass*)

$$M_{\rm ch} = 5.8\, Y_{\rm e}^2 M_{\rm solar} \simeq 1.2\text{–}1.5\, M_\odot \tag{4.27}$$

where $M_\odot$ is the solar mass unit and the *electron fraction* $Y_{\rm e}$ is defined by

$$Y_{\rm e} = \frac{\text{Number of Electrons}}{\text{Number of Baryons}}. \tag{4.28}$$

For the cases that will interest us, we may take the number of baryons to be the sum of the number of protons and the number of neutrons. Thus, electron degeneracy pressure is capable of stabilizing a gravitating mass against collapse, but only up to a limiting mass that is of order a solar mass. This will be crucial for our subsequent discussion of stellar core collapse.

5. The Hertzsprung–Russell Diagram.

If one plots the luminosity versus the surface temperature for observable stars, the resulting *spectrum–luminosity diagram* or Hertzsprung–Russell (HR) diagram indicates that stars are not scattered randomly, but are strongly grouped into a few regions. Such an HR diagram is illustrated in Fig. 1, for a population of stars like that in the neighborhood of the sun. Most stars cluster in the S-shaped region termed the *main sequence*. These represent stars in the stable period of their lives that are producing energy by hydrogen fusion. Since this period occupies the largest portion of a normal star's life, most stars at any one time will be main sequence stars. The giant and supergiant regions correspond to stars late in their lives that have greatly extended outer envelopes. This expansion of the outer envelope cools the surface and greatly increases the surface area relative to the original main-sequence star. Thus, these stars appear on the HR diagram above the main sequence and shifted to lower surface temperatures.

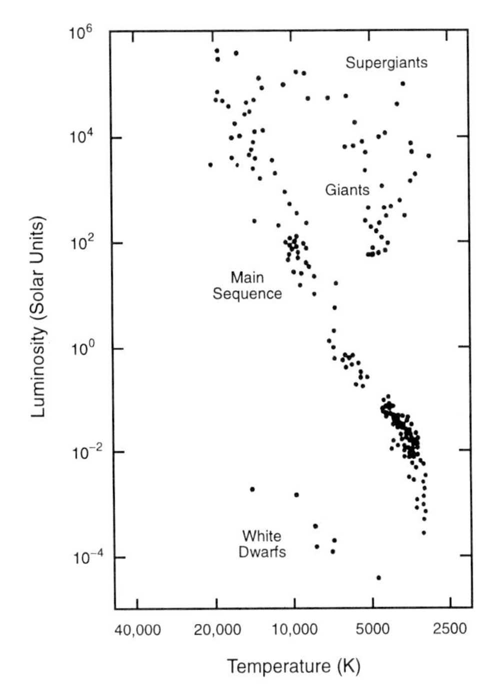

FIGURE 1. Hertzsprung–Russell diagram for stars near the Sun.

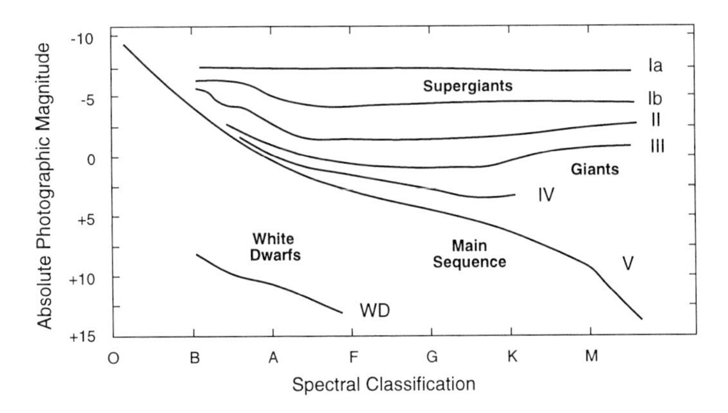

FIGURE 2. The primary luminosity classes.

5.1. *Luminosity Classes*

The qualitative groupings according to luminosity that we have discussed may be made more quantitative by introducing the *luminosity classes* illustrated in Fig. 2. In this classification, the main sequence stars (also sometimes called "dwarfs"—not to be confused with white dwarfs) are in luminosity class V, the white dwarfs are luminosity class WD, the giants are distributed in three luminosity classes (II, III, IV) according to brightness, and the supergiants are divided into two categories (Ia and Ib) according to their intrinsic brightness.

We may then classify stars according to both their spectral class and luminosity class; this classification confines them to a rather localized region of the HR diagram. For example Rigel (β-Orionis) may be classified as B8 (Ia), which indicates that it is a

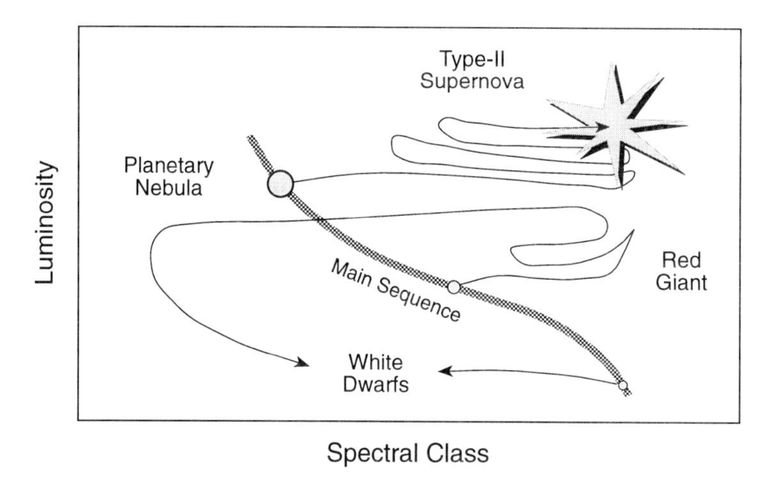

FIGURE 3. Three general categories of post main sequence evolution.

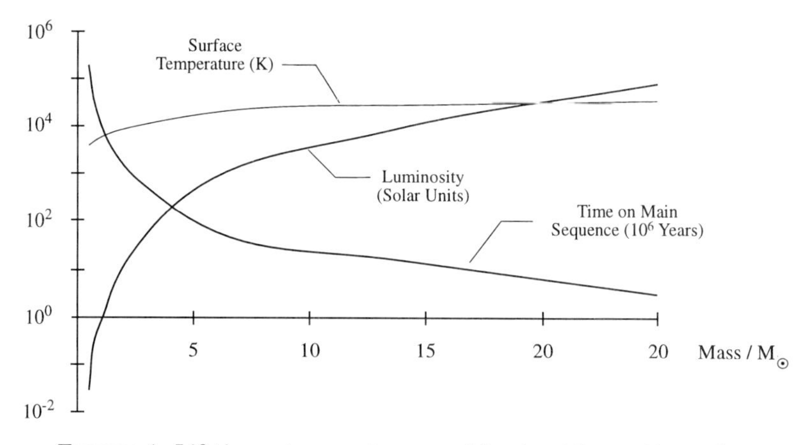

FIGURE 4. Lifetimes, temperatures, and luminosities on the main sequence.

spectral class B8, luminosity class Ia star: a luminous, blue supergiant. We shall be very interested in such luminous supergiants, for they are prime candidates to become supernovae.

5.2. *The HR Diagram as an Evolutionary Sequence*

Stars evolve from one region of the HR diagram to another during the course of their lives. The length of time spent in various regions of the HR diagram in the course of a star's life, and the nature of the evolutionary track, is primarily determined by the mass of the star. Generally, more massive stars evolve faster than less massive stars because the higher densities and pressures in the central regions of massive stars leads to rapid consumption of their fuel. Their ultimate fate is also often different, with their lives terminated by a supernova explosion after evolving to the supergiant region, leaving behind a neutron star or a black hole. Three general classes of post main-sequence evolutionary tracks are illustrated in Fig. 3, and some average times spent on the main sequence (and selected other properties) as a function of initial mass for the star are illustrated in Fig. 4.

6. Some Timescales of Significance

In discussing the fate of stars that will become supernovae, there are some timescales that play an important role.

6.1. *Nuclear Burning Timescales*

The length of time that a star spends on the main sequence and in other stages of its life is governed primarily by the rate at which it consumes its nuclear fuel. As we have just noted, this timescale is much shorter for massive than for less massive stars. Thus, we shall find that the massive stars that are destined to become type II supernovae race through all stages of their lives at breakneck speeds compared with normal timescales for stellar evolution.

6.2. *Hydrodynamical Timescales*

In our previous discussion of hydrostatics we have assumed that fluid elements are not accelerated. If fluid elements experience net forces the corresponding motion is described by the equations of hydrodynamics [Shore (1992), Bowers & Wilson (1991)]. For the motion of such fluid elements in a gravitational field, we can define a dynamical timescale by asking the following question: if the pressure support for the star were suddenly taken away, what is the characteristic timescale on which the star would collapse? Since there is no restoring force, the only possible dependence of this timescale is on the mass of the star M, the radius of the star R, and the gravitational constant G. The only combination of these quantities having the dimension of time is

$$t_{\mathrm{ff}} = \sqrt{\frac{R^3}{GM}}, \tag{6.29}$$

which we take to be the characterisic timescale for free fall. Introducing a mean density $\bar{\rho} = M/(\frac{4}{3}\pi R^3)$, this may be written as

$$t_{\mathrm{ff}} \approx \sqrt{\frac{1}{G\bar{\rho}}} \approx \sqrt{\frac{R}{g}} \tag{6.30}$$

where $g = GM/R^2$ is the gravitational acceleration.

We may introduce a second dynamical timescale by considering the opposite extreme: if gravity were taken away, how fast would the star expand by virtue of its pressure? This timescale can depend only on R, $\bar{\rho}$, and $\bar{P}$, and the only combination of these quantities having time units is

$$t_{\mathrm{exp}} \approx R\sqrt{\frac{\bar{\rho}}{\bar{P}}}, \tag{6.31}$$

which we take to define a characteristic expansion timescale. This has a simple physical interpretation: $(\rho/P)^{1/2}$ is roughly the inverse mean sound speed; thus, t_{exp} is approximately the time for a pressure wave to travel from the center of the star to its surface.

Hydrostatic equilibrium will clearly be precarious unless these two timescales are comparable; therefore, we define a *hydrodynamical timescale* for the system through

$$\tau_{\mathrm{hydro}} \approx t_{\mathrm{exp}} \approx t_{\mathrm{ff}} \approx \sqrt{\frac{1}{G\bar{\rho}}}. \tag{6.32}$$

In Table 1 we illustrate the hydrodynamical timescale for several kinds of stars calculated using this formula (with appropriate constants of proportionality inserted).

Object	$\sim M/M_\odot$	$\sim R/R_\odot$	τ_{hydro}
Red Giant	1	100	36 days
Sun	1	1	55 minutes
White Dwarf	1	1/50	9 seconds
Iron Core	1	10^{-2}	100 ms
Neutron Star	1	10^{-5}	1 ms

TABLE 1. Hydrodynamical Timescales

Nuclear Fuel	Nuclear Products	Ignition Temperature	Minimum Main Sequence Mass	Period in $25M_\odot$ Star
H	He	4×10^6 K	$0.1M_\odot$	7×10^6 years
He	C, O	1.2×10^8 K	$0.4M_\odot$	5×10^5 years
C	Ne, Na, Mg, O	6×10^8 K	$4M_\odot$	600 years
Ne	O, Mg	1.2×10^9 K	$\sim 8M_\odot$	1 years
O	Si,S, P	1.5×10^9 K	$\sim 8M_\odot$	~ 0.5 years
Si	Ni–Fe	2.7×10^9 K	$\sim 8M_\odot$	~ 1 day

TABLE 2. Burning stages in massive stars (Woosley)

6.3. *Timescale Set by Random Walk of Diffusing Particles*

For a random walk, the distance traversed after Z scatterings is

$$\Delta x \approx \sqrt{Z}\,\lambda, \tag{6.33}$$

where λ is the average mean free path for the particle. To escape a star of radius R a particle must undergo approximately

$$Z = \left(\frac{\Delta x}{\lambda}\right)^2 = \left(\frac{R}{\lambda}\right)^2 \tag{6.34}$$

scatterings. From the center of the Sun, assuming an average mean free path of 0.5 cm for photons, this corresponds to 10^{22} scatterings before reaching the surface. We may attach a timescale to this by estimating the average lifetime of the state formed with each scattering.

7. Advanced Stellar Burning Stages

If a star is massive enough, burnings beyond H $\to$ He $\to$ C are possible by virtue of the high temperatures and pressures that result as the core contracts after exhausting its fuel. Typical burning stages in massive stars and their characteristics are listed in Table 2.

7.1. *Carbon Burning*

Carbon burns at $T \sim 5 \times 10^8$ K and $\rho \sim 3 \times 10^6$ g cm^{-3}, primarily through

$$^{12}\mathrm{C} + {}^{12}\mathrm{C} \to {}^{20}\mathrm{Ne} + \alpha \qquad {}^{12}\mathrm{C} + {}^{12}\mathrm{C} \to {}^{23}\mathrm{Na} + p \qquad {}^{12}\mathrm{C} + {}^{12}\mathrm{C} \to {}^{23}\mathrm{Mg} + p$$

As indicated in Table 2, such reactions are possible for stars having masses $\gtrsim 4M_\odot$. Subsequent burning stages require conditions that are probably only realized for stars having $M \gtrsim 8M_\odot$ or so.

7.2. *Neon Burning*

At $T \sim 10^9$ K, neon burns by a two-step sequence. (1) Neon is photodisintegrated by a high-energy photon (which become more plentiful as the temperature is increased): $\gamma + {}^{20}\mathrm{Ne} \longrightarrow {}^{16}\mathrm{O} + \alpha$. (2) The α-particle thus produced can initiate a radiative capture $\alpha + {}^{20}\mathrm{Ne} \rightarrow {}^{24}\mathrm{Mg} + \gamma$. This sequence produces a core of ${}^{16}\mathrm{O}$ and ${}^{24}\mathrm{Mg}$.

7.3. *Oxygen Burning*

At a temperature of 2×10^9 K, oxygen can fuse through the reaction ${}^{16}\mathrm{O} + {}^{16}\mathrm{O} \longrightarrow {}^{28}\mathrm{Si} + \alpha$. The silicon thus produced can only react at temperatures where the photon spectrum becomes sufficiently hard that photodissociation reactions begin to play a central role.

7.4. *Silicon Burning*

At $T \sim 3 \times 10^9$ K, silicon may be burned to heavier elements. At these temperatures the photons are quite energetic and those in the high-energy tail of the distribution can readily photodissociate nuclei. A network of photodisintegration and capture reactions develops that evolves preferentially to those elements that are the most bound [*Nuclear Statistical Equilibrium* or (NSE)]. Since the most stable nuclei are in the iron group, silicon burning carried to completion under equilibrium conditions produces iron-group nuclei. The initial step in silicon burning is a photodisintegration such as, $\gamma + {}^{28}\mathrm{Si} \longrightarrow {}^{24}\mathrm{Mg} + \alpha$, which requires a photon energy of 9.98 MeV. The α particles thus liberated can now undergo radiative capture reactions; a representative sequence is

$$
\begin{aligned}
\alpha + {}^{28}\mathrm{Si} &\longleftrightarrow {}^{32}\mathrm{S} + \gamma \\
\alpha + {}^{32}\mathrm{S} &\longleftrightarrow {}^{36}\mathrm{Ar} + \gamma \\
&\vdots \\
\alpha + {}^{56}\mathrm{Fe} &\longleftrightarrow {}^{56}\mathrm{Ni} + \gamma
\end{aligned}
$$

The reactions in this series are typically in equilibrium or quasiequilibrium, and are much faster than the initial photodisintegration; thus the photodisintegration of silicon becomes the rate-controlling step in silicon burning.

Since the iron group nuclei are the most stable in the Universe, silicon burning represents the last stage by which fusion and radiative capture reactions can build heavier elements under equilibrium conditions. In principle, one could make heavier elements by increasing the temperature so that the required extra energy could be absorbed from the kinetic energy of the gas, but this becomes self-defeating in statistical equilibrium because the higher temperatures will also lead to increased photodissociation and one is still left with iron-group nuclides as the most evolved end product.

7.5. *Timescales for Advanced Burning*

As is apparent from Table 2, the timescales for advanced burning are greatly compressed relative to earlier burning stages. These differences are particularly striking for very massive stars that are our primary concern; they rush through all phases of their lives in times that are fleeting compared with the usual evolutionary timescales. For example, the $25M_\odot$ example used for Table 2 takes almost 10 million years to advance through its hydrogen and helium burning phases, but completes its burning of oxygen in only 6 months and transforms its newly minted silicon into iron group nuclei in an astonishing 1 day.

These timescales are set by the amount of fuel available, the energy per reaction derived from burning the fuel, and the rate of energy loss from the star (which ultimately governs the burning rate). The last factor is particularly important because energy losses are

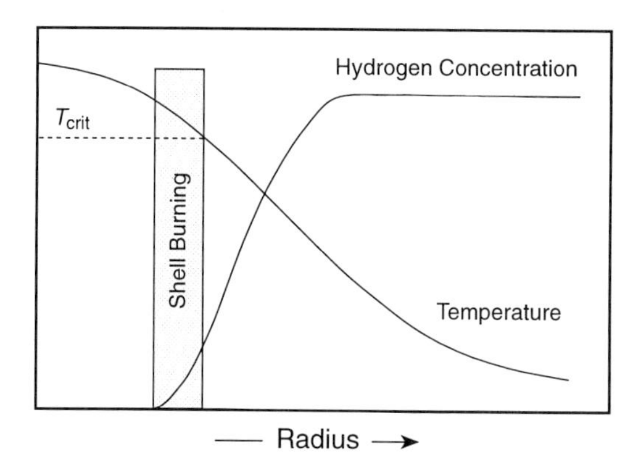

FIGURE 5. Conditions for hydrogen shell burning.

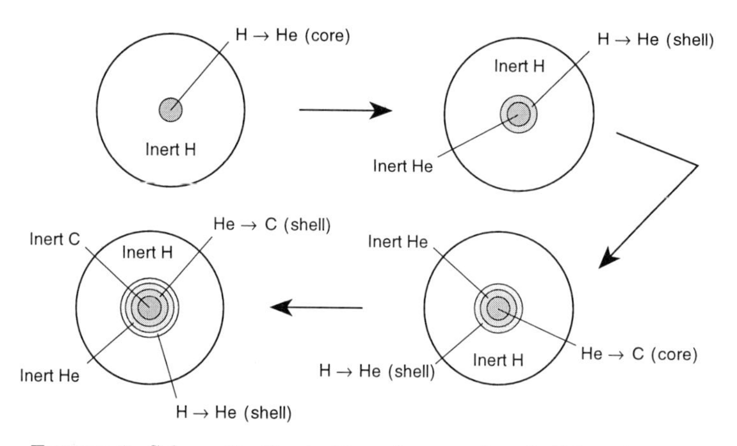

FIGURE 6. Schematic illustration of successive shell burnings.

large when the reaction must run at high temperature. Each factor separately shortens the timescale for advanced burning; taken together they make the times for the most advanced burning almost instantaneous on the scale set by the hydrogen burning.

To get a perspective, if the lifetime of the $25M_\odot$ star of Table 2 were compressed to a single year, the hydrogen fuel would be gone by around December 7, the helium would burn over the next 24 days, the carbon would burn in the 42 minutes before midnight, the neon and oxygen would burn in the last several seconds before midnight, and the silicon would be converted to iron in the last 1/100 second of the year (with a quite impressive New Year's Eve fireworks display in the offing—see the subsequent discussion!).

7.6. *Shell Burning*

An important aspect of advanced burning in post main-sequence evolution is the establishment of *shell burning sources*, as illustrated in Fig. 5 and Fig. 6. As the initial core hydrogen is depleted, a thermonuclear ash of helium builds up in its place. This ash is inert at hydrogen fusion temperatures because much higher temperatures and densities are necessary to initiate helium fusion. However, as the core becomes depleted in hydrogen there may remain a concentric shell in which the hydrogen concentration and the temperature are both sufficiently high to support hydrogen fusion (Fig. 5). This is termed a hydrogen shell source (Fig. 6).

As the core contracts after exhausting its hydrogen fuel the temperature and density rise and this may eventually ignite helium in the core. As the helium burns in the core a central ash of carbon is left behind that is inert because much higher temperatures are required to fuse it to heavier elements. This is termed core helium burning. However, similar to the case of hydrogen, once sufficient carbon ash has accumulated in the core the helium burning will be confined to a concentric shell surrounding the inert core; this is termed a helium shell source. If the star is sufficiently massive, the preceding scenario may be repeated for successively heavier core and shell sources, as illustrated in Fig. 6. For more massive stars there may exist at any particular time only a core source, only a shell source, or a core source and one or more shell sources burning simultaneously. In the schematic example illustrated in Fig. 6, the third step involves both a helium core source and a hydrogen shell source.

8. Thermal Equilibrium.

The condition for thermal equilibrium in a star is

$$\frac{dL(r)}{dr} = 4\pi r^2 \epsilon(r)\rho(r) \tag{8.35}$$

where $L(r)$ is the energy flux through a shell at radius r and $\epsilon(r)$ is the energy released per gram per second in the region enclosed by r. For stars, there are three ways to transport energy in establishing thermal equilibrium:

(*a*) *Conduction* is not important in normal stars because the mean free path of electrons and ions is short compared with the stellar radius. In very dense degenerate matter the mean free paths become much longer and conduction may become important as a means of transporting energy.

(*b*) *Radiation* plays a central role in stellar energy transport. We include in this category the transport of energy both by photons and by (massless or nearly massless) neutrinos and antineutrinos. In the sun, radiation of photons is the dominant energy transport mechanism below about 100,000 km from the surface.

(*c*) *Convection,* which involves the vertical motion of macroscopic regions of a fluid, is an important means of energy transport in many stars. The region just below the surface of the sun is dominated by convective energy transport.

In all three of these modes of energy transport, the radial energy flux in the star is primarily determined by the temperature gradient $\Delta T/\Delta r$, unless the mean free path of the radiation is much larger than the radius of the star. In non-degenerate stellar matter, radiative transport generally dominates energy transfer unless the temperature gradient (or equivalently the opacity of the stellar material) is large. Then energy transfer is dominated by convection.

9. Energy Transport in Stars

Both conduction and radiative transport may be viewed as resulting from random thermal motion of constituent particles (electrons in the first case and photons in the second), while convection is a macroscopic or collective phenomenon.

9.1. *Diffusion of Energy*

Consider the small cube illustrated in Fig. 7, for which we introduce a random velocity distribution with a small temperature gradient in the x direction. Therefore, let us assume that 1/6 of the particles move in the x direction with average velocity $\langle v \rangle$ and

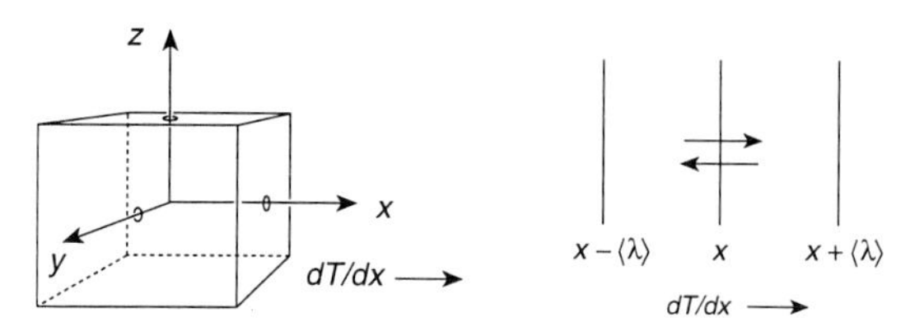

FIGURE 7. Diffusion of energy.

average mean free path $\langle\lambda\rangle$. Let $u(x)$ be the thermal energy per unit volume; because of the temperature gradient, particles crossing a plane at x from left to right have a different thermal energy than those crossing from right to left (see Fig. 7). Thus, energy is transported across the surface by virtue of the temperature gradient. This transport is associated with a current

$$\begin{aligned} j(x) &\approx \tfrac{1}{6}\langle v\rangle u(x-\langle\lambda\rangle) - \tfrac{1}{6}\langle v\rangle u(x+\langle\lambda\rangle) \\ &\approx -\tfrac{1}{3}\langle v\rangle\langle\lambda\rangle\frac{du}{dx} = -\tfrac{1}{3}\langle v\rangle\langle\lambda\rangle\frac{du}{dT}\frac{dT}{dx} \approx -\tfrac{1}{3}\langle v\rangle\langle\lambda\rangle C\frac{dT}{dx}, \end{aligned}$$

where the heat capacity $C = du/dT$ has been used. Thus, the current across the surface may be written as

$$j(x) = -K\frac{dT}{dx}, \qquad K \equiv \tfrac{1}{3}\langle v\rangle\langle\lambda\rangle C. \tag{9.36}$$

where K is termed the *coefficient of thermal conductivity.* Equation (9.36) is characteristic of diffusive processes. As noted above, in normal stars heat conduction by either electrons or ions is negligible compared with radiative transport. However, degenerate matter behaves much like a metal and transport of energy by conduction becomes important in stars containing such matter.

9.2. *Radiative Energy Transport by Photons*

Assuming stars to be blackbody radiators, the photons constitute a gas with

$$\langle v\rangle = c \qquad u = aT^4 \qquad C = \frac{du}{dT} = 4aT^3. \tag{9.37}$$

Thus for radiative diffusion we may write

$$j(x) = -K_{\mathrm{r}}\frac{dT}{dx} \qquad K_{\mathrm{r}} \equiv \tfrac{4}{3}c\langle\lambda\rangle aT^3, \tag{9.38}$$

where K_{r} is the *coefficient of radiative diffusion.* The mean free path λ may also be expressed as

$$\langle\lambda\rangle = \frac{1}{\rho\kappa}, \tag{9.39}$$

where κ is termed the *opacity* and clearly has units of area over mass.

9.3. *Energy Transport by Convection*

I some situations, the energy to be transported in a stellar interior is too large to be efficiently carried by radiative transport or conduction. In those instances the system can become unstable to macroscopic overturn of material; this process is called *convection.* Because convection is a collective process that moves entire blobs of material up and down in the gravitational field, it can transport energy very efficiently. In the next section we consider the conditions that can lead to convective instability in a fluid.

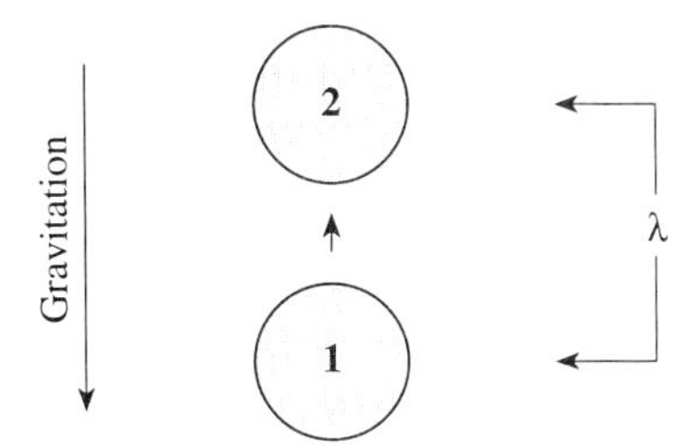

FIGURE 8. Convective motion.

Schwarzchild Instability

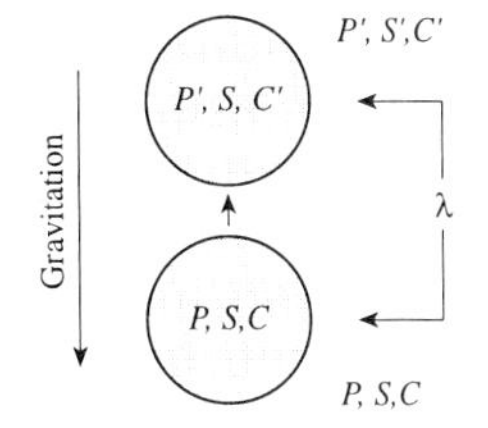

The blob moves adiabatically, but in both pressure and composition equilibrium with surroundings.

The region is unstable if $\rho(P', S', C') - \rho(P', S, C') \geq 0$

FIGURE 9. Convective instability by the Schwarzschild criterion.

10. Conditions for Convective Instability

Consider the following thought experiment [Wilson & Mayle (1988)]: imagine a blob of matter in a fluid (in a gravitational field) which moves upward a distance λ from position 1 to position 2 because of some infinitesimal stimulus (see Fig. 8). If the blob of material at position 2 is less dense than the surrounding material, it will be driven upward by buoyancy forces and the region is said to be convectively unstable.

10.1. *Schwarzschild Instability*

Let us suppose that the blob moves adiabatically (constant entropy), but in pressure and composition equilibrium with the surrounding medium. Denote the pressure, entropy, and composition of the medium at position 1 by P, S, and C, and at position 2 by the corresponding variables with a prime, as illustrated in Fig. 9. The condition for convective instability is $\rho(P', S', C') - \rho(P', S, C') \geq 0$. Expanding in a Taylor series,

$$\rho(P', S', C') - \rho(P', S, C') = \left.\frac{\partial \rho}{\partial S}\right|_{P,C} \lambda \frac{dS}{dr}. \tag{10.40}$$

By introducing the specific heat at constant pressure $C_{\rm p}$, we may exchange the entropy S for the temperature T as a variable and obtain

$$\frac{T}{C_{\rm p}} \left.\frac{\partial \rho}{\partial T}\right|_{P,C} \lambda \frac{dS}{dr} \geq 0. \tag{10.41}$$

For normal equations of state the partial derivative is negative and Eq. (10.41) is equivalent to the *Schwarzschild condition* for convective instability,

$$dS/dr \leq 0 \qquad \text{(Schwarzschild condition)}. \tag{10.42}$$

Thus, a region is unstable against Schwarzschild convection if there is a *negative entropy gradient,* as illustrated schematically in Fig. 10.

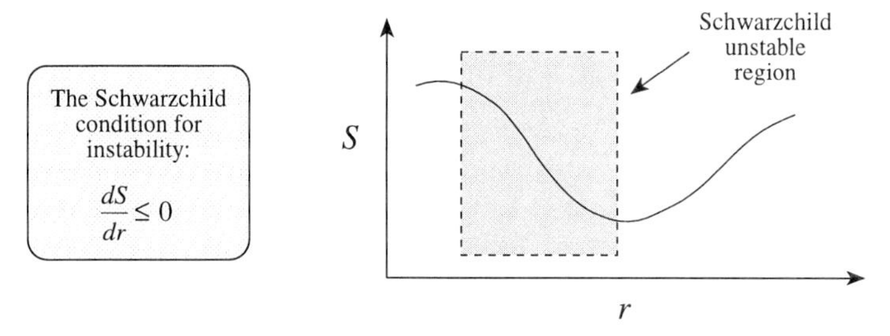

FIGURE 10. A Schwarzchild unstable region.

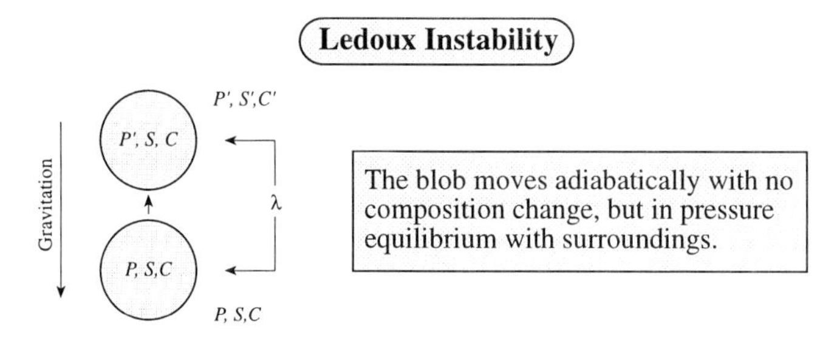

FIGURE 11. Convective instability according to the Ledoux criterion.

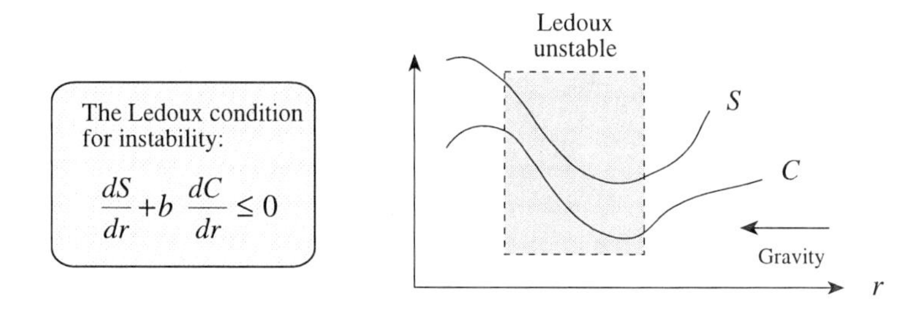

FIGURE 12. A Ledoux unstable region.

10.2. *Ledoux Instability*

Now suppose that the blob moves adiabatically with no composition change, but in pressure equilibrium, as illustrated in Fig. 11. The condition for convective instability may be expressed in this case as

$$\frac{T}{C_{\mathrm{p}}} \frac{\partial \rho}{\partial T}\bigg|_{P,C} \lambda \frac{dS}{dr} + \frac{\partial \rho}{\partial C}\bigg|_{P,S} \lambda \frac{dC}{dr} \geq 0, \tag{10.43}$$

where the first term is the same as for the Schwarzschild instability and the second term arises because of the assumption that there is no composition change. For most cases of interest, both partial derivatives are negative and the Ledoux condition for instability is

$$\frac{dS}{dr} + k\frac{dC}{dr} \leq 0 \qquad \text{(Ledoux condition)} \tag{10.44}$$

where k is a positive constant. Thus, a region is unstable against Ledoux convection if both the entropy and the concentration variable have a negative gradient, as illustrated schematically in Fig. 12. If the entropy and concentration gradients have opposite signs, the stability depends on the relative sizes of the two terms in Eq. (10.44).

11. Critical Temperature Gradient for Convection

For the structure and evolution of normal stars, the most important mode of convective instability is typically that set by the Schwarzchild condition (10.42) and driven by entropy gradients. The instability criterion for Schwarzchild convection may also be expressed in terms of a critical temperature gradient

$$\frac{dT}{dr} < \left(1 - \frac{1}{\gamma}\right) \frac{T}{P} \frac{dP}{dr}, \tag{11.45}$$

where γ is the adiabatic index and where generally both derivatives are negative. Dividing both sides by dT/dr, we may also express the instability condition (11.45) in the form

$$\frac{d \ln P}{d \ln T} < \frac{\gamma}{\gamma - 1}. \tag{11.46}$$

Thus, convection requires the temperature to fall off sufficiently fast with height that the actual temperature gradient satisfies Eq. (11.45). The two most important factors governing the behavior of the right side of (11.45) are the adiabatic index γ and the pressure gradient dP/dr.

11.1. *Role of the Adiabatic Index in Convection*

For an ideal gas the adiabatic index is $\gamma = C_P/C_V = [1 + s/2)]/(s/2)$, where s is the number of classical degrees of freedom per particle, each carrying average thermal energy $E = \frac{1}{2}kT$. Therefore, for a monatomic gas with only 3 translational degrees of freedom the adiabatic index is $\gamma = (1 + 3/2)/(3/2) = \frac{5}{3}$, and the condition (11.46) for convective instability is that $d(\ln P)/d(\ln T) < 2.5$. But if the gas has additional degrees of freedom the adiabatic index will decrease and for sufficiently many degrees of freedom γ will approach unity. Notice that as $\gamma \to 1$ the factor $(\gamma - 1)/\gamma \to 0$ and the critical temperature gradient of Eq. (11.45) becomes less steep. Thus, a decrease in the adiabatic index γ, which for example can result from an increase in the effective degrees of freedom for a gas, will generally enhance the possibility of convective instability. Since nonrelativistic and ultrarelativistic ideal gases have γ values of $\frac{5}{3}$ and $\frac{4}{3}$, respectively, more relativistic gases will tend to have adiabatic indices lower than corresponding nonrelativistic gases.

11.2. *Role of the Pressure Gradient in Convection*

In hydrostatic equilibrium, the pressure gradient is given by Eq. (2.1). Therefore, pressure falls off more gradually where $g(r)$ is small and the smaller values of dP/dr make the condition (11.45) easier to satisfy, thereby favoring convective instability. Regions with weak local gravity will, other things being equal, be more susceptible to convective instabilities.

12. Stellar Temperature Gradients

The condition (11.45) defines a critical temperature gradient for convective instability. Therefore, we must investigate the actual temperature gradients of stars in order to assess their stability against convection. Generally, stars will choose from among the available modes of energy transport the one that leads to the smallest temperature gradient and largest luminosity. Furthermore, from previous considerations we may assume that the temperature gradients of normal stars that are not convective are determined by the rate of radiative energy transport.

Let $L(r)$ denote the rate of energy flow through a shell of thickness dr at a radius r, and let $\epsilon(r)$ denote the nuclear power per unit volume generated at radius r. Then the power generated in the shell of thickness dr at radius r is given by $4\pi r^2 \epsilon(r) dr$. This is added to the outward power flow from interior shells and be may write

$$\frac{dL}{dr} = 4\pi r^2 \epsilon(r). \tag{12.47}$$

Outside the central power generating regions we may expect that $L(r)$ becomes constant and equal to the surface luminosity of the star. If we assume radiative energy flow,

$$L(r) = 4\pi r^2 j(r), \tag{12.48}$$

where for radiative transport

$$j(r) = -\frac{4ac}{3} \frac{T^3}{\rho\kappa} \frac{dT}{dr}, \tag{12.49}$$

which is (9.38) rewritten in terms of the opacity κ using (9.39). Inserting (12.49) in (12.48), we may solve for the temperature gradient associated with radiative diffusion,

$$\left(\frac{dT}{dr}\right)_{\rm rad} = -\frac{3\rho(r)\kappa(r)}{4acT^3(r)} \frac{L(r)}{4\pi r^2} \tag{12.50}$$

If this gradient becomes steeper than the critical gradient defined by Eq. (11.45), the system will become convectively unstable.

13. Examples of Convective Regions in Normal Stars

Because it is efficient, convection will normally dominate radiative transport as soon as the critical temperature gradient (11.45) is reached in a region of a star. Generally, it is believed that the most massive stars are centrally convective and radiative in the outer envelope, that stars of a solar mass or so have subsurface convection zones but the central region is not convective, and that in the least massive stars essentially the entire star is convective. Figure 13 illustrates. To get a feeling for the nature of convection in normal stars as preparation for our discussion of convection in supernovae, let us examine two such convective regions in more detail: (1) ionization zones in the surface layers of stars, and (2) the cores of massive main sequence stars.

13.1. *Surface Ionization Zones*

Convection is favored in the surface layers of stars where constant ionization and recombination transitions are taking place for two basic reasons: (1) the opacity is large, making the temperature gradient for radiative transport steep [see Eq. (12.50)], and (2) the critical convective temperature gradient is not very steep because there are many degrees of freedom associated with the ionization–recombination transitions and the adiabatic index is decreased toward unity. More physically, convection is favored because electron recombination can supply part of the energy to expand the rising cells of gas. Thus, they do not cool much and are more likely to remain buoyant. In the Sun, there is a convective layer from a distance of about $0.1R_\odot$ to $0.2R_\odot$ below the surface associated with such ionization zones that is responsible for the *granules* observed on the solar surface.

13.2. *Convection in the Cores of Stars*

Convection in stellar cores is favored if the power is generated in a compact central region: (1) there is a large energy flow through a region with small gravitational acceleration,

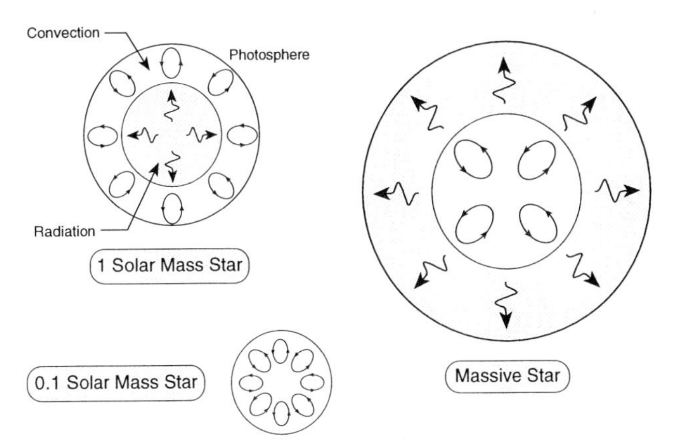

FIGURE 13. Schematic illustration of convective regions in stars of different mass.

and (2) because gravity is weak, pressure falls off gradually and rising gas tends to remain buoyant because it need not expand much. If we set the radiation temperature gradient (12.50) equal to the critical temperature gradient (11.45), introduce the radiation pressure $P_{\mathrm{r}} = aT^4/3$, and rearrange the resulting expression we obtain

$$\frac{L(r)}{m(r)} = \frac{16\pi Gc}{\kappa}\left(\frac{\gamma-1}{\gamma}\right)\left(\frac{P_{\mathrm{r}}}{P}\right). \tag{13.51}$$

This defines a critical value of $L(r)/m(r)$ favoring convection over radiative diffusion. Thus, we expect convective cores of radius r and enclosed mass $m(r)$ to develop in stars if the critical value of $L(r)/m(r)$ defined by Eq. (13.51) is exceeded. Such convective cores are common in main sequence stars more massive than the Sun where the CNO cycle is the dominant energy production mechanism and its strong temperature dependence (Power $\sim T^{17}$) confines power production to a small central region.

14. Critical Luminosities

The pressure due to photons grows as the fourth power of the temperature and thus will be most important for very hot stars. It is instructive to ask what radiation luminosity is required such that the magnitude of the force associated with the radiation field is equivalent to the magnitude of the gravitational force. For photons, this critical luminosity is termed the *Eddington Luminosity.*

14.1. *Eddington Luminosity*

The force per unit volume associated with a photon gas is given by the gradient of the radiation pressure,

$$\frac{1}{V}F_{\mathrm{r}} = -\frac{dP_{\mathrm{r}}}{dr} = \frac{4}{3}aT^3\frac{dT}{dr} \tag{14.52}$$

If the magnitude of this force is equated to the magnitude of the gravitational force, we obtain an expression for the Eddington luminosity

$$L_{\mathrm{Edd}} = \frac{4\pi cGM}{\kappa}. \tag{14.53}$$

We may expect that stars exceeding the luminosity (14.53) can eject surface layers by radiation pressure. Thus, the Eddington luminosity places an upper limit on the masses of stars.

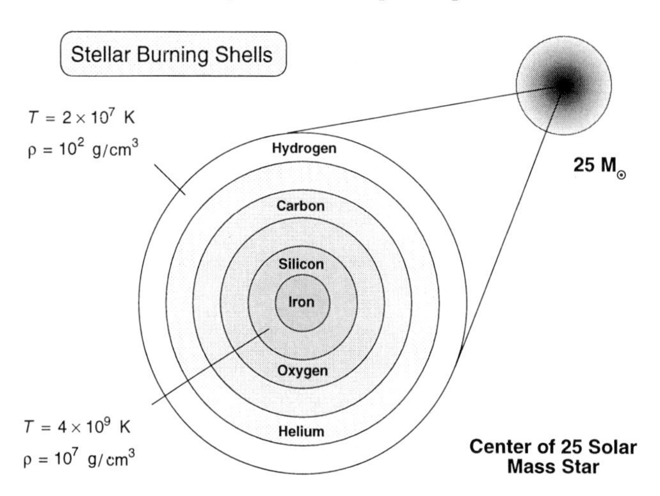

FIGURE 14. Center of a 25 solar mass star late in its life.

14.2. *Limiting Neutrino Luminosities*

A flux of neutrinos also exerts a pressure. By a similar derivation as for Eq. (14.53), we can find a limiting "Eddington luminosity" for neutrinos,

$$L_{\text{Edd}}(\nu) = \frac{4\pi c G M}{\kappa_\nu}. \tag{14.54}$$

where κ_ν is a mean opacity associated with neutrino scattering. In normal stars any neutrinos produced interact weakly so this is not very important. As we shall see below, the interaction of neutrinos with the core material of an exploding supernova may be crucial, both through the radiation pressure implied by Eq. (14.54), and (probably more important) through increases of the gas pressure induced by neutrino heating.

15. The Death of Massive Stars

Massive stars near the ends of their lives build up the layered structure depicted in Fig. 14 as a consequence of the advanced shell burning described in §7.6. The iron core cannot produce energy by fusion, so it must ultimately be supported by electron degeneracy pressure. As the silicon layer undergoes reactions the central iron core becomes more massive. Electron degeneracy can support the iron core against gravitational collapse only if its mass remains below the Chandrasekhar limit (4.27), which depends on the electron fraction but is approximately 1.4 solar masses for the typical case. When the iron core exceeds this critical mass it begins to collapse because the electron degeneracy pressure can no longer balance the gravitational forces. At the point where the collapse begins, the iron core of a 25 solar mass star has a mass of about 1.4 solar masses and a diameter of maybe several thousand kilometers.† The core density is about 6×10^9 g/cm^3, the core temperature is approximately 6×10^9 K, and the dimensionless entropy per baryon per Boltzmann constant of the core is about 1.

This entropy is remarkably low: the entropy of the original main-sequence star that produced this iron core is perhaps 15–20 in these units. At first glance, it may seem contradictory for the entropy to decrease as the star burns its fuel. However, the star is

† This is but a tiny fraction of the total diameter. The massive star is typically a supergiant by this time (see Fig. 2), with its tenuous outer layers spread over a volume that would encompass much of the inner solar system if it were placed at the position of the sun.

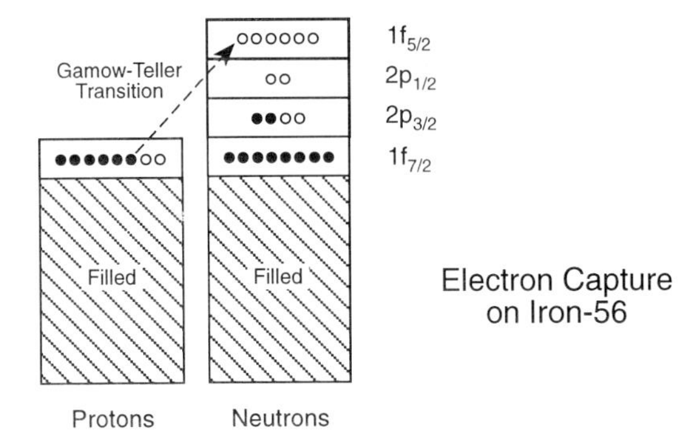

FIGURE 15. Electron capture in ^{56}Fe.

not a closed system: as nuclear fuel is consumed, energy leaves the star in the form of photons and neutrinos and the nucleons in the original main-sequence star are converted to iron nuclei. In ^{56}Fe, the 26 protons and 30 neutrons are highly ordered compared with 56 free nucleons in the original star. Thus, the core of the star becomes *more ordered* compared with the original star as the nuclear fuel is consumed. The entire universe becomes more disordered, as required by the second law, because the star radiates photons and neutrinos as it builds its ordered core.

16. Sequence of Events in Core Collapse

The collapse of the iron core as the Chandrasekhar limit is exceeded triggers a remarkable sequence of events that will occur in an elapsed time of less than a second.

(1) When the mass of the iron core exceeds about 1.4 solar masses, it begins to collapse under the influence of gravity. This collapse is rapidly accelerated by two factors that are accentuated as the temperature and density of the collapsing core rise:

(*a*) As the core heats up, high-energy γ-rays are produced in copious amounts. These photodisintegrate some of the iron-peak nuclei into α-particles. The corresponding process is highly endothermic; for example, the reaction $^{56}\mathrm{Fe} \longrightarrow 13\alpha + 4\mathrm{n}$ has a Q-value of -124.4 MeV. This decreases the kinetic energy of the particles in the core, which lowers the pressure and hastens the collapse.

(*b*) As the density and temperature increase in the core, the rate for the electron capture reaction $\mathrm{p}^+ + \mathrm{e}^- \rightarrow \mathrm{n} + \nu$ (see Fig. 15) is greatly enhanced. This reaction is called *neutronization;* because it converts protons and electrons into neutrons and neutrinos, it decreases the electron fraction Y_e of the core and thus the (dominant) electronic contribution to the pressure. The neutrinos produced in this reaction easily escape the core during the initial phases of the collapse because their mean free path is much larger than the initial radius of the core. These neutrinos carry energy with them, decreasing the core pressure and accelerating the collapse even further.

(2) The accelerated core collapse proceeds on a timescale of milliseconds, with velocities that are significant fractions of the free-fall velocities (see §6.2). The core separates (conceptually) into an *inner core* that collapses subsonically and homologously,† and an *outer core* that collapses in approximate free fall with a velocity exceeding the local velocity of sound in the medium (it is supersonic)—see the left side of Fig. 17 below.

† Subsonic means that the velocities are below the local speed of sound in the medium. Homologous means that the collapse is "self-similar", in that it can be described by changing a scale factor.

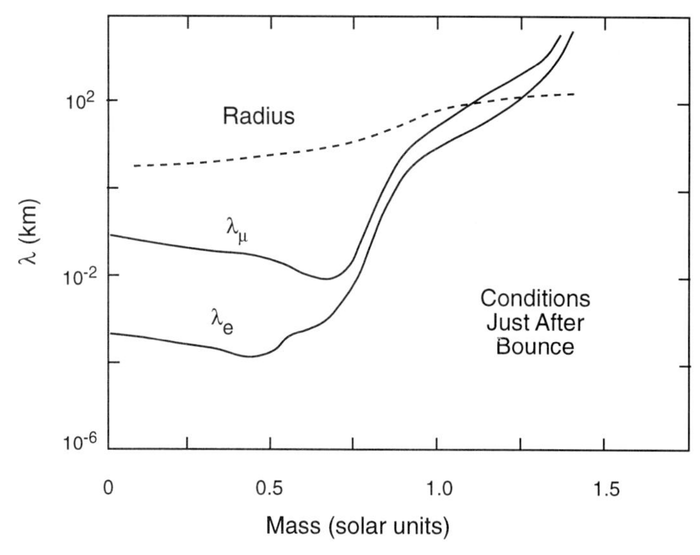

FIGURE 16. Mean free path for neutrinos shortly after core bounce (Burrows & Lattimer).

(3) This collapse is rapid on the timescales characteristic of most stellar evolution, but it is slow compared with the reaction rates and the core is approximately in equilibrium during all phases of the collapse. This implies that the *entropy is constant*, and the highly ordered iron core before collapse ($S \approx 1$) remains ordered during the collapse, with the entropy during the collapse shared approximately equally between the nuclei and nearly degenerate electrons.

(4) As the collapse proceeds and the temperature and density rise, a point is reached where the neutrino interactions become sufficiently strong that the mean free path of the neutrinos becomes less than the radius of the core (see §17.1). Shortly thereafter, the time for neutrinos to diffuse outward becomes longer than the characteristic time of the collapse and the neutrinos are effectively trapped in the collapsing core (see Fig. 16, where we see that neutrino mean free paths in the collapsed core can become as short as a fraction of a meter). The radius at which this occurs is termed the *trapping radius*, and it is closely related to the neutrinosphere defined below.

(5) Because the collapse proceeds with low entropy, there is little nuclear excitation and nucleons remain in nuclei until densities where nuclei begin to touch [Bethe, et al (1979)]. At this stage of the collapse, the collapsing core begins to resemble a "macroscopic nucleus".

(6) This core of extended nuclear matter is a nearly degenerate fermi gas and has a very stiff equation of state (nuclear matter is highly incompressible). At this point, the pressure of the nucleons begins to dominate that from the electrons and neutrinos.

(7) Somewhat beyond nuclear density, the incompressible core of nearly degenerate nuclear matter rebounds violently as a pressure wave reflects from the center of the star and proceeds outward. This wave steepens into a shock wave as it moves outward through material of decreasing density, with the shock wave forming near the boundary between the subsonic inner core and supersonic outer core (this point is called the *sonic point;* see Fig. 17). In the simplest picture this shock wave would eject the outer layers of the star, resulting in a supernova explosion. This is called the *prompt shock mechanism.* The gravitational binding energy of the core at rebound is about 10^{53} ergs, and the

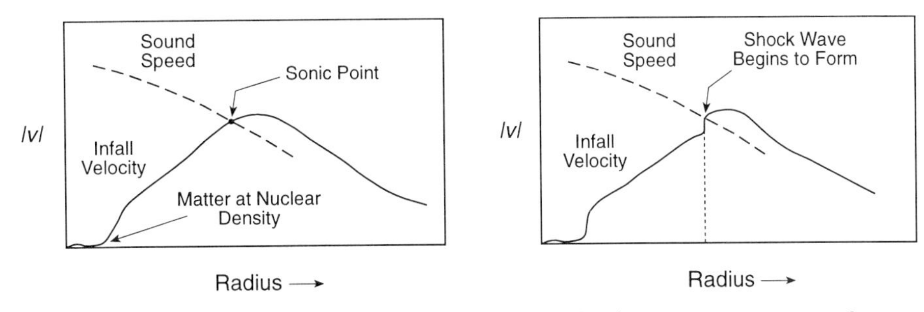

FIGURE 17. The sonic point during collapse (left) and the beginning of shock-wave formation following the bounce (right).

typical observed energy of a supernova (the expanding remnant plus photons) is about 10^{51} ergs;† thus, only about 1% of the gravitational energy need be released in the form of light and kinetic energy to account for the observed properties of supernovae.

(8) Unfortunately, a simple prompt-shock mechanism runs afoul of the details: realistic calculations suggest that the prompt shock dissipates energy rapidly as it progresses through the outer core. This energy loss derives from two primary sources:

(*a*) The shock wave dissociates Fe nuclei as it passes through the outer core, and this highly endothermic reaction saps it of a large amount of energy.

(*b*) As the shock wave passes into increasingly less dense material, the local mean free path for the trapped neutrinos increases until the neutrinos are again freely radiated from the core. The radius at which the neutrinos change from diffusive to radiative behavior is termed the *neutrinosphere* (by analogy with the photosphere of a regular star; see Fig. 18. More precisely, the neutrinosphere is the radius at which the neutrinos have an optical depth of $\frac{2}{3}$)). When the shock wave penetrates the neutrinosphere, a burst of neutrinos is emitted, carrying with it large amounts of energy (most of the energy released in the gravitational collapse resides in the neutrinos). This further deprives the shock of its vitality by lowering the pressure behind it, and also depletes the lepton fraction near the neutrinosphere.

The most realistic calculations indicate that the shock wave stalls into an *accretion shock* (a standing shock wave at a constant radius) before it can exit the core, unless the original iron core contains less than about 1.1 solar masses. In a typical calculation, the accretion shock forms at about 100–300 km from the center of the star within about 10 ms of core bounce. Figure 19 illustrates the conditions characteristic of a stalled accretion shock in a modern calculation of shock propagation in a core-collapse supernova. Since there is considerable agreement that SN1987A resulted from the collapse of a core having 1.3–1.4 solar masses, the prompt shock mechanism is not viable as a generic explanation of type II supernovae.

17. Neutrino Reheating

The idea that neutrinos might play a significant role in supplying energy and momentum to eject the outer layers of a star in a supernova event is an old one (see §14.2), but the failure of the prompt mechanism to consistently lead to supernovae led to a revival of interest in such mechanisms [Wilson(1985), Bethe & Wilson (1985]. This evolved into what is generally termed the *delayed shock* or *neutrino reheating mechanism*. In this

† This defines a unit of energy commonly used in supernova physics that is termed the *foe:* 1 foe $\equiv 10^{51}$ ergs, with the name foe deriving from the first letters of fifty-one ergs.

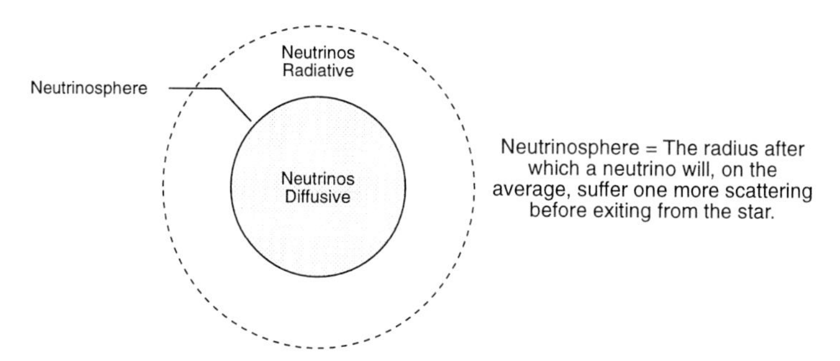

FIGURE 18. The neutrinosphere.

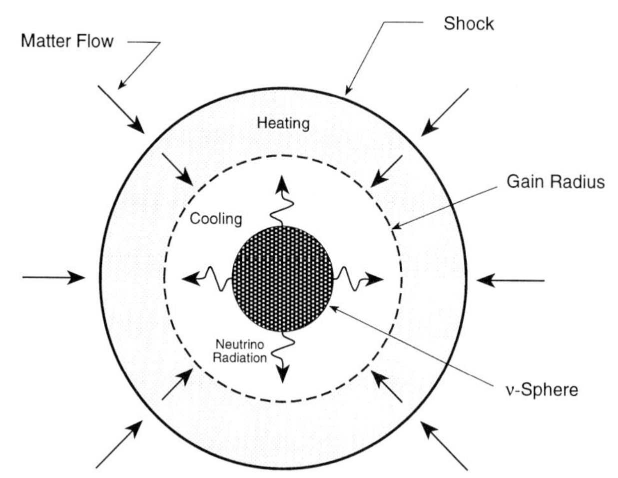

FIGURE 19. Conditions at time of shock stagnation in a core-collapse supernova. The neutrinosphere and the gain radius (defined in §17.2) are indicated.

picture, the stalled accretion shock is re-energized by heating of matter behind the shock through interactions with neutrinos produced in the region interior to the shock. This raises the pressure sufficently to impart an outward velocity to the stalled shock on a timescale of several hundred milliseconds, and the reborn shock then proceeds through the outer envelope of the star. The schematic mechanism for the supernova event thus becomes the two-stage process depicted in Fig. 20.

17.1. *Neutrino Interactions*

Neutrinos and antineutrinos are produced copiously in the hot, dense region near the center of the collapsed core. Expected luminosities for electron neutrinos, and for neutrinos and antineutrinos of all flavors, are shown in Fig. 21. The basic interactions of the neutrinos and antineutrinos with leptons and with nuclei are illustrated in terms of Feynman diagrams in Fig. 22. These may be divided into neutral-current interactions involving the Z boson and charged-current interactions mediated by the $W^{\pm}$ bosons, and may be further classified according to whether hadrons are involved. Generally, the interactions involving nucleons (either free, or bound in nuclei) have larger cross sections and are the most important contributions to neutrino opacity. Of the interactions shown in Fig. 22, all but the last neutral current interaction with nucleons or nuclei (diagram g) involve momentum and energy exchange. Diagram g, for kinematic reasons, exchanges momen-

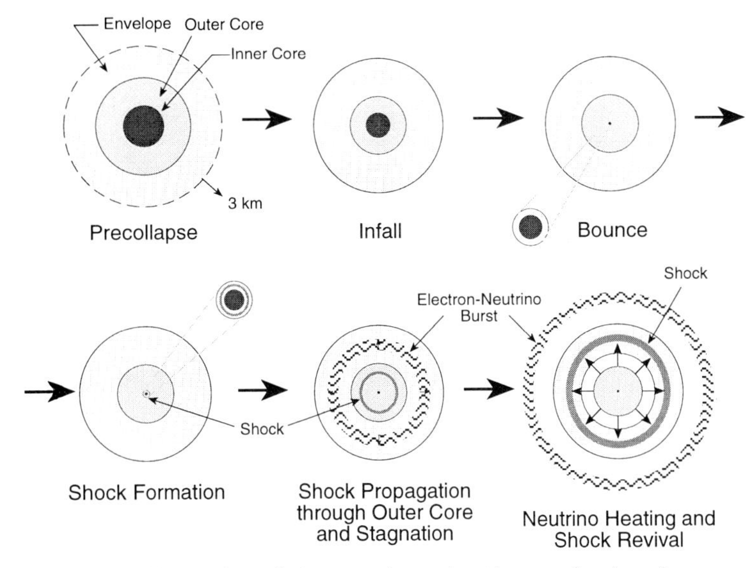

FIGURE 20. Illustration of the neutrino reheating mechanism for a supernova explosion [after Bruenn(1993)]. Figures are approximately to scale, except that the surface of the star would lie some 3 km from the center if represented on this scale.

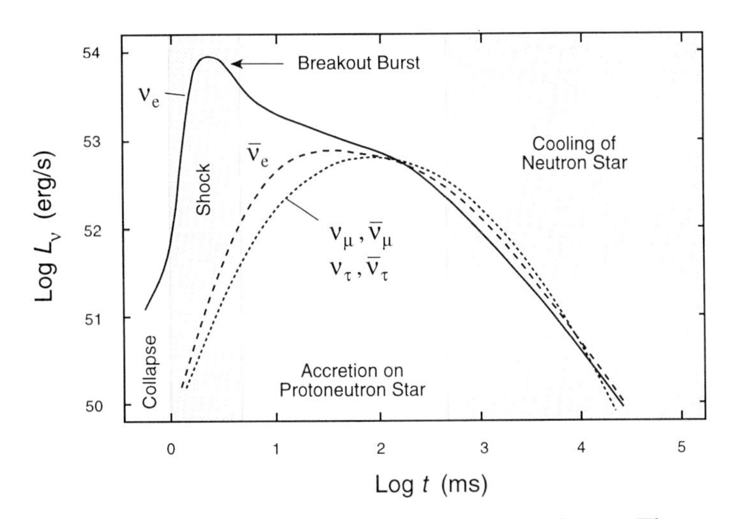

FIGURE 21. Neutrino luminosities in a supernova explosion. The zero of the time scale is at core bounce.

tum but little energy. Neutrino interactions relevant for the supernova are discussed in Bruenn (1985), Bruenn & Haxton (1991), Cooperstein (1988), and Myra (1988).

Calculations indicate that in the supernova core, *coherent scattering* of long wavelength neutrinos from nuclei through the neutral-current diagram g is the dominant interaction of the neutrinos [Freedman (1974), Freedman, Schramm, and Tubbs (1977)]. If we assume that the mean free path λ of the neutrinos in the core is determined by such coherent scattering,

$$\lambda = \frac{10 \text{ km}}{\rho_{12}(\epsilon_\nu/10)^2(N^2/6A)}, \tag{17.55}$$

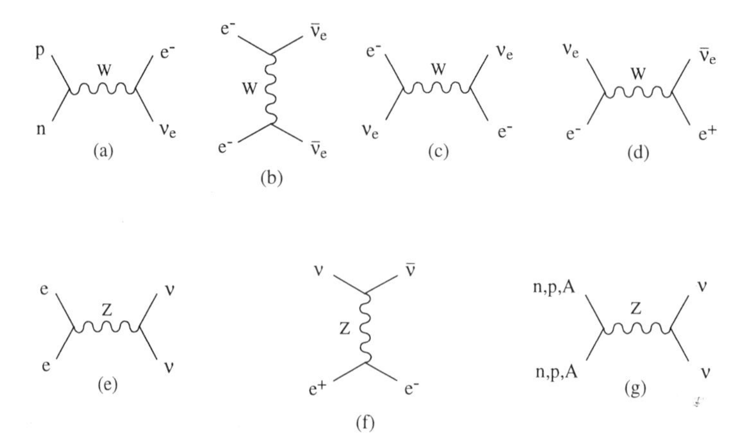

FIGURE 22. Neutrino interactions relevant in collapsed supernova cores.

where ρ_{12} is the density in units of 10^{12} g/cm^3, ϵ_ν is the neutrino energy, N is the neutron number, and A is the atomic mass number of the heavy nucleus. For a random walk in a uniform sphere of radius R, the neutrino diffusion time is

$$t_{\text{diff}} = \frac{3}{\pi^2}\left(\frac{R}{\lambda}\right)^2 \frac{\lambda}{c}. \tag{17.56}$$

If the sphere collapses with a fraction f of the free-fall velocity, the time to shrink to a singular point defines a dynamical time scale

$$t_{\text{dyn}} = \left(f\sqrt{6\pi G\rho}\right)^{-1}. \tag{17.57}$$

The ratio of these timescales is

$$\frac{t_{\text{diff}}}{t_{\text{dyn}}} = \frac{7 \text{ km}}{\lambda} f M^{2/3} \rho_{12}^{-1/6}, \tag{17.58}$$

where M is the mass in units of $M_\odot$. The ratio (17.58) becomes unity for $\rho \simeq 10^{12}$ g/cm$^3 \equiv \rho_{12}$, which defines a trapping density for neutrinos: above this density, the rate of outward diffusion of the neutrinos is slower than the rate of collapse, and the neutrinos are effectively trapped in the imploding core [Cooperstein (1988), Myra (1988)].

17.2. *Reheating of Shocked Matter*

In the post-bounce phase, the neutrinos can impart energy to the shocked matter behind the stalled shock in three ways:

(*a*) The electron neutrinos and antineutrinos can undergo charged current interactions with nuclei and free nucleons, $e^+ + p \leftrightarrow n + \nu_e$ and $e^+ + n \leftrightarrow p + \bar{\nu}_e$.

(*b*) Each flavor of neutrino can undergo the charged and neutral current interactions $\nu + e \rightarrow \nu + e$.

(*c*) Each flavor of neutrino can participate in the pair interaction $\nu + \bar{\nu} \leftrightarrow e^+ + e^-$. For the conditions found in supernova cores, this latter interaction is typically less important in the collapse phase.

By these types of interactions, the neutrinos that are produced in the core can interact with the matter behind the shock wave. In discussing the details of that interaction, it is useful to introduce two characteristic radii. The first we have already described in conjunction with Fig. 18: The *neutrinosphere* is a sphere defined by a radius beyond which an average neutrino will suffer one more scattering before it leaves the star (it lies at optical depth $\frac{2}{3}$). The second is associated with the observation that the neutrino interactions with the shocked matter could cool the matter rather than heat it. General considerations suggest that there is always a radius outside of which the net effect of the neutrino interactions is to heat the matter. This break-even radius, beyond which the neutrino interactions become effective in increasing the pressure behind the shock is termed the *gain radius*. These radii are indicated schematically in Fig. 19.

17.3. *Calculations with Neutrino Reheating*

The result of a large number of calculations is that the neutrino reheating helps, and can often turn a failed supernova into a supernova. However, such calculations do not succeed without artificial boosts of the neutrino luminosities. These persistent results suggest that there still are missing ingredients in the supernova mechanism that must be included to obtain a quantitative description. One suggestion, which is not new but has until recently been avoided because of the computational complexity that it implies, is that convection in the region interior to the shock wave alters the neutrino reheating in a non-negligible fashion. Another suggestion, also not new but not yet fully implemented because of the computational difficulty that it presents is that a more complete treatment of neutrino transport in the supernova core is required. Let us now turn to a general discussion of what role neutrino transport and convection might play in supernova explosions.

18. Neutrino Transport

Neutrinos are central to the supernova mechanism and they generally are not in equilibrium. Thus, we must use a transport theory [Mihalas & Mihalas (1984)] to describe their propagation and influence. The full solution of the corresponding Boltzmann equations is a large computational task that has not been completed, even for spherical symmetry [Mezzacappa(1993), Bruenn(1993), Mezzacappa & Bruenn (1993)].

We shall return to the issue of Boltzmann transport in the final section, but all complete calculations to date have (probably rather severely) approximated the neutrino transport. The most ambitious approximation scheme employed thus far is termed *Multi-Group Flux-Limited Diffusion* (MGFLD), which may be viewed as interpolating semi-empirically between diffusion and free-streaming regimes for the neutrinos. Since MGFLD is the method of neutrino transport to be employed in the calculations that I will describe, let us briefly outline the MGFLD approximation [Bruenn (1985)].

18.1. *The Diffusion Approximation*

In optically thick regions a particle has a mean free path λ that is short with respect to a characteristic length scale ℓ_0 for the physical system: $\lambda << \ell_0$. The number of collisions over a distance ℓ is then approximately given by the value $(\ell/\lambda)^2$ expected for a random walk. The average time between collisions is then $\tau = \lambda/c$, if we assume the particles to travel at the speed of light c, and thus the characteristic timescale τ_d to diffuse a distance ℓ is

$$\tau_\mathrm{d} \approx \left(\frac{\lambda}{c}\right) \times \left(\frac{\ell^2}{\lambda^2}\right) = \frac{\ell^2}{c\lambda}. \tag{18.59}$$

In a *diffusion approximation* the particle is assumed to *always* travel a distance λ between collisions. One obvious problem with this assumption is that it can lead to violations of causality.

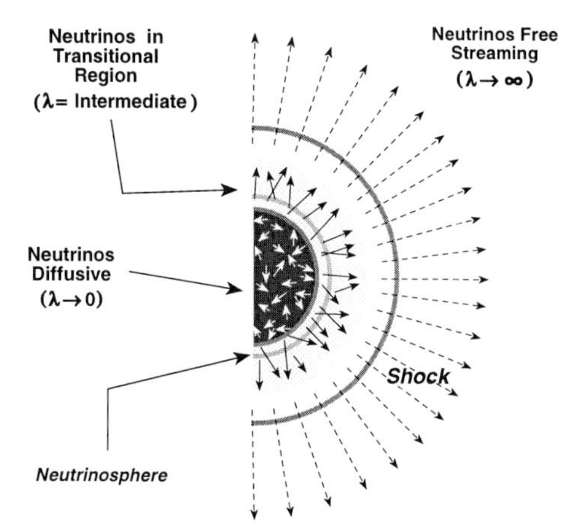

FIGURE 23. Schematic illustation of diffusive, intermediate transparency, and free-streaming regions for neutrino transport.

18.2. *Causality Violation*

In optically thin regimes $\lambda \to \infty$, but this implies that eventually we can reach a point where $\lambda > c\Delta t$, where Δt is a characteristic timestep in the solution of the problem. But this implies a violation of causality because flux from an optically thin region can then exceed the value

$$|\boldsymbol{F}_{\text{max}}| = \text{speed of light} \times \text{energy density} = cE, \tag{18.60}$$

which is the flux carried by particles streaming freely with $v = c$.

18.3. *Flux Limiting*

We may expect that the range of validity for a diffusion approximation might be extended if we prevented the approximation from violating causality. In particular, we could accomplish this by modifying the diffusion formula (18.59) such that it yields (1) the usual diffusion formula

$$\boldsymbol{F}_{\text{diff}} = -\tfrac{1}{3}c\lambda\boldsymbol{\nabla}E \tag{18.61}$$

for high opacity (optically thick) regions, and (2) the free-streaming limit

$$\boldsymbol{F}_{\text{free}} = cE\mathbf{n} \tag{18.62}$$

in optically thin regions ($\boldsymbol{n}$ is a unit vector directed opposite $\boldsymbol{\nabla}E$). For example,

$$\begin{aligned}\boldsymbol{F} = -c\left(\frac{\boldsymbol{\nabla}E}{3/\lambda + |\boldsymbol{\nabla}E|/E}\right) &= \left(\frac{-c\,\boldsymbol{\nabla}E}{3/\lambda - \boldsymbol{\nabla}E/E\boldsymbol{n}}\right) \\ \longrightarrow &\begin{cases} -\tfrac{1}{3}c\lambda\boldsymbol{\nabla}E & \text{diffusion } (\lambda \text{ small}) \\ cE\boldsymbol{n} & \text{free-streaming } (\lambda \to \infty) \end{cases}\end{aligned} \tag{18.63}$$

preserves causality and is a simple example of a *flux limiter*. Many other flux limiters are possible since various expressions could reduce to the two limits of the preceding equation. In most cases flux limiters are (1) empirical and (2) not unique.

18.4. *Multigroup Diffusion*

In multigroup diffusion particles are binned into energy groups and the diffusion is calculated separately for each energy group. This can be important for neutrinos because

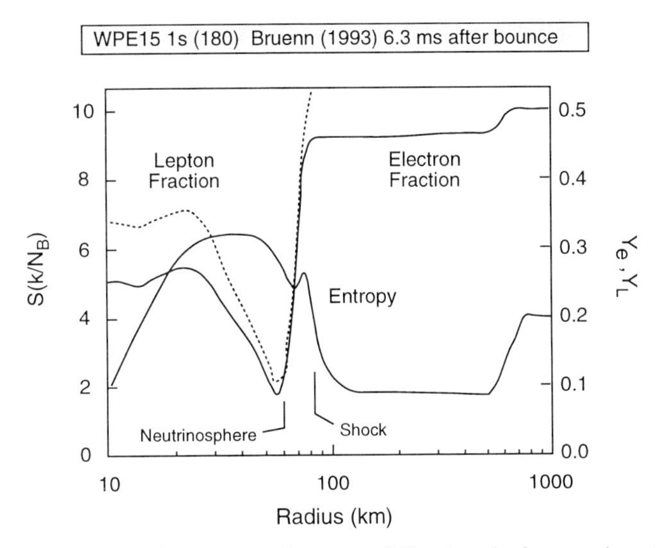

FIGURE 24. Lepton fractions and entropy following the bounce in a typical supernova calculation. Regions that may be expected to be particularly favorable for convective motion are marked. The progenitor had a mass of 15 $M_\odot$, and the calculation is described in Bruenn(1993).

generally the weak interaction cross sections scale quadratically with energy and the the mean free path is $\lambda_\nu \approx 1/\epsilon_\nu^2$.

18.5. *Multigroup Flux-Limited Diffusion*

To summarize, Multi-Group Flux-Limited Diffusion (MGFLD) is an approximation to full Boltzmann transport in which (1) a diffusion approximation is used; (2) a flux limiter (generally empirical and not unique) is used to give a smooth transition from diffusion in opaque regimes to free streaming in transparent regimes; (3) the energy dependence of the diffusion process is retained. We may expect such an approximation to be excellent for the deep interior where mean free paths are very short and at large radii where the mean free paths are very long. However, it is precisely the interpolated intermediate regime that is of most interest in the neutrino reheating mechanism and this is where MGFLD is only approximately valid.

19. Convection and Neutrino Reheating

In Fig. 24 we show results due to Bruenn for a 15 $M_\odot$ progenitor star at 6 ms after bounce. There are two obvious regions where we may expect the possibility of significant convective instability. In the supernova 6 ms after the bounce there is a Schwarzschild unstable region lying inside the shock front at about 100 km where there is a large negative entropy gradient, and there is a region near the neutrinospheres where both the entropy and the lepton fraction exhibit strong negative gradients and thus favor an instability against Ledoux convection.

Our arguments to this point only identify regions that are favorable for convective motion. Whether such regions develop convection, the timescale for that convection, and the quantitative implications for supernova explosions may only be settled by detailed calculations (see the next section). Nevertheless, we may conjecture that convection inside the stalled shock could have significant influence on the possibility of neutrinos reenergizing the shock, and on the quantitative characteristics of a reenergized shock. In particular, we note that to boost the stalled shock it is necessary for the neutrinos to deposit energy behind the shock front (outside the gain radius, but inside the shock). Convective motion inside the shock front could, by overturning hot and cooler matter, cause more neutrino production and alter the neutrino spectrum. The convection could also move neutrino-producing matter beyond the neutrinosphere, so that the neutrinos that are produced would have a better chance to propagate into the region closely behind the shock where deposition of energy would have the most favorable influence in increasing the pressure and reenergizing the shock. This would provide a possible method to produce a supernova explosion with the required $\mathcal{O}\left(10^{51}\right)$ ergs of energy, thereby solving the "supernova problem".

20. Convection in Multidimensional Hydrodynamics

In this section, we summarize some recent calculations that have used multidimensional hydrodynamical calculations to test the preceding conjectures about convective instabilities and the role that they might have in reenergizing the stalled shock of a type II supernova. There are at least three modes of convection that may develop during the shock reheating phase. (1) Prompt convection near and below the neutrinospheres, which may occur immediately after the formation of the shock, (2) Doubly diffusive instabilities below the neutrinospheres, and (3) Neutrino-driven convection below the shock. We shall the discuss the first and third of these possibilities. Recent calculations that account for convection using multidimensional hydrodynamics may be found in Miller, Wilson, & Mayle (1993), Herant, et al (1994), Burrows, Hayes, & Fryxell (1995), Burrows (1997), Janka & Müller (1995), Janka & Müller (1996), Keil et al (1996), Mezzacappa, et al (1997a), and Mezzacappa, et al (1997b). I will describe the last two of these.

20.1. *Method of Calculation*

In the calculations to be described the initial models (provided by Woosley) were evolved through core collapse and bounce using MGFLD neutrino transport and Lagrangian hydrodynamics. The one-dimensional data shortly after bounce were mapped onto a two-dimensional Eulerian grid, with the inner and outer boundaries lying at radii of approximately 25 km and 1000 km, respectively. We use 128 nonuniform radial spatial zones and 128 uniform angular zones spanning a range of 180 degrees with reflecting boundary conditions. Spherically symmetric,time-dependent boundary data for the two-dimensional hydrodynamics were supplied by one-dimensional simulations, and at each time step the boundary data at the fixed inner and outer Eulerian radii were interpolated from the one-dimensional data.

The two-dimensional hydrodynamics was evolved using an extended version of the PPM [Woodward & Colella (1984)] hydrodynamics code VH-1 due to Blondin and collaborators. The matter was in nuclear statistical equilibrium for most of the calculations,† and we employed the Lattimer & Swesty (1991) equation of state. Convection was seeded by applying random velocity perturbations to the radial and angular velocities between $\pm 5\%$ of the local sound speed. In our two-dimensional simulations, gravity was assumed to be spherically symmetric.

The neutrino heating and cooling, and the change in the electron fraction, were com-

† The special steps taken when nuclear statistical equilibrium conditions were not satisfied are discussed in Mezzacappa, et al (1997a).

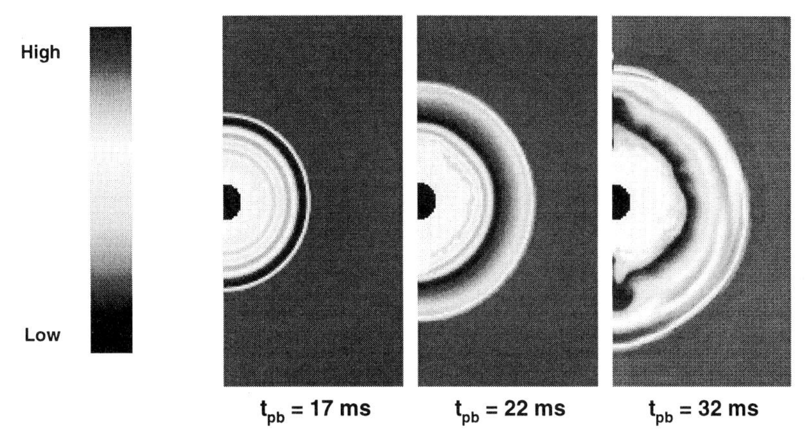

FIGURE 25. Entropy distribution in 15 solar mass model with neutrino transport turned off.

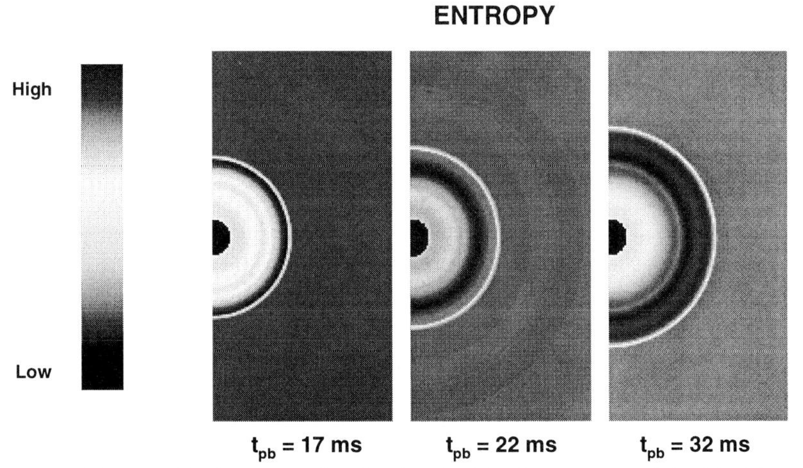

FIGURE 26. Entropy distribution in 15 solar mass model with full MGFLD neutrino transport.

puted using

$$\begin{aligned}\frac{d\epsilon}{dt} &= c\sum_{i=1}^{2}\int E_\nu^3 dE_\nu[\psi_i^0/\lambda_i^{(a)} - j_i(1-\psi_i^0)]/\rho(hc)^3 \\ \frac{dY_{\mathrm{e}}}{dt} &= cm_{\mathrm{B}}\sum_{i=1}^{2}\alpha_i\int E_\nu^2 dE_\nu[\psi_i^0/\lambda_i^{(a)} - j_i(1-\psi_i^0)]/\rho(hc)^3,\end{aligned} \tag{20.64}$$

where ϵ is the internal energy per gram; E_ν, ψ_i^0, $\lambda_i^{(a)}$, and j_i are the electron neutrino or antineutrino energy, zeroth distribution function moment, absorption mean free path, and emissivity, respectively; m_{B} is the baryon mass; $i = 1, \alpha_1 = 1$ corresponds to electron neutrinos, and $i = 2, \alpha_2 = -1$ corresponds to electron antineutrinos. The time-dependent ψ_i^0's are obtained from tables in r and t constructed from one-dimensional MGFLD simulations.

20.2. *Results for Prompt Convection*

Prompt convection may be initiated by negative entropy gradients produced by the weakening shock and may be sustained by the negative lepton gradient produced near the electron neutrinosphere by electron neutrino losses. Previous calculations using 2-dimensional hydrodynamics and simplified neutrino transport have suggested that prompt convection may increase the neutrinosphere luminosities and push the shock to larger radius.

Prompt convection was investigated for both 15 and 25 $M_\odot$ models, representative of stars with compact and extended iron cores, respectively. In the absence of neutrino transport, prompt convection develops and dissipates on a time scale ~15 ms for both models (Fig. 25). Prompt convection seeds further convection behind the shock that distorts it, but on average the shock radius is not increased significantly over that found from 1-dimensional simulations.

With MGFLD neutrino transport turned on (Fig. 26), we find that "prompt convection" essentially never develops, because near the neutrinosphere neutrino transport equilibrates both the entropy and electron fraction of a convecting fluid element with its surroundings in a fraction of a millisecond (see the discussion in §10). As a result, convection velocities become too small relative to the bulk inflow to result in any significant convective transport of entropy and leptons to the neutrinospheres. Therefore, we conclude that prompt convection will have little effect on boosting the neutrinosphere luminosities or neutrino reheating of the stalled supernova shock wave.

20.3. *Analytical Model of "Leaky Convection"*

In effect, convection becomes "leaky" in the presence of neutrino transport and differences between the entropy and lepton fraction of a convecting fluid element and that of the background are reduced. We may construct a simple model of such convection by assuming it to be driven by a constant negative entropy gradient, with neutrino transport equilibrating the entropy of a fluid element and the background entropy in a characteristic time τ_s. If the fluid element and the background are in pressure balance and viscosity is neglected, an element of the fluid is governed by the equations of motion

$$\dot{v} = \frac{g}{\rho}\alpha_s\theta_s \qquad \dot{\theta}_s = -\frac{\theta_s}{\tau_s} - \frac{d\bar{s}}{dr}v \tag{20.65}$$

where $\theta_s = s - \bar{s}$, with s and $\bar{s}$ being the fluid element entropy and the background entropy, respectively; g is the local acceleration of gravity; v is the radial velocity; and

$$\alpha_s \equiv -\left(\partial \ln \rho / \partial \ln s\right)_{P,Y_\ell} > 0, \tag{20.66}$$

where Y_ℓ is the common lepton fraction. The first expression in Eq. (20.65) equates the fluid element acceleration to the buoyancy force arising from the difference between its entropy and the background entropy, while the second equates $\dot{\theta}_s$ to $\dot{s}$ minus $\dot{\bar{s}}$, where $\dot{s}$ results from equilibration of the fluid element with the background and $\dot{\bar{s}}$ results from its motion through the gradient in $\bar{s}$.

If we neglect neutrino effects ($\tau_s = \infty$), the solutions to equations (20.65) indicate that (a) if $d\bar{s}/dr > 0$, the fluid element oscillates with the Brunt-Väisälä frequency

$$\omega_{\rm BV} \equiv \left[(g\alpha_s d\bar{s}/dr)/\rho\right]^{1/2}, \tag{20.67}$$

and (b) if $d\bar{s}/dr < 0$, it convects (its velocity increases exponentially), with a convection growth time given by

$$\tau = \tau_{\rm BV} \equiv \left[-(g\alpha_s d\bar{s}/dr)/\rho\right]^{1/2}. \tag{20.68}$$

When neutrino transport effects are included in the convectively unstable case ($d\bar{s}/dr <$

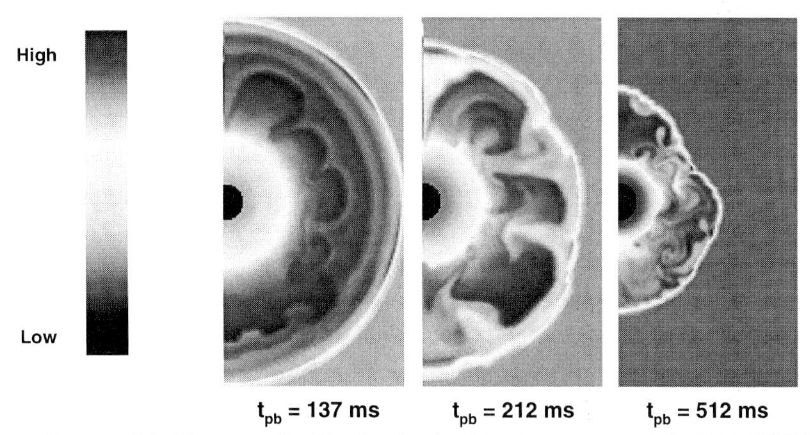

FIGURE 27. Entropy distribution in the 15 solar mass model with MGFLD neutrino transport in the neutrino-driven convection region. Times are measured from bounce.

0), the fluid element convects, but the convection growth time scale $\tau > \tau_{\rm BV}$ is now

$$\frac{1}{\tau} = \left[\frac{1}{\tau_{\rm BV}^2} + \frac{1}{4\tau_s^2}\right]^{1/2} - \frac{1}{2\tau_s}. \tag{20.69}$$

In the limit $\tau_s \ll \tau_{\rm BV}$, the growth time scale increases by $\tau_{\rm BV}/\tau_s$ such that $\tau \simeq \tau_{\rm BV}^2/\tau_s$. Neutrino transport also reduces the asymptotic convective velocities: in the limit $\tau_s \ll \tau_{\rm BV}$, the solutions to equations (20.65) show that after moving a distance ℓ from rest a fluid element's velocity is reduced by a factor $\tau_s/\tau_{\rm BV}$.

In our protoneutron star convection simulations, $\tau_{\rm BV}$ is 2 – 3 ms in the region between 10^{11} and 10^{12} g cm^{-3} immediately after shock propagation and our $\dot{\epsilon}$ from neutrino heating and cooling give estimates for τ_s that are of order 1/100 to 1 millisecond, depending on conditions. This implies that neutrino transport should reduce the growth rate and asymptotic velocities of entropy-driven convection by a factor $\sim$ 4 at the neutrinosphere and a factor $\sim$100 at $\rho = 10^{12}$ g cm^{-3}, for both the 15 and 25 solar mass models. Our detailed simulations are consistent with these model estimates, and with similar estimates for convection driven by lepton gradients.

20.4. *Neutrino-Driven Convection*

We have also investigated neutrino-driven convection in supernova cores. As in the preceding calculations for prompt convection, we couple two-dimensional (PPM) hydrodynamics to one-dimensional MGFLD neutrino transport. Results for the 15 solar mass initial model are illustrated in Fig. 27. By approximately 200 ms after bounce, large-scale and violent convection develops behind the shock, characterized by high-entropy upflows and low-entropy downflows. The upflows distort the shock and radial convection velocities become supersonic just below the shock, as illustrated in Fig. 29. However, the shock eventually recedes to smaller radii and at $\sim$500 ms after bounce there is no evidence of a developing explosion.

Thus, with two-dimensional (PPM) hydrodynamics coupled to one-dimensional multi-group flux-limited diffusion neutrino transport, we see vigorous — in some regions supersonic — neutrino-driven convection develop behind the shock. However, do not obtain explosions for what should be an "optimistic" 15 $M_\odot$ model with a compact core.

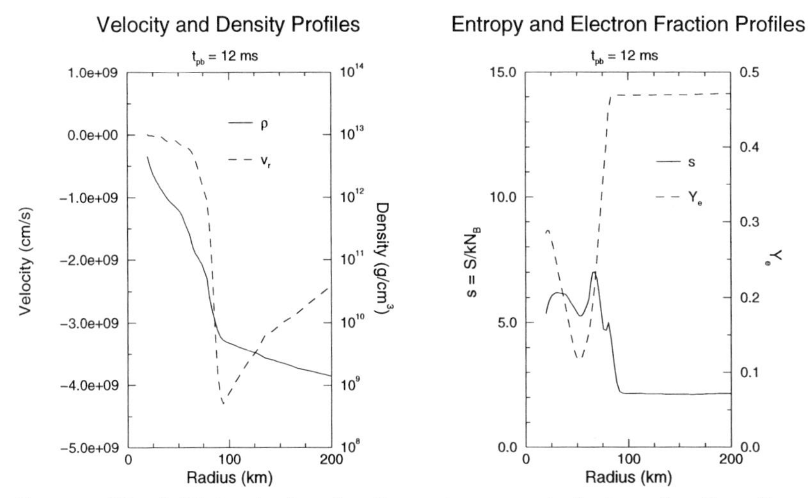

FIGURE 28. Initial velocity, density, entropy, and electron fraction distribution in the 15 solar mass model for the neutrino-driven convection calculation.

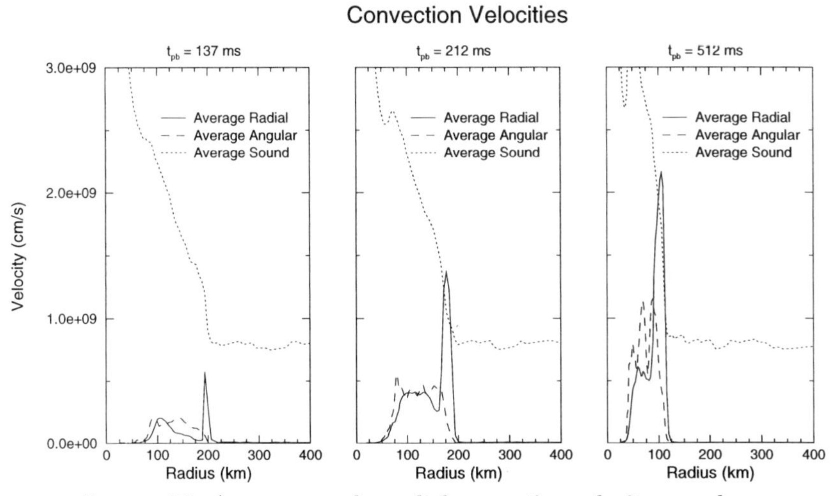

FIGURE 29. Average angular radial convection velocity, angular convection velocity, and local sound speed for the three times shown in Fig. 27.

20.5. *Summary of 2D Hydro + MGFLD Results*

The preceding results indicate that 1-dimensional MGFLD neutrino transport coupled with 2-dimensional hydrodynamics does not generally ensure supernova explosions. Detailed investigation of these results indicates that the reason that our models do not explode, when some previous multidimensional simulations have produced explosions, lies in the different treatment of neutrino transport in the present calculations. Since the MGFLD neutrino transport that we have implemented is arguably the most complete yet employed in a multidimensional supernova simulation, our results indicate that convection alone is not sufficient to ensure a supernova explosion. Rather, these results suggest that the neutrino transport, not the convection, is the most important factor governing the success of explosions.

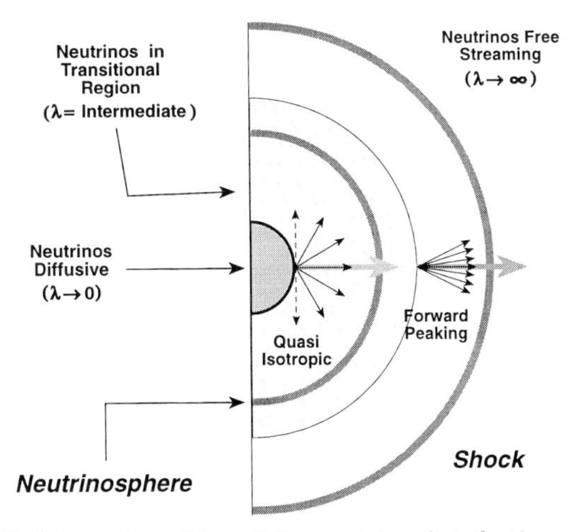

FIGURE 30. Forward peaking of the neutrino distribution near the neutrinosphere.

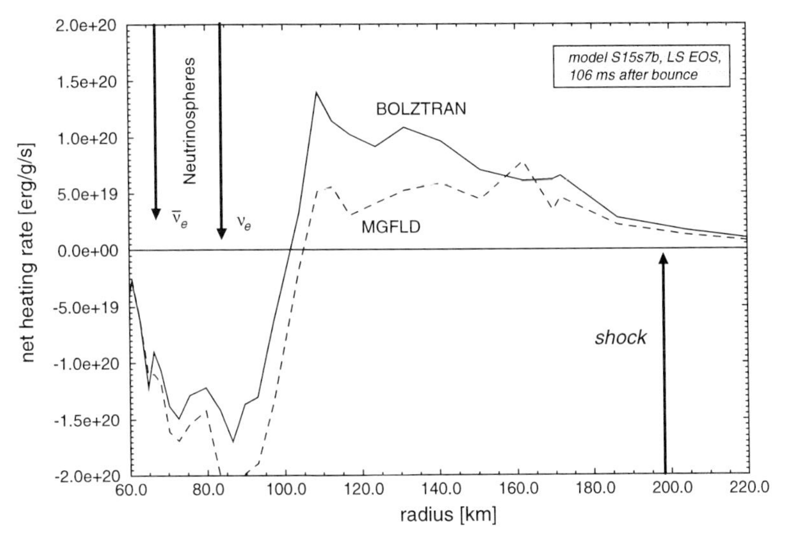

FIGURE 31. Comparison of net heating rate for Boltzmann transport and MGFLD.

Although the calculations presented above represent state of the art neutrino transport for multi-dimensional simulations, the MGFLD method still represents an approximation to the underlying Boltzmann transport theory that reduces to the full results in the diffusive and free-streaming limits, but is an empirical approximation precisely in the transition regime that is of most interest in the present discussion. Accordingly, let us conclude by examining an approximate Boltmann transport solution and its effect on neutrino reheating behind the stalled shock.

21. Boltzmann Neutrino Transport

We have compared three-flavor Boltzmann neutrino transport and three-flavor multi-group flux-limited diffusion (MGFLD) in a postbounce core collapse supernova environment by developing stationary state neutrino distributions in thermally and hydrodynamically frozen time slices obtained from Lagrangian–MGFLD core collapse and bounce simulations. We find that both the electron neutrino and antineutrino inverse flux factors, which describe how forward peaked these neutrinos are in the regions between their respective neutrinospheres and the shock (see Fig. 30), differ by as much as ~20–30%, which implies an equivalent difference in their heating rates. These results may have important implications for shock revival and neutrino-driven convection when full Boltzmann transport of neutrinos is included in a realistic calculation.

22. Conclusions

Two-dimensional simulations of convection in core collapse supernovae by various groups have produced mixed results that run the full gamut from robust explosions to complete duds. These disparate outcomes appear to result primarily from differences in numerical hydrodynamics methods and neutrino transport approximations used, with differences in equations of state, neutrino opacities, etc. also making smaller contributions. It is important to note that no group has yet implemented neutrino transport that simultaneously (a) is multidimensional, (b) is multigroup (neutrino-energy dependent), and (c) simulates with sufficient realism the transport of neutrinos in all three important regions: opaque, semitransparent, and transparent.

Our results point to the need for either improved neutrino transport (relative to MGFLD) or new physics in order to obtain consistently robust explosions. Recent results suggest that Boltzmann transport will yield greater neutrino heating and more vigorous neutrino-driven convection than MGFLD; each would increase the chances of reviving the stalled shock. Although these results are promising, firm conclusions await the completion of postbounce supernova simulations with full Boltzmann transport.

I thank John Blondin, Steve Bruenn, Alan Calder, Bronson Messer, Tony Mezzacappa, and Friedel Thielemann for useful discussions. Oak Ridge National Laboratory is managed by Lockheed Martin Energy Research Corp. for the U. S. Department of Energy under Contract No. DE–AC05–96OR22464.

REFERENCES

Arnett, D. 1996 *Supernovae and Nucleosynthesis.* Princeton University Press.

Bethe, H. A., Brown, G. E., Applegate, J., & Lattimer, J. M. 1979 Equation of state in the gravitational collapse of stars. *Nucl. Phys.* **A324**, 487–533.

Bethe, H. A. 1990 Supernova mechanisms. *Rev. Mod. Phys.* **62**, 801–867.

Bethe, H. A. and Wilson, J. R. 1985 Revival of a stalled supernova shock by neutrino heating. *Ap. J.* **295**, 14–23.

Bowers, R. L. and Wilson, J. R. 1984 *Numerical Modeling in Applied and Astrophysics.* Jones and Bartlett.

Brown, G. E. (ed.) 1988 Volume **163** of *Phys. Rep.*: Theory of Supernovae.

Bruenn, S. W. 1985 Stellar core collapse: numerical model and infall epoch. *Ap. J. Suppl.* **58**, 771–841.

Bruenn, S. W. & Haxton, W. C. 1991 Neutrino–nucleus interactions in core-collapse supernovae. *Ap. J. Suppl.* **58**, 771–841.

Bruenn, S. W. 1993 Type II supernovae: stellar core collapse and Boltzmann neutrino transport. In *First Symposium on Nuclear Physics in the Universe* (ed. M. W. Guidry & M. R. Strayer) p. 31. IOP Publishing.

BURROWS, A., HAYES, J., & FRYXELL, B. 1995 On the nature of core-collapse supernova explosions. *Ap. J.* **450**, 830–850.

BURROWS, A. 1997 New insights into core-collapse supernova theory. Preprint astro-ph/9706137.

COOPERSTEIN, J. 1988 Neutrinos in supernovae. *Phys. Rep.* **163**, 95–126.

FREEDMAN, D. Z. 1974 Coherent effects of a weak neutral current. *Phys. Rev.* **D9**, 1389–1392.

FREEDMAN, D. Z., SCHRAMM, D. N., & TUBBS, D. L. 1977 The weak neutral current and its effects in stellar collapse. *Ann. Rev. Nucl. Sci.* **27**, 167–207.

HANSEN, C. J. AND KAWALER, S. D. 1994 *Stellar Interiors: Physical Principles, Structure, and Evolution.* Springer–Verlag.

HERANT, M., BENZ, W., HIX, W. R., FRYER, C. L., AND COLGATE, S. A. 1994 Inside the supernova: a powerful convective engine. *Ap. J.* **435**, 339–361.

JANKA, H.-T. & MÜLLER, E. 1995 The first second of a type II supernova: convection, accretion, and shock propagation. *Ap. J.* **448**, L109–L113.

JANKA, H.-T. & MÜLLER, E. 1996 *Astron. Astrophys.* **306**, 167.

KEIL, W., JANKA, H.-T., & MÜLLER, E. 1996 Ledoux convection in protoneutron stars—a clue to supernova nucleosynthesis? *Ap. J.* **473**, L111–L114.

KIPPENHAHN, R. AND WIEGERT, A. 1994 *Stellar Structure and Evolution.* Springer–Verlag.

LATTIMER, J. M. AND SWESTY, F. D. 1991 A generalized equation of state for hot, dense matter. *Nucl. Phys.* **A535**, 331-376.

MCCRAY, R. 1993 Supernova 1987A revisisted. *Annu. Rev. Astron. Astrophys.* **31**, 175–216.

MEZZACAPPA, A. 1993 Type II supernovae: stellar core collapse and Boltzmann neutrino transport. In *First Symposium on Nuclear Physics in the Universe* (ed. M. W. Guidry & M. R. Strayer) pp. 51–62. IOP Publishing.

MEZZACAPPA, A., & BRUENN, S. W. 1993 Type II supernovae and Boltzmann neutrino transport: the infall phase. *Ap. J.* **405**, 637-668.

MEZZACAPPA, A., CALDER, A. C., BRUENN, S. W., BLONDIN, J. M., GUIDRY, M. W., STRAYER, M. R., & UMAR, A. S. An investigation of neutrino-driven convection and the core collapse supernova mechanism using multigroup neutrino transport. *Ap. J.*, in press, 1997.

MEZZACAPPA, A., CALDER, A. C., BRUENN, S. W., BLONDIN, J. M., GUIDRY, M. W., STRAYER, M. R., & UMAR, A. S. The competition between protoneutron star convection and neutrino transport in core collapse supernovae. *Ap. J.*, in press, 1997.

MEZZACAPPA, A., MESSER, O. E. B., BRUENN, S. W., & GUIDRY, M. W., A comparison of Boltzmann and multigroup flux-limited diffusion neutrino transport in core-collapse supernovae. In preparation for *Ap. J.*, 1997.

MIHALAS, D. AND MIHALAS, B. W. 1984 *Foundations of Radiation Hydrodynamics.* Oxford University Press.

MILLER, D. S., WILSON, J. R., AND MAYLE, R. W. 1993 Convection above the neutrinosphere in type II supernovae. *Phys. Rep.* **163**, 127-136.

MYRA, E. S. 1988 Neutrino transport in stellar collapse. *Phys. Rep.* **163**, 127-136.

SHORE, S. N. 1992 *An Introduction to Astrophysical Hydrodynamics.* Academic Press.

WILSON, J. R. 1985 In *Numerical Astrophysics* (ed. J. Centrella, J. LeBlanc, and R. Bowers) p. 422. Jones and Bartlett.

WILSON, J. R. AND MAYLE, R. W. 1988 Convection in core-collapse supernovae. *Phys. Rep.* **163**, 63–77.

WOODWARD, P., AND COLELLA, P. 1984 The numerical simulation of two-dimensional fluid flow with strong shocks. *Comp. Phys.* **54**, 115-173.

Neutron Stars

By MADAPPA PRAKASH

Department of Physics and Astronomy, SUNY at Stony Brook, Stony Brook, NY 11794, USA

The structure of neutron stars is discussed with a view to explore (1) the extent to which stringent constraints may be placed on the equation of state of dense matter by a comparison of calculations with the available data on some basic neutron star properties; and (2) some astrophysical consequences of the possible presence of strangeness, in the form of baryons, notably the Λ and Σ^-, or as a Bose condensate, such as a K^- condensate, or in the form of strange quarks.

1. Introduction

Almost every physical aspect of a neutron star tends to the extreme when compared to similar traits of other commonly observed objects in the universe. Stable matter containing $A \sim 10^{57}$ baryons and with a mass in the range of $(1-2)\ M_\odot$ ($M_\odot \cong 2\times10^{33}$ g) confined to a sphere of radius $R \sim 10$ km (recall that $R_\odot = 6.96\times10^5$ km) represents one of the densest forms of matter in the observable universe. Depending on the equation of state (EOS) of matter at the core of a neutron star, the central density could reach as high as $(5-10)\rho_0$, where $\rho_0 \cong 2.65\times10^{14}$ g cm^{-3} (corresponding to a number density of $n_0 \cong 0.16$ fm^{-3}) is the central mass density of heavy laboratory nuclei (compare this to $\rho_\odot = 1.4$ g cm^{-3}). Observations of electromagnetic emissions from the surface imply surface magnetic fields of $B_{\rm ns} \sim 10^{12\pm3}$ gauss to be compared with the highest magnetic fields of $B_{\rm lab} \sim 10^5$ gauss currently achievable in the laboratory. The interior magnetic fields are expected to be about three orders of magnitude larger than the surface fields.

Neutron stars are believed to be made in the aftermath of type II supernova explosions which result from the gravitational core collapse of massive ($\geq 8M_\odot$) stars. All known forces of nature, strong, weak, electromagnetic and gravitational, play key roles in the formation, evolution and the composition of neutron stars. Thus, intriguing aspects of nearly all branches of physics come to the fore in observing and understanding these fascinating objects. For pedagogical accounts of the physics of neutron stars and for more recent developments in understanding the many facets of its composition, structure and evolution, the reader may refer to Weinberg (1972), Misner *et al.* (1973), Baym & Pethick (1975; 1979), Shapiro & Teukolsky (1983), Ogelman (1989), Ventura & Pines (1991), Pines (1992), Pethick (1992), van Riper *et al.* (1993), Prakash (1994), Alpar *et al.* (1995), Pethick & Ravenhall (1995), Prakash (1996).

Fig. 1 shows a cross-sectional view of the structure of a neutron star of mass $1.45M_\odot$ and radius $R \simeq 11$ km. Beginning from the outer crust in the mass density regime $10^7 \leq \rho \leq 4\times10^{11}$ g cm^{-3}, the equilibrium configuration of matter consists of nuclei set in a lattice against a uniform background of electrons. Near the upper limit of this density range, neutrons begin to drip out of nuclei. The nuclei encountered in this regime closely resemble those found in the laboratory with neutron excess $\delta = (N-Z)/(N+Z) \sim 0.3$. (In astrophysics, the proton fraction $x = Z/A$ is often used denote the neutron-proton asymmetry.) Typically, the width of the solid-like outer crust is $\sim 1/3$ km. The physical properties of the outer crust may be understood in terms of the interplay between electrostatic and nuclear energies, knowledge about the latter taken from the many studies of nuclear systematics.

Above the neutron drip density, $4\times10^{11} \leq \rho \leq 10^{14}$g cm^{-3}, the inner crust region

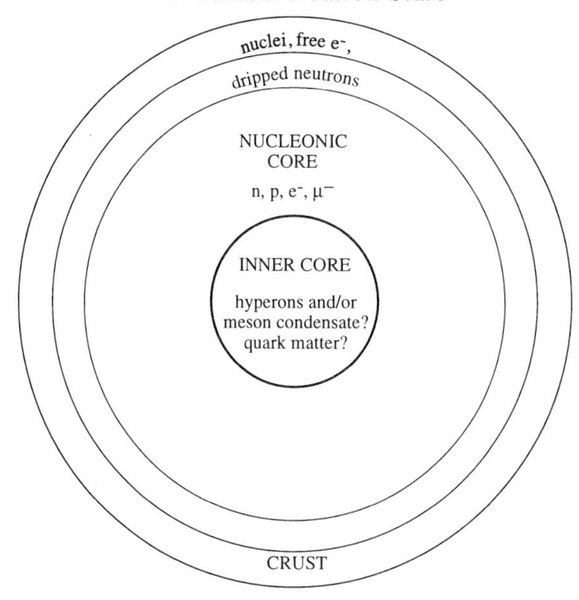

FIGURE 1. Schematic cross-sectional view of a neutron star.

consists of nuclei, dripped neutrons and electrons, and has a width of ~ 0.5 km. Due to the presence of a gas-like environment of dripped neutrons, the equilibrium nuclei encountered in the inner crust are highly neutron-rich with $\delta \sim 0.7$, unlike laboratory nuclei for which an adequate understanding of the energetics is available, both experimentally and theoretically. Hence, the physical properties of the inner crust are largely based on theoretical studies, which use extrapolations of existing mass formulae to highly neutron-rich systems. Current developments in the study of highly neutron-rich nuclei using radio-active beam techniques are expected to play an important role in constraining such extrapolations.

With increasing density, the possibility of more exotic non-spherical shapes of nuclear matter (rods, plates and tubes) have been theoretically predicted to exist in a surrounding of neutrons and electrons. Such non-spherical shapes are favored over the more common spherical shapes due to the balance between Coulomb and nuclear energies. The width of this surrounding region is about ~ 0.1 km. As the density is further increased up to about half nuclear equilibrium density, a phase of uniform nuclear matter coexisting with nuclear bubbles has also been predicted.

The star's crust extends up to about half nuclear density and is made up of matter at large neutron excess. The crust is typically about 1 km thick and contributes $\sim 0.01 M_\odot$, about 1% of the total $1.45 M_\odot$. The dependence of the crustal mass and thickness on the total mass and radius of the star are discussed in more detail by Lorenz *et al.* (1993), and Pethick *et al.* (1995), to which the reader's attention is drawn. The properties of the crust are important in understanding the thermal emissions from the star's surface, of X-ray bursts occurring upon accretion of ambient matter on to the star's surface, and in the interpretation of neutron star glitches.

For $10^{14} \leq \rho \leq (4-6)10^{14}$ g cm^{-3}, which corresponds to about one-half to 2-3 times nuclear density, matter is more liquid-like and is comprised of neutrons, protons, electrons, and, beginning at roughly nuclear density, also muons. The EOS of matter in this density regime has long been a subject of intensive study using traditional many-body techniques of nuclear physics. Many approaches, including both potential models and field theoretical models, have been employed. For higher densities, many intriguing possibilities exist. These include the appearance of strangeness either in the form of hyperons, or in the form of a Bose condensate likely made of K^- mesons. The nature of strong interactions may also allow a quark phase to appear in the central regions of the star. These possibilities are the subject matter of active research, some of which we will consider in these lectures.

1.1. *Some observed neutron star properties*

1.1.1. *Masses*

Observed neutron star masses (Thorsett em et al. (1993), van Kerkwijk *et al.* (1996), Brown *et al.* (1996)) are shown in Fig. 2. The smallest range that is consistent with all of the data has an upper limit of $1.44 M_\odot$ from PSR1913+16 and a lower limit of $1.36 M_\odot$ from the precisely measured total mass of PSR2127+11C and its companion. The upper limit, $M = 1.44 M_\odot$, provides constraints on the neutron star EOS. A conservative upper limit is shown by the dashed line in Fig. 2 at $M = 1.5 M_\odot$. Any theoretical neutron star EOS has to support a maximum mass of at least this value. The fact that all the measured neutron star masses consistently lie within a narrow range of $1.4 M_\odot$ is intriguing. Since neutron stars are formed in the gravitational core collapse of massive stars, their masses may depend on the structure of the progenitor star. The cores of stars which evolve into neutron stars have precollapse masses of about $1.4 M_\odot$, (which is the Chandrashekar limit for stars with electron/baryon ratios of about 0.5) which introduces a natural mass scale for possible neutron star masses. The final neutron star mass may, however, depend on the amount of accretion subsequent to a neutron star's birth. Thus, rigorous arguments for the happenstance of observations are not yet available. Later, we will return to the question of whether or not it is the nature of strong interactions which restricts the formation of stars in this mass range.

1.1.2. *Thermal emissions and surface temperatures*

Through recent advances in imaging X-ray telescopes, it has been possible to detect X-rays from some 14 rotation-powered neutron stars (Ogelman (1993)). Fig. 3 shows the inferred surface temperatures of some X-ray emitters. Of these, 4 appear to have the signatures of neutron stars on the initial cooling curve. These pulsars, Vela (PSR0833-45), Geminga (PSR0630+18) and PSR's 0656+14, 1055-52 span an age bracket of 10^4 to 5×10^5 yr. Their X-ray luminosity is currently being compared with calculations of both the standard cooling and the rapid cooling scenarios. It must be borne in mind, however, that the surface temperature of a pulsar may be determined not entirely by the cooling of the neutron star but also by irradiation of its surface by magnetopsheric X-rays (Halpern & Ruderman (1993)) so that all temperature estimates or "measurements" must be considered as upper limits.

1.1.3. *Pulsar periods*

The periods of over 600 presently known pulsars range between 1.56 ms to several s (Taylor *et al.* (1993)) with the median period around 0.7 s (see Fig. 4). The millisecond pulsars form a distinct tail in the pulsar period distribution. Pulsar periods increase slowly with time except for occasional decreases known as glitches (see below). The pulsar slow down is caused by a steady loss of the rotational kinetic energy, either in the form of electromagnetic waves and/or energetic particles. The period derivative $\dot{P}$ (most observed pulsars have $\dot{P} \sim 10^{-15}$ s s^{-1}) may be used to define the pulsar characteristic time (also referred to as the spin-down age) by $\tau = P/2\dot{P}$, which is an indicator of the

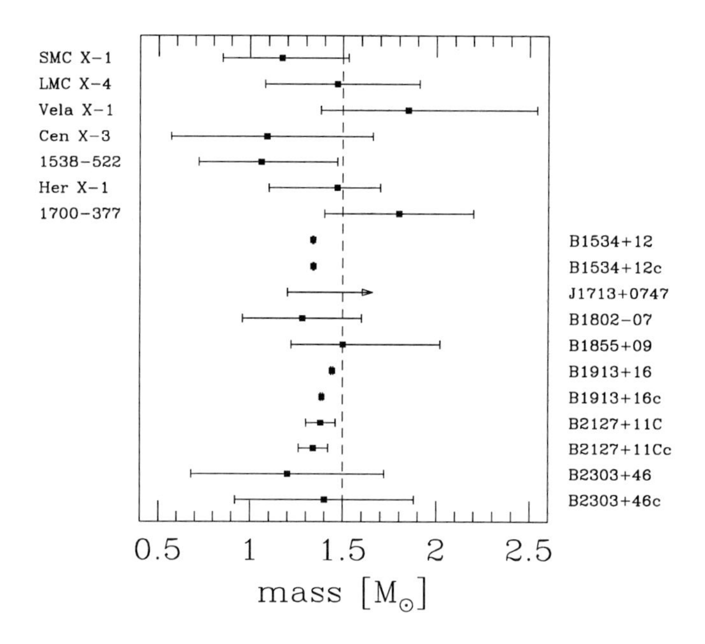

FIGURE 2. Measured neutron star masses with 95% confidence limits.

pulsar age. Pulsars associated with supernova remnants are all only about 20,000 years old. In contrast, all millisecond and binary pulsars are old with $P/\dot{P} \sim 10^9$ yr.

1.1.4. *Pulsar glitches*

Although pulsar periods generally increase with time, several pulsars are known to exhibit small, discontinuous decreases in their periods which are termed pulsar glitches and are believed to originate in the interior of neutron stars. The pulsars PSR0355+54, PSR0531+21 (Crab), PSR0833-45 (Vela) and PSR1737-30 have been observed to glitch more than once. The Vela pulsar has so far exhibited eight giant glitches and one minor glitch. The basic glitch characteristics are the change $\Delta\nu$ in the pulsar rotation frequency $\nu = 1/P$, and its time derivatives such as $\Delta\dot{\nu}/\dot{\nu}$ (Manchester (1992)), for the observed glitch characteristics of the above pulsars). The glitch size $\Delta\nu/\nu$ varies between pulsars, but is typically greater than 10^{-6}. For the pulsars mentioned above, $\Delta\dot{\nu}/\dot{\nu}$ varies between $(1 - 105) \times 10^{-3}$. For the same pulsars, the time period, τ_d, over which the preglitch behavior is recovered (after the pulsar period jump) ranges from about 1 to 44 days. Pulsars with even longer periods of recovery, up to a year, are known. The long time scales of the recovery imply that the core rotation is weakly coupled to that of the crust. Successful models (Alpar (1992)) of the postglitch behavior require that the weakly coupled component consists of superfluid neutrons in the interior of the star. Since the observed glitch phenomena exhibit a great variety, prediction of pulsar glitches has not met with much success so far.

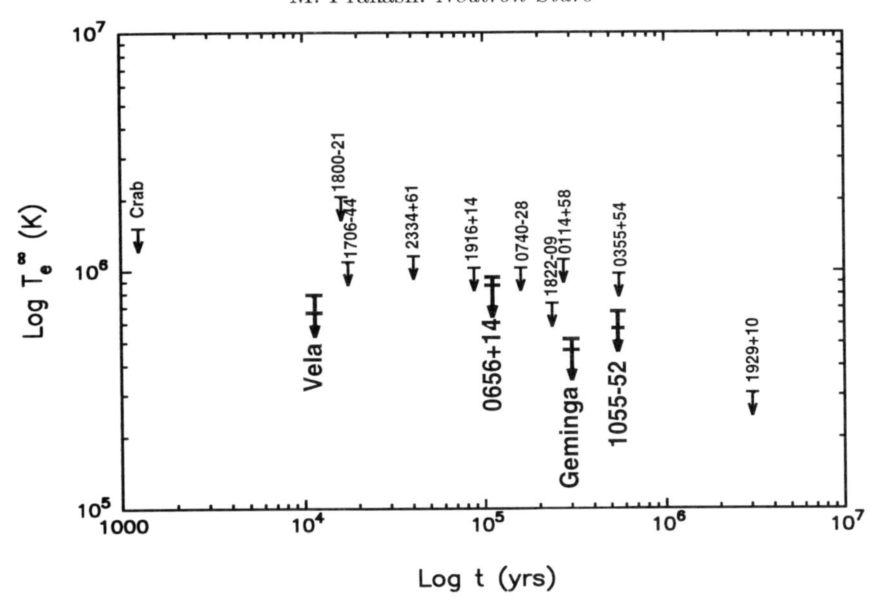

FIGURE 3. Measurement, estimates or upper limits for the surface temperature of fourteen pulsars. I am grateful to Dany Page for preparing this figure.

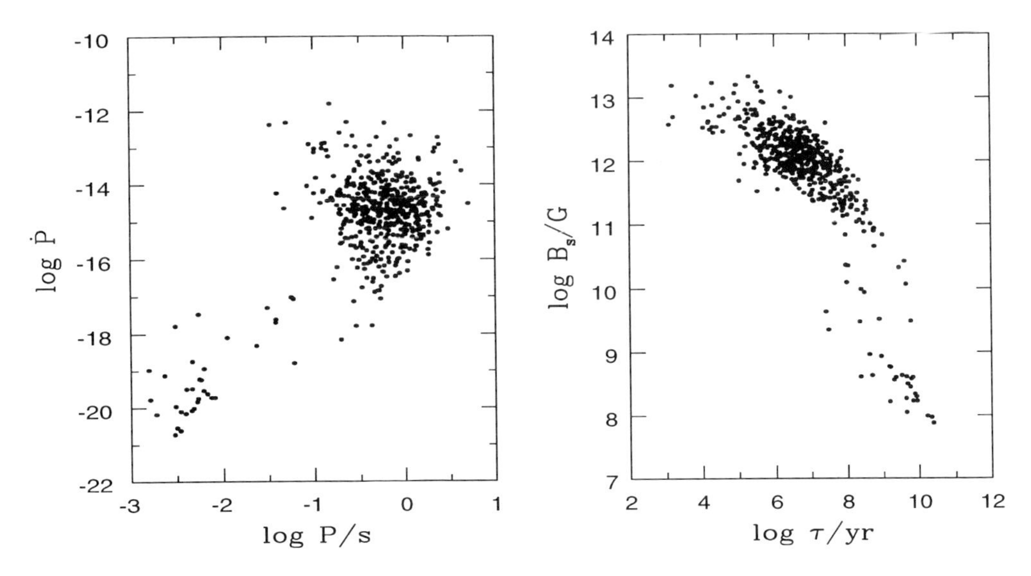

FIGURE 4. Left panel: Period derivative vs. period. Right panel: Inferred surface magnetic field vs. spin down age. I thank J. Holmer for preparing this figure using the Princeton pulsar data base.

2. Neutron star structure

2.1. *Equations of stellar structure*

In hydrostatic equilibrium, the structure of a spherically symmetric neutron star is determined by the Tolman-Oppenheimer-Volkov (TOV) equations (Tolman (1934), Tolman

(1939), Oppenheimer & Volkoff (1939)):

$$\frac{dM}{dr} = 4\pi r^2 \epsilon\,, \quad \frac{dP}{dr} = -\frac{GM\epsilon}{c^2r^2}\left[1+\frac{P}{\epsilon}\right]\left[1+\frac{4\pi r^3 P}{Mc^2}\right]\left[1-\frac{2GM}{c^2r}\right]^{-1} \tag{2.1}$$

Above, G is the gravitational constant, P is the pressure, ϵ is the energy density inclusive of the rest mass density and M is the enclosed gravitational mass. The quantity $R_s = 2GM/c^2$ has the units of a length and is known as the Schwarzschild radius.

The gravitational and baryon masses of the star are defined by

$$M_G c^2 = \int_0^R dr\, 4\pi r^2\, \epsilon\,, \quad M_A c^2 = m_A \int_0^R dr\, 4\pi r^2\, n \left[1-\frac{2GM}{c^2r}\right]^{-1/2} \tag{2.2}$$

where m_A is the baryonic mass and n is the baryon number density. The binding energy of the star is then $B.E. = (M_A - M_G)c^2$.

By specifying the EOS of enclosed matter, $P = P(\epsilon)$, the structure of the star is determined by choosing a central pressure $P_c = P(\epsilon_c)$ at $r = 0$ and integrating the coupled differential equations Eqs. (2.1) out to the star surface at $r = R$ determined by the condition $P(r = R) = 0$. Two limiting situations are of special interest. If the quantities in brackets of the right hand side of Eqs. (2.1) are all close to unity, as is the case when $2GM/c^2R = R_s/R \ll 1$, the structure is determined by Newtonian gravity. Thus, the significance of general relativity may be gauged by the magnitude of R_s/R. The surface approaches the event horizon as $R_s/R \to 1$; larger values result in black hole configurations.

The gravitational potential satisfies

$$-\frac{d\phi}{dr} = \frac{1}{\epsilon + P}\frac{dP}{dr} \tag{2.3}$$

with the metric given by

$$ds^2 = -e^{2\phi}dt^2 + e^{2\Lambda}dr^2 + r^2(d\theta^2 + \sin^2\theta\, d\phi^2)\,, \tag{2.4}$$

where the function Λ and the enclosed mass are related through $M = (r/2)(1-e^{-2\Lambda})$. For slow rigid rotation with angular velocity Ω, the moment of inertia I is given by (Hartle (1973))

$$I = \frac{8\pi}{3c^2}\int_0^R dr\, \frac{r^4[\epsilon + P]e^{\phi}}{\left[1-\dfrac{2GM}{c^2r}\right]^{1/2}}\frac{\overline{\omega}}{\Omega} \tag{2.5}$$

where the angular velocity $\overline{\omega}$ is obtained from

$$\frac{d}{dr}\left(r^4 j \frac{d\overline{\omega}}{dr}\right) + 4r^3\frac{dj}{dr}\overline{\omega} = 0\,, \quad j = \left[1-\frac{2GM}{c^2r}\right]e^{-\phi}\,. \tag{2.6}$$

The non-rotating gravitational mass distribution M and the unperturbed EOSs are used to solve the above equations. The boundary conditions are $d\overline{\omega}/dr = 0$ at $r = 0$, $\phi(\infty) = 0$, and $\overline{\omega} = \Omega - (R/3)(d\overline{\omega}/dr)_R$. An arbitrary value of $\overline{\omega}_c = \overline{\omega}(0)$ is chosen which determines Ω; a new value of $\overline{\omega}_c$ corresponding to the observed Ω' may be obtained by the scaling relation $(\overline{\omega}_c/\Omega) = (\overline{\omega}'_c/\Omega')$. Finally, the angular momentum is given by $J = I\Omega$.

2.2. *Macroscopic properties*

A comparison of the results obtained by solving the TOV equations for two different EOSs which differ in their high density behavior offers useful insights concerning the physical attributes of a star. Fig. 5 shows some of the essential features. The top left panel contains P vs ϵ for two representative EOSs . The dash–dotted curve shows the causal EOS $P = \epsilon$. The solid curve is referred to as the soft EOS since pressure varies less rapidly with the energy density than the dashed curve which is termed the stiff EOS. The EOS directly influences the physical attributes of a star as seen from the results shown in the other panels of this figure. For example, the limiting or maximum mass for the stiff EOS is larger than that for the soft EOS. This is seen from the mass versus radius plots (top right panel) and also from the mass versus central baryon density (center left panel). Also, stiffer the EOS, lower is the central density of the maximum mass star. Further, the radius (top right panel) and the binding energy (center right panel) of the maximum mass star are larger for the stiff EOS than for the soft EOS. Finally, the bottom panels contrast the moments of inertia (bottom left) and the surface red shift (bottom right) for the soft and stiff EOS's.

Stellar configurations with central densities $n_c > n_c(M_{max})$ are unstable towards small perturbations, since the gravitational attraction in such stars overwhelms the repulsive forces in matter. Such stars are thought to form black holes. Stars with central densities $n_c < n_c(M_{max})$ represent stable configurations and are candidates for stable neutron stars. A detailed discussion of neutron structure and a compendium of various EOSs (prior to 1977) may be found in the article by Arnett & Bowers (1977).

2.3. *Constraints on the EOS*

Stringent constraints onthe EOS may be placed if measurements of stars' masses, radii, pulsar frequencies, moments of inertia, etc. are available. To see how this works, consider the dependence of these properties on the mass and radius in a so-called $M - R$ diagram.

(*a*) Eq. (2.1) requires that the radius $R > R_s = 2GM/c^2$. This yields $M/M_\odot \leq R/R_s(\odot)$, where $R_s(\odot) = 2GM_\odot/c^2 \cong 2.95$ km.

(*b*) Since the pressure in the center is finite, $Pc < \infty$, $R > (9/8)R_s$ (Weinberg (1972)). This translates to $M/M_\odot \leq (8/9)R/R_s(\odot)$.

(*c*) Since the adiabtic sound speed $c_s = (dP/d\epsilon)^{1/2} \leq c$, where c is the speed of light, one has $R > 1.39R_s$, giving $M/M_\odot \leq R/(1.39 \times R_s(\odot))$.

(*d*) If, instead of employing the causal EOS $P = \epsilon$ at all densities, one requires it to hold above a fiducial density $n_t \cong 2n_0$ (below which the EOS is presumed to be known), the limit $R > 1.52R_s$ is obtained (Lattimer *et al.* (1990)) yielding $M/M_\odot \leq R/(1.52 \times R_s(\odot))$.

In practice, however, these restrictions allow a large class of EOSs to be consistent with observations. One is thus forced to utilize additional constraints that are based on observations. In the hope that continued measurements will prescribe stringent limits, we will note the constraints employed currently.

(*a*) The limiting or maximum mass M_{max} should exceed the largest of the observed neutron star masses. Currently, this condition is taken to imply $M_{max} \geq 1.44M_\odot$, the most accurately measured neutron star mass (Thorsett *et al.* (1993)) in the binary pulsar system PSR1913+16.

(*b*) Nearly all (up to 99%) of the binding energy of a neutron star is released in the form of neutrinos during the birth of a neutron star, after the gravitational core collapse of a massive ($\geq 8M_\odot$) star which results in a type II supernova. Estimates (Burrows & Lattimer (1986)) of the energy released in neutrinos from the SN 1987A explosion lie in the range $(1 - 2) \times 10^{53}$ ergs. This places a restriction on the EOS that the B.E $\geq (1 - 2) \times 10^{53}$ ergs.

(*c*) The Keplerian frequency of the star (this is the rotational frequency Ω_K beyond which the star will begin to shed mass at the equator) should exceed the spin period of the fastest spinning pulsar, namely, that of PSR1957+20. This translates to $P_K \geq 1.56$ ms. (In reality, a star may spin at a frequency lower than Ω_K due to viscous effects. Choosing

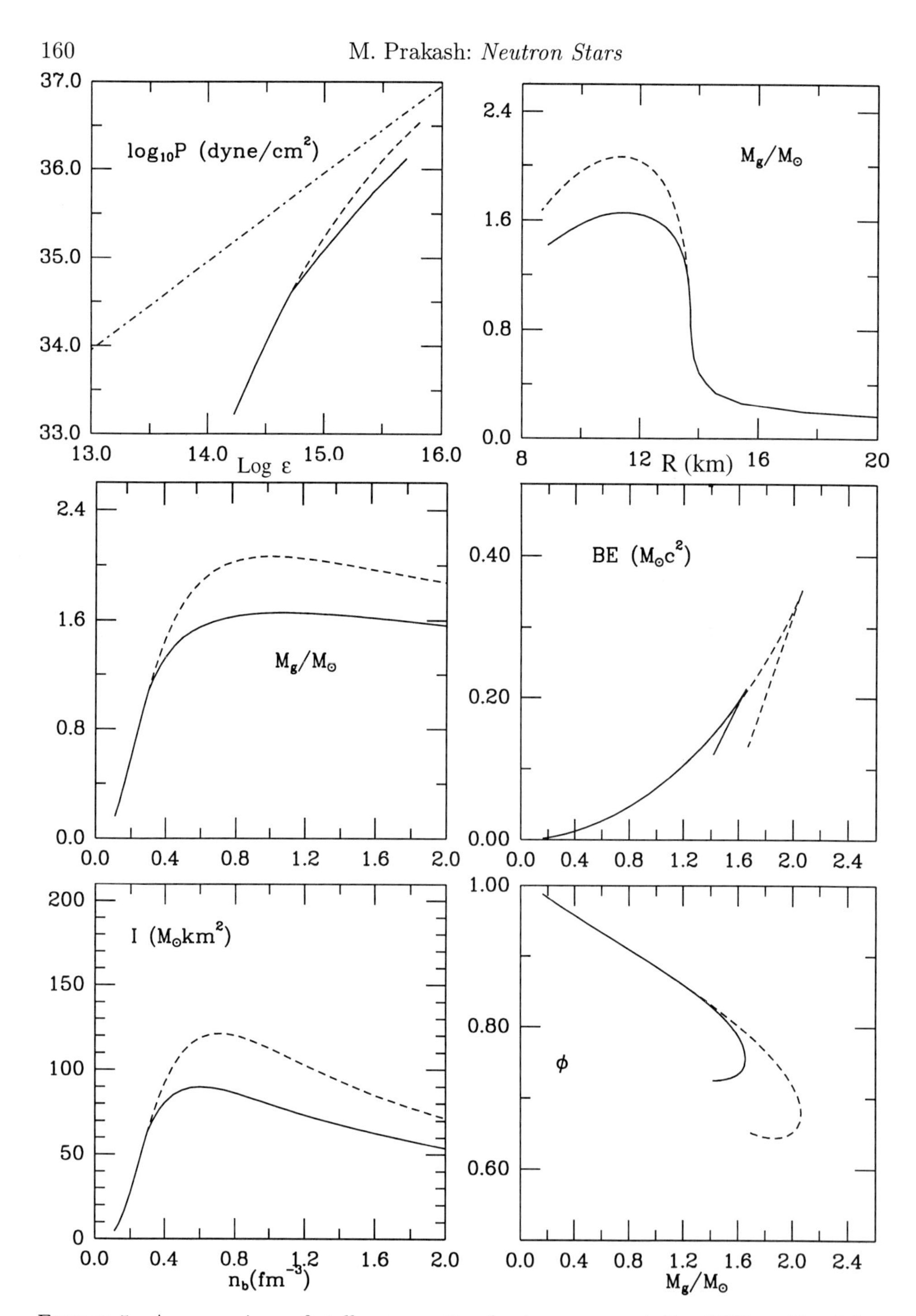

FIGURE 5. A comparison of stellar properties for two representative EOSs . Top left panel: The causal limit EOS $P = \epsilon$ is shown by the dash-dotted line. The solid and dashed curves show a soft and stiff (by comparison) EOS. Top right panel: Gravitational mass (in units of the solar mass $M_\odot$) versus radius. Center left panel: Gravitational mass versus central density. Center right panel: The binding energy versus gravitational mass. Bottom left panel: Moment of inertia versus central density. Bottom right panel: The surface red shift as a function of the gravitational mass.

the larger Keplerian frequency thus gives an upper limit.) GR calculations of rapidly rotating stars give (Lattimer *et al.* (1990))

$$\Omega_K \cong 7.7 \times 10^3 (M_{max}/M_\odot)^{1/2} (R_{max}/10 \text{ km})^{-3/2} \text{ s}^{-1}, \tag{2.7}$$

where M_{max} and R_{max} refer to the mass and radius of the non-rotating spherical configuration. It is worthwhile to note that the discovery of a sub-millisecond pulsar (say of 0.5 ms) as was purported (Kristian *et al.* (1989)) to be the case in the wake of the SN1987A explosion and which was later retracted (Pennypacker (1990)) as an erroneous measurement, would place rather severe limits on the EOS.

(*d*) Another limit (Haensel (1990),Pethick *et al.* (1995)), employs the maximum moment of inertia, expressed in terms of M_{max} and R_{max} of the non-rotating configurations:

$$I_{max} = 0.6 \times 10^{45} \frac{(M_{max}/M_\odot)(R_{max}/10 \text{ km})^2}{1 - 0.295(M_{max}/M_\odot)/(R_{max}/10 \text{ km})} \text{ g cm}^2. \tag{2.8}$$

Limits on I would severely limit the allowed region in the $M - R$ plane.

(*e*) X-ray and γ-ray observations of a new class of objects (Walter *et al.* (1996)), isolated old neutron stars with the prototype RXJ185635-3754, promises to yield accurate information about their radii, temperatures and ages. Assuming blackbody emission from the entire surface of a neutron star, an observed flux f_∞ and temperature T_∞ (the subscript ∞ denotes observations made at a very large distance from the star) implies that the local stellar radius R and mass M are related by (Lewin *et al.* (1993))

$$R_\infty \equiv D\sqrt{f_\infty/(\sigma T_\infty^4)} = R/\sqrt{1 - R_s/R}, \tag{2.9}$$

where D is the distance to the star, σ is the Stefan-Boltzmann constant and the Schwarzschild radius $R_s = 2GM/c^2 = 2.954(M/\text{M}_\odot)$ km. The largest permitted mass, $R_\infty c^2/(3^{3/2}G)$, occurs when $R = 1.5R_s$.

In Fig. 6 (left panel), the implications of the various restrictions mentioned above are considered. Besides the theoretical constraints, observational constraints imposed by mass, moments of inertia and pulsar periods are also illustrated in this figure. Also shown are the mass-radius relationships of two representative EOSs . To date, data are consistent with a wide variety of EOSs , which highlights the need for continuing observations. The right panel in this figure, where four values of R_∞ are shown, illustrates the possibility for a radius determination. An upper limit to R_∞ would constrain the mass and radius of the star to lie under that R_∞ curve and to the right of the line $R = 1.52R_s$.

3. Equation of state

3.1. *Introduction*

The many attempts to determine the EOS for dense nucleon matter can be conveniently grouped into three broad categories.

Nonrelativistic potential models: Here one starts from a Hamiltonian with a two-nucleon potential that fits nucleon-nucleon scattering data and the properties of deuteron. The quantum many-body problem is traditionally handled either by a selective summation of diagrams in perturbation theory (the Brueckner-Bethe-Goldstone approach) or by using a variational method with correlation operators (Pandharipande (1971)). Nuclear matter with two-body forces alone consistently saturates near $2n_s$ (n_s is the empirical equilibrium density) rather than n_s (Day & Wiringa (1985)). This result holds for all phase shift equivalent potentials and for the different methods (Brueckner-Bethe-Goldstone or variational) used to calculate the energy. To achieve saturation at the empirical density,

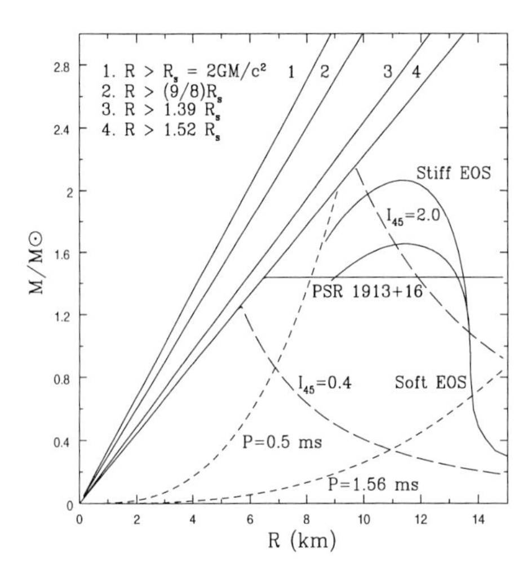

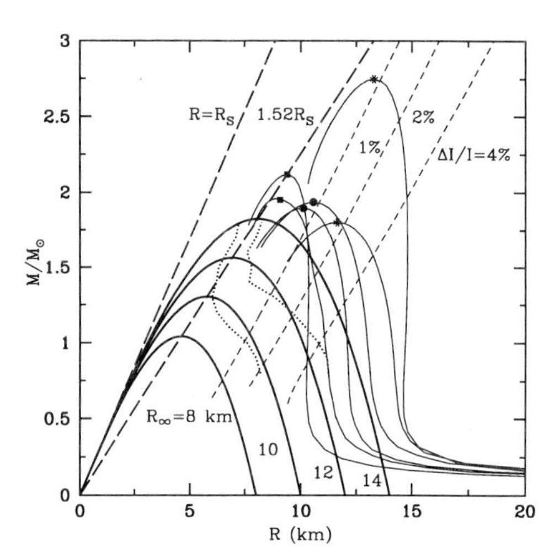

FIGURE 6. Left panel: Valid EOSs must produce neutron stars which lie to the right of curve labeled 4. Current observational constrains include $M > M_{PSR1913+16}$, $I > 0.4 \times 10^{45}$ g cm^2 (Crab pulsar) and $P_{min} < 1.6$ ms (PSR1957+20). The promise of future observations is illustrated with curves of other values of I and P. Right panel: $M - R$ curves for several recent EOSs which permit only nucleons and leptons to be present. The curves labeled $R = R_s$ and $R = 1.52R_s$ show the limits imposed by GR and GR + causality, respectively. Also displayed are contours of $R_\infty = 8, 10, 12$, and 14 km, and $\Delta I/I = 1, 2$, and 4 %. The dotted lines indicate the allowed region inferred from analyses of X-ray bursts.

Friedman & Pandharipande (1981) added a density dependent three-nucleon interaction, which effectively incorporates the suppression of non-nucleonic degrees of freedom, *e.g.*, Δ isobar resonances, in the construction of a two-body potential. Wiringa *et al.* (1988) updated such variational calculations with explicit three-nucleon terms whose form is suggested by theory and whose parameters are determined by the binding of few–body nuclei and the saturation properties of nuclear matter. This work also includes several technical improvements in the calculation of energy expectation values and a more thorough search for the best variational wave functions. These represent definite improvements over earlier calculations inasmuch as they are able to come closer to the empirical properties of nuclear matter. However, a serious shortcoming of many potential models is that the speed of sound exceeds the speed of light at densities relevant for maximum mass neutron stars. Furthermore, the high density behavior of the EOS, particularly, the symmetry energy, depends sensitively on the choice of the three-body interactions.

Field-theoretical models: Here one starts from local, Lagrangian densities with baryon and meson degrees of freedom (Serot & Walecka (1986), Serot (1992)). The models rigorously satisfy causality but sacrifice the connection to NN scattering data since the coupling constants and masses are chosen to fit only the nuclear matter saturation properties. The traditional approach is to start with a mean-field approximation and then include effects of vacuum fluctuations and correlations. At the mean-field level, calculations with and without scalar self-interactions have been reported (Serot (1979), Boguta (1981), Glendenning (1985)). Inclusion of nucleon and meson vacuum fluctuation terms at the 1-loop level generally improves fits to nuclear matter saturation properties (Chin (1977), Jackson *et al.*(1985), Prakash & Ainsworth (1987)). Recently, detailed two-loop corrections in the basic scalar-vector model have been computed (Furnstahl *et al.* (1989)). The immense size and nature of these corrections led the authors to conclude that the loop expansion is not convergent in any sense. Since the coupling constants are on the order of 10 and the model is not asymptotically free, this should have been no surprise. Prakash *et al.* (1992) argued that the composite structure of nucleons ought to soften these contributions. This was demonstrated in a model calculation, in which non-local interactions were incorporated through strong form factors at the vertices of meson-nucleon couplings. At nuclear densities and beyond, the total two-loop result was then found to be perturbatively small in comparison to the one-loop calculation.

Relativistic Dirac-Brueckner approach: This approach (Brockman & Machleidt (1984), ter Haar & Malfliet (1986), ter Haar & Malfliet (1987), Horowitz & Serot (1987)) is based on the idea that the self-energy of a nucleon in medium is made up from a large attractive scalar potential and a repulsive vector potential of comparable magnitude. The magnitude of these large self-energy terms indicate that relativistic effects are important even at low energies. The nucleon spinors that are solutions of the Dirac equation are characterized by an effective mass equal to the sum of the free mass and the scalar potential. In a self-consistent calculation these spinors are used to evaluate the matrix element of the NN potential, *e.g.*, the one-boson-exchange potential, in the nuclear medium. Following the lines of Brueckner theory, the Bethe-Goldstone equation is solved to account for the effects of NN correlations. The nucleon self-energy is evaluated in the Brueckner-Hartree-Fock approximation. There are a few calculations (ter Haar & Malfliet (1986), ter Haar & Malfliet (1987), Horowitz & Serot (1987), Muether *et al.* (1987), Engvik *et al.* (1994)) which have been carried out to high enough densities for neutron star structures to be calculated.

It is beyond the scope of these lectures to describe the details of these methods. Each method has it's own merit, but is not without serious shortcomings to address the issue of the high density EOS. We will, therefore, concentrate on the basic ingredients that enter the calculations of neutron star structure. Towards this end, a phenomenological approach that reproduces the pressure-density relationships of the more microscopic approaches will be utilized to highlight the physics issues.

3.2. *Nuclear matter*

Let us begin with the state variables for matter with equal numbers of neutrons (N) and protons (Z) contained in a large volume V at zero temperature. Let $n = (N+Z)/V = n_n + n_p$ denote the number density (the subscripts n and p stand for neutron and proton, respectively) of nucleons. The nucleon density is related to the Fermi momentum k_F by $n = 2k_F^3/(3\pi^2)$. Given the energy density $\epsilon(n)$ inclusive of the rest mass density mn, denote the energy per particle by $E/A = \epsilon/n$, where $A = N + Z$. In terms of $\epsilon(n)$, the other state variables are easily derived.

Pressure: From thermodynamics, we have

$$
\begin{aligned}
P = & -\frac{\partial E}{\partial V} = - \frac{dE}{d(A/n)} \\
& = n^2 \frac{d(\epsilon/n)}{dn} = n \frac{d\epsilon}{dn} - \epsilon = n\mu - \epsilon \,,
\end{aligned}
\tag{3.10}
$$

where $\mu = d\epsilon/dn$ is the chemical potential inclusive of the rest mass m. At the equilibrium density n_0, where $P(n_0) = 0$, one has the identity $\mu = \epsilon/n = E/A$.

Incompressibility: The compressibility χ is usually defined by

$$
\chi = -\frac{1}{V}\frac{\partial V}{\partial P} = \frac{1}{n}\left(\frac{dP}{dn}\right)^{-1}
\tag{3.11}
$$

However, in nuclear physics applications, the incompressibility factor

$$K(n) = 9\frac{dP}{dn} = 9\,n\frac{d^2\epsilon}{dn^2}\,, \qquad \text{or}$$
$$= 9\,\frac{d}{dn}\left[n^2\frac{d(E/A)}{dn}\right] = 9\,\left[n^2\frac{d^2(E/A)}{dn^2} + 2n\frac{d(E/A)}{dn}\right] \tag{3.12}$$

is more commonly used. At the equilibrium density n_0, where $d(E/A)/dn = 0$, this is referred to as the compression modulus:

$$K(n_0) = 9n_0^2\left.\frac{d^2(E/A)}{dn^2}\right|_{n_0} = k_F^{0^2}\left.\frac{d^2(E/A)}{dk_F^2}\right|_{k_F^0}\,. \tag{3.13}$$

Above, $k_F^0 = (3\pi^2 n_0/2)^{1/3}$ denotes the equilibrium Fermi momentum. Note that the factor 9 above serves to make Eq. (3.12) consistent with the standard definition of the compression modulus given in the rightmost factor of Eq. (3.13). For densities $n \gg n_0$, the second term in Eq. (3.12) gives sizeable contributions to $K(n)$ as we shall see later.

Adiabatic sound speed: The propagation of small scale density fluctuations occurs at the sound speed obtained from the relation

$$\left(\frac{c_s}{c}\right)^2 = \frac{dP}{d\epsilon} = \frac{dP/dn}{d\epsilon/dn} = \frac{1}{\mu}\frac{dP}{dn} = \frac{d\ln\mu}{d\ln n}\,. \tag{3.14}$$

Alternative relations for the sound speed squared are

$$\left(\frac{c_s}{c}\right)^2 = \frac{K}{9\mu} = \Gamma\frac{P}{P+\epsilon}\,, \tag{3.15}$$

where $\Gamma = d\ln P/d\ln\epsilon$ is the adiabatic index. It is desirable to require that the sound speed does not exceed that of light.

As a simple example, consider the schematic equation of state (EOS) given by

$$\epsilon_{nm} = mn_0u + \frac{3}{5}E_F^0 n_0 u^{5/3} + V(u) \tag{3.16}$$

where m is the nucleon mass and $u = n/n_0$ denotes the density compression ratio. The second term in the brackets above is the mean kinetic energy of isospin symmetric matter. The potential contribution $V(u)$ is given by (Prakash *et al.* (1988))

$$V(u) = \frac{1}{2}An_0u^2 + \frac{Bn_0u^{\sigma+1}}{1+B'u^{\sigma-1}} + u\sum_{i=1,2}C_i4\int\frac{d^3k}{(2\pi)^3}\,g(k,\Lambda_i)\theta(k_F-k)\,. \tag{3.17}$$

The function $g(k,\Lambda_i)$ is suitably chosen to simulate finite range effects. Following Gale *et al.* (1987), we choose the simple form $g(k,\Lambda_i) = [1+(k/\Lambda_i)^2]^{-1}$, which leads to a closed form expression (at zero temperature) for the last term:

$$V^{fr} = 3n_0u\sum_{i=1,2}C_i\left(\frac{\Lambda_i}{p_F^0}\right)^3\left(\frac{p_F}{\Lambda_i} - \arctan\frac{p_F}{\Lambda_i}\right)\,, \tag{3.18}$$

where p_F is the Fermi momentum, related to p_F^0 by $p_F = p_F^0u^{1/3}$. The parameters Λ_1 and Λ_2 parametrize the finite range forces between nucleons. Specifically, the values used were $\Lambda_1 = 1.5p_F^0$ and $\Lambda_2 = 3p_F^0$. The parameters A, B, σ, C_1, C_2, and B', a small parameter introduced to maintain causality, are determined from constraints provided by the properties of nuclear matter at the saturation: $E/A - m = -16$ MeV, $n_0 = 0.16$ fm^{-3}, nucleon effective mass $m^* = 0.7m$, the single particle potential depth at zero momentum $U(n_0) = -76.34$ MeV. With appropriate choices of the parameters, it is possible to vary the nuclear incompressibility K_0 so that the dependence on the stiffness of the EOS

K_0	A	B	B'	σ	C_1	C_2
120	75.94	−30.88	0	0.498	−83.84	23
180	440.94	−213.41	0	0.927	−83.84	23
240	−46.65	39.45	0.3	1.663	−83.84	23

TABLE 1. Parameters in Eq. (3.17) for some input values of the compression modulus K_0 of symmetric nuclear matter. All energies in MeV.

may be explored. Some numerical values of the parameters, appropriate for symmetric nuclear matter, are given in Table 1.

3.3. *Neutron–rich matter*

We turn now to matter with unequal numbers of neutrons (N) and protons (Z). Let $\alpha = (n_n - n_p)/n$ denote the excess neutron density and $n = n_n + n_p$ denote the total baryon density. In astrophysical problems, the state variables are commonly expressed in terms of the proton fraction $x = n_p/n = (1-\alpha)/2$. The neutron and proton densities are given by

$$n_n = \frac{(1+\alpha)}{2}\, n = (1-x)\, n\,, \quad n_p = \frac{(1-\alpha)}{2}\, n = x\, n\,. \tag{3.19}$$

For nuclear matter, $\alpha = 0$, while, for pure neutron matter, $\alpha = 1$. Correspondingly, $x = 0.5$ for nuclear matter and $x = 0$ for pure neutron matter. It will be useful to write the energy per particle (by simplifying the notation from E/A to E) as

$$\begin{aligned} E(n,\alpha) &= \ E(n,\alpha=0) \ + \Delta E_{kin}(n,\alpha) + \Delta E_{pot}(n,\alpha)\,, \quad \text{or} \\ E(n,x) &= E(n,x=1/2) + \Delta E_{kin}(n,x) + \Delta E_{pot}(n,x)\,, \end{aligned} \tag{3.20}$$

where the first terms above refer to the energy of symmetric nuclear matter, and the second and third terms respectively denote the contributions of the isospin asymmetric parts arising from the kinetic and interaction terms in the many–body hamiltonian. In a non–relativistic description, the kinetic energy density of neutron–rich matter is

$$\begin{aligned} \epsilon_{kin}(n,\alpha) &= \frac{3}{5}\frac{\hbar^2}{2m}\left[(3\pi^2 n_n)^{2/3}\; n_n + (3\pi^2 n_p)^{2/3}\; n_p\right] \\ &= n\,\langle E_F\rangle \cdot \frac{1}{2}\left[(1+\alpha)^{5/3} + (1-\alpha)^{5/3}\right]\,, \end{aligned} \tag{3.21}$$

where $\langle E_F\rangle = (3/5)(\hbar^2/2m)(3\pi^2 n/2)^{2/3}$ is the mean kinetic energy of nuclear matter. Thus, the second term in Eq. (3.20) takes the form

$$\begin{aligned} \Delta E_{kin}(n,\alpha) &= E_{kin}(n,\alpha) - E_{kin}(n,\alpha=0) \\ &= \langle E_F\rangle \cdot \left[\frac{1}{2}\left\{(1+\alpha)^{5/3} + (1-\alpha)^{5/3}\right\} - 1\right]\,, \end{aligned} \tag{3.22}$$

or, using $\alpha = (1-2x)$,

$$\begin{aligned} \Delta E_{kin}(n,x) &= E_{kin}(n,x) - E_{kin}(n,x=0) \\ &= \langle E_F\rangle \cdot \left[2^{2/3}\left\{(1-x)^{5/3} + x^{5/3}\right\} - 1\right]\,. \end{aligned} \tag{3.23}$$

Thus, at least for the kinetic energy per particle, the isospin asymmetry dependence may be readily separated from the baryon density dependence. Specializing to pure neutron matter, we have the useful result that

$$\Delta E_{kin}(n,\alpha=1) = \Delta E_{kin}(n,x=0) = \langle E_F\rangle \cdot (2^{2/3}-1)\,. \tag{3.24}$$

It is instructive to examine $\Delta E_{kin}(n,\alpha)$ expressed as a power series in α^2:

$$\Delta E_{kin}(n,\alpha) = \langle E_F\rangle \cdot \frac{5}{9}\alpha^2\left(1+\frac{\alpha^2}{27}+\cdots\right) = \frac{1}{3}\;E_F\cdot\alpha^2\left(1+\frac{\alpha^2}{27}+\cdots\right), \tag{3.25}$$

which shows that the quadratic term above offers a useful approximation to explore the isospin dependence of the state variables.

Given that the bulk symmetry energy deduced from experiments on laboratory nuclei is of order 30 MeV, and that the contribution from kinetic energy amounts to $E_F^0/3 \simeq (12-13)$ MeV, the potential interactions must give a sizeable contribution to the total bulk symmetry energy. Potential contributions calculated from the many–body theory of neutron–rich matter indicate that the quadratic approximation in α or $(1-2x)$ is generally adequate for $\Delta E_{pot}(n,\alpha)$ also. However, for densities much greater than the equilibrium density, there is not, as yet, general agreement on the density dependence of the symmetry energy arising from the potential interactions. This, therefore, suggests that the energy per particle of neutron–rich matter may be written as

$$E(n,x) = E(n,1/2) + S_2(n)\;(1-2x)^2 + S_4(n)\;(1-2x)^4+\cdots\,, \tag{3.26}$$

where the density dependence of the terms $S_2(n), S_4(n),\cdots$ may be adjusted to agree with the results of microscopic calculations. To a very good approximation, it is sufficient to retain in the above expansion only the quadratic term and write the total nuclear symmetry energy $S(u)$ as the sum of kinetic and potential terms:

$$S(u) = (2^{2/3}-1)\;\frac{3}{5}E_F^0\left(u^{2/3}-F(u)\right)+S_0F(u) \tag{3.27}$$

where $S_0 \simeq 30$ MeV is the bulk symmetry energy parameter. The function $F(u)$ parametrizes the potential contribution to the symmetry energy and satisfies $F(0)=0$ and $F(1)=1$. For example, three representative forms that mimic the results of more microscopic models are:

$$F(u)=u\,,\qquad F(u)=\frac{2u^2}{1+u}\qquad\text{and}\qquad F(u)=\sqrt{u}\,. \tag{3.28}$$

We will now establish some general relationships involving the chemical potentials. Given $\epsilon \equiv \epsilon(n_n,n_p)$, the neutron and proton chemical potentials are given by

$$\mu_n = \left.\frac{\partial\epsilon}{\partial n_n}\right|_{n_p}\,,\qquad \mu_p = \left.\frac{\partial\epsilon}{\partial n_p}\right|_{n_n}\,. \tag{3.29}$$

The partial derivatives above may be expressed in terms of the baryon density n and the proton fraction x by noting that

$$\frac{\partial}{\partial n_n} = \frac{\partial}{\partial n}\left.\frac{\partial n}{\partial n_n}\right|_{n_p} + \frac{\partial}{\partial x}\left.\frac{\partial x}{\partial n_n}\right|_{n_p}\,,\qquad \frac{\partial}{\partial n_p} = \frac{\partial}{\partial n}\left.\frac{\partial n}{\partial n_p}\right|_{n_n} + \frac{\partial}{\partial x}\left.\frac{\partial x}{\partial n_p}\right|_{n_n}\,, \tag{3.30}$$

where

$$\left.\frac{\partial n}{\partial n_n}\right|_{n_p} = 1 = \left.\frac{\partial n}{\partial n_p}\right|_{n_n}\,,\qquad \left.\frac{\partial x}{\partial n_n}\right|_{n_p} = -\frac{x}{n}\,,\qquad \left.\frac{\partial x}{\partial n_p}\right|_{n_n} = \frac{(1-x)}{n}\,, \tag{3.31}$$

which follows from the definitions $n=n_n+n_p$ and $x=n_p/n$. Utilizing the definitions $E=\epsilon/n$ and $u=n/n_0$, we obtain the relations

$$\mu_n = E + u\left.\frac{\partial E}{\partial u}\right|_x - x\left.\frac{\partial E}{\partial x}\right|_n\,,$$

$$\mu_p = \mu_n + \left.\frac{\partial E}{\partial x}\right|_n , \quad \widehat{\mu} = \mu_n - \mu_p = - \left.\frac{\partial E}{\partial x}\right|_n . \tag{3.32}$$

The quantity $\widehat{\mu}$ will play an important role in determining the composition of charge neutral neutron star matter. Employing the parametrization in Eq. (3.26) for the energy per particle of baryons, we notice that

$$\widehat{\mu} = 4(1-2x)\ \left[S_2(n) + 2S_4(n)\ (1-2x)^2 + \cdots\right] , \tag{3.33}$$

which shows that the magnitude of $\widehat{\mu}$ is in large part governed by the magnitude and the density dependence of the symmetry energy.

3.4. *Charge neutral neutron–rich matter*

Cold catalyzed neutron stars are in equilibrium with respect to weak interactions. We are thus led to find the ground state of a system of strongly interacting hadrons and weakly interacting leptons that are generated in both β-decay and inverse β-decay processes of the hadrons. The energy must be minimized, simultaneously satisfying local charge neutrality and chemical equilibrium. The chemical equilibrium conditions are enumerated below, assuming, to begin with, that matter consists of only neutrons, protons and electrons. For the densities to be considered, the processes $n \to p + e^- + \overline{\nu}_e$ and $p + e^- \to n + \nu_e$ take place simultaneously. In cold catalyzed neutron star matter, we may also assume that neutrinos generated in these reactions have left the system. This implies that

$$\widehat{\mu} = \mu_n - \mu_p = \mu_e , \tag{3.34}$$

where at the temperatures of interest here, the chemical potentials of the participating Fermions are close to their respective Fermi energies. The total energy per particle is obtained by adding the lepton energy $E_\ell(x) = E_e(x)$ to the specific energy of baryons $E_b(n,x)$, and in beta equilibrium, one has

$$\frac{\partial}{\partial x}\left[E_b(n,x) + E_e(x)\right] = 0 . \tag{3.35}$$

The charge neutrality condition implies that $n_e = n_p = nx$, or, $k_{F_e} = k_{F_p}$. Combining the results in Eq. (3.33) and Eq. (3.34), we arrive at the relation

$$4(1-2x)\ \left[S_2(n) + 2S_4(n)\ (1-2x)^2 + \cdots\right] = \hbar c\left(3\pi^2 nx\right)^{1/3} , \tag{3.36}$$

which determines the equilibrium proton fraction $\widetilde{x}(n)$, once the density dependent symmetry energy terms $S_2(n), S_4(n), \cdots$ are specified. When higher order terms such as $S_4(n), \cdots$ may be neglected in comparison with $S_2(n)$, Eq. (3.36) takes the form of a cubic equation in $\widetilde{x}$:

$$\beta\,\widetilde{x} - (1-2\widetilde{x})^3 = 0 , \tag{3.37}$$

where $\beta = 3\pi^2 n\ (\hbar c/4S_2)^3$. The physically relevant solution of this equation is

$$\widetilde{x} = \frac{1}{2} - \frac{1}{4}\left\{\left[2\beta(\gamma+1)\right]^{1/3} - \left[2\beta(\gamma-1)\right]^{1/3}\right\} , \quad \gamma = \left(1 + \frac{2\beta}{27}\right)^{1/2} . \tag{3.38}$$

For $u \le 1$, the equilibrium proton fraction is small and to a good approximation is given by $\widetilde{x} \simeq (\beta+6)^{-1}$, from which the high sensitivity to the density dependent symmetry energy, which favors the addition of protons to matter, is evident.

When the electron Fermi energy is large enough, *i.e.* $\mu_e \ge m_\mu c^2$, it is energetically favorable for the electrons to convert to muons $e^- \to \mu^- + \overline{\nu}_\mu + \nu_e$. Denoting the muon chemical potential by μ_μ, the chemical equilibrium established by the above process and its inverse is given by $\mu_\mu = \mu_e$. At the threshold for muons to appear, $\mu_\mu = m_\mu c^2 \sim 105$

MeV. By inspecting Eqs. (3.33) and (3.34) and noting that the proton fraction at nuclear density is small, one has the approximate relationship $4S_2(u)/m_\mu c^2 \sim 1$. Using a typical value $S(u=1) \simeq 30$ MeV, one may expect muons to appear roughly at nuclear density $n_0 = 0.16$ fm^{-3}. Above the threshold density,

$$\mu_\mu = \sqrt{k_{F_\mu}^2 + m_\mu^2 c^4} = \sqrt{(\hbar c)^2(3\pi^2 n x_\mu)^{2/3} + m_\mu^2 c^4}\,, \tag{3.39}$$

where $x_\mu = n_\mu/n_b$ is the muon fraction in matter. The charge neutrality condition now takes the form $n_e + n_\mu = n_p$, which, together with the relation of chemical equilibrium in Eq. (3.39), establishes the lepton and proton fractions in matter. The appearance of muons has the consequence that the electron fraction x_e is lower than its value without the presence of muons.

Since the electromagnetic interactions give negligible contributions, it is sufficient to consider the non–interacting forms for the energy density and pressure of the leptons:

$$\epsilon_\ell = 2\int \frac{d^3k}{(2\pi)^3}\sqrt{k^2+m_\ell^2}\,, \quad P_\ell = \frac{1}{3}\cdot 2\int \frac{d^3k}{(2\pi)^3}\frac{k^2}{\sqrt{k^2+m_\ell^2}} \tag{3.40}$$

The total energy density and pressure of charge neutral and chemically equilibrated matter is then

$$\epsilon_{tot} = \epsilon_b + \sum_{\ell=e^-,\mu^-} \epsilon_\ell\,, \quad P_{tot} = P_b + \sum_{\ell=e^-,\mu^-} P_\ell\,, \tag{3.41}$$

where ϵ_b and P_b are the energy density and pressure, respectively, of the baryons. For the parametrization employed in Eqs. (3.16) and (3.26), one has

$$\epsilon_b = mn_0u + \left\{\frac{3}{5}E_F^0 n_0 u^{5/3} + V(u)\right\} + n_0(1-2x)^2 uS(u)\,, \tag{3.42}$$

$$P_b = \left\{\frac{2}{5}E_F^0 n_0 u^{5/3} + \left(u\frac{dV}{du} - V\right)\right\} + n_0(1-2x)^2 u^2\frac{dS}{du}\,. \tag{3.43}$$

3.5. *State variables at nuclear density*

Let us estimate the magnitudes of the energy, pressure and the chemical potentials of stellar matter at the equilibrium density of nuclear matter: $u=1$. These may be contrasted with the (more familiar) corresponding quantities for symmetric nuclear matter.

Start with an approximate form for the density dependence of the symmetry energy

$$S_2(u) = (13u^{2/3} + 17u)\ \text{MeV}\,. \tag{3.44}$$

Above, the first term gives the kinetic contribution and the second term arises from potential interactions. The linear density dependence of the potential interactions holds for $u \simeq 1$. At nuclear density, the muon concentration is small. Thus, when electrons account for the bulk of the leptons, we may use $\tilde{x} = (\beta+6)^{-1}$ to estimate the equilibrium proton fraction $\tilde{x} \simeq x_e$. For $S_2(1) = 30$ MeV, $\tilde{x} = 0.037$. Utilizing this value and Eq. (3.44) in Eqs. (3.42) and (3.43), the energy and pressure of baryons is easily obtained. Notice that for $u=1$, the terms in braces in Eq. (3.42) give a contribution of -16 MeV to the specific energy, while the terms in braces in Eq. (3.43) vanishes for the pressure. Alternatively, the result in Eq. (3.36) may be employed to get (in units of MeV)

$$\frac{\epsilon_b}{n} - m = -16 + \frac{\hbar c}{4}(3\pi^2 n_0)^{1/3}\tilde{x}^{1/3}(1-2\tilde{x})^{1/3} \tag{3.45}$$

$$P_b = \frac{\hbar c}{4}(3\pi^2 n_0)^{1/3}\tilde{x}^{1/3}(1-2\tilde{x})^{1/3}\,. \tag{3.46}$$

For the chemical potentials, the simple density dependence in Eq. (3.44) and the

Quantity	$\tilde{x}$	$\epsilon_b/n-m$	ϵ_e/n	P_b	P_e	μ_n-m	μ_p-m	μ_e
Nuclear	0.5	−16	0	0	0	−16	−16	0
Stellar	0.037	9.6	3.18	3.5	0.17	35.74	−75.14	110.88

TABLE 2. State variables at nuclear density. Energies in MeV and pressure in MeV fm^{-3}. The numerical estimates are based on an assumed symmetry energy $S_2(u) = 13u^{2/3} + 17u$.

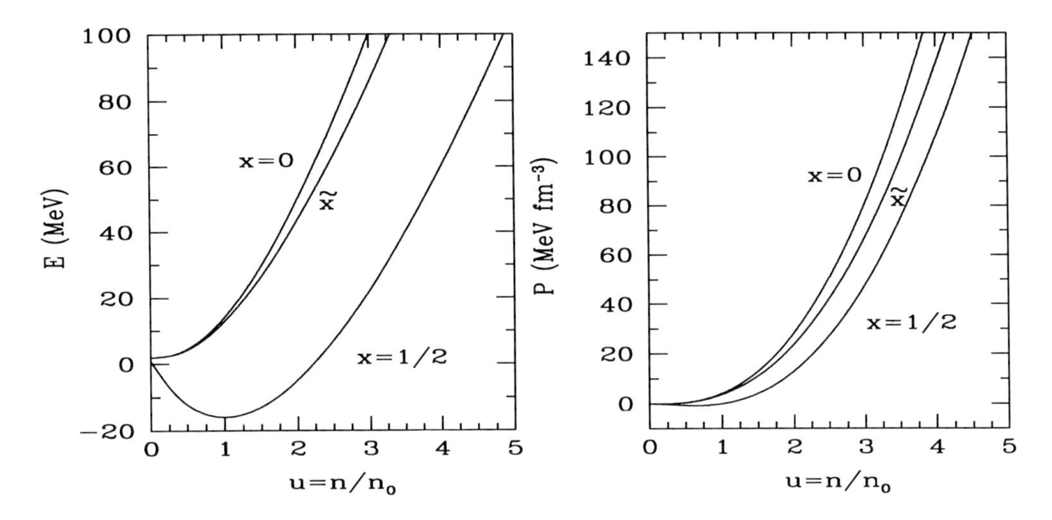

FIGURE 7. The energy and pressure of nuclear matter ($x = 1/2$), pure neutron matter ($x = 0$) and of stellar matter in β-equilibrium ($x = \tilde{x}$).

quadratic behavior of the energy with $(1-2x)$ allow the derivatives

$$\left(\frac{\partial E}{\partial u}\right)_x = (1-2x)^2\frac{\partial S_2}{\partial u}\,; \quad \left(\frac{\partial E}{\partial x}\right)_u = -4S_2(1-2x) \tag{3.47}$$

to be readily evaluated.

Compared to the baryons, electrons (assumed massless here) give relatively small contributions to the energy density and pressure:

$$\epsilon_e = 3P_e = \frac{3}{4}(3\pi^2 n_0)^{1/3} n_0 Y_e^{4/3} = 0.51\times 10^{-3}\ \text{MeV fm}^{-3}\,, \tag{3.48}$$

since their abundance $Y_e = n_e/n_b = \tilde{x}$ is rather small at $u = 1$.

Table 2 provides a comparison of the various physical quantities of interest between symmetric nuclear matter and stellar matter in beta equilibrium. In Fig. 7, we show the energy and pressure of stellar matter in β-equilibrium. For comparison, results for pure neutron matter and isospin symmetric nuclear matter are also shown. Results shown are for the EOS in Eq. (3.20) with parameters in the third row of Table 1. It is evident that the state variables of stellar matter are bracketed between those of pure neutron matter and isospin symmetric nuclear matter.

3.6. *EOS and the maximum mass of neutron stars*

The maximum neutron star mass (M_{max}), subject only to the constraints of causality and general relativity, has been shown (Nauenberg & Chapline (1973)) to be $\sim 3\ M_\odot$ ($M_\odot \simeq 2 \times 10^{33}$ gm). Observed neutron star masses generally lie well below this value. Can the observed masses limit the EOS of dense matter? Can the parameters of symmetric nuclear matter, which have the potential of being measured in other contexts, be constrained? The answers hinge on whether or not the bulk of the matter in maximum mass neutron stars lies close to symmetric matter saturation density, $n_0 \simeq 0.16$ fm^{-3}. If the bulk lies at much higher values the limits to symmetric matter parameters cannot be effectively constrained.

In what follows, we examine the dependence of neutron star structure on the EOS. The importance of this sensitivity has bearing on other issues as well. The EOS so obtained must be consistent with the EOS inferred from the analysis of giant resonances in laboratory nuclei (Blaizot (1980)). It must also correspond to what is obtained in the studies of particle multiplicities and the matter, momentum and energy flows in heavy ion collisions (Bertsch & Dasgupta (1988)). Finally, there may be a significant effect of the EOS on supernova simulations, as the strength of a supernova shock is correlated (van Riper (1980), Baron *et al.* (1985)), with the behavior of the EOS near n_0.

The properties of maximum mass neutron stars, obtained by integrating the TOV equation, are displayed in Table 3 for the parameters shown in Table 1 and $S_0 = 30$ MeV. The choices for $F(u)$ mimic the results of the potential and hybrid models referred to earlier. Results of M_{max} of a star with pure neutrons are shown in parentheses. For a given parametrization, M_{max} roughly scales (Bludman (1973), Cooperstein (1988)) as $K_0^{1/2}$, and the stiffer the EOS the less important the symmetry energy (Prakash & Ainsworth (1987)). However, even in the case $K_0 = 120$ MeV, relatively large neutron stars (with $M_{max} \sim 1.5\ M_\odot$) are possible. In addition, the more rapidly rising is the symmetry energy, the stiffer is the overall EOS. Changing the parameter S_0 to 35 MeV has an almost negligible effect on beta–equilibrium stars, although the M_{max} of a pure neutron star is somewhat increased.

It is instructive to examine the fraction of the gravitational massexterior to a given radius versus the density at that radius for the maximum mass configurations. Independently of the EOS employed here, most of the mass lies at densities higher than 3 n_0. Only for very stiff EOS's is this finding modified. Therefore, it is mostly the high density properties that play an important role in neutron star structure Prakash *et al.* (1988).

These results demonstrate that observed neutron star masses are consistent with K_0 smaller than 140 MeV, using a plausible EOS that is consistent with empirical nuclear matter properties and causality. Other equally plausible EOS's have set a relatively larger lower limit to K_0, e.g. , $M_{max} = (1.36,\ 1.45,\ 1.50) M_\odot$ were found for $K_0 = (199,\ 225,\ 240)$ MeV by Prakash & Ainsworth (1987) and $M_{max} = (0.8,\ 1.3,\ 1.5,\ 1.8) M_\odot$ were found for $K_0 = (100,\ 200,\ 240,\ 285)$ MeV by Glendenning (1986). In both these (field theoretic) calculations the potential contribution to the symmetry energy varied linearly with density. The calculations of Glendenning (1986) included hyperons, which substantially reduced M_{max}. However, there are considerable uncertainties regarding the importance and abundances of hyperons in dense neutron star matter. The fact is that very little is known about hyperonic potentials from experiments. A further uncertainty, at the densities where hyperons might be abundant, is a possible transition to quark matter, in which case the role of hyperons may be reduced. The results shown above imply

$F(u)$	K_0 (MeV)	$\frac{M_{max}}{M_\odot}$	R (km)	$\frac{n_c}{n_0}$	I ($M_\odot$ km^2)	B.E (10^{53} ergs)
	120	1.458(1.70)	9.114	10.841	43.83	3.431
u	180	1.722(1.90)	9.879	8.680	66.73	4.733
	240	1.935(2.07)	10.57	7.269	90.93	5.890
	120	1.470(1.95)	9.895	9.631	49.36	3.260
$2u^2/(1+u)$	180	1.738(2.10)	10.318	8.166	70.82	4.572
	240	1.952(2.24)	10.933	6.953	95.32	5.734
	120	1.404(1.45)	8.435	12.28	37.75	3.476
$\sqrt{u}$	180	1.679(1.71)	9.324	9.46	60.07	4.812
	240	1.895(1.92)	10.112	7.740	83.83	5.954

TABLE 3. Neutron star properties. The mass (M_{max}), radius (R), central density (n_c), moment of inertia (I) and gravitational binding energy ($B.E.$) of the maximum mass neutron star for different choices of the potential contribution to the symmetry energy $F(u)$ and compression modulus K_0. Results for the maximum mass with pure neutrons are shown in parentheses.

that it is possible to construct an EOS at high densities, with acceptable saturation and causality properties, and including hyperons, which nevertheless has $M_{max} \geq 1.5M_\odot$.

The above discussion shows that the relationship of M_{max} to K_0 is extremely model dependent, and this is the main point. The compression modulus K_0 of symmetric matter by itself does not provide a good model-independent basis for contrasting the structure of neutron stars. This is because the high density EOS is uncertain, and there is no unique way to link the high density EOS to properties around n_0. Evidently, star properties may not yet be firmly linked to equilibrium properties of symmetric nuclear matter.

4. Exotica

The physical state and internal constitution of neutron stars chiefly depends on the nature of strong interactions. Although the composition and the EOS of neutron star matter are not yet known with certainty, QCD based effective Lagrangians have opened up intriguing possibilities. Among these is the possible existence of matter with strangeness to baryon ratio, $|S|/B$, of order unity. Strangeness may occur in the form of fermions, notably the Λ and Σ^- hyperons (Glendenning (1985), Kapusta & Olive (1990), Ellis *et al.* (1991)), or as a Bose condensate, such as a K^- meson condensate (Kaplan & Nelson (1986), Brown *et al.* (1987), Politzer & Wise (1991), Brown *et al.* (1992)), or in the form of strange quarks in a mixed phase of hadrons and quarks (Freedman & McLerran (1978), Glendenning (1992)). *All these alternatives involve negatively charged matter*, which, if present in dense matter, results in important consequences for neutron stars. For example, the appearance of strangeness-bearing components results in protoneutron (newly born) stars having larger maximum masses than catalyzed (older, neutrino-free) neutron stars, a reversal from ordinary nucleons-only matter (Thorsson *et al.* (1994), Keil & Janka (1994), Prakash *et al.* (1995)). This permits the existence of metastable protoneutron stars that could collapse to black holes during their deleptonization (Prakash (1993), Brown & Bethe (1994)). In older stars, the presence of such components also implies rapid cooling of the star's interior via the direct Urca processes (Lattimer *et al.* (1991),

	M	J	I	I_3	q	s	Y	Quark content
N	939	$\frac{1}{2}$	$\frac{1}{2}$	$-\frac{1}{2}$	0	0	1	udd
				$\frac{1}{2}$	1	0		uud
Δ	1232	$\frac{3}{2}$	$\frac{3}{2}$	$-\frac{3}{2}$	-1	0	1	ddd
				$-\frac{1}{2}$	0			udd
				$\frac{1}{2}$	1			uud
				$\frac{3}{2}$	2			uuu
Λ	1115	$\frac{1}{2}$	0	0	0	-1	0	uds
Σ	1190	$\frac{1}{2}$	1	-1	-1	-1	0	dds
				0	0			uds
				1	1			uus
Ξ	1315	$\frac{1}{2}$	$\frac{1}{2}$	$-\frac{1}{2}$	-1	-2	-1	dss
				$\frac{1}{2}$	0			uss
Ω	1673	$\frac{3}{2}$	0	0	-1	-3	-2	sss

TABLE 4. Baryons and their charge states. Mass M in MeV. Spin is J, isospin is I, its third component I_3, charge is q, strangeness is s, and $Y = B + s$ is hypercharge, where B is the baryon number.

Pethick (1992), Prakash (1994)). Interpretation of the surface temperatures of neutron stars in conjunction with different possibilities for the stars' core cooling is currently a topic of much interest (Ogelman (1993)).

4.1. *Matter with strangeness–rich baryons*

Many calculations of the composition of dense matter lead to the conclusion that baryons more massive than the nucleon, namely hyperons and nucleon isobars, should be present in dense matter. We turn now to examine some general relationships that are satisfied in such a multi–component system.

The simplest possible neutrino emitting processes are of the type

$$B_1 \to B_2 + \ell + \bar{\nu}_\ell\,; \quad B_2 + \ell \to B_1 + \nu_\ell\,, \tag{4.49}$$

where B_1 and B_2 are baryons, and ℓ is a lepton, either an electron or a muon. Here we shall consider processes where the baryons can be strange particles (Λ's, Σ's, and Ξ's) or non–strange nucleon isobars (Δ^0, Δ^-), as well as nucleons. The basic properties of these baryons are listed in Table 4.

Beta equilibrium implies that the chemical potentials satisfy the condition $\mu_{B_1} = \mu_{B_2} + \mu_\ell$, which is equivalent to the general condition for chemical equilibrium

$$\mu_i = b_i\mu_n - q_i\mu_\ell, \tag{4.50}$$

where b_i is the baryon number of particle i and q_i is its charge. This implies the equalities

$$\mu_{\Sigma^0} = \mu_\Lambda = \mu_{\Xi^0} = \mu_n\,, \quad \mu_{\Sigma^-} = \mu_{\Xi^-} = \mu_n + \mu_e\,, \quad \mu_{\Sigma^+} = \mu_p = \mu_n - \mu_e\,, \tag{4.51}$$

for the hyperons that are likely to be of relevance. These relations also show that, in equilibrium, there exist only two independent chemical potentials, μ_n and μ_e, corresponding to the conservation of baryon number and electric charge.

4.1.1. *Model calculations*

The field-theoretical model to be described below, studied extensively by Glendenning Glendenning (1985), illustrates the influence of hyperons in dense matter. Further work can be found, e.g., in Glendenning & Moszkowski (1991), Ellis *et al.* (1995), Prakash *et al.* (1997). Here, interactions between the baryons are mediated by the exchange of σ, ω and ρ mesons. The full Lagrangian density is given by

$$\begin{aligned}\mathcal{L} = & \sum_B \bar{B}(-i\gamma^\mu\partial_\mu - g_{\omega B}\gamma^\mu\omega_\mu - g_{\rho B}\gamma^\mu\mathbf{b}_\mu\cdot\mathbf{t} - M_B + g_{\sigma B}\sigma)B \\ & - \frac{1}{4}F_{\mu\nu}F^{\mu\nu} + \frac{1}{2}m_\omega^2\omega_\mu\omega^\mu - \frac{1}{4}\mathbf{B}_{\mu\nu}\mathbf{B}^{\mu\nu} + \frac{1}{2}m_\rho^2\mathbf{b}_\mu\cdot\mathbf{b}^\mu \\ & + \frac{1}{2}\partial_\mu\sigma\partial^\mu\sigma + \frac{1}{2}m_\sigma^2\sigma^2 - U(\sigma) + \sum_l \bar{\Psi}(-i\gamma^\mu\partial_\mu - m_l)\Psi\,.\end{aligned}$$

Here, B are the Dirac spinors for baryons and $\mathbf{t}$ is the isospin operator. The sums include baryons $B = n, p, \Lambda, \Sigma$, and Ξ, and leptons $l = e^-$ and μ^-. The field strength tensors for the ω and ρ mesons are $F_{\mu\nu} = \partial_\mu\omega_\nu - \partial_\nu\omega_\mu$ and $\mathbf{B}_{\mu\nu} = \partial_\mu\mathbf{b}_\nu - \partial_\nu\mathbf{b}_\mu$.

The potential $U(\sigma)$ represents the self-interactions of the scalar field and has the form

$$U(\sigma) = \frac{1}{3}bM(g_{\sigma N}\sigma)^3 + \frac{1}{4}c(g_{\sigma N}\sigma)^4\,, \tag{4.52}$$

where the nucleon mass M renders the constant b dimensionless. Electrons and muons are included as noninteracting particles, since the magnitudes of their interactions are small compared to those of the free Fermi gas parts.

In the mean field approximation, the baryon source currents and meson fields are replaced by their ground-state expectation values. The resulting nonlinear equations have to be solved self-consistently for the meson field expectation values. The ground state is defined to be uniform and isotropic. The sources for the charged ρ mesons vanish, so that of the three charged states of the ρ, only the neutral charge remains. It is denoted by $b_\mu^{(3)}$ according to its isospin. For ease of notation, the superscript denoting the isospin component will be omitted, so that b_μ will be used for $b_\mu^{(3)}$.

In the Euler–Lagrange equations for the meson fields,

$$\partial_\mu \frac{\partial\mathcal{L}}{\partial(\partial_\mu q)} = \frac{\partial\mathcal{L}}{\partial q}\,, \tag{4.53}$$

where q denotes any of the fields; the fields are replaced by their ground state expectation values. Hereafter, $\sigma = \langle\sigma\rangle$, $\omega_\mu = \langle\omega_\mu\rangle$ and $b_\mu = \langle b_\mu\rangle$ denote expectation values. For static solutions in infinite, uniform and isotropic matter, all space-time derivatives of the fields vanish. The resulting meson field equations are

$$\begin{aligned} m_\sigma^2\sigma &= -\frac{dU}{d\sigma} + \sum_B g_{\sigma B}\langle\bar{B}B\rangle \\ m_\omega^2\omega_\mu &= \sum_B g_{\omega B}\langle\bar{B}\gamma_\mu B\rangle\,, \quad m_\rho^2 b_\mu = \sum_B g_{\rho B}t_{3B}\langle\bar{B}\gamma_\mu B\rangle\,,\end{aligned} \tag{4.54}$$

where t_{3B} is the 3-component of isospin of baryon B.

In momentum-space, the Dirac equation for the baryon B is

$$(\gamma^\mu p_\mu - g_{\omega B}\gamma^\mu\omega_\mu - g_{\rho B}t_{3B}\gamma^\mu b_\mu - M_B + g_{\sigma B}\sigma)B = 0\,,$$

which can be solved for the energy eigenvalues. In infinite isotropic matter, the space-components of ω_μ and b_μ vanish, i.e. $\omega_\mu = b_\mu = 0$ for $\mu \neq 0$. The energy eigenvalues then take the form

$$e_B(\mathbf{k}) = g_{\omega B}\omega_0 + g_{\rho B}t_{3B}b_0 \pm E_B^*(k) \tag{4.55}$$

$$E_B^*(k) = \sqrt{\mathbf{k}^2 + M_B^{*\,2}}\,, \quad M_B^* = M_B - g_{\sigma B}\sigma\,. \tag{4.56}$$

The expectation values for the source currents in the meson field equations are the baryon number and scalar density given by

$$n_B = \langle B^\dagger B\rangle = \frac{\gamma}{2\pi^2}\int_0^{k_B} dk\; k^2 = \frac{\gamma k_B^3}{6\pi^2} \qquad n_B^s = \langle \bar{B}B\rangle = \frac{\gamma}{2\pi^2}\int_0^{k_B} dk\; k^2\frac{M_B^*}{E_B^*}\,, \tag{4.57}$$

where the spin degeneracy has been denoted by γ.

Finally, energy density and pressure are obtained from the energy-momentum tensor, and are given by

$$\varepsilon = U(\sigma) + \frac{1}{2}m_\sigma^2\sigma^2 + \frac{1}{2}m_\omega^2\omega_0^2 + \frac{1}{2}m_\rho^2 b_0^2 + \sum_B \varepsilon_B + \sum_l \varepsilon_l \tag{4.58}$$

$$P = -U(\sigma) - \frac{1}{2}m_\sigma^2\sigma^2 + \frac{1}{2}m_\omega^2\omega_0^2 + \frac{1}{2}m_\rho^2 b_0^2 + \sum_B P_B + \sum_l P_l\,. \tag{4.59}$$

Here, ε_i and P_i are

$$\varepsilon_i = \frac{\gamma}{2\pi^2}\int_0^{k_{F_i}} dk\; k^2 E_i^*(k)\,, \quad P_i = \frac{\gamma}{6\pi^2}\int_0^{k_{F_i}} dk \frac{k^4}{E_i^*(k)}\,. \tag{4.60}$$

The energies E_i^* above are, $E_B^*(k) = \sqrt{k^2 + M_B^{*\,2}}$ and $E_l(k) = \sqrt{k^2 + m_l^2}$ for baryons and leptons, respectively.

At any given baryon density, the above equations are solved self-consistently for the mean meson fields and the particle fractions under the constraints of charge neutrality and β-equilibrium. At zero temperature, the particle fractions are given in terms of the Fermi momenta k_{Fi}. The general β-equilibrium condition Eq. (4.50) determines whether a particular baryon will be present at a given density. This will be the case if the lowest-lying energy state of that baryon in matter, which is implicitly density-dependent through the values of the meson fields, is less than its chemical potential as dictated by β-equilibrium. The value of k_{FB} is thus determined by the requirement that $\mu_B = e_B(k_{FB})$.

4.1.2. *Coupling Constants*

The effective field theoretical model described above contains coupling constants which are chosen to reproduce empirically observable properties of saturated nuclear matter, of nuclei off the stability line, and of hypernuclei.

Nucleon Couplings

In the nucleon sector, the constants $b, c, g_{\sigma N}, g_{\omega N}$ and $g_{\rho N}$ are determined by reproducing the properties of nucleonic matter. These properties are the saturation density, the binding energy per nucleon, the symmetry energy coefficient, the compression modulus, and the Dirac effective mass $M^* = M - g_{\sigma N}\sigma$. Models from Glendenning & Moszkowski (1991) are termed GM and from Horowitz & Serot (1981) are termed HS81. For the compression modulus K and the effective mass M^*, different choices are made to span a range of uncertainty in their values. The values of the nucleon–meson coupling constants and the constants b and c in the scalar self-interaction adjusted to reproduce the properties listed above are given in Table 5.

Hyperon Couplings

The hyperon couplings are undetermined by nuclear matter properties. One way to obtain the hyperon couplings is to require the model to reproduce the binding energy of the Λ hyperon in nuclear matter at saturation (Glendenning & Moszkowski (1991)). Millener *et al.* (1988) have determined that $(B/A)_\Lambda \cong -28$ MeV. The couplings of hyperons to mesons may then be parametrized in terms of the ratios of hyperon to nucleon

	$g_{\sigma N}/m_\sigma$ (fm)	$g_{\omega N}/m_\omega$ (fm)	$g_{\rho N}/m_\rho$ (fm)	b	c
GM1	3.434	2.674	2.100	0.002947	-0.001070
GM2	3.025	2.195	2.189	0.003478	0.01328
GM3	3.151	2.195	2.189	0.008659	-0.002421
HS81	3.974	3.477	2.069	0.0	0.0

TABLE 5. Nucleon coupling constants for models GM1–GM3 and HS81.

couplings to the respective mesons,

$$x_{\sigma H} = g_{\sigma H}/g_{\sigma N} \ , \qquad x_{\omega H} = g_{\omega H}/g_{\omega N} \ , \qquad x_{\rho H} = g_{\rho H}/g_{\rho N} \ .$$

The simplest choice is to assume that the coupling constant ratios of all the different hyperons are the same. That is $x_\sigma = x_{\sigma\Lambda} = x_{\sigma\Sigma} = x_{\sigma\Xi}$; and similarly for the other mesons, ω and ρ. In the present model, the Λ binding energy is given by

$$(B/A)_\Lambda = x_\omega V - x_\sigma S = -28 \text{ MeV} \ , \tag{4.61}$$

where $V = g_{\omega N}\omega$ and $S = g_{\sigma N}\sigma$ are the vector and scalar field strengths at saturation. For a particular choice of x_σ, the value of x_ω may then be adjusted to reproduce the Λ binding energy. The value of x_ρ is undetermined by the Λ binding, since Λ has isospin zero. To begin with, one may set $x_\rho = x_\sigma$. The range of x_σ compatible with observed neutron star masses and hypernuclear levels is approximately $0.5 - 0.7$ (Glendenning & Moszkowski (1991)). The analysis of level spectra of Λ hypernuclei in the framework of the relativistic mean field model sets the upper limit of $x_{\sigma\Lambda} = 0.72$. The lower limit is obtained by the requirement that an EOS obtained with a particular choice of x_σ must yield a maximum mass larger than the largest measured neutron star mass.

4.1.3. *Stellar structure with hyperons*

Fig. 8 shows the maximum mass as a function of x_σ for the three models used by Glendenning & Moszkowski (1991). (A similar figure is given in this reference. The maximum masses have been recalculated here, and differ slightly from the values shown in Glendenning & Moszkowski (1991).) For the stiffest EOS (GM1), the lower limit for the mass of $1.5M_\odot$ suggested by observation sets the lower limit for the coupling ratio to $x_\sigma = 0.5$. (A mass of $1.44M_\odot$ gives $x_\sigma = 0.43$.)

The information contained in Fig. 8 clarifies the role of hyperon couplings on structure. The weaker the couplings, the lower is the maximum mass; or, in other words, the greater the effect of the hyperons. This seems paradoxical at first, but is easily understood in terms of the pressure generated by the hyperons. The main part of the pressure is contained in the meson fields, which are considerably reduced in the presence of hyperons. This is due to the fact that the source terms in the field equations are proportional to $g_{\sigma B}$. The smaller these couplings, the less hyperons contribute to the fields, and the less is the interaction pressure. The lower pressure leads to a reduction in the maximum mass.

How sensitive is the maximum mass to the choice of the coupling constants? In the range $x_\sigma = 0.5 - 0.7$, the maximum mass changes roughly by $0.01M_\odot$ for every change of 0.05 in x_σ. Taking the central value of $x_\sigma = 0.6$ with an uncertainty of ± 0.1, which is

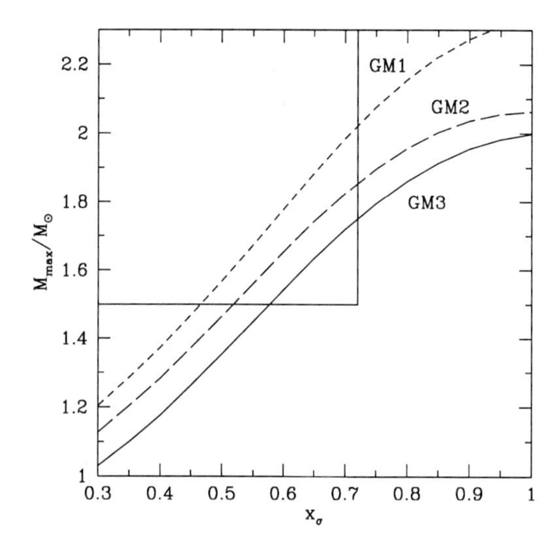

FIGURE 8. Maximum neutron star mass as function of hyperon to nucleon coupling constant ratio x_σ for models GM1–GM3.

probably an underestimation of the uncertainty, leads to an uncertainty in the maximum mass of about $0.4M_\odot$. *This is quite large and significant.* In the results to be discussed below, the central value of $x_\sigma = 0.6$ will be used, unless otherwise indicated. The hyperon coupling ratios for different sets of constants in the nucleon sector are determined as described above.

The influence of hyperons may be gauged from the results in Fig. 9. Comparison of the particle fractions in matter without and with hyperons (panels (1) and (2), respectively) shows that as soon as hyperons start to appear, the lepton fractions are dramatically reduced. Comparison of the meson field strengths (panels (3) and (4)) reveals that in the presence of hyperons, the values of ω_0 and b_0 are reduced, while the value of σ is slightly increased. Panel (5) of Fig. 9 illustrates that in matter with hyperons, both energy density and pressure are lowered as compared to matter without hyperons. The reduction in the pressure is more significant than the reduction in the energy density. The pressure as a function of energy density is lowered by the presence of hyperons. This softening of the EOS is shown in panel (6).

The impact of the presence of hyperons on neutron star properties is shown in Table 6. A comparison with the corresponding results for matter without hyperons reveals that the presence of hyperons in neutron star matter considerably reduces the maximum mass. The reduction is on the order of $0.5M_\odot$, which is about 20% of the maximum mass of a hyperon-free star.

4.2. *Matter with Bose condensates*

The idea that, above some critical density, the ground state of baryonic matter might contain a Bose–Einstein condensate of negatively charged kaons is due to Kaplan & Nelson (1986). Subsequently, the formulation, in terms of chiral perturbation theory, was discussed by Brown *et al.* (1987), Politzer & Wise (1991) and Brown *et al.* (1992), and a number of calculations exploring the astrophysical consequences have been carried out, e.g. by Thorsson *et al.* (1994) and Muto *et al.* (1988; 1993). Physically, the strong attraction between K^- mesons and nucleons increases with density and lowers the energy

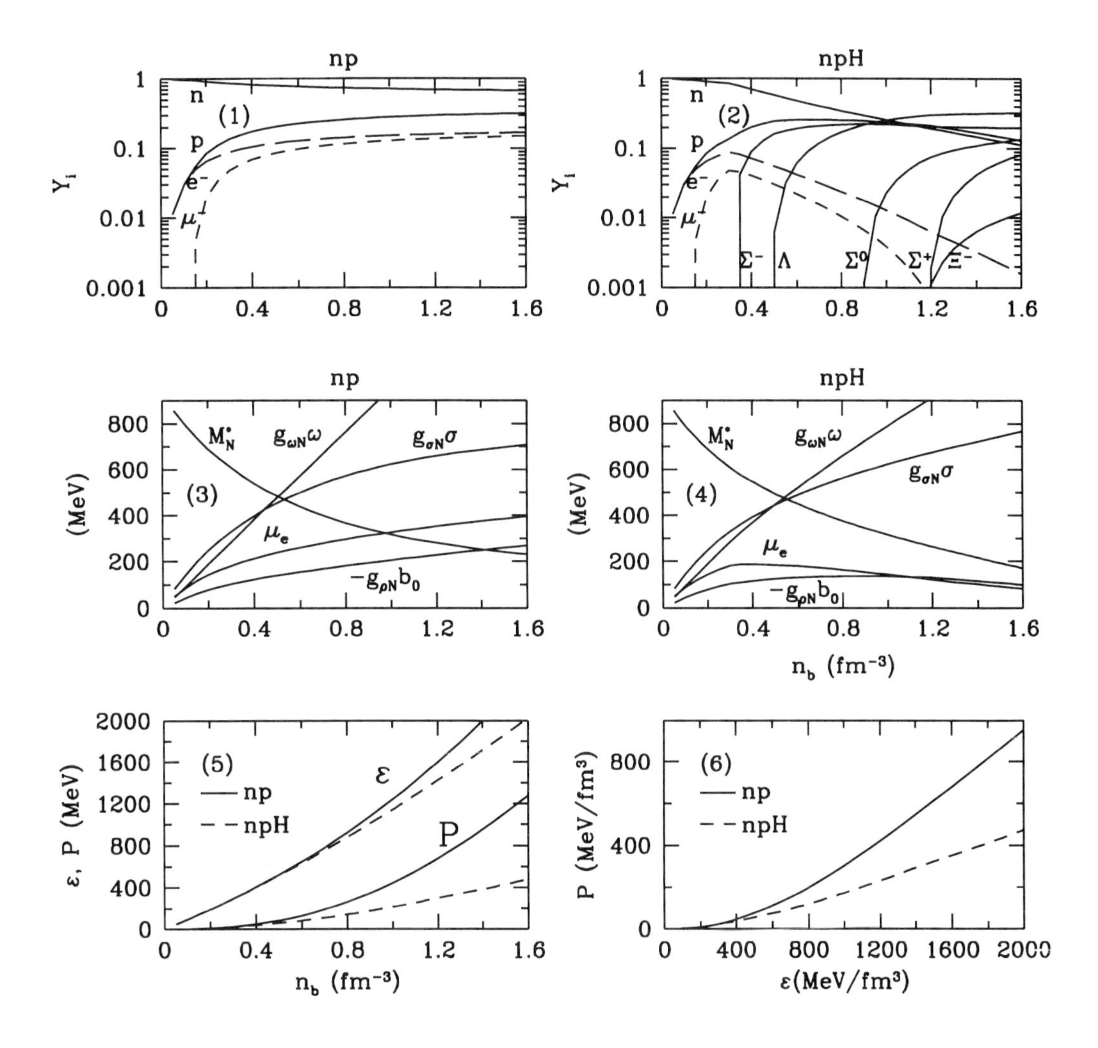

FIGURE 9. Results for model GM3. The total baryon number density is denoted by n_b. The symbol np refers to nucleons-only matter, npH to matter with additional hyperons. Panels (1) and (2): particle fractions $Y_i = n_i/n_b$. Panels (3) and (4): field strengths $g_{\sigma N}\sigma$, $g_{\omega N}\omega_0$, $-g_{\rho N}b_0$, nucleon effective mass M_N^* and electron chemical potential μ_e. Panel (5): energy density and pressure versus n_b. Panel (6): pressure versus energy density.

of the zero-momentum state. A condensate forms when this energy becomes equal to the kaon chemical potential, μ. In dense neutron star matter, μ is related to the electron and nucleon chemical potentials by $\mu = \mu_e = \mu_n - \mu_p$ due to chemical equilibrium in the reactions $n \leftrightarrow p+e^- + \bar{\nu}_e$ and $n \leftrightarrow p+K^-$. The density at which this takes place is model and parameter dependent, but is typically $\sim 4n_0$, where n_0 denotes equilibrium nuclear density. Since this may be less than the central density in neutron stars, a K^- condensate is expected to be present in the core region. Apart from the softening effect on the EOS, which lowers the maximum mass, the proton abundance is dramatically increased, so that the star might be dubbed a "nucleon", rather than a neutron, star. Based on the Kaplan-Nelson Lagrangian, Thorsson *et al.* (1994) have considered the astrophysical con-

Model	Matter	$\frac{M_{max}}{M_\odot}$	R (km)	$\frac{n_c}{n_0}$	P_c MeV fm^{-3}	I $M_\odot$ km^2
GM1	np	2.346	11.93	5.70	512.7	152.0
	npH	1.776	11.79	6.53	266.0	91.26
GM2	np	2.064	11.36	6.58	469.9	112.2
	npH	1.655	11.46	6.97	268.2	77.53
GM3	np	2.005	10.92	7.14	545.8	100.6
	npH	1.544	10.90	7.98	309.4	63.31

TABLE 6. Star properties for the maximum mass star. The symbol np denotes matter containing nucleons and npH denotes matter which includes hyperons. Other symbols are: M_{max} for the maximum gravitational mass, R for the radius, n_c/n_0 for the ratio of central density to nuclear matter saturation density, P_c for the central pressure, and I the moment of inertia.

sequences in some detail. In the subsequent discussion, we follow Knorren *et al.* (1995) who offered a complementary approach based on a traditional Yukawa exchange model to kaon-baryon interactions.

4.3. *Model*

The total hadron Lagrangian is written as the sum of the baryon and the kaon Lagrangians, $\mathcal{L}_H = \mathcal{L}_B + \mathcal{L}_K$. In the baryon sector, we employ the field-theoretical model described earlier. For the kaon sector, we take a Lagrangian which contains the usual kinetic energy and mass terms, along with the meson interactions,

$$\mathcal{L}_K = \partial_\mu K^+ \partial^\mu K^- - (m_K^2 - g_{\sigma K} m_K \sigma) K^+ K^- \\ + i \left[g_{\omega K} \omega^\mu + g_{\rho K} b^\mu \right] (K^+ \partial_\mu K^- - K^- \partial_\mu K^+) \, . \tag{4.62}$$

Here b^μ denotes the ρ^0 field and m_K is the vacuum kaon mass (which is present in the fourth term so that $g_{\sigma K}$ is dimensionless). The scalar interaction term can be combined with the kaon mass into an effective kaon mass defined by

$$m_K^{*\,2} = m_K^2 - g_{\sigma K} m_K \sigma \, . \tag{4.63}$$

We shall treat the kaons in the mean field approximation, writing the time dependence of the fields $K^\pm = \frac{1}{\sqrt{2}} f \theta e^{\pm i \mu t}$; thus, θ gives the condensate amplitude. For the baryons, we shall consider calculations at the mean field level. We need to calculate the potential, Ω, of the grand canonical ensemble at zero temperature. It is straightforward to obtain

$$\frac{\Omega}{V} = \tfrac{1}{2}(f\theta)^2 [m_K^{*\,2} - 2\mu(g_{\omega K}\omega_0 + g_{\rho K} b_0) - \mu^2] + \tfrac{1}{2} m_\sigma^2 \sigma^2 + \tfrac{1}{3} b M (g_{\sigma N}\sigma)^3 \\ + \tfrac{1}{4} c (g_{\sigma N}\sigma)^4 - \tfrac{1}{2} m_\omega^2 \omega_0^2 - \frac{\zeta}{4!}(g_{\omega N}\omega_0)^4 - \tfrac{1}{2} m_\rho^2 b_0^2 + \sum_B \frac{1}{\pi^2} \int_0^{k_{FB}} dk \, k^2 (E_B^* - \nu_B) \tag{4.64}$$

Here V is the volume, $E_B^* = \sqrt{k^2 + M_B^{*2}}$, and the baryon effective masses are $M_B^* = M_B - g_{\sigma B}\sigma$. The chemical potentials μ_B are given in terms of the effective chemical potentials, ν_B, by

$$\mu_B = \nu_B + g_{\omega B}\omega_0 + g_{\rho B} t_{3B} b_0 \, , \tag{4.65}$$

where t_{3B} is the z-component of the isospin of the baryon. The relation to the Fermi momentum k_{FB} is provided by $\nu_B = \sqrt{k_{FB}^2 + M_B^{*2}}$.

The thermodynamic quantities can be obtained from the grand potential in Eq. (4.64) in the standard way; thus the baryon number density $n_B = k_{FB}^3/(3\pi^2)$, while for kaons

$$n_K = (f\theta)^2(\mu + g_{\omega K}\omega_0 + g_{\rho K}b_0) \,. \tag{4.66}$$

The pressure $P = -\Omega/V$ and the energy density $\varepsilon = -P + \sum_B \mu_B n_B + \mu n_K$. The meson fields are obtained by extremizing Ω, giving

$$\begin{aligned} m_\omega^2 \omega_0 &= -\frac{\zeta}{6} g_{\omega N}^4 \omega_0^3 + \sum_B g_{\omega B} n_B - (f\theta)^2 \mu g_{\omega K} \\ m_\rho^2 b_0 &= \sum_B g_{\rho B} t_{3B} n_B - (f\theta)^2 \mu g_{\rho K} \\ m_\sigma^2 \sigma &= -bM g_{\sigma N}^3 \sigma^2 - c g_{\sigma N}^4 \sigma^3 + \sum_B g_{\sigma B} n_B^s + \tfrac{1}{2}(f\theta)^2 g_{\sigma K} m_K \,. \end{aligned} \tag{4.67}$$

Here n_B^s denotes the baryon scalar density

$$n_B^s = \frac{1}{\pi^2} \int_0^{k_{FB}} dk\, k^2 \frac{M_B^*}{E_B^*} \,. \tag{4.68}$$

Notice that the condensate contributes directly to the equations of motion (4.67), whereas in chiral models the contribution appears in the effective chemical potentials and effective masses.

The condensate amplitude, θ, is also found by extremizing Ω. This yields the solutions $\theta = 0$ (no condensate), or, if a condensate exists, the equation

$$\mu^2 + 2\mu(g_{\omega K}\omega_0 + g_{\rho K}b_0) - {m_K^*}^2 = 0 \,. \tag{4.69}$$

The roots of this equation are the energies of the zero-momentum K^- and K^+ states,

$$\omega^\pm = \sqrt{(g_{\omega K}\omega_0 + g_{\rho K}b_0)^2 + {m_K^*}^2} \pm (g_{\omega K}\omega_0 + g_{\rho K}b_0) \,, \tag{4.70}$$

so Eq. (4.69) amounts to setting the chemical potential equal to the energy of the lowest (K^-) state.

Eq. (4.69) can be used to simplify the expressions for pressure and energy density:

$$P = -\tfrac{1}{2}m_\sigma^2\sigma^2 - \tfrac{1}{3}bM(g_{\sigma N}\sigma)^3 - \tfrac{1}{4}c(g_{\sigma N}\sigma)^4 + \tfrac{1}{2}m_\omega^2\omega_0^2 + \frac{\zeta}{4!}(g_{\omega N}\omega_0)^4 + \tfrac{1}{2}m_\rho^2 b_0^2 + \sum_B \frac{1}{3\pi^2}\int_0^{k_{FP}} dk\, \frac{k^4}{E_B^*} \tag{4.71}$$

$$\varepsilon = (f\theta)^2 {m_K^*}^2 + \tfrac{1}{2}m_\sigma^2\sigma^2 + \tfrac{1}{3}bM(g_{\sigma N}\sigma)^3 + \tfrac{1}{4}c(g_{\sigma N}\sigma)^4 + \tfrac{1}{2}m_\omega^2\omega_0^2 + \frac{\zeta}{8}(g_{\omega N}\omega_0)^4 + \tfrac{1}{2}m_\rho^2 b_0^2 + \sum_B \frac{1}{\pi^2}\int_0^{k_{FB}} dk\, k^2 E_B^* \,. \tag{4.72}$$

Note that by virtue of Eq. (4.69), the first term in the expression for the thermodynamical potential Eq. (4.64) vanishes, and so the pressure due to the kaons is contained entirely in the meson fields via their field equations (4.67).

To complete the thermodynamics, leptonic contributions to the total energy density and pressure, which are given adequately by the standard free gas expressions, must be added to Eq. (4.71) and Eq. (4.72).

	$g_{\sigma K}/m_\sigma$ (fm)	$g_{\omega K}/m_\omega$ (fm)	$g_{\rho K}/m_\rho$ (fm)	$-S^K_{opt}$ (MeV)	$-V^K_{opt}$ (MeV)	$-U^K_{opt}$ (MeV)
GM1	0.712	1.863	0.619	29	151	180
GM2	0.808	2.269	0.594	28	151	179
GM3	0.776	2.269	0.594	25	151	176

TABLE 7. Kaon-baryon coupling constants that reproduce the phase shift data. Also shown are the Scalar and vector contributions and the kaon optical potential in equilibrium nuclear matter using the kaon-baryon couplings from the phase shifts.

4.3.1. *Kaon couplings*

In order to investigate the effect of a kaon condensate on the equation of state in high-density baryonic matter, the kaon-meson coupling constants have to be specified. Empirically-known quantities can be used to determine these constants, but it is important to keep in mind that laboratory experiments give information only about the kaon-nucleon interaction in free space or in nuclear matter. The physical setting here, however, is matter in the dense interiors of neutron stars, i.e. infinite matter containing baryons and leptons in β-equilibrium, that has a different composition and spans a wide range in densities (up to central densities $\sim 8n_0$). As a consequence, kaon-meson couplings as determined from experiments might not be appropriate to describe the kaon-nucleon interaction in neutron star matter, and the particular choices of coupling constants should be regarded as parameters that have a range of uncertainty.

One possibility of experimentally determining the strength of the kaon-nucleon interaction is the analysis of phase shift data. An analysis of KN scattering data using a meson-exchange model (Buttgen *et al.* (1990)) was used to determine couplings of nucleons and kaons to σ, ω, and ρ mesons. This yielded $G^\sigma_{KN} = g_{\sigma N}g_{\sigma K}/m_\sigma^2 = 2.44$ fm^2, $G^\omega_{KN} = g_{\omega N}g_{\omega K}/m_\omega^2 = 4.98$ fm^2, and $G^\rho_{KN} = g_{\rho N}g_{\rho K}/m_\rho^2 = 1.30$ fm^2 with our Lagrangian conventions. Since only the ratio g/m enters the formalism, it is not necessary to specify the masses, and the kaon ratios are listed in Table 7.

With these couplings and the field strengths in nuclear matter at saturation, we can determine the value of the optical potential felt by a single kaon in infinite nuclear matter for the present model. Lagrange's equation for an s-wave K^- with a time dependence $K^- = k^-(\mathbf{x})\ e^{-iEt}$, where $E = \sqrt{p^2 + m_K^2}$ is the asymptotic energy, is obtained from Eq. (4.62) as

$$\begin{aligned}[\nabla^2 + E^2 - m_K^2]\ k^-(\mathbf{x}) &= [-2(g_{\omega K}\omega_0 + g_{\rho K}b_0)E - g_{\sigma K}m_k\sigma]\ k^-(\mathbf{x}) \\ &= 2\ m_K\ U^K_{opt}\ k^-(\mathbf{x})\ . \end{aligned} \tag{4.73}$$

In nuclear matter, $b_0 = 0$, so for a kaon with zero momentum ($E = m_K$) the optical potential is

$$U^K_{opt} \equiv S^K_{opt} + V^K_{opt} = -\tfrac{1}{2}g_{\sigma K}\sigma - g_{\omega K}\omega_0\ . \tag{4.74}$$

The value of U^K_{opt} for the different models are contained in Table 7. Friedman *et al.* (1994) have recently reanalyzed the kaonic atom data examining a more general parameterization of U^K_{opt} in nuclear matter than the standard $t_{eff}\rho$ approximation. They were able to obtain a better fit with a kaon optical potential whose real part had a depth of -200 ± 20 MeV. The coupling constants adjusted to the parameters obtained from phase shift measurements lead to a value of the kaon optical potential close to this value.

Note that the ratio of the $K\omega$ coupling to the $N\omega$ coupling, $x_{\omega K} = g_{\omega K}/g_{\omega N}$, is close to one, whereas $x_{\sigma K}$ and $x_{\rho K}$ are close to the value of $\frac{1}{3}$ which is suggested by naive quark counting.

4.3.2. *Condensation in nucleons-only matter*

In Fig. 10 we display our results for matter containing nucleons and leptons for a representative case, parameter set GM2 with kaon couplings from Table 7. The particle fractions, $Y_i = n_i/n$, are shown in panel 1 of Fig 10. The proton fraction becomes much closer to the neutron fraction once kaons are present, and for high u they are essentially equal. It can be seen that the threshold condition for kaon condensation, Eq. (4.69), is fulfilled at a density of $2.6n_0$. (The dashed lines in Fig. 10 show the behavior if kaons are excluded.) In panel 2 of this figure, the energies of the zero-momentum kaon states, $\omega^{\pm}$, are plotted as a function of the ratio of baryon density to equilibrium nuclear matter density, i.e. $u = \sum_B n_B/n_0 \equiv n/n_0$. We see that the ω^- energy drops with increasing density and meets the chemical potential μ at threshold. The effective kaon mass m_K^* does not vary greatly; so, referring to Eq. (4.70), the density-dependent contributions, dominated by the term containing the ω meson, are critical in obtaining condensation. Panel 3 (right scale) shows that the condensate amplitude, θ, rises rapidly at threshold and then slowly approaches a maximum value of $\sim 40°$; this is smaller than in chiral models. The effect of the condensate on the ω and σ fields (panel 2) follows from the field equations (4.67), whereas the behavior of the ρ field is dominated by the changes in the neutron–proton ratio. The proton charge is balanced by an approximately equal number of K^- mesons beyond threshold, since the lepton contributions rapidly become negligible. Thus the magnitude of the strangeness/baryon, $|S|/B$, in panel 3 is $\sim \frac{1}{2}$ once kaons condense. Panel 4 of Fig. 10 shows that the total pressure and energy density are reduced when kaons are present.

Finally, in the upper part of Table 8, the gross properties of neutron stars are given for the various EOSs. The critical density ratios lie in a narrow range, $u_{crit} \sim 2.5 - 3$, so that a significant region of the star will contain kaons. This softens the EOS causing a reduction in the maximum mass by 4–10%. The precise value depends on the magnitude of the ω repulsion at high density, which is governed by the coupling constants of Table 5. This also affects the changes in the central density.

4.3.3. *Condensation in matter with hyperons*

We now consider the case where hyperons are allowed to be present in addition to nucleons. The first strange particle to appear is the Σ^-, since the somewhat higher mass of the Σ^- is compensated by the electron chemical potential in the equilibrium condition of the Σ^-. Since the Σ^- carries a negative charge, it causes the lepton fractions to drop. This means that the chemical potential μ is reduced, requiring a smaller value of ω^- for kaon condensation which results in a higher threshold density. This arises partly from the reduction in the condensate amplitude and partly from changes in the baryon fractions.

Immediately above threshold the kaon fraction rises dramatically to reach a maximum of $0.1 - 0.2$ per baryon. Since the kaons carry negative charge, charge neutrality for the system leads to a small drop in the Σ^- fraction, and the lepton concentrations become even smaller. By contrast, the fraction of the neutral Λ is little influenced by kaon condensation. In fact this is the largest fraction at large values of u, with roughly comparable amounts of n, p, Σ^- and a relatively small kaon presence. Thus, the hyperons dominate the strangeness/baryon, $|S|/B$, of ~ 0.6 at the highest density considered.

The neutron star properties for matter containing nucleons, hyperons and leptons are given in the lower part of Table 8. Excluding kaons for the moment, we see that the

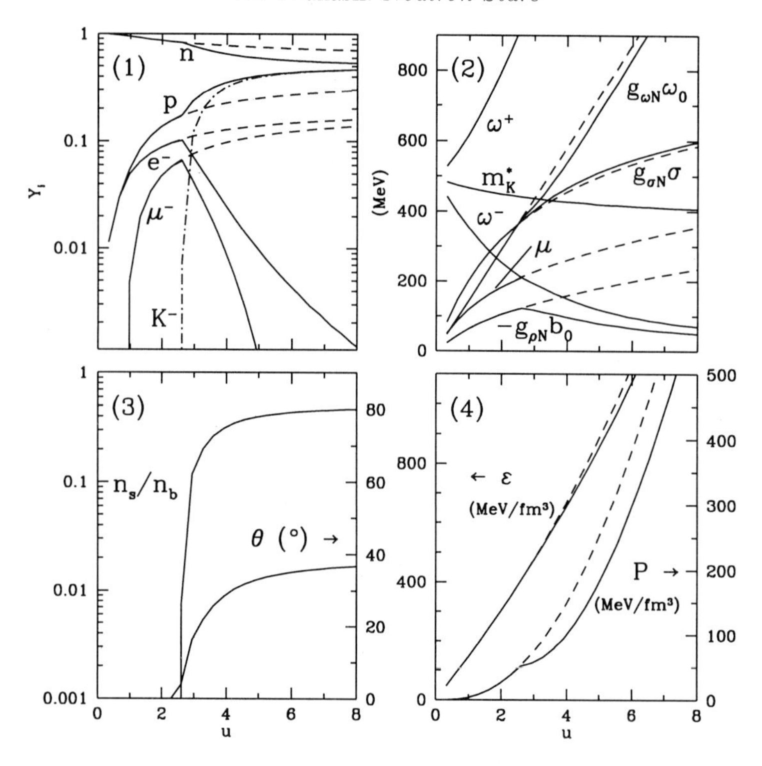

FIGURE 10. Matter containing nucleons and leptons with parameter set GM2. Solid (dashed) lines show quantities in matter with (without) kaons, as a function of the baryon density ratio $u = n/n_0$. Panel (1): Particle fractions $Y_i = n_i/n$. Panel (2): Kaon energies $\omega^{\pm}$ and effective mass m_K^*, meson field strengths and electron chemical potential μ. Panel (3): Kaon condensate amplitude, θ and strangeness/baryon, $|S|/B$. Panel (4): Pressure and energy density.

softening effect of hyperons causes a reduction of $\sim 0.5M_\odot$ and a corresponding increase in the central density. If we include kaons in the calculation, condensation takes place within the star only for models GM2 and GM3, and it does so at a higher density than when hyperons are absent. The reduction in the maximum neutron star mass due to the presence of kaons amounts to only about $0.01M_\odot$. The change in the central density is likewise small.

Thus, in this model, the influence of the hyperons is decisive. In nucleons-only matter, the pressure is significantly decreased by a kaon condensate, which lowers the maximum mass. An even larger reduction in the maximum mass occurs when hyperons are present in matter. The additional presence of condensed kaons in hyperonic matter induces relatively small changes in the EOS so that there is little influence of the condensate on the gross stellar properties.

It must be borne in mind, however, that due to limited guidance about the couplings of the strange particles, be they kaons or hyperons, the densities at which they appear in dense matter are uncertain. The importance of individual hyperon species likewise remains unclear. For example, with the choice of couplings made above (suggested by the Λ binding in nuclei), hyperons play a dominant role, while kaons, which appear at a higher density, are of lesser importance. However, these roles may be altered by other suitable choices of the coupling constants. If, in addition to the Λ couplings implied by

		without kaons			with kaons	
		$\frac{M_{max}}{M_\odot}$	u_{cent}	u_{crit}	$\frac{M_{max}}{M_\odot}$	u_{cent}
np	GM1	2.346	5.70	2.49	2.185	6.46
	GM2	2.064	6.58	2.60	1.854	8.37
	GM3	2.005	7.14	2.59	1.809	9.18
npH	GM1	1.776	6.53	**	—	—
	GM2	1.655	6.96	3.37	1.645	7.36
	GM3	1.544	7.98	3.20	1.536	8.46

TABLE 8. Gravitational mass and central density of the maximum mass neutron stars for matter with and without kaon condensates. The critical density ratio for condensation is given in the middle column. The symbol np denotes matter containing nucleons and leptons, and npH denotes matter containing nucleons, hyperons and leptons. The kaon coupling constants are taken from Table 7. The symbol ** indicates that for this choice of constants no condensation takes place up to the maximum density considered ($u = 10$).

hypernuclei, the couplings implied by Σ^- atoms are employed, then, depending upon the couplings of the Ξ, it is possible that the K^- is the first negatively charged hadron to appear in matter (Knorren *et al.* (1995)). This highlights the importance of further work in this area using inputs from hypernuclear physics and advances in both theory and techniques for calculating the energy of interacting systems.

4.4. *Matter with Quarks*

Depending on the pressure–density relationship, the density of matter in the core of a neutron star may be $n \sim 1 - 2$ fm^{-3}. At such high density, the energy of quark matter may well lie below that of baryonic matter. Here, we explore the conditions under which a phase transition to quark matter may take place.

4.4.1. *Quark phase EOS*

For the EOS in the quark phase, we follow Freedman & McLerran (1978), in which the thermodynamic potential of a quark gas to fourth order in the quark–gluon coupling, g, was calculated.

The Fermi gas contribution to the thermodynamic potential from a quark species i ($= u$, d or s) of mass m_i and chemical potential μ_i is

$$\Omega_i^{\rm FG} = -\frac{1}{4\pi^2}\left\{\mu_i\mu_i^*\left(\mu_i^2 - \frac{5}{2}m_i^2\right) + \frac{3}{2}m_i^4 \ln\left(\frac{\mu_i + \mu_i^*}{m_i}\right)\right\}, \tag{4.75}$$

where $\mu_i^* = (\mu_i^2 - m_i^2)^{1/2}$. For massless quarks (e.g., u and d quarks), the Fermi gas result simplifies to $\Omega_i^{\rm FG} = -(1/4\pi^2)\mu_i^4$.

The two–loop or exchange contributions are given by (Baym & Chin (1976)

$$\Omega_i^{\rm ex} = \frac{1}{4\pi^2}\frac{2\alpha_c}{\pi}\left\{3\left[\mu_i\mu_i^* - m_i^2 \ln\left(\frac{\mu_i + \mu_i^*}{m_i}\right)\right]^2 - 2(\mu_i^2 - m_i^2)^2\right\}, \tag{4.76}$$

which reduces to $\Omega_i^{\rm ex} = (1/4\pi^2)(2\alpha_c/\pi)\mu_i^4$ for massless quarks. Here, $\alpha_c \equiv g^2/(4\pi)$.

Higher order correlation corrections (from ring diagrams, etc.) have only been evaluated for the massless case, for which

$$\Omega_i^{\rm corr} = \frac{1}{4\pi^2}\mu_i^4\left[(\alpha_c/\pi)^2\ln(\alpha_c/4\pi) + 31.1\,(\alpha_c/4\pi)^2\right]. \tag{4.77}$$

To the same order, a further contribution associated with the interference between up and down quarks also exists (Freedman & McLerran (1978)). The screened charge α_c falls off with the chemical potential μ of the quark according to the Gell–Mann–Low equation. To the order considered above, the precise way in which this fall off occurs is described in Freedman & McLerran (1978).

The production of electrons by the beta decays

$$d \leftrightarrow u + e^- + \overline{\nu}_e\,, \qquad s \leftrightarrow u + e^- + \overline{\nu}_e \tag{4.78}$$

leads to the additional contribution $\Omega_e \cong -\mu_e^4/(12\pi^2)$ to the thermodynamic potential, where, as before, the neutrinos are taken to have left the system. The beta–equilibrium and charge neutrality conditions are

$$\mu_d = \mu_u + \mu_e = \mu_s \quad \text{and} \quad Q = e\left(\frac{2}{3}n_u - \frac{1}{3}n_d - \frac{1}{3}n_s - n_e\right) = 0\,. \tag{4.79}$$

respectively. In the above, the number densities are given by $n_i = -(\partial/\partial\mu_i)\Omega$, with the baryon number density $n = (n_u + n_d + n_s)/3$. The existence of muons is contingent on the electron Fermi energies exceeding the muon mass. However, as is clear from Eq. (4.79), quarks, in equal amounts of each flavor, maintain charge neutrality amongst themselves, needing the aid of electrons only when the strange quarks are massive. Thus, electron concentrations in quark matter are rather small, so small in fact as to preclude the existence of muons.

The total pressure and energy density of the electro–neutral quark matter system are thus

$$P = -\Omega = -(\Omega_u + \Omega_d + \Omega_s + \Omega_{int} + \Omega_e) \quad \text{and} \quad \mathcal{E} = \Omega + \sum_i \mu_i n_i\,. \tag{4.80}$$

Fahri & Jaffe (1984) retain terms up to order α_c in the thermodynamic potential of Eq. (4.80) and choose a renormalization procedure in which α_c is held fixed. The renormalization point, ρ_R, for the massive s quark is chosen at $\rho_R = 313$ MeV to minimize its (unphysical) impact on physical variables. In this scheme, the exchange contribution from the strange quarks is given by

$$\begin{aligned}\Omega_s^{\rm ex} = \frac{1}{4\pi^2}\frac{2\alpha_c}{\pi}\Bigg\{ & 3\left[\mu_s\mu_s^* - m_s^2\ln\left(\frac{\mu_s+\mu_s^*}{\mu_s}\right)\right]^2 - 2(\mu_s^2 - m_s^2)^2 \\ & - 3m_s^4\ln^2\frac{m_s}{\mu_s} + 6\ln\frac{\rho_R}{\mu_s}\left[\mu_s\mu_s^* m_s^2 - m_s^4\ln\left(\frac{\mu_s+\mu_s^*}{m_s}\right)\right]\Bigg\}\,,\end{aligned} \tag{4.81}$$

where $\mu_s^* = (\mu_s^2 - m_s^2)^{1/2}$.

To complete the description of the quark matter phase, the results in Eq. (4.80) are supplemented with the dynamics of the MIT bag by adding a positive constant contribution, B, to the energy density $\mathcal{E}$, whence

$$\Omega \to \Omega + B\,, \qquad P \to P - B\,, \qquad \text{and} \qquad \mathcal{E} \to \mathcal{E} + B. \tag{4.82}$$

The constant B has the simple interpretation as the thermodynamic potential of the vacuum, and in comparisons of energy and pressure of quark–matter to nuclear matter calculations, is regarded as a phenomenological parameter.

4.4.2. *Quark-hadron phase transition*

Let us begin by considering the equilibrium conditions for bulk baryonic and quark matter coexisting in a uniform background of leptons (e^- and μ^-). For the influence of complicated finite size structures possible due to Coulomb and surface effects, see Heiselberg *et al.* (1993). Following Glendenning (1992), let us require only *global* charge neutrality of bulk matter, considering the separately conserved charges: baryon number and electric charge. An important consequence of this treatment is that baryonic and quark matter can coexist for a finite range of pressures.

For the pure phase in which the strongly interacting particles are baryons (hereafter termed phase 1), the composition is determined by the requirements of charge neutrality and equilibrium under the weak processes

$$B_1 \to B_2 + \ell + \overline{\nu}_\ell \quad \text{and} \quad B_2 + \ell \to B_1 + \nu_\ell \,, \tag{4.83}$$

where B_1 and B_2 are baryons, and ℓ is a lepton, either an electron or a muon. Under conditions when the neutrinos have left the system, these two requirements imply that the relations

$$Q_1 = \sum_i q_i n_{B_i} + \sum_{\ell=e,\mu} q_\ell n_\ell = 0 \quad \text{and} \quad \mu_i = b_i \mu_n - q_i \mu_\ell \,, \tag{4.84}$$

are satisfied. Above, q and n denote the charge and number density, respectively, and the subscript i runs over all the baryons considered. The symbol μ_i refers to the chemical potential of baryon i, b_i is its baryon number and q_i is its charge. The chemical potential of the neutron is denoted by μ_n.

In the pure quark phase (hereafter referred to as phase 2), the relevant weak decay processes are similar to Eq. (4.83), but with B_i replaced by q_f, where f runs over the quark flavors u , d and s. In neutrino-free matter, charge neutrality and chemical equilibrium under the weak processes imply

$$Q_2 = \sum_f q_f n_f + \sum_{\ell=e,\mu} q_\ell n_\ell = 0 \quad \text{and} \quad \mu_d = \mu_u + \mu_\ell = \mu_s \,. \tag{4.85}$$

To determine the equilibrium concentrations in each phase, we employ a field theoretic model at the mean field level for hadronic matter and a bag model for quark matter. Specific details of these models are kept close to the work of Glendenning (1992). The pressure P_1 of the hadronic phase is obtained from the Lagrangian proposed by Zimanyi & Moszkowski (1990), in which the baryons B interact through the exchange of σ, ω and ρ mesons. All charge states of the baryon octet $B = n, p, \Lambda, \Sigma^+, \Sigma^-, \Sigma^0, \Xi^-, \Xi^0$, as well as the Δ quartet are considered. The contributions of the leptons, $\ell = e^-$ and μ^-, are adequately given by their non-interacting expressions. In our calculations, we assume that all the baryon couplings mediated by a given meson field are the same, which limits the number of unknown constants to three: $g_\sigma/m_\sigma, g_\omega/m_\omega$ and g_ρ/m_ρ. These may be determined by fitting the empirical properties of nuclear matter at the equilibrium density of $n_0 = 0.16$ fm^{-3} (see Glendenning (1992) for more details). Specifically, their values are $(g_\sigma/m_\sigma)^2 = 7.487$ fm^2, $(g_\omega/m_\omega)^2 = 2.615$ fm^2 and $(g_\rho/m_\rho)^2 = 4.774$ fm^2. The energy density in phase 1 is $\epsilon_1 = -P_1 + \sum_B n_B \mu_B + \sum_\ell n_\ell \mu_\ell$.

The pressure of quarks in phase 2 is given by $P_2 = -B + \sum_{f=u,d,s} P_f + P_\ell$, where the first term accounts for the cavity pressure, and the second and third terms are the Fermi degeneracy pressures of quarks and leptons, respectively. The constant B is regarded as a phenomenological parameter in the range $(100 - 250)$ MeV fm^{-3}. The chemical potential of free quarks in the cavity is $\mu_f = \sqrt{k_{F_f}^2 + m_f^2}$, where k_{F_f} is the Fermi momentum of quarks of flavor f. For numerical calculations, we take the u and d quarks as massless

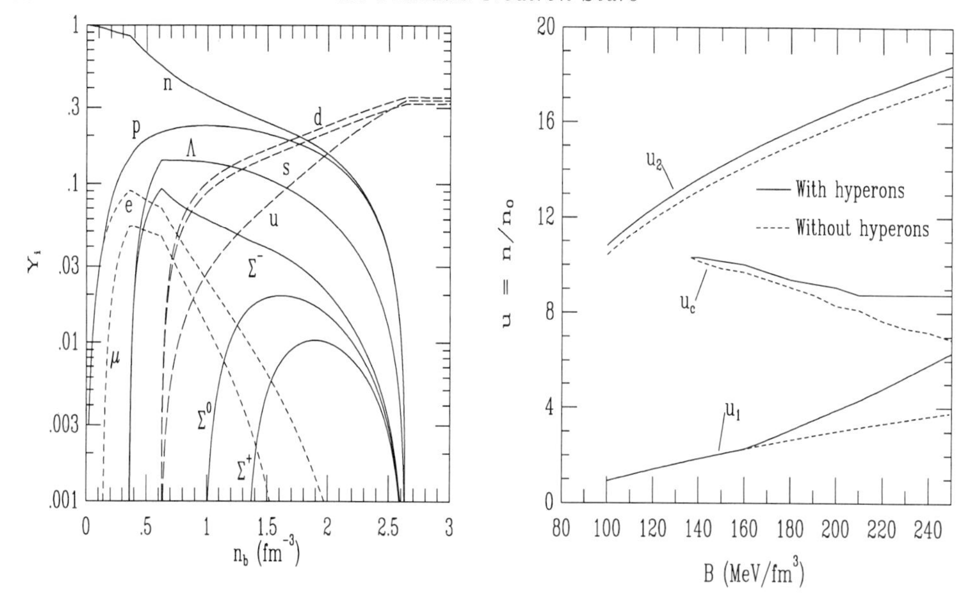

FIGURE 11. Left panel: Composition of neutrino-free matter. The quark phase cavity pressure $B = 200$ MeV fm^{-3}. Right panel: Quark-hadron phase transition boundaries in matter with nucleons and leptons (top) and nucleons, hyperons, and leptons (bottom). The equilibrium nuclear density $n_0 = 0.16$ fm^{-3}.

and $m_s = 150$ MeV. The baryon density is $n_{b_2} = (n_u + n_d + n_s)/3$ and the energy density is $\epsilon_2 = -P_2 + \sum_q n_q\mu_q + \sum_\ell n_\ell\mu_\ell$.

The description of the mixed phase of hadrons and quarks is achieved by satisfying Gibbs' phase rules: $P_1 = P_2$ and $\mu_n = \mu_u + 2\mu_d$. Further, the conditions of global charge neutrality and baryon number conservation are imposed through the relations

$$Q = fQ_1 + (1-f)Q_2 = 0 \qquad \text{and} \qquad n_b = fn_{b_1} + (1-f)n_{b_2}\,, \tag{4.86}$$

where f represents the fractional volume occupied by phase 1. Notice that unlike in the pure phases (Eqs. (4.84) and (4.85)), Q_1 and Q_2 do not separately vanish in the mixed phase. The total energy density is $\epsilon = f\epsilon_1 + (1-f)\epsilon_2$.

Fig. 11 shows the composition of neutrino-free matter (left panel). One expects that Λ, with a mass of 1116 MeV, and the Σ^-, with a mass of 1193 MeV, first appear roughly at the same density, because the somewhat higher mass of the Σ^- is compensated by the presence of μ_e in the equilibrium condition of the Σ^-. More massive and more positively charged particles than these appear at higher densities. Notice that with the appearance of the negatively charged Σ^- hyperon, which competes with the leptons in maintaining charge neutrality, the lepton concentrations begin to fall. With the appearance of quarks, which occurs around $4n_0$ for $B = 200$ MeV fm^{-3}, the neutral and negative particle abundances begin to fall, since quarks furnish both negative charge and baryon number.

The right panel of in Fig. 11 shows the phase boundaries as a function of the bag pressure B. The onset of the transition is at density $n_1 = u_1 n_0$, and a pure quark phase begins at density $n_2 = u_2 n_0$. Also shown are the central densities $n_c = u_c n_0$ of the maximum mass stars. When hyperons are present, the mixed phase occurs over a smaller range in density than that found in the absence of hyperons. The abrupt change in the onset of the transition around $B = 140$ MeV fm^{-3} is related to the appearance of hyperons prior to that of quarks. Whether or not hyperons are present, the mixed phase is present over a wide range of density inside the star. Note also that, since the central

$M_{\max}/M_{\odot}$		
	Without hyperons	With hyperons
B (MeV fm^{-3})	$Y_\nu = 0$	$Y_\nu = 0$
136.6	1.440	1.434
150	1.444	1.436
200	1.493	1.471
250	1.562	1.506
No quarks	1.711	1.516

TABLE 9. Maximum masses of stars with baryonic matter that undergoes a phase transition to quark matter. Results are for a mean-field model of baryons and a bag model of quarks. B denotes the bag pressure in the quark EOS.

density of the star $u_c < u_2$ for all cases considered, the presence of a pure quark phase is precluded.

4.4.3. *Stellar structure results*

Table 9 shows the maximum masses of stars as a function of the composition of matter. For each bag constant B shown in the table, the limiting mass of a star with hyperons is similar to, but slightly lower than, that of a star with nucleons only. The influence of quarks in a mixed phase is readily seen when we compare the maximum masses of stars with and without hyperons, but without quarks in a mixed phase. In both cases, a significant reduction in the maximum mass is achieved, due to the presence of quarks at high density.

5. Neutrino trapped matter

5.1. *The Fate of a Newborn Neutron Star*

After a supernova explosion, the gravitational mass of the remnant is less than 1 $M_{\odot}$. It is lepton rich and has an entropy per baryon of $S \simeq 1$ (in units of Boltzmann's constant k_B). The leptons include both electrons and neutrinos, the latter being trapped in the star because their mean free paths in the dense matter are of order 1 cm, whereas the stellar radius is about 15 km. Accretion onto the neutron star increases its mass to the 1.3–1.5 $M_{\odot}$ range, and should mostly cease after a second. It then takes about 10–15 s (Burrows & Lattimer (1986), Burrows (1990)) for the trapped neutrinos to diffuse out, and in the diffusion process they leave behind most of their energy, heating the protoneutron star to fairly uniform entropy values of about $S = 2$. Cooling continues as thermally-produced neutrinos diffuse out and are emitted. After about 50 s, the star becomes completely transparent to neutrinos, and the neutrino luminosity drops precipitously (Burrows & Lattimer (1986)).

Denoting the maximum mass of a cold, catalyzed neutron star by M_{max} and the maximum mass of the protoneutron star with abundant trapped leptons by M^L_{max}, there are two possible ways that a black hole could form after a supernova explosion. First, accretion of sufficient material could increase the remnant's mass to a value greater than either M_{max} or M^L_{max} and produce a black hole, which then appears on the accretion time scale (Brown *et al.* (1992)). Second, if exotic matter plays a role and if accretion is insignificant after a few seconds, then for $M^L_{max} > M > M_{max}$, where M is the final remnant mass, a black hole will form as the neutrinos diffuse out (Prakash *et al.* (1997), Thorsson *et al.* (1994), Keil & Janka (1994), Prakash *et al.* (1995), Glendenning (1995)) on the deleptonization time scale of 10–15 s.

The existence of metastable neutron stars has some interesting implications. First, it could explain why no neutron star is readily apparent in the remnant of SN1987A despite our knowledge that one existed until at least 12 s after the supernova's explosion. Second, it would suggest that a significant population of relatively low mass black holes exists (Brown & Bethe (1994)), one of which could be the compact object in the X-ray binary 4U1700-37 (Brown *et al.* (1992)).

How is the stellar structure, particularly the maximum mass, influenced by the trapped neutrinos? (The finite entropy plays a lesser role (Prakash *et al.* (1997)). In order to investigate this question, one needs the EOS up to $\sim$ 10 times the baryon density encountered in the center of a nucleus. The EOS at such high densities is not known with any certainty. Nevertheless, recent work (Glendenning (1985), Kaplan & Nelson (1986), Glendenning (1992)) has emphasized the possibility that hyperons, a condensate of K^- mesons, or u, d, and s quarks, may be present in addition to nucleons and leptons. These additional components can appear separately or in combination with one another. Compared to a star containing just plain-vanilla nucleons and leptons, the presence of these additional components qualitatively changes the way in which the structure of the star depends upon neutrino trapping (Prakash *et al.* (1997)).

5.2. *Equilibrium conditions*

The composition of the star is constrained by three important physical principles: baryon conservation, charge neutrality and beta equilibrium. The third exists because the time scales of weak interactions, including those of strangeness-violating processes, are short compared to the dynamical time scales of evolution. For example, the process $p + e^- \leftrightarrow n + \nu_e$ in equilibrium establishes the relation

$$\mu \equiv \mu_n - \mu_p = \mu_e - \mu_{\nu_e} \,, \tag{5.87}$$

allowing the proton chemical potential to be expressed in terms of three independent chemical potentials: μ_n, μ_e, and μ_{ν_e}.

At densities where μ exceeds the muon mass, muons can be formed by $e^- \leftrightarrow \mu^- + \overline{\nu}_\mu + \nu_e$, hence the muon chemical potential is

$$\mu_\mu = \mu_e - \mu_{\nu_e} + \mu_{\nu_\mu} \,, \tag{5.88}$$

requiring the specification of an additional chemical potential μ_{ν_μ}. However, unless $\mu > m_\mu c^2$, the net number of μ's and ν_μ's per baryon (designated by Y_μ and Y_{ν_μ}, respectively) is zero, because no muon-flavor leptons are present at the onset of trapping, so $Y_{\nu_\mu} = -Y_\mu$ determines μ_{ν_μ}. Following deleptonization, $Y_{\nu_\mu} = 0$, and Y_μ is determined by $\mu_\mu = \mu_e$ for $\mu_e > m_\mu c^2$, or by $Y_\mu = 0$ otherwise.

Negatively charged kaons can be formed in the process $n + e^- \leftrightarrow n + K^- + \nu_e$ when $\mu_{K^-} = \mu$ becomes equal to the energy of the lowest eigenstate of a K^- in matter. In addition, weak reactions for the Λ, Σ, and Ξ hyperons are all of the form $B_1 + \ell \leftrightarrow B_2 + \nu_\ell$, where B_1 and B_2 are baryons, ℓ is a lepton, and ν_ℓ is a neutrino of the corresponding flavor. The chemical potential for a baryon B with baryon number b_B and electric charge

q_B is then given by the general relation

$$\mu_B = b_B \mu_n - q_B \mu \,, \tag{5.89}$$

which leads to

$$\mu_\Lambda = \mu_{\Sigma^0} = \mu_{\Xi^0} = \mu_n \quad ; \quad \mu_{\Sigma^-} = \mu_{\Xi^-} = \mu_n + \mu \quad ; \quad \mu_p = \mu_{\Sigma^+} = \mu_n - \mu \,. \tag{5.90}$$

The same considerations apply to quarks, for which Eq. (5.89) gives

$$\mu_d = \mu_s = (\mu_n + \mu)/3 \quad ; \quad \mu_u = (\mu_n - 2\mu)/3 \,. \tag{5.91}$$

Therefore, if there are no trapped neutrinos present, so that $\mu_{\nu_\ell} = 0$, there are two independent chemical potentials (μ_n, μ_e) representing conservation of baryon number and charge. If trapped neutrinos are present $(\mu_{\nu_\ell} \neq 0)$, further constraints, due to conservation of the various lepton numbers over the dynamical time scale of evolution, must be specified. At the onset of trapping, during the initial inner core collapse, $Y_\ell = Y_e + Y_{\nu_e} \approx 0.4$ and $Y_e/Y_{\nu_e} \sim 5-7$, depending upon the density (Bethe *et al.* (1979)). These numbers are not significantly affected by variations in the EOS. Following deleptonization, $Y_{\nu_e} = 0$, and Y_e can vary widely, both with the density and EOS.

As long as both weak and strong interactions are in equilibrium, the above general relationships determine the constituents of the star during its evolution. Since electromagnetic interactions give negligible contributions, it is sufficient to consider the non-interacting (Fermi gas) forms for the partition functions of the leptons. Hadrons, on the other hand, receive significant contributions at high density from the less well known strong interactions.

5.3. *Neutrino-poor versus neutrino-rich stars*

A detailed discussion of the composition and structure of protoneutron stars may be found in Prakash *et al.* (1997), and Ellis *et al.* (1996). The main findings were that the structure depends more sensitively on the compostion of the star than its entropy and that the trapped neutrinos play an important role in determining the composition. Since the structure is chiefly determined by the pressure of the strongly interacting constituents and the nature of the strong interactions is poorly understood at high density, several models of dense matter, including matter with strangeness-rich hyperons, a kaon condensate and quark matter were studied there. For the purpose of illustration, we show here only the cases in which hyperons and quarks appear at high densities. The qualitative trends when other forms of strangeness, including kaon condensates, appear are very similar.

Figure 12 shows the various concentrations, Y_i (the number of particles of species i per baryon), as a function of density when the only hadrons allowed are nucleons. The arrows indicate the central density of the maximum mass stars. The left hand panel refers to the case in which the neutrinos have left the star. At high density the proton concentration is about 30%, charge neutrality ensuring an equal number of negatively charged leptons. This relatively large value is the result of the symmetry energy increasing nearly linearly with density in this model. Many non-relativistic potential models (Wiringa *et al.* (1988)) predict a maximum proton concentration of only 10%. However, the generic results discussed here are not sensitive to the behavior of the nuclear symmetry energy. The effects of neutrino trapping are displayed in the right hand panel. The fact that $\mu_{\nu_e} \neq 0$ in Eq. (5.87) results in larger values for μ_e and Y_e. Because of charge neutrality, Y_p is also larger, and it approaches 40% at high density. It is clear that the maximum mass configuration has a much lower density when neutrinos are trapped than when they are not. As is evident from the bottom row of Table 10, neutrino-trapping reduces the maximum mass M^L_{max} from the value found in neutrino-free matter M_{max};

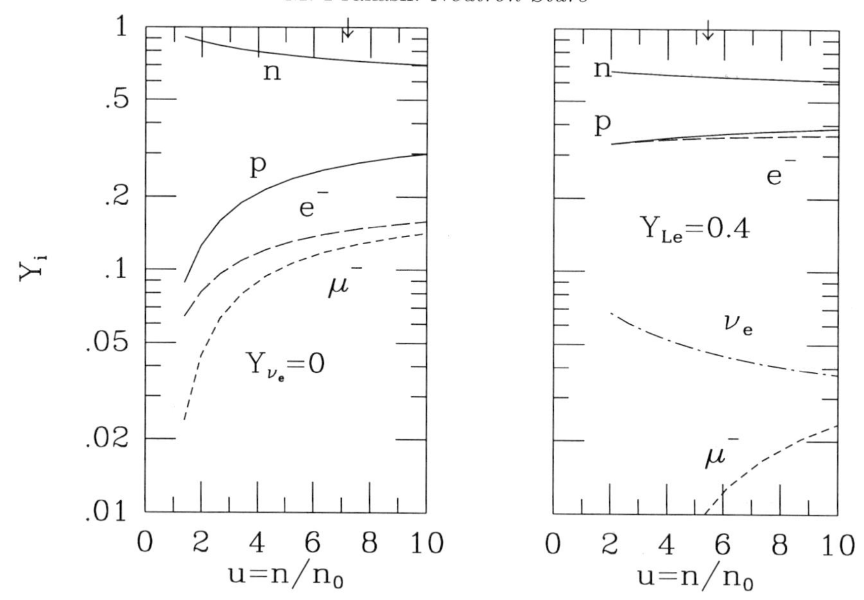

FIGURE 12. Individual concentrations, Y_i, as a function of the baryon density ratio $u = n/n_0$, where n_0 is the density of equilibrium nuclear matter. The arrows indicate the central density of the maximum mass stars. Left panel: neutrino free. Right panel: with trapped neutrinos ($Y_\ell = 0.4$).

although neutrino-trapped nucleons-only matter contains more leptons and more leptonic pressure, it also contains more protons and, therefore, less baryonic pressure. Thermal effects increase the pressure and therefore the maximum mass, but only slightly. Even for $S = 2$, the central temperature is only $\sim$ 50 MeV, which is much less than the nucleon Fermi energies. Thus, because $M^L_{max} \lesssim M_{max}$, a black hole could only form promptly after bounce from nucleons-only stars, in the absence of significant accretion at late times.

For comparison, Fig. 13 shows the compositions of neutrino-free matter (left panel) and neutrino-trapped matter (right panel) in the event that strange particles are allowed to appear. In neutrino-free matter, one expects that the Λ, with a mass of 1116 MeV, and the Σ^-, with a mass of 1193 MeV, both first appear at roughly the same density, because the somewhat higher mass of the Σ^- is compensated by the presence of μ_e in the equilibrium condition of the Σ^-. More massive and more positively charged particles than these appear at higher densities. Notice that with the appearance of the negatively charged Σ^- hyperon, which competes with the leptons in maintaining charge neutrality, the lepton concentrations begin to fall. When quarks appear, at around $4n_0$ (for $B = 200$ MeV fm^{-3}), the neutral and negative particle abundances begin to fall, since quarks furnish both negative charge and baryon number.

Trapped neutrinos again increase the proton and electron abundances, and this strongly influences the threshold for the appearance of hyperons. The Λ and the Σ's now appear at densities higher than those found in the absence of neutrinos. In addition, the transition to a mixed phase with quarks is delayed to about $10n_0$, which is beyond the central density of the maximum mass star (see arrow in Fig. 13). This makes the overall EOS stiffer, so that when matter contains strangeness the behavior is opposite to that of the nucleons-only case. Specifically, the maximum mass is *larger* in the neutrino-trapped case. This behavior is summarized in Table 10 for different assumptions about the com-

	$M_{\max}/M_\odot$			
	Without hyperons		With hyperons	
B (MeV fm^{-3})	$Y_\nu = 0$	$Y_{Le} = 0.4$	$Y_\nu = 0$	$Y_{Le} = 0.4$
136.6	1.440	1.610	1.434	1.595
150	1.444	1.616	1.436	1.597
200	1.493	1.632	1.471	1.597
250	1.562	1.640	1.506	1.597
No quarks	1.711	1.645	1.516	1.597

TABLE 10. Maximum masses of stars with baryonic matter that undergoes a phase transition to quark matter without ($Y_\nu = 0$) and with ($Y_{Le} = 0.4$) trapped neutrinos. Results are for a mean field model of baryons and a bag model of quarks. B denotes the bag pressure in the quark EOS.

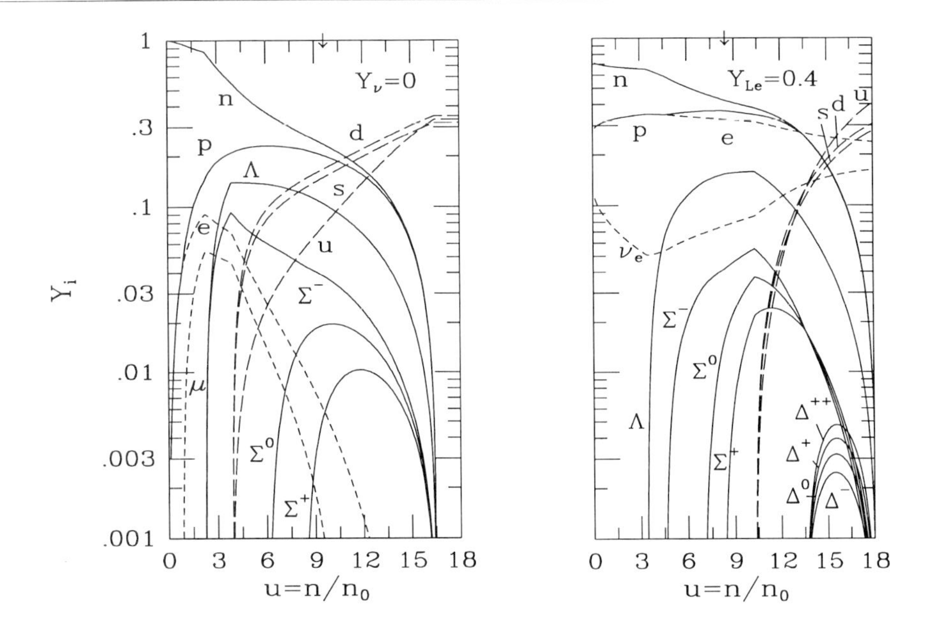

FIGURE 13. As for Fig. 12, but for matter which contains strangeness-bearing hyperons and quarks, as well as nucleons and leptons.

position and was first noted by Thorsson *et al.* (1994) in the study of stars with kaon condensates and subsequently by Keil & Janka (1994) of stars with hyperons.

Evolutionary calculations (Burrows & Lattimer (1986), Keil & Janka (1994)) without accretion show that it takes on the order of 10–15 s for the trapped neutrino fraction to vanish. for a nucleons-only EOS To see qualitatively what might transpire during the early evolution, we show in Fig. 14 the dependence of the maximum stellar mass upon the trapped neutrino fraction Y_{ν_e}, which decreases during the evolution. When the only

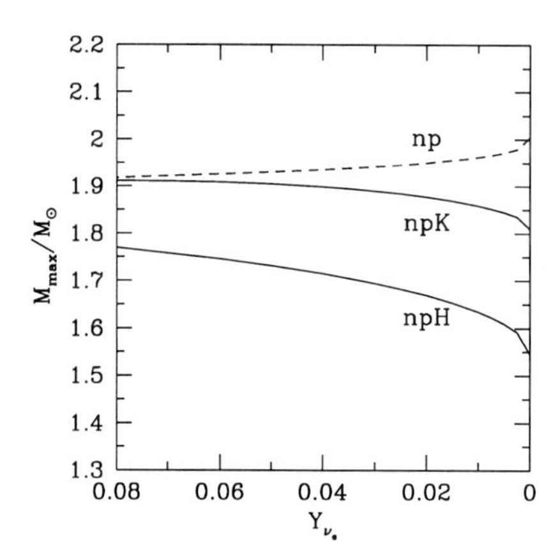

FIGURE 14. Maximum neutron star mass as a function of Y_{ν_e} for hadronic matter with only nucleons (np) or with nucleons and hyperons (npH) or kaons (npK).

hadrons are nucleons (np), the maximum mass increases with decreasing Y_{ν_e}, whereas when hyperons (npH) or kaons (npK) are also present, it decreases. Further, the rate of decrease accelerates for rather small values of Y_{ν_e}. Coupled with this is the fact that the central density of stars will tend to increase during deleptonization. The implication is clear. *If* hyperons, kaons, or other negatively-charged hadronic species are present, an initially stable star can change into a black hole after most of the trapped neutrinos have left, and this takes 10 – 15 s. This happens only if the remnant mass M satisfies $M^L_{max} > M > M_{max}$.

It must be emphasized that the maximum mass of the cold catalyzed star still remains uncertain due to the uncertainty in strong interactions at high density. At present, all nuclear models can only be effectively constrained at nuclear density and by the condition of causality at high density. The resulting uncertainty is evident from the range of possible maximum masses predicted by the different models. Notwithstanding this uncertainty, our findings concerning the effects of neutrino trapping offer intriguing possibilities for distinguishing between the different physical states of matter. These possibilities include both black hole formation in supernovae and the signature of neutrinos to be expected from supernovae.

5.4. *Supernova SN1987A*

On February 23 of 1987, neutrinos were observed from the explosion of supernova SN1987A (Bionta *et al.* (1987), Hirata *et al.* (1987)), indicating that a neutron star, not a black hole, was initially present. (The appearance of a black hole would have caused an abrupt cessation of any neutrino signal (Burrows (1988).)) The neutrino signal was observed for a period of at least 12 s, after which counting statistics fell below measurable limits. From the handful of events observed, only the average neutrino energy, $\sim$ 10 MeV, and the total binding energy release of $\sim (0.1 - 0.2)\mathrm{M}_\odot$ could be estimated.

These estimates, however, do not shed much light on the composition of the neutron star. This is because, to lowest order, the average neutrino energy is fixed by the neutrino mean free path in the outer regions of the protoneutron star. Further, the binding energy exhibits a universal relationship (Prakash *et al.* (1997)) for a wide class of EOS's, including those with strangeness bearing components, namely

$$B.E. = (0.065 \pm 0.01)(M_B/\mathrm{M}_\odot)^2 \mathrm{M}_\odot\,, \tag{5.92}$$

where M_B is the baryonic mass. This allows us only to determine a remnant gravitational mass of $(1.14 - 1.55)\mathrm{M}_\odot$, but not the composition.

The ever-decreasing optical luminosity (light curve) (Kumagai *et al.* (1991)) of the remnant of SN1987A suggests two arguments against the continued presence of a neutron star. First, accretion onto a neutron star at the Eddington limit is already ruled out for the usual hydrogen-dominated Thomson electron scattering opacity. (However, if the atmosphere surrounding the remnant contains a sufficient amount of iron-like elements, as Chen & Colgate (1995) suggest, the appropriate Eddington limit is much lower.) Second, a Crab-like pulsar cannot exist in SN1987A, since the emitted magnetic dipole radiation would be observed in the light curve. Either the magnetic field or the spin rate of the neutron star remnant would have to be much less than in the case of the Crab and what is inferred from other young neutron stars. The spin rate of a newly formed neutron star is expected to be high; however, the time scale for the generation of a significant magnetic field is not well known and could be greater than 10 years.

Although most of the binding energy is released during the initial accretion and collapse stage in about a second after bounce, the neutrino signal continued for a period of at least 12 s. The compositionally-induced changes in the structure of the star occur on the deleptonization time scale, which we have estimated to be of order 10–15 s (Prakash *et al.* (1997)), not on the binding energy release time scale. Thus, the duration of the neutrino signal from SN1987A was comparable to the time required for the neutrinos initially trapped in the star to leave. However, counting statistics prevented measurement of a longer duration, and this unfortunate happenstance prevents one from distinguishing a model in which negatively-charged matter appears and a black hole forms from a less exotic model, in which a neutron star still exists. As we have pointed out, the maximum stable mass drops by as much as $0.2\mathrm{M}_\odot$ when the trapped neutrinos depart if negatively charged particles are present, which could be enough to cause collapse to a black hole.

Observed neutron stars lie in a very small range of gravitational masses (see Fig. 2). Thielemann *et al.* (1990), and Bethe & Brown (1995) have estimated the gravitational mass of the remnant of SN1987A to be in the range $(1.40 - 1.56)\mathrm{M}_\odot$, using arguments based on the observed amounts of ejected ^{56}Ni and/or the total explosion energy. This range extends above the largest accurately known value for a neutron star mass, 1.44 $\mathrm{M}_\odot$, so the possibility exists that the neutron star initially produced in SN1987A could be unstable in the cold, deleptonized state. In this case, SN1987A would have become a black hole once it had deleptonized, and no further signal would be expected. Should this scenario be observationally verified, it would provide strong evidence for the appearance of strange matter.

5.5. *Future Directions*

The emitted neutrinos, of all flavors, are the only direct probe of the mechanism of supernova explosions and the structure of newly formed neutron stars. The cooling of the star can yield information on the stellar composition. The two most important microphysical ingredients for detailed simulations of the cooling of a newborn neutron star are the EOS of dense matter and the neutrino opacities.

Future efforts must address the crucial question of the strong interactions of strange particles in dense matter – even near nuclear equilibrium density, our knowledge is sketchy at present.

What can be expected in future detections? In an optimistic scenario, several thousand neutrinos from a typical galactic supernova might be seen in upgraded neutrino detectors, such as SNO in Canada and Super Kamiokande in Japan. (For rough characteristics of present and future neutrino detectors, see Burrows, Klein & Gandhi (1992). Among the interesting features that could be sought are:

(*a*) Possible cessation of a neutrino signal, due to black hole formation.

(*b*) Possible burst or light curve feature associated with the onset of negatively-charged, strongly interacting matter near the end of deleptonization, whether or not a black hole is formed.

(*c*) Identification of the deleptonization/cooling epochs by changes in luminosity evolution or neutrino flavor distribution.

(*d*) Determination of a radius-mean free path correlation from the luminosity decay time or the onset of neutrino transparency.

(*e*) Determination of the neutron star mass from the universal binding energy-mass relation.

6. Cooling of neutron stars

The thermal evolution of neutron stars has long been regarded as the source of information about the possible physical states of dense matter. Neutron stars are born with interior temperatures of order 20–50 MeV, but cool via neutrino emission to temperatures of less than 1 MeV within minutes (Burrows & Lattimer (1986)). The subsequent cooling consists of two phases: a neutrino-dominated cooling epoch followed by a photon-dominated cooling epoch. Even during the neutrino epoch, thermal photons are radiated from the neutron star's surface. The temperature and luminosity of this thermal radiation is controlled by the interior temperature evolution of the star. Until recently, the general view was that the interiors of newly-formed neutron stars would cool relatively slowly (Chiu & Salpeter (1964), Bahcall & Wolf (1965), Tsuruta (1986)), unless they contained non-standard or exotic matter, such as a pion condensate, a kaon condensate, or quark matter (Maxwell *et al.* (1977), Muto & Tatsumi (1988), Iwamoto (1982)). In the standard model, the interior cooling is slow enough that the surface temperatures of neutron stars remain above 10^6 K for about 10^5 yr, and they are potentially observable for this length of time in the X-ray or UV bands. When exotic matter with enhanced neutrino emissivity is present, the core cools so rapidly that a temperature inversion develops. The size of the cooler interior grows as the energy from the hot crust is conducted to the core. After about 1 to 100 years, depending upon the star's structure, this cooling wave reaches the surface and the surface temperature plummets, perhaps to unobservably low values.

6.1. *Thermal evolution*

The equations governing the thermal evolution of a spherical, non-rotating and non-magnetic star in hydrostatic equilibrium are (Tsuruta (1986)):

$$\frac{dM}{dr} = 4\pi r^2 \epsilon\,, \qquad \frac{dP}{dr} = -\frac{GM\epsilon}{c^2 r^2}\left[1+\frac{P}{\epsilon}\right]\left[1+\frac{4\pi r^3 P}{Mc^2}\right]e^{2\Lambda} \tag{6.93}$$

$$\frac{d}{dr}\left(Te^{\Phi/c^2}\right) = -\frac{3}{16\sigma}\frac{\kappa\rho}{T^3}\frac{L_d}{4\pi r^2}e^{\Phi/c^2}e^{\Lambda}\,, \qquad \frac{d\Phi}{dr} = \frac{G\left(M+4\pi r^3 P/c^2\right)}{r^2}e^{2\Lambda} \tag{6.94}$$

$$\frac{d}{dr}\left(L_\nu e^{2\Phi/c^2}\right) = \epsilon_\nu e^{2\Phi/c^2} 4\pi r^2 e^\Lambda \,, \quad \frac{d}{dr}\left(Le^{2\Phi/c^2}\right) = -c_v \frac{dT}{dt} e^{\Phi/c^2} 4\pi r^2 e^\Lambda \,, \quad (6.95)$$

with $\Lambda = (1 - 2GM/c^2r)^{-1/2}$. Above, G is the gravitational constant, P is the pressure, ϵ is the energy density inclusive of the rest mass density, and M is the enclosed gravitational mass. The first two equations are the TOV equations of hydrostatic equilibrium. The third equation describes heat transport within the star. Symbols are: T is the temperature, κ is the total opacity of stellar matter, σ is the Stefan-Boltzmann constant, L_d is the luminosity due to thermal conductivity and radiation, and Φ is the gravitational potential, which is determined by the fourth equation. There is no corresponding heat transport equation for neutrinos, since for degenerate conditions ($T << T_F$), the mean free paths of neutrinos are typically a few times the radius of the neutron star. Thus, neutrinos are only a sink of energy and do not transport energy from one part of the star to another. The fifth equation describes the spatial variation of the neutrino luminosity L_ν for stars in which nuclear burning has ceased. Here, ϵ_ν denotes the neutrino emisssivity. The sixth equation is a statement of energy conservation. Symbols are: $L = L_d + L_\nu$ is the net luminosity and c_v is the specific heat per unit volume. The time variable t is the time measured by an observer at $r = \infty$, who is at rest with respect to the star. The quantities $P, \epsilon, T, \kappa, L, L_d, L_\nu, c_v$, and ϵ_ν are local quantities and are evaluated in a proper reference frame comoving with the stellar material.

The solution of the six ordinary differential equations governing the thermal evolution of the star are facilitated by provding appropriate inner and outer boundary conditions. The inner boundary conditions are

$$M(0) = L(0) = L_\nu(0) = 0 \,. \quad (6.96)$$

The outer boundary conditions are

$$P_s = \tfrac{2}{3} g_s/\kappa_s \quad (6.97)$$

$$L_s = L_d(R) = 4\pi R^2 \sigma T_s^4 \quad (6.98)$$

$$e^{\Phi/c^2} = \left(1 - \tfrac{2GM}{c^2 r}\right)^{-1/2} = e^{-\Lambda} \,, \quad (6.99)$$

where $g_s = (GM/R^2)e^{\Lambda_s}$ is the surface gravity, κ_s is the opacity at the surface, and T_s is the surface temperature. In the last relation for the red-shift factor, which holds for $r >> R$, M is the total mass of the star.

The physics ingredients required to establish the thermal evolution are: (1) the equation of state $P = P(\epsilon)$, (2) the opacity κ and the specific heat c_v, and (3) photon and neutrino emissivites. We turn now to the various sources leading to loss of energy via neutrino emission.

6.2. *Neutrino emission processes*

The so–called standard model of neutron star cooling is based on neutrino emission from the interior that is dominated by the modified Urca† process (Chiu & Salpeter (1964), Bahcall & Wolf (1965))

$$(n,p) + p + e^- \to (n,p) + n + \nu_e \,, \quad (n,p) + n \to (n,p) + p + e^- + \bar{\nu}_e \,. \quad (6.100)$$

In the standard model, the surface temperatures of neutron stars remain above 10^6 K for about 10^5 years, so they are potentially observable in the X–ray or UV bands. Nevertheless, the thermal radiation from a neutron star has yet to be identified unambiguously. All positive observations to date are for pulsars, and it is unclear how much of the observed emissions is due to the pulsar phenomenon, to a synchrotron–emitting nebula, or

† Named after the now extinct Urca casino in Rio de Janeiro, which was reputedly a perfect sink for money.

to the neutron star itself (Tsuruta (1986)). If matter were in an exotic state, the cooling would be faster – so fast, in fact, that thermal emission from the star's surface would be too small to be observable.

The main observation by Lattimer *et al.* (1991) and Prakash *et al.* (1992) was that it is likely that all neutron stars will cool rapidly, whether they contain the so–called exotic matter or not. The considerations that have necessitated a revision of the general view, held for about thirty years now, are summarized below.

6.3. *The direct Urca process revisited*

The direct Urca process

$$n \to p + e^- + \bar{\nu}_e\,, \quad p + e^- \to n + \nu_e\,, \tag{6.101}$$

was not usually considered, because the proton abundance was thought to be too small to allow simultaneous energy and momentum conservation (Chiu & Salpeter (1964), Bahcall & Wolf (1965)). At temperatures well below typical Fermi temperatures ($T_F \sim 10^{12}$ K), massive fermions of species $i = (n,p)$ participating in the process must have momenta close to their respective Fermi momenta, p_{F_i}. The neutrino and antineutrino momenta are $\sim kT/c \ll p_{F_i}$. Since matter is very close to beta equilibrium, the chemical potentials of the constituents satisfy the condition $\mu_n = \mu_p + \mu_e$. Thus, the condition of energy conservation is easily satisfied for some states close to the respective Fermi surfaces. The condition for momentum conservation is that it must be possible to construct a triangle from the Fermi momenta of the two baryons and the electron, and therefore the three triangle inequalities $p_{F_i} + p_{F_j} \geq p_{F_k}$, where i, j, and k are p, e, and n, and cyclic permutations of them, must be satisfied. When the inequalities are not satisfied, the dominant neutrino emission processes are the modified Urca processes, which differ from Eq. (6.101) by the presence in the initial and final states of a bystander particle, whose sole purpose is to make possible conservation of momentum for particles close to the Fermi surfaces.

6.3.1. *Threshold proton fraction and density for the direct Urca process*

The number density of particle species i (n, p or e) is given by $n_i = p_{Fi}^3/(3\pi^2\hbar^3)$. Thus, the proton fraction $x = n_p/(n_p + n_n)$ may be rewritten, using the momentum conservation condition at threshold, as

$$x_c = \frac{p_{Fp}^3}{p_{Fp}^3 + (p_{Fp} + p_{Fe})^3} = \frac{1}{1 + (1 + p_{Fe}/p_{Fp})^3}\,. \tag{6.102}$$

If matter consists only of neutrons, protons and electrons, charge neutrality requires that $n_p = n_e$, or $p_{Fp} = p_{Fe}$. Hence, the proton fraction at threshold is $x_c = 1/9$.

To explore the dependence of the proton fraction on nuclear properties, the energy per baryon ϵ may be expanded quadratically in the proton concentration x about its value for symmetric matter ($x = 1/2$):

$$\epsilon(n, x) = \epsilon(n, 1/2) + S_v(n)(1 - 2x)^2 + \cdots, \tag{6.103}$$

where $n = n_n + n_p$ is the baryon density and S_v is the bulk symmetry energy, which is density dependent. At nuclear saturation density, $n_s \simeq 0.16$ fm^{-3}, $S_v(n_s) \equiv S_o \approx 27-36$ MeV (Möller *et al.* (1988), Pearson *et al.* (1991)). The matter we are concerned with is degenerate, and therefore the temperature dependence of Eq. (6.103) may be neglected. The condition for beta equilibrium is

$$\mu_e = \mu_n - \mu_p = -(\partial\epsilon/\partial x)\,. \tag{6.104}$$

Studies of pure neutron matter strongly suggest that the expansion Eq. (6.103), with

q	1/3	2/3	1	4/3
n_c/n_s	25(9.7)	5.0(3.1)	2.2(1.8)	1.71(1.46)

TABLE 11. Urca threshold densities. The quantity n_c/n_s was calculated for power law symmetry energies $S_v \propto n^q$ using $S_o = 30(35)$ MeV.

only the quadratic term, is a good approximation for all x, at any density (Prakash *et al.* (1988)). If muons and other charged species are ignored, x is then given by (Muether *et al.* (1987))

$$\hbar c(3\pi^2 nx)^{1/3} = 4S_v(n)(1-2x)\,, \tag{6.105}$$

where the electrons are assumed ultrarelativistic and degenerate. The density, n_c, at which x equals the critical value, $x_c = 1/9$, is found from

$$S_v(n_c) = 51.2\left(\frac{S_o}{30\text{ MeV}}\right)\left(\frac{n_c}{n_s}\right)^{1/3}\text{ MeV}. \tag{6.106}$$

If S_v has a power law dependence on the density, $S_v \propto n^q$, we find $n_c/n_s = [1.71(30\text{ MeV}/S_o)]^{1/(q-1/3)}$. The case $q = 2/3$ and $S_0 = (1/3)(\hbar^2 k_{F_s}^2/2m) \cong 12.28$ MeV corresponds to free non–relativistic nucleons for which $n_c/n_s \simeq 73$! Table 11 shows the critical density for a select choice of q and $S_o = 30(35)$ MeV. Clearly, the critical density is sensitive to interactions and the magnitude of the symmetry energy. This situation had, in fact, been anticipated by Boguta (1981), who pointed out that large symmetry energies at high densities would allow rapid cooling via the direct Urca process to take place.

Muons will be present when $\mu_e > m_\mu c^2 = 105.7$ MeV, which generally is the case for $n \geq n_s$. In the presence of muons, the charge neutrality condition, $n_e + n_\mu = n_p$, and the energy conservation condition, $\mu_e = \mu_\mu$, may be combined to yield

$$p_{Fe}^3 + \left(p_{Fe}^2 - m_\mu^2 c^2\right)^{3/2} = p_{Fp}^3\,. \tag{6.107}$$

To estimate the effects of muons, consider the case $\mu_e = p_{Fe}c >> m_\mu c^2$. In this case, $p_{Fe} = p_{F\mu} = (1/2)^{1/3} p_{Fp}$, which gives

$$x_c = \frac{1}{1 + (1 + 1/2^{1/3})^3} \simeq 0.148\,. \tag{6.108}$$

In general, when the effects of the muon mass are included, $p_{F\mu} < p_{Fe}$. Hence, x_c will be altered from the above estimate, but the difference is not large.

Although x_c is now larger than before, the critical density (for the Urca process with electrons) is smaller, provided $q > 1/3$; the equation determining x, Eq. (6.105), has the x on its left–hand side replaced by $x/2$, and the critical density becomes $n_c/n_s = [1.65(30\text{ MeV}/S_o)]^{1/(q-1/3)}$, in the case of power law symmetry energy and ultrarelativistic muons. Note that if $p_{Fp} + p_{F\mu} > p_{Fn}$, the direct Urca process with muons will occur. The threshold concentration and density for this process are higher than that for electrons, since $m_\mu > m_e$. In the discussion that follows, we shall consider symmetry energies from microscopic calculations.

6.3.2. *Models of dense matter*

The calculation most firmly grounded in available nuclear data is that of Wiringa *et al.* (1988), which is based on a two–body potential fitted to nucleon–nucleon scattering and a three–body term whose form is suggested by theory and whose parameters are determined by the binding of few–body nuclei and the saturation properties of nuclear matter. The symmetry energies for two choices of the three–body forces are shown in Fig. 15(a). Because of Eq. (6.105), suitably modified to include muons, x mimics the behavior of $S_v(n)$, as displayed in Fig. 15(b). The proton concentration attains values very close to those required for the direct Urca process to occur, but never quite reaches them. Note, however, the large spread in x, which is consistent with available data. This largely reflects uncertainties in the three–body interaction. Also shown in Fig. 15 are results for two other types of models: a field–theoretical model (Chin (1977)), with baryon and meson degrees of freedom calculated up to the one–loop level, and the relativistic Dirac–Brueckner approach (Muether *et al.* (1987), Horowitz & Serot (1987)), in which the matrix elements of the boson exchange potentials are calculated using in–medium nucleon spinors, and the effects of correlations are calculated using the Bethe–Goldstone equation. In these models, x can be large enough for the direct Urca process to occur. It is evident that the more rapidly the symmetry energy increases with density, the lower the density at which the direct Urca process begins to operate.

For the direct Urca process to occur in a neutron star, its central density, which depends on the pressure rather than on the symmetry energy, must exceed the critical density. Fig. 15(b) shows the central densities of a neutron star with a mass of 1.4 $M_\odot$ and a neutron star with the maximum mass for the particular equation of state (EOS). Of the models selected, only that of Horowitz & Serot (1987) allows the direct Urca process in a 1.4 $M_\odot$ star. However, in each of the models displayed in Fig. 15, the central density of the maximum mass neutron star considerably exceeds that for the 1.4 $M_\odot$ star.

6.3.3. *Neutrino emissivity from the direct Urca process*

The emissivity due to the direct Urca process may be derived from Fermi's Golden Rule. The antineutrino energy emission rate from neutron decay is given by

$$\epsilon_\beta = \frac{2\pi}{\hbar} 2 \sum_i G_F^2 (1+3g_A)^2 n_1(1-n_2)(1-n_3)E_4 \delta^{(4)}(p_1 - p_2 - p_3 - p_4)\,, \quad (6.109)$$

where n_i is the Fermi function and the subscripts $i = 1-4$ refer to the neutron, proton, electron and antineutrino, respectively. The p_i's are four-momenta and E_4 is the antineutrino energy. The sum over states is to be performed only over three-momenta $\vec{p_i}$ and the prefactor 2 takes into account the initial spin states of the neutron. The square of the neutron beta–decay matrix element, summed over spins of final particles and averaged over angles, is $G_F^2(1+3g_A^2)$, where $G_F \simeq 1.436 \times 10^{-49}$ erg cm^{-3} is the weak coupling constant and $g_A \simeq -1.261$ is the axial vector coupling constant. The factors $1 - n_2$ and $1 - n_3$ are final state blocking factors. The phase space sums in Eq. (6.109) may be simply performed using the methods of Fermi liquid theory. Electron capture gives the same luminosity as neutron decay, but in neutrinos, and thus the total luminosity for the Urca process is $\epsilon_{Urca} = 2\epsilon_\beta$, or

$$\begin{aligned} \epsilon_{Urca} &= \frac{457\pi}{10080} \frac{G_F^2(1+3g_A^2)}{\hbar^{10}c^5} m_n m_p \mu_e (kT)^6 \Theta_t \\ &= 4.00 \times 10^{27} (Y_e n/n_s)^{1/3} T_9^6 \Theta_t \text{ erg cm}^{-3}\text{ s}^{-1}\,, \end{aligned} \quad (6.110)$$

where T_9 is the temperature in units of 10^9K, $n_s = 0.16$ fm^{-3}, $Y_e = n_e/n$ is the electron fraction, and $\Theta_t = \theta(p_{Fe} + p_{Fp} - p_{Fn})$ is the threshold factor, $\theta(x)$ being +1 for $x > 0$ and zero otherwise. If the muon Urca process can occur, the emissivity is increased by a factor of 2, irrespective of the value of $\mu_e/(m_\mu c^2)$.

The estimates above were made assuming that the participating particles are free.

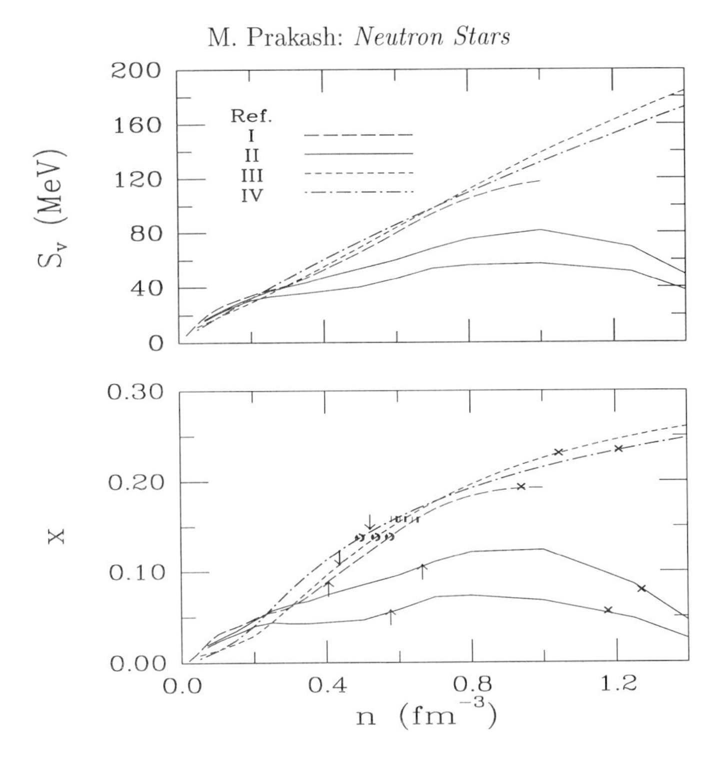

FIGURE 15. (a) Nuclear symmetry energy as a function of density for several different EOSs. EOSs for the different curves are taken from I: Muether *et al.* (1987), II: Wiringa *et al.* (1988), III: Chin (1977), and IV: Horowitz & Serot (1987). (b) Equilibrium proton fraction for the EOS shown in (a), including the presence of muons. Solid circles (squares) denote the critical density for the direct URCA process for electrons (muons). Arrows(crosses) denote the central density of $1.4M_\odot$ (maximum–mass) neutron stars.

Interactions give rise to a number of changes. First, the neutron and proton densities of states at the Fermi surfaces are renormalized, which results in the factor $m_n m_p$ in Eq. (6.110) being replaced by $m_n^* m_p^*$, where m^* is the effective mass (in the sense of Landau Fermi liquid theory). This factor may well be of order 0.5–0.2. A second effect is that in a nuclear medium, the effective value of $|g_A|$ is quenched (Wilkinson (1973), Rho (1974). At the saturation density, $|g_A| \simeq 1$, and it is expected to remain at approximately this value at higher densities. Third, final state interactions will modify the effective weak–interaction matrix element, but this is a small effect, since $n-p$ interactions are small at the momentum transfers of importance ($\sim p_{Fp}$). Thus, the total reduction of the Urca rates due to interaction effects may amount to a factor of 10, but similar factors must be applied to other neutrino emission processes involving nucleons.

Let us now compare the rate of the direct Urca process with that of other processes. The modified Urca process, Eq. (6.100), may be regarded as a correction to the direct Urca process due to damping of participating nucleon states by collisions. The neutrino emissivity from the modified Urca process, for free particles, is Friman & Maxwell (1979)

$$\epsilon_{mod.Urca} \approx 10^{22}(Y_e n/n_s)^{1/3} T_9^8 \text{ erg cm}^{-3}\text{ s}^{-1}. \tag{6.111}$$

A small correction to this result should be applied to take into account reactions in which the bystander nucleon is a proton. From Eq. (6.110) and Eq. (6.111), we find

$$\epsilon_{Urca}/\epsilon_{mod.Urca} \approx 5\times 10^5 T_9^{-2}. \tag{6.112}$$

Roughly speaking, this is a factor $(T/T_F)^{-2}$, reflecting the fact that the bystander particles in the initial and final states of the modified Urca process each lead to a factor T/T_F. This indicates that the rate of the modified Urca process would be comparable to that of the direct Urca process at $T \simeq T_F \simeq 10^{12}$ K.

The time for a star's center to cool by the direct Urca process, Eq. (6.110) to a temperature T_9 may be estimated using Baym & Pethick (1975)

$$\Delta t = -\int \frac{c_v}{\epsilon_{urca}} dT \,, \tag{6.113}$$

where T is the temperature and c_v is the specific heat per unit volume, which yields $\Delta t \sim 30\ T_9^{-4}$ seconds.

6.3.4. *Consequences of the direct Urca process*

The threshold for the direct Urca process is expected to lie in the range of central densities for neutron stars. In view of the step function character of the threshold, neutron stars rather close in mass could have very different cooling histories. This expectation is borne out in the detailed cooling calculations of Page & Applegate (1992). At some densities, neutrons may be superfluid and/or protons superconducting. The direct Urca rate is then reduced by a factor $\sim \exp(-\Delta/kT)$, since for the process to occur, the total energy of particles in the initial or final states must exceed Δ, the larger of the neutron and proton gaps. Calculated gaps are uncertain, but are typically of the order of a few hundred keV (Takatsuka (1971), Chen *et al.* (1986), Ainsworth *et al.* (1989). Thus, when $\Delta >> kT$, neutrino emission rates are significantly reduced. However, it should be remembered that the modified Urca rates are reduced by a factor $\sim \exp(-2\Delta/kT)$. The precipitous drop in the surface temperature caused by cooling due to the direct Urca process is halted by the effects due to superfluidity in the core of the star. Comparisons with observations of surface temperatures thus hold the potential for pinning down superfluid gaps in dense matter.

When the direct Urca process can occur, the bulk viscosity of neutron star matter will be increased by a factor $(T_F/T)^2$ compared to that for the modified Urca process (Haensel & Schaeffer (1992). This would strongly damp radial pulsations and differential rotation in young neutron stars.

6.4. *Direct Urca processes with hyperons and Δ-isobars*

Many calculations of the composition of dense matter lead to the conclusion that baryons more massive than the nucleon, namely hyperons and nucleon isobars, should be present in dense matter. We turn now to consider neutrino emission from matter containing these baryons. One of our most important conclusions is that even minuscule concentrations of Λ's can give rise to a luminosity in neutrinos exceeding that of exotic states. This provides additional channels for rapid cooling of neutron stars which do not involve either exotic states or a large proton fraction.

The simplest possible neutrino emitting processes are of the type

$$B_1 \to B_2 + \ell + \bar{\nu}_\ell\,; \quad B_2 + \ell \to B_1 + \nu_\ell\,, \tag{6.114}$$

where B_1 and B_2 are baryons, and ℓ is a lepton, either an electron or a muon. Here we shall consider processes in which the baryons can be strange particles (Λ's, Σ's, and Ξ's) or non–strange nucleon isobars (Δ^0, Δ^-), as well as nucleons. Beta equilibrium implies that the chemical potentials satisfy the condition

$$\mu_{B_1} = \mu_{B_2} + \mu_\ell\,, \tag{6.115}$$

which is equivalent to the general condition for chemical equilibrium $\mu_i = b_i\mu_n - q_i\mu_\ell$, where b_i is the baryon number of particle i and q_i is its charge. The condition for momentum conservation, expressed through the three triangle inequalities

$$p_{F_i} + p_{F_j} \geq p_{F_k}\,, \tag{6.116}$$

where i, j, and k are B_2, ℓ, and B_1, and cyclic permutations of them, must be satisfied. When the inequalities are not satisfied, the dominant neutrino emission processes are the modified Urca processes. For hyperons, these modified Urca processes have been considered by Maxwell (1987).

6.4.1. *Neutrino emissivity from hyperon Urca processes*

The rate of antineutrino energy emission from baryon (spin half) beta decay is given by

$$\epsilon = \frac{2\pi}{\hbar} 2 \sum_{\vec{p}_i} \overline{\mathcal{M}^2} n_1(1-n_2)(1-n_3)E_4 \delta^{(4)}(p_1 - p_2 - p_3 - p_4)\,, \tag{6.117}$$

where the subscripts $i = 1, 2$ refer to the baryons B_1 and B_2, and the subscripts i=3 and 4 refer to the lepton and the antineutrino, respectively. The other symbols have the same meaning as in Eq. (6.109). Assuming SU(3) symmetry, the matrix element for the process $B_1 \to B_2 \ell \bar{\nu}_\ell$ is given by (see for example, Gaillard & Sauvage (1984))

$$M = (G_F/\sqrt{2}) C \overline{u}_{B_2} [\gamma_\mu f_1 + \gamma_\mu \gamma_5 g_1] u_{B_1}\ \overline{u}_\ell \gamma^\mu (1 + \gamma_5) u_\nu\,, \tag{6.118}$$

where very small tensor terms have been ignored. In the above, G_F is the weak coupling constant, $C = \cos\theta_c (\sin\theta_c)$ for a change of strangeness $|\Delta S| = 0(1)$, and θ_c is the Cabibbo angle. In the non–relativistic limit, the squared matrix element is given by

$$\overline{\mathcal{M}^2} = \overline{M^2}/(\Pi_{i=1}^4 2E_i) = G_F^2 C^2 [(f_1^2 + 3g_1^2) + \vec{v}_\ell \cdot \vec{v}_\nu (f_1^2 - g_1^2)]\,, \tag{6.119}$$

where $\vec{v} = \vec{p}/E$ is the velocity. The parameters f_1 and g_1 of the baryon weak decay process are given in Table 12.

If the neutrino momentum is neglected in the momentum conserving delta function, the contribution of the term involving $\vec{v}_\ell \cdot \vec{v}_\nu$ vanishes on integrating over angles. The energy emission in neutrinos of the Urca pair by process (6.114) is equal to that in antineutrinos by process (6.114), so the total emissivity in both neutrinos and antineutrinos is

$$\epsilon = \frac{457\pi}{10080} \frac{G_F^2 C^2 (f_1^2 + 3g_1^2)}{\hbar^{10} c^5} m_{B_1} m_{B_2} \mu_\ell (kT)^6 \Theta_t \tag{6.120}$$

$$= 4.00 \times 10^{27} (Y_e n/n_s)^{1/3} \frac{m_{B_1} m_{B_2}}{m_n^2} R T_9{}^6 \Theta_t \text{ erg cm}^{-3}\ s^{-1}\,, \tag{6.121}$$

where Θ_t is the threshold factor, which is +1 if the triangle inequalities are satisfied and zero otherwise. The mass of species i is denoted by m_i, and T_9 is the temperature in units of 10^9 K. The electron fraction is $Y_e = n_e/n$, and $n_s = 0.16\ \text{fm}^{-3}$ is the equilibrium density of nuclear matter. The baryon squared matrix element normalized to the corresponding neutron quantity is denoted by R in Eq. (6.121). Explicitly,

$$R = \frac{[C^2(f_1^2 + 3g_1^2)]_{B_1 \to B_2 \ell \nu_\ell}}{[C^2(f_1^2 + 3g_1^2)]_{n \to p \ell \nu_\ell}}\,. \tag{6.122}$$

In comparison with the nucleon direct Urca process, the emissivities of hyperon processes are suppressed, because they have smaller matrix elements, as shown in the last column of Table 12, which gives R for the various processes. Apart from the threshold factor Θ, the luminosity for Urca processes with muons is the same as that for the corresponding processes with electrons, since in equilibrium $\mu_e = \mu_\mu$. Note that characteristic hyperon direct Urca emissivities, although smaller than that for nucleons, are larger

Transition	C	f_1	g_1	R
$n \to p\ell\bar{\nu}_\ell$	$\cos\theta_c$	1	$F+D$	1
$\Lambda \to p\ell\bar{\nu}_\ell$	$\sin\theta_c$	$-\sqrt{3/2}$	$-\sqrt{3/2}(F+D/3)$	0.0394
$\Sigma^- \to n\ell\bar{\nu}_\ell$	$\sin\theta_c$	-1	$-(F-D)$	0.0125
$\Sigma^- \to \Lambda\ell\bar{\nu}_\ell$	$\cos\theta_c$	0	$\sqrt{2/3}D$	0.2055
$\Sigma^- \to \Sigma^0\ell\bar{\nu}_\ell$	$\cos\theta_c$	$\sqrt{2}$	$\sqrt{2}F$	0.6052
$\Xi^- \to \Lambda\ell\bar{\nu}_\ell$	$\sin\theta_c$	$\sqrt{3/2}$	$\sqrt{3/2}(F-D/3)$	0.0175
$\Xi^- \to \Sigma^0\ell\bar{\nu}_\ell$	$\sin\theta_c$	$\sqrt{1/2}$	$(F+D)/\sqrt{2}$	0.0282
$\Xi^0 \to \Sigma^+\ell\bar{\nu}_\ell$	$\sin\theta_c$	1	$F+D$	0.0564
$\Xi^- \to \Xi^0\ell\bar{\nu}_\ell$	$\cos\theta_c$	1	$F-D$	0.2218

TABLE 12. Weak processes for nucleons and hyperons. The quantity $R = [C^2(f_1^2+3g_1^2)]_{B_1\to B_2\ell\nu_\ell}/[C^2(f_1^2+3g_1^2)]_{n\to p\ell\nu_\ell}$ was calculated using the central values of $\sin\theta_c = 0.231 \pm 0.003, F = 0.477 \pm 0.012$ and $D = 0.756 \pm 0.011$.

by a factor of order $(T_F/T)^2 \sim 10^6 T_9{}^2$ than those of hyperon modified Urca processes (Maxwell (1987)). Interactions will affect neutrino emission in a number of ways, but they are not expected to reduce the emission by more than a factor of ten. Superconductivity and superfluidity of particles participating in an Urca process will reduce neutrino emission rates in a manner similar to that for the nucleon process.

Among other particles that we have not yet considered are the Δ isobars, which are non–strange spin 3/2 baryons with masses of about 1232 MeV, only a little greater than that of the Σ^-. Should any process with Δ's be kinematically allowed, it would result in neutrino energy emission comparable to the characteristic rate for the nucleon direct Urca process, and much greater than that for reactions in which there is a change of strangeness.

6.4.2. *Threshold concentrations for hyperon Urca processes*

The threshold concentration for an Urca process to occur is determined by the triangle inequalities in Eq. (6.116), and one general consequence of these is that, under conditions of neutrons being the most abundant baryons, the threshold concentrations for reactions in which neutrons participate will generally be higher than those for reactions without neutrons. For the first process listed, the one for nucleons, the threshold proton/nucleon ratio lies in the range $x_c = 0.11 - 0.15$, the exact value of x_c depending upon the ratio of muons to electrons. However, it is not known if this condition is ever satisfied in neutron stars, because the proton concentration is determined by the symmetry energy of matter above nuclear density, and this is poorly known. In models such as relativistic mean field ones, in which the symmetry energy increases linearly with density, the condition is generally satisfied in neutron stars with masses exceeding $1M_\odot$. However, if the symmetry energy increases less rapidly with density, the critical density (if it exists at all) may never be attained in a given neutron star.

We turn now to hyperons and nucleon isobars. The composition of matter is determined by solving the chemical equilibrium conditions Eq. (6.115), subject to the conditions of charge neutrality and baryon number conservation. One expects that Λ, with a mass of 1116 MeV, the Σ^-, with a mass of 1197 MeV, and Δ^-, with a mass of 1232 MeV, first appear at roughly the same density, because the somewhat higher masses of the Σ^- and Δ^- are compensated by the presence of the electron chemical potential in the equilibrium condition for the Σ^-. More massive, and more positively charged, particles than these appear only at higher densities, if they do so at all. Most model calculations show that positively charged particles appear at densities in excess of those found in $1.4M_\odot$ neutron stars, and in the following discussion we shall assume that they are not present. In that case, the individual concentrations of Σ^- and Δ^- will always be less than that of protons.

The question of whether any of the hyperon Urca processes are allowed when the nucleon Urca processes are forbidden is most conveniently addressed in terms of the proton to nucleon ratio $x = x_p/(x_n + x_p)$, where x_i are the number fractions of species i. Recall that nucleon Urca processes are forbidden for $x < x_c$. If $x > x_c$, the Urca process will be permitted with nucleons, unless the concentrations of hyperons and/or nucleon isobars is large enough to reduce substantially the electron concentration. *In any event, if the nucleon Urca process is forbidden, the Urca processes* $\Sigma^- \to n + \ell + \bar{\nu}$ *and* $\Delta^- \to n + \ell + \bar{\nu}$ *are also forbidden, because* $x_{\Sigma^-}, x_{\Delta^-} < x_p$. Under these circumstances, the only Urca processes that could still be allowed are those involving Λ's: $\Lambda \to p + \ell + \bar{\nu}$, $\Sigma^- \to \Lambda + \ell + \bar{\nu}$ and $\Delta^- \to \Lambda + \ell + \bar{\nu}$.

The Urca threshold concentration of Λ's, x_{Λ_c}, can be roughly estimated from the triangle inequalities, Eq. (6.116), by its value when the abundances of Σ^- and Δ^- are zero, namely $x_{\Lambda_c} = |\, {x_p}^{1/3} - {x_e}^{1/3} \,|^3$. This threshold concentration is zero if the numbers of protons and electrons are equal, as they are if electrons and protons are the only charged particles present. Including the presence of muons, but no additional particles, an upper limit for x_{Λ_c} when $x_{\Sigma^-} = x_{\Delta^-} = 0$ is found to be

$$x_{\Lambda_c} = \frac{x(2^{1/3}-1)^3}{2 + x(2^{1/3}-1)^3} \leq \frac{(2^{1/3}-1)^3}{6(2^{1/3}+1)} \simeq 0.0013\,, \tag{6.123}$$

where the numerical value is obtained assuming $x < x_c$. (Recall that if $x \geq x_c$, the more efficient nucleon Urca process is permitted when no hyperons or nucleon isobars are present.) A more realistic estimate of x_{Λ_c} is obtained by relaxing the condition that $x_{\Sigma^-} = x_{\Delta^-} = 0$. When Σ^-'s and/or Δ^-'s are present, x_{Λ_c} may be larger or smaller than the value in Eq. (6.123) depending on the relative concentrations, but it is possible to show that an upper limit still exists:

$$x_{\Lambda_c} < \frac{x}{(1+\sqrt{x})(2 + 4\sqrt{x} + 5x + 4x^{3/2} + 2x^2)} < 0.032\,. \tag{6.124}$$

Thus the Urca process *must* occur either when $x \geq x_c$ and no hyperons or isobars are present, or, for any x (and independent of x_{Σ^-} and x_{Δ^-}), when $x_\Lambda \geq x_{\Lambda_c}$, which is, typically, a few parts per mille. Note that these conditions have been derived completely independently of any knowledge of nucleon and hyperon interactions. *It is important to appreciate that the* x_{Λ_c}*'s in Eq. (6.123) and Eq. (6.124) are exceedingly small.* Once a hyperon is present, its concentration increases rapidly with density, and, consequently, the threshold density for the Urca process for Λ differs little from the threshold density for its appearance. As a result of such low Λ Urca thresholds, at high temperatures there could be regions in the star where the process could occur with the Λ's being nondegenerate. The emissivity then varies as $T^{9/2}$.

6.5. *Neutrino emission from exotic states*

Neutrino emission from pion and kaon condensates and quark matter all have the same temperature dependence as the direct Urca process, because the phase space considerations are essentially identical.

For a pion condensate, the neutrino emissivity is estimated to be (Maxwell *et al.* (1977))

$$\epsilon_\pi = \epsilon_{urca}\frac{\theta_\pi^2}{4}\left[1+\left(\frac{g_A k}{p_{F_e}}\right)^2\right]\frac{p_{F_e}}{k}, \tag{6.125}$$

where k is the condensate momentum and θ_π is the pion condensate angle, which is expected to be considerably less than unity. Estimates of the pion emissivity are typically at least a factor of 10 less than that of the direct Urca process. It is now considered likely that in–medium modifications are strong and repulsive in spin–isospin channels, which effectively prevents pion condensation in neutron star matter.

The attractive $s-$wave interaction of kaons with nucleons is thought to be the main driving force of kaon condensation (Kaplan & Nelson (1986), Brown *et al.* (1987), Brown *et al.* (1994), Thorsson *et al.* (1994)). For a spatially uniform condensate, the kaon emissivity has been estimated for small amplitudes of the condensate angle θ_K to be (Brown *et al.* (1988), Thorsson *et al.* (1995))

$$\epsilon_K = \epsilon_{Urca}\frac{\theta_K^2}{8}\sin^2\theta_c\,, \tag{6.126}$$

where θ_c is the Cabibbo angle. For $\theta_K^2 \simeq 0.1$, the kaon emissivity is less than that of the direct Urca process by a factor of about 1000.

The equilibrium composition of a Bose condensed state of matter with a net negative charge is generally characterized by the presence of large proton concentrations (Thorsson *et al.* (1994)). This leads to regions in the star where the condensation serves as a catalyst for rapid neutron star cooling via direct nucleon Urca processes. However, the region of the star where such catalyzed Urca processes occur is expected to be somewhat smaller than the size of the condensate core, since the electron concentrations drop with increasing negative charge in the condensate.

Quark matter, if present, would give rise to direct Urca processes involving u and d quarks with an emissivity (Iwamoto (1982))

$$\begin{aligned}\epsilon_{q.Urca} &= \frac{914}{315}\;\frac{G_F^2\cos^2\theta_C}{\hbar^{10}c^7}\;\alpha_c p_{F_d} p_{F_u}\mu_e\;(kT)^6\\ &= 8.8\times10^{26}\alpha_c(n/n_s)Y_e^{1/3}T_9^6 \text{ erg cm}^{-3}\text{ s}^{-1}\,,\end{aligned} \tag{6.127}$$

where p_{F_u} and p_{F_d} are the Fermi momenta of u and d quarks. With the standard value of the QCD coupling constant $\alpha_c \simeq 0.1$, and choosing $n = 4n_s$ and $Y_e = 10^{-4}$ (the equilibrium electron fraction in quark matter would be zero if quarks were massless), one finds that $\epsilon_{q.Urca}/\epsilon_{Urca} \sim 10^{-3}$.

6.6. *Conclusions*

We have considered the implications of neutron stars cooling very rapidly compared to the standard case, in which cooling is via modified Urca processes. Such stars will undergo a sharp decrease in surface temperature at the time t_w, which is essentially given by the thermal diffusion time through the crust (Lattimer *et al.* (1994)). This time primarily depends on the square of the thickness of the crust, but is modified by relativistic effects and is also influenced by the presence of neutron superfluidity in the crust. The time is not very sensitive to the details of the accelerated emissivity, such as the density threshhold or the net rate, including the question of whether or not superfluidity quenches the rapid cooling. The surface temperature during the isothermal phase following the time t_w, on the other hand, depends strongly on the details of the superfluid gaps above nuclear density and on the central density of the star, both of which are uncertain.

The crustal thickness, appropiately defined, depends only on the mass and radius of a neutron star with a relatively small uncertainty arising from the EOS. An observational determination of t_w, and the subsequent inference of the crustal thickness, would, therefore, constrain the structure of the star. An estimate of the neutron star's mass would, in addition, constrain the high-density EOS.

Nearly all potential candiates for thermal emission (see, e.g., Tsuruta (1986), Ogelman (1993), Becker, Trumper & Ogelman (1993)) are pulsars, and it is unclear how much of the observed emission is due to the pulsar phenomenon, to a synchrotron-emitting nebula, or to thermal emission from the neutron star itself. Improved spectral resolution in recent observations show strong evidence for a thermal spectrum in sources such as PSR 0656+14 (Córdova, Hjellming, Mason & Middleditch (1989)) and PSR 1055-52 (Brinkman & Ogelman). Periodicity in the thermal emission from PSR 0656+14 is interpreted as a hotter region on the surface sweeping past the field of view as the star rotates (Finley *et al.* (1992), Anderson *et al.* (1993)). All objects observed to date, with the exception of SN 1987A, are more than 100 years old, and the rapid cooling paradigm suggests that no thermal emission will be observed from them. In the case of SN 1987A, no thermal emission is seen as yet. Rapid cooling may be the reason for the absence of thermal emission from the young neutron stars which may be present in the center of the Kepler (Helfand, Chanan & Novick (1980)), Tycho (Gorenstein, Seward & Tucker (1983)) or SN 1006 (Pye *et al.* & Willingale (1981)) supernova remnants.

6.7. *Model calculations of cooling*

All neutron stars suspected of having surface thermal emission show evidence of non-uniform surface temperatures. Observations of several pulsars with ROSAT have shown X-ray emission varying with the pulse period. This has been interpreted as being due to emission from hot spots on the rotating surface. Non-uniform emission complicates the extraction of a surface temperature. Adding to the confusion are uncertainties concerning the atmospheric compositions and the strength of surface magnetic fields, both of which affect the temperature deduced from observations. In the case of various proposed atmospheric composition, the uncertainty in temperatures can surpass a factor of three.

In terms of theoretical models (see Page & Applegate (1992), Lattimer *et al.* (1994), Page (1994), Page & Sarmiento (1994), Schaab et al (1992), Page (1997), Page (1998)), standard cooling due to modified Urca processes or rapid cooling due to direct Urca processes are complicated at late times by superfluidity and by possible sources of heat in the star's crust. The latter include the release of energy by the decay of metastable heavy nuclei in the crust, diffusion of neutrons into the interior from shells of non-equilibrium matter containing heavy nuclei and friction between the differentially rotating superfluid of the core and the crust. Even a small source of heat is sufficient to counter the rapid cooling of the interior at late times. Page (Page 1994) has concluded that the thermal emission suspected from Geminga can be adequately explained by either a rapid cooling or a standard cooling model, given certain assumptions about superfluidity. More recently, it has also been suggested that within a kaon condensate scenario, pairing is not necessary if (1) the EOS is stiff enough for the star to have a thick crust in which sufficient friction can occur to heat the star and (2) a thin layer, of mass ΔM larger than $\sim 10^{-12}\mathrm{M}_{\odot}$, of light elements (H and He) is present at the stellar surface. Both these features can be confirmed or refuted by future observations. Planned new observations with the ROSAT, ASCA, AXAF, and the XMM facilities should provide a wealth of much required data.

REFERENCES

AINSWORTH, T. L., WAMBACH, J. & PINES D. 1989 Effective interactions and superfluid energy gaps for low density neutron matter. *Phys. Lett.*, **B222** (1989) 173-178.

ALPAR, M. A. 1992 In *The Structure and Evolution of Neutron Stars* p. 148-155, (ed. Pines, D., Tamagaki, R. & Tsuruta, S.) Addison-Wesley.

ALPAR, M. A., KIZILOĞLU, U. & VAN PARADIJS, J. 1995 *The Lives of the Neutron Stars* (ed. Alpar, M. A., Kiziloğlu, U. & van Paradijs, J.) Dordrecht: Kluwer.

ANDERSON, S. B., CÓRDOVA, F. A., PAVLOV, G. G., ROBINSON, C. R. & THOMPSON JR., R. J. 1993 Rosat high resolution imager observations of PSR 0656+14. *Ap. J.*, **414** (1993) 867-871.

ARNETT, W. D. & BOWERS, R. L. 1977 A microscopic interpretation of neutron star structure. *Ap. J.*, **33** (1977) 415-436.

BAHCALL, J. N. , & WOLF, R. A. 1965 Neutron stars. *Phys. Rev. Lett.* **14** (1965) 343-346.

BARON, E., COOPERSTEIN, J. & KAHANA, S. 1985 Type-II supernovae in $12M_\odot$ and $15M_\odot$ stars: The equation of state and general relativity. *Phys. Rev. Lett.*, **55** (1985) 126-129.

BAYM, G. & PETHICK, C. J. 1975 Neutron stars. *Ann. Rev. Nucl. Sci.*, **25** (1975) 27-77.

BAYM,G. & CHIN, S. A. 1976 Landau theory of relativistic Fermi liquids. *Nucl. Phys.*, **A262** (1976) 527-538.

BAYM, G. & PETHICK, C. J. 1979 Physics of neutron stars. *Ann. Rev. Astron. Astrophys.*, **17** (1979) 415-443.

BECKER, W., TRUMPER, J. & BRAZIER, K. T. S. 1993 I *Isolated Pulsars*, (ed. van Riper, K. A., Epstein, R. & Ho, C.) p 104, Cambridge.

BERTSCH, G. F. & DAS GUPTA, S. 1988 A guide to microscopic models for intermediate energy heavy ion collisions. *Phys. Rep.*, **160** (1988) 189-233.

BETHE, H. A., BROWN, G. E., APPLEGATE, J. & LATTIMER, J. M. 1979 Equation of state in the gravitational collapse of stars. *Nucl. Phys.*, **A324** (1979) 487-533.

BETHE, H. A. & BROWN, G. E. 1995 Observational constraints on the maximum neutron star mass. *Ap. J. Lett.*, **445** (1995) L129-L132.

BIONTA *et al.* 1987 Observation of a neutrino burst in coincidence with supernova SN 1987A in the large Magellanic cloud. *Phys. Rev. Lett.*, **58** (1987) 1494-1497.

BLAIZOT, J. P. 1989 Nuclear incompressibilities. *Phys. Rep.*, **64** (1980) 171-248.

BLUDMAN, S. A. 1973 Simple calculation of critical parameters of neutron stars. *Ap. J.*, **183** (1973) 649-656.

BLUDMAN, S. A. & RUDERMAN, M. A. 1970 Noncausality and instability in ultradense matter. *Phys. Rev.*, **D1** (1970) 3243-3246.

BOGUTA, J. 1981 Remarks on beta stability in neutron stars. *Phys. Lett.* **B106** (1981) 255-258.

BRINKMANN, W. & OGELMAN, H. 1987 Soft X-ray observations of the radio pulsar PSR 1055-22. *Astron. & Astrophys.*, **182** (1987) 71-74.

BROCKMAN, R. & MACHLEIDT, R. 1984 Nuclear saturation in a relativistic Brueckner-Hartree-Fock approach. *Phys. Lett.*, **B149** (1984) 283-287.

BROWN, G. E., KUBODERA, K. & RHO, M. 1987 Strangeness condensation and "clearing" of the vacuum. *Phys. Lett.*, **B192** (1987) 273-278.

BROWN, G. E., KUBODERA, K, PAGE, D. & PIZZOCHERO, P. 1988 Strangeness condensation and cooling neutron stars. *Phys. Rev.*, **D37** (1988) 2042-2046.

BROWN, G. E., BRUENN, S. & WHEELER, C. 1992 Is there a black hole in supernova 1987A? *Comments Astrophys.* **16** (1992) 153.

BROWN, G. E., KUBODERA, K. RHO, M. & THORSSON, V. 1992 A novel mechanism for kaon condensation in neutron star matter. *Phys. Lett.*, **B291** (1992) 355-362.

BROWN, G. E. & BETHE, H. A. 1994 A scenario for a large number of low mass black holes in the galaxy. *Ap. J.,* **423** (1994) 659-664.

BROWN, G. E. LEE, C-H., RHO, M. & THORSSON, V. 1994 From kaon-nuclear interactions to kaon condensation. *Nucl. Phys.,* **A567** (1994) 937-956.

BROWN, G. E., WEINGARTNER, J. C. & WIJERS, R. A. M. J. 1996 On the formation of low-mass black holes in massive binary stars. *Ap. J.,* **463** (1996) 297-304.

BURROWS, A & LATTIMER, J. M. 1986 The birth of neutron stars. *Ap. J.,* **307** (1986) 178-196.

BURROWS, A. 1988 Supernova neutrinos. *Ap. J.,* **334** (1988) 891-908.

BURROWS, A. 1990 Neutrinos from supernova explosions. *Ann. Rev. Nucl. Sci.,* **40** (1990) 181-212.

BURROWS, A., KLEIN, D. & GANDHI, R. 1992 The future of supernova neutrino detection. *Phys. Rev.,* **D45** (1992) 3361-3385.

BUTTGEN, R., HOLINDE, K., MULLER-GROELING, A., SPETH, J. & WYBORNY, P. (1990) A meson exchange model for the K^+N interaction. *Nucl. Phys.,* **A506** (1990) 586-614.

CHEN, J. M. C., CLARK, J. W., KROTSCHEK, E., SMITH, R. A. 1986 Nucleonic superfluidity in neutron stars: $^1S_0$ neutron pairing in the inner crust. *Nucl. Phys.* **A 451** (1986) 509-540.

CHEN, K. & COLGATE, S. A. 1995 Formation of inner ring around SN 1987A and remnant neutron star. *Los Alamos Preprint* LA-UR-95-2972.

CHIN, S. A. 1977 A relativistic many-body theory of high density matter *Ann. Phys.,* **108** (1977) 301-367.

CHIU, H. & SALPETER, E. E. 1964 Surface X-ray emission from neutron stars. *Phys. Rev. Lett.* **12** (1964) 413-415.

COOPERSTEIN, J. 1988 Neutron stars and the equation of state. *Phys. Rev.,* **C37** (1988) 786-796.

CÓRDOVA, F. A., HJELLMING, R. M., MASON, K. O. & MIDDLEDITCH, J. 1989 Soft X-ray emission from the radio pulsar PSR 0656+14. *Ap. J.,* **345** (1989) 451-463.

DAY, B. & WIRINGA, R. B. 1985 Brueckner-Bethe and variational calculations of nuclear matter. *Phys. Rev.,* **C32** (1985) 1057-1062..

ELLIS, J., KAPUSTA, J. I., & OLIVE, K. A. 1991 Strangeness, glue and quark matter content of neutron stars. *Nucl. Phys.* **B348** (1991) 345-372.

ELLIS, P. J., KNORREN, R. & PRAKASH, M. 1995 Kaon condensation in neutron star matter with hyperons. *Phys. Lett.,* **B349** (1995) 11-15.

ELLIS, P. J., LATTIMER, J. M. & PRAKASH, M. 1996 Strangeness and metastable neutron stars: What might have happened to SN 1987A. *Comments Nucl. Part. Phys.,* **22** (1996) 63-75.

ENGVIK, L., HJORTH-JENSEN, M., OSNES, E, BAO, G & ØSTGAARD, E. 1994 Asymmetric nuclear matter and neutron star properties. *Phys. Rev. Lett.,* **73** (1994) 2650-2653.

FAHRI, E. & JAFFE, R. L. 1984 Strange matter. *Phys. Rev.,* **D30** (1984) 2379-2390.

FINLEY, J. P., OGELMAN, H. & KIZILOĞLU, U. 1992 Rosat observations of PSR 0656+14: A pulsating and cooling neutron star. *Ap. J. Lett.,* **394** (1992) L21-L24.

FREEDMAN, B. & MCLERRAN, L. 1978 Quark star phenomenology. *Phys. Rev.,* **D17** (1978) 1109-1122.

FRIEDMAN, B. & PANDHARIPANDE, V. R. 1981 Hot and cold, nuclear and neutron matter. *Nucl. Phys.,* **A361** (1981) 502-520.

FRIEDMAN, E., GAL, A. & BATTY, C. J. (1994) Density-dependent K^- nuclear optical potentials from kaonic atoms. *Nucl. Phys.,* **A579** (1994) 518-538.

FRIMAN, B. & MAXWELL, O. V. 1979 Neutrino emissivities of neutron stars. *Ap. J.,* **232** (1979) 541-557.

FURNSTAHL, R. J., PERRY, R. J., & SEROT, B. D. 1989 Two-loop corrections for nuclear matter in the Walecka model. *Phys. Rev.,* **C40** (1989) 321-353.

GAILLARD, J.-M. & SAUVAGE, G. 1984 Hyperon beta decays. *Ann. Rev. Nucl. Part. Sci.,* **34** (1984) 351-402.

GALE, C., BERTSCH, G. F. & DAS GUPTA, S. 1987 Heavy-ion collision thoery with momentum-dependent interactions. *Phys. Rev.,* **C35** (1987) 1666-1671.

GASSER, J. & LEUTWYLER, H. 1982 Quark masses. *Phys. Rep.,* **C87** (1982) 77-169.

GLENDENNING, N. K. 1985 Neutron stars are giant hypernuclei? *Ap. J.,* **293** (1985) 470-493.

GLENDENNING, N. K. 1986 Neutron- star masses as a constraint on the nuclear equation of state. *Phys. Rev. Lett.,* **57** (1986) 1120-1123.

GLENDENNING, N. K. & MOSZKOWSKI, S. A. 1991 Reconciliation of neutron-star masses and binding of the Λ in hypernuclei. *Phys. Rev. Lett.* **67** (1991) 2414-2417.

GLENDENNING, N. K. 1992 First-order phase transitions with more than one conserved charge: Consequences for neutron stars. *Phys. Rev.,* **D46** (1992) 1274-1287.

GLENDENNING, N. K. 1995 Prompt subsidence of a protoneutron star into a black hole. *Ap. J.,* **448** (1995) 797-802.

GORENSTEIN, P., SEWARD, F. & TUCKER, W., 1983 In *Supernova Remnants and their X-Ray Emission,* (ed. Danziger, J. & Gorenstein, P.) p. 1., Dordrecht: Reidel.

HAENSEL, P. 1990 *Copernicus Astronomical Center Preprint*

HAENSEL, P. & SCHAEFFER, R. F. 1992 Bulk viscosity of hot-neutron-star matter from direct Urca processes. *Phys. Rev.,* **D45** (1992) 4708-4712.

HALPERN, J. P. & RUDERMAN, M. 1993 Soft X-ray properties of the Geminga pulsar. *Ap. J.,* **415** (1993) 286-297.

HARTLE, J. B. 1973 Slowly rotating relativistic stars. *Ap. Space. Sci.,* **24** (1973) 385-405.

HEISELBERG, H., PETHICK, C. J. & STAUBO, E. F. 1993 Quark matter droplets in neutron stars. *Phys. Rev. Lett.,* **70** (1993) 1355-1359.

HELFAND, D. J., CHANAN, G. A. & NOVICK, R. 1980 Thermal X-ray emission from neutron stars. *Nature.,* **283** (1980) 337-343.

HELFAND, D. J. & BECKER, R. H. 1984 Observation of stellar remnants from recent supernovae. *Nature.,* **307** (1984) 215-221.

HIRATA, K. S., *et al.* 1987 Observation of a neutrino burst from the supernova SN1987A. *Phys. Rev. Lett.,* **58** (1987) 1490-1493.

HOROWITZ, C. & SEROT, B. D. 1981 Self-consistent Hartree description of finite nuclei in a relativistic quantum field theory. *Nucl. Phys.,* **A368** (1981) 503-528.

HOROWITZ, C. & SEROT, B. D. 1987 The relativistic two-nucleon problem in nuclear matter. *Nucl. Phys.,* **A464** (1987) 613-699.

IWAMOTO, N. 1982 Neutrino emissivities and mean free paths of degenerate quark matter. *Ann. Phys.,* **141** (1982) 1-49.

JACKSON, A. D., RHO, M. & KROTSCHEK, E. 1985 The σ-model and the binding energy of nuclear matter. *Nucl. Phys.,* **407** (1985) 495-506.

KAPLAN, D. B. & NELSON, A. E. 1986 Strange goings on in dense nucleonic matter. *Phys. Lett.,* **B175** (1986) 57; *Phys. Lett.,* **B179** (1986) 409 (E).

KAPUSTA, J. I. & OLIVE, K. A. 1990 Effects of strange particles on neutron-star cores. *Phys. Rev. Lett.,* **64** (1990) 13-15.

KEIL, W. & JANKA, H. T. 1994 Hadronic phase transitions at supranuclear densities and the delayed collapse of newly formed neutron stars. *Astron. & Astrophys.,* **296** (1994) 145-163.

KNORREN, R., PRAKASH, M. & ELLIS, P. J. 1995 Strangeness in hadronic stellar matter. *Phys. Rev.,* **C52** (1995) 3470-3482.

KRISTIAN, C., *et al.* 1989 Submillisecond optical pulsar in supernova 1987A. *Nature.,* 338 (1989) 234-236.

KUMAGAI, S., SIGEYAMA, T., HASHIMOTO, M. & NOMOTO, K. (1991) Implications of the recent light curve of SN 1987A. *Astron. & Astrophys.,* **243** (1991) L13-L16.

LATTIMER, J. M., PRAKASH, M., MASAK, D. & YAHIL, A. 1990 Rapidly rotating pulsars and the equation of state. *Ap. J.,* **355** (1990) 241-254.

LATTIMER, J. M., PETHICK, C. J., PRAKASH, M. & HAENSEL, P. 1991 Direct Urca process in neutron stars. *Phys. Rev. Lett.,* **66** (1991) 2701-2704.

LATTIMER, J. M., VAN RIPER, K. A., PRAKASH, M. & PRAKASH, MANJU 1994 Rapid cooling and the structure of neutron stars. *Ap. J.,* **425** (1994) 802-813.

LEWIN, W. H. G., VAN PARADIJS, J. & TAAM, R. E. (1993) X-ray Bursts. *Space. Sci. Rev.,* **62** (1993) 223-389.

LORENZ, C. P. , RAVENHALL, D. G. & PETHICK, C. J. 1993 Neutron star crusts. *Phys. Rev. Lett.,* **70** (1993) 379-382.

MANCHESTER, R. N. 1992 In *The Structure and Evolution of Neutron Stars* p. 32-49, (ed. Pines, D., Tamagaki, R. & Tsuruta, S.) Addison-Wesley.

MAXWELL, O., BROWN, G. E., CAMPBELL, D. K., DASHEN, R. F. & MANASSAH, J. T. 1977 Beta decay of pion condensates as a cooling mechanism for neutron stars. *Ap. J.,* **216** (1977) 77-85.

MAXWELL, O. V. 1987 Neutrino emission processes in hyperon populated stars. *Ap. J.,* **316** (1987) 691-707.

MILLENER, J., DOVER, C. & GAL, A. 1988 Λ-nucleus single particle potentials. *Phys. Rev.,* **C38** (1988) 2700-2708.

MISNER, C. W., THORNE, K. S. & WHEELER, J. A. 1973 *Gravitation* W. H. Freeman and Company.

MÖLLER, P. , MYERS, W. D., SWIATECKI, W. J. & TREINER, J. 1988 Nuclear mass formula with a finite-range droplet model and a filded-Yukawa single-particle potential. *At. Data and Nucl. Data Tables,* **39** (1988) 225-233.

MUETHER, H., PRAKASH, M. & AINSWORTH, T. L. 1987 The nuclear symmetry energy in relativistic Brueckner-Hartree-Fock calculations. *Phys. Lett.,* **199B** (1987) 469-474.

MUTO, T. & TATSUMI, T. 1988 Exact neutrino emissivity of neutron stars under the combined condensation of neutral and charged pions. *Prog. Theor. Phys.,* **79** (1988) 461-479.

MUTO, T. 1993 Role of weak interactions on kaon condensation in neutron matter. *Prog. Theor. Phys.,* **89** (1993) 415-435.

MUTO, T., TAKATSUKA, T., TAMAGAKI, R. & TATSUMI, T. 1993 Implications of various hadron phases to neutron star phenomena. *Prog. Theor. Phys. Suppl.,* **112** (1993) 221-275.

MUTO, T. & TAMAGAKI, R. 1993 A chiral symmetry approach to meson condensations. *Prog. Theor. Phys. Suppl.,* **112** (1993) 159-196.

NAUENBERG, M. & CHAPLINE, G. 1973 Determination of properties of cold stars in general relativity by a variational method. *Ap. J.,* **179** (1973) 277-287.

OGELMAN, H. 1989 In *Timing Neutron Stars,* (ed. Ogelman, H. & van den Heuvel, E. P. J) Dordrecht: Kluwer.

OGELMAN, H. 1993 In *Isolated Pulsars,* (ed. van Riper, K, Epstein, R. & Ho, C.) p. 96, Cambridge.

OPPENHEIMER, J. R. & VOLKOFF, G. M. 1939 On massive neutron star cores. *Phys. Rev.,* **55** (1939) 374-381.

PAGE, D. & APPLEGATE, J. 1992 The cooling of neutron stars by the direct Urca process. *Ap. J. Lett.,* **394** (1992) L17-L20.

PAGE, D. 1994 Geminga: A cooling superfluid neutron star. *Ap. J.,* **428** (1994) 250-260.

PAGE, D. & SARMIENTO, A. 1996 Surface temperature of a magnetized neutron star and interpretation of the ROSAT data: II. *Ap. J.,* **473** (1996) 1067-1078.

PAGE, D. 1997 Fast cooling of neutron stars: Superfluidity versus heating and accreted envelope. *Ap. J. Lett.,* **479** (1997) L43-L46.

PAGE, D. 1998 In *The Many Faces of Neutron Stars* (ed. Alpar, A., Buccheri, R. & van Paradijs, J.) Dordrecht: Kluwer, in press.

PANDHARIPANDE, V. R. 1971 Dense neutron matter with realistic interactions. *Nucl. Phys.,* **A174** (1971) 641-656.

PEARSON, J. M., ABOUSSIR, Y., DUTTA, A. K., NAYAK, R. C., FARINE, M. & TONDEUR, F. 1991 Thomas Fermi approach to nuclear mass formula. *Nucl. Phys.* **A528** (1991) 1-47.

PENNYPACKER, C. 1990 Private communication.

PETHICK, C. J. 1992 Cooling of neutron stars. *Rev. Mod. Phys.*, **64** (1992) 1133-1140.

PETHICK. C. J. & RAVENHALL, D. G. 1995 Matter at large neutron excess and the physics of neutron-star crusts. *Ann. Rev. Nucl. Sci.*, **45** (1995) 429-484.

PETHICK, C. J., RAVENHALL, D. G. & LORENZ, C. P. The inner boundary of a neutron-star crust. *Nucl. Phys.*, **A584** (1995) 675-703.

PINES, D. 1992 In *The Structure and Evolution of Neutron Stars* (ed. Pines, D., Tamagaki, R. & Tsuruta, S.) Addison-Wesley.

POLITZER, H. D. & WISE, M. B. 1991 Kaon condensation in nuclear matter. *Phys. Lett.*, **B273** (1991) 156.

PRAKASH, M. & AINSWORTH, T. L. 1987 Sigma model calculations of neutron-rich nuclear matter. *Phys. Rev.*, **C36** (1987) 346-353.

PRAKASH, M., AINSWORTH, T. L. & LATTIMER, J. M. 1988 Equation of state and the maximum mass of neutron stars. *Phys. Rev. Lett.*, **61** (1988) 2518-2521.

PRAKASH, M., ELLIS, P. J. & KAPUSTA, J. I. 1992 Relativistic nuclear matter with composite nucleons. *Phys. Rev.*, **C45** (1992) 2518-2521.

PRAKASH, M., PRAKASH, MANJU, LATTIMER, J. M. & PETHICK, C. J. 1992 Rapid cooling of neutron stars by hyperons and Δ isobars. *Ap. J. Lett.*, **390** (1992) L77-L80.

PRAKASH, M. 1993 The existence of a large window for nucleosynthesis in the case of a neutron star with strangeness condensates, with later collapse into a black hole, was first pointed by this author.

PRAKASH, M. 1994 Rapid cooling of neutron stars. *Phys. Rep.*, **242** (1994) 297-312.

PRAKASH, M., COOKE, J. R. & LATTIMER, J. M. 1995 The quark-hadron phase transition in protoneutron stars. *Phys. Rev.*, **D52** (1995) 661-665.

PRAKASH, M. 1996 The equation of state and neutron stars. In *The Nuclear Equation of State* p. 83-208, (ed. Ansari, A. & Satpathy, L.) World Scientific.

PRAKASH, M., BOMBACI, I., PRAKASH, MANJU, ELLIS, P. J., LATTIMER, J. M. & KNORREN, R. 1997 Composition and structure of protoneutron stars. *Phys. Rep.*, **280** (1997) 1-77.

PYE, J. P., POUNDS, K. A., ROLF, D. P., SEWARD, F. D., SMITH, A. & WILLINGALE, R. 1981 An X-ray map of SN 1006 from the Einstein observatory. *M. N. R. A. S.*, **194** (1981) 569-582.

RHO, M. 1974 Quenching of axial-vector coupling constant in β-decay and pion-nucleus optical potential. Nucl. Phys. **A231** (1974) 493-503.

SCHAAB, C., WEBER, F., WEIGEL, M. K. & GLENDENNING, N. K. 1996 Thermal evolution of compact stars *Nucl. Phys.*, bf A605 (1996) 531-565.

SEROT, B. D. 1979 A relativistic nuclear field theory with π and ρ mesons. *Phys. Lett.*, **B86** (1979) 146-150.

SEROT, B. D. & WALECKA, J. D. 1986 The relativistic nuclear many-body problem. *Advances in Nuclear Physics* **16** (1986) 1.

SEROT, B. D. 1992 Quantum hadrodynamics. *Rep. Prog. Phys.* **55** (1992) 1855-1946.

SHAPIRO, S. L. & TEUKOLSKY, S. A. 1983 *Black Holes, White Dwarfs, and Neutron Stars* Wiley.

TAKATSUKA, T. & TAMAGAKI, R. 1971 Superfluid state in neutron matter: II. Properties of anisotropic energy gap of $^3P_2$ pairing. *Prog. Theor. Phys.*, **46** (1971) 114-134.

TAYLOR, J. H., MANCHESTER, R. N. & LYNE, A. G. 1993 Catalog of 558 pulsars. *Ap. J. Suppl.*, **88** (1993) 529-568.

TER HAAR, B. & MALFLIET, R. 1986 Equation of state of nuclear matter in the relativistic Dirac-Brueckner appraoch. *Phys. Rev. Lett.*, **56** (1986) 1237-1240.

TER HAAR, B. & MALFLIET, R. 1987 Equation of state of dense asymmetric matter. *Phys. Rev. Lett.*, **59** (1987) 1652-1655.

THIELEMANN, F.-K., HASHIMOTO, M. & NOMOTO, K. 1990 Explosive nucleosynthesis in SN 1987A. II. Composition, radioactivities and the neutron star mass. *Ap. J.*, **349** (1990)

222-240.

THORSETT, S. E., ARZOUMANIAN, Z., MCKINNON, M. M., & TAYLOR, J. H. 1993 The masses of two binary neutron star systems. *Ap. J. Lett.*, **405** (1993) L29-L32.

THORSSON, V., PRAKASH, M. & LATTIMER, J. M. 1994 Composition, structure and evolution neutron stars with kaon condensates. *Nucl. Phys.*, **A572** (1994) 693-731.

THORSSON, V., PRAKASH, M., TATSUMI, T. & PETHICK, C. J. 1995 Neutrino emission from dense matter containing meson condensates. *Phys. Rev.*, **D52** (1995) 3739-3741.

TOLMAN, R. C. 1934 *Proc. Nat. Acad. Sci. U. S. A.*, **20** (1934) 3.

TOLMAN, R. C. 1939 Static solutions of Einstein's field equations for spheres of fluid. *Phys. Rev.*, **15** (1939) 364-373.

TSURUTA, S. 1986 Neutron stars: Current cooling theories and observational results. *Comments Astrophys.* **11** (1986) 151-192.

VAN KERKWIJK, M. H., VAN PARADIJS, J. & ZUIDERWIJK, E. J. On the masses of neutron stars. *Astron. & Astrophys.*, **303** (1995) 497-501.

VAN RIPER, K. A. 1980 Stellar core collapse: II. Inner core bounce and shock propagation. *Ap. J.*, **257** (1980) 793-820.

VAN RIPER, K. A., EPSTEIN, R. & HO, C. 1993 *Isolated Pulsars* (ed. van Riper, K. A., Epstein, R. & Ho, C.) Cambridge.

VENTURA, J. & PINES, D. (1991) *Neutron Stars: Theory and Observation* (ed. Ventura, J. & Pines, D.) Dordrecht: Kluwer.

WALTER, F. K., WOLK, S. J., & NEUHÄUSER, R. 1996 Discovery of a nearby isolated neutron star. *Nature.*, **379** (1996) 233-235.

WEINBERG, S. 1972 *Gravitation and Cosmology: Principles and Applications of the General Theory of Relativity*, John Wiley & Sons.

WILKINSON, D. H. 1973 Renormalization of the axial-vector coupling constant in nuclear β-deacy. *Phys. Rev.* **C7** (1973) 930-936.

WIRINGA, R. B., FIKS, V., & FABROCINE, A. 1988 Equation of state of dense nucleon matter. *Phys. Rev.*, **C38** (1988) 1010-1037.

WIRINGA, R. B. 1993 From deuterons to neutron stars: Variations in nuclear many-body theory. *Rev. Mod. Phys.*, **63** (1993) 231-242.

ZIMANYI, J. & MOSZKOWSKI, S. A. 1990 Nuclear equation of state with derivative scalar coupling. *Phys. Rev.*, **C42** (1990) 1416-1421.

Massive Neutrinos

By PETR VOGEL

Department of Physics, California Institute of Technology, Pasadena, CA 91125, USA

---

1. Introduction, formalism, and cosmological bound

In these four lectures I will present a brief and rather elementary description of the physics of massive neutrinos as it emerges from studies involving nuclear physics, particle physics, astrophysics and cosmology. The lectures are meant for physicists who are not experts in this field, which I believe covers most of the participants in this School, and many potential readers elsewhere. I hope that such readers can find here enough information that they will be able to understand and appreciate the connection between the hunt for neutrino mass and mixing described here, and their own field of expertize.

Throughout I will use original references sparingly. Instead I refer to several monographs, written and published during the last decade [Boehm & Vogel (1992), Kayser, Gibrat-Debu & Perrier (1989), Winter (1991), Mohapatra & Pal (1991), Kim & Pevsner (1993), Klapdor-Kleingrothaus & Staudt (1995)] where an interested reader can find references to the original papers. When appropriate I will also refer to review papers on various aspects of the neutrino mass or related topics. For the experimental data, including the list of thc most recent original experimental papers, the best source is the Review of Particle Physics, periodically updated, with the latest printed version in PDG (1996) . The update of this very useful publication is available even between printed editions on the World-Wide Web at http://pdg.lbl.gov/.

As is well known, the existence of neutrino has been proposed by Pauli in 1930, in a desperate attempt to save the law of energy and angular momentum conservation, seemingly violated in nuclear β decay. The proposal was quickly followed by realization, due to Bethe & Peierls in 1934, that neutrinos interact with matter with an extremely small cross section. Hence they are very difficult to detect.

Proof that the electron neutrinos (actually electron antineutrinos) really exist took more than twenty years. Reines & Cowan accomplished that in 1953; Reines had to wait longer, until 1995, for the Nobel prize for this accomplishment. Relatively soon afterwards, in 1962 Danby et al. have shown that muon and electron neutrinos are distinct particles. (That work was awarded Nobel prize already in 1988.) Now we know (from the invisible width of Z to be discussed later) that there are precisely three neutrino flavors. But we do not know whether neutrinos are massless, as the standard electroweak theory suggests, or whether they have a nonvanishing mass. Related to this is the question of the lepton number conservation. Are all three lapton flavors conserved separately, or is there just a conservation of the total lepton number. Or, perhaps, even the total lepton number is not conserved at some level. These questions are the central topics of our discussion here.

1.1. *Neutrino electron (or neutrino quark) interaction cross section (schematic)*

The basic practical difficulty of all of the neutrino physics is the small cross section which makes any experiments involving neutrinos complicated and costly. To appreciate how small the cross section really is, I present here a schematic derivation of its magnitude.

Consider the neutrino-electron scattering

$$\nu(k_1) + e(p_1) \to \nu(k_2) + e(p_2) \ . \tag{1.1}$$

The only invariant quantity of dimension *energy* is the center-of-mass energy, which when squared equals

$$s = (k_1 + p_1)^2 = 2k_1 \cdot p_1 \ + \ p_1^2 = 2m_e E_\nu^{lab} + m_e^2 \ , \tag{1.2}$$

where the last equality is valid for the target lectrons at rest. Since the neutrino-electron interaction is weak, the cross section must be proportional to the Fermi coupling constant squared, G_F^2, of dimension $energy^{-4}$. At the same time the cross section must have dimension $length^2$; it is then obvious from dimensional arguments that (this is valid for moderate energies, at energies comparable to the W mass modifications occur)

$$\sigma \ \approx \ \frac{G_F^2}{\pi} s(\hbar c)^2 \simeq 1.68 \times 10^{-44} E_\nu^{lab} \ \mathrm{MeVcm}^2 \ . \tag{1.3}$$

where I have guessed the numerical factor π ($G_F = 1.17 \times 10^{-11}$ MeV^{-2} , $\hbar c = 197 \times 10^{-13}$ MeV cm). This is a very small cross section indeed.

1.2. *Motivation for neutrino mass*

Why do we expect that neutrinos are massive? In the standard electroweak theory neutrinos are postulated to be exactly massless, and each lepton flavor is exactly conserved.

However, in grand unified theories, where one tries to unify the description of the weak, electromagnetic, and strong interactions, neutrinos are *naturally* massive. In a generic theory quarks q^+ (charge +2/3), q^- (charge -1/3), charged leptons and neutrinos of one "family" (say u, d, e, ν_e) are in one multiplet:

$$\begin{bmatrix} q^+ \\ q^- \\ l^- \\ \nu \end{bmatrix}$$

Since charged leptons and quarks are massive, it is perhaps more natural that neutrinos are massive as well. But why is their mass so much smaller than their charged partners?

Maybe that small neutrino masses have to do with the fact neutrinos are neutral, while the quarks and other leptons are charged. So, neutrinos can be the ultimate neutral particles, i.e. Majorana particles (particle = antiparticle), while the others are clearly Dirac particles (particle $\neq$ antiparticle). We will see later, in the subsection devoted to the so-called see-saw mechanism, that in that case, at least, the small neutrino mass is a natural consequence of the model. But what about the lepton number conservation, which we know is obeyed very well?

Recall the origin of the lepton number. The neutrino produced in, say, π^+ decay (called ν_μ or muon neutrino), when interacting with nucleons, is known to produce μ^- but never μ^+. Similarly, the neutrino produced in the decay of π^- (called $\bar{\nu}_\mu$ or muon antineutrino) produces μ^+ only.

Conventional explanation of this phenomenon involves the concept of the conserved lepton quantum number:

(*a*) ν_μ and $\bar{\nu}_\mu$ are distinct, hence they cannot be Majorana particles.

(*b*) There is a conserved lepton quantum number (+1 for μ^- and ν_μ, and -1 for μ^+ and $\bar{\nu}_\mu$).

But there is an alternative explanation:

(*a*) The neutral particle produced in the decay of π^+ is left-handed, and the one produced in the decay of π^- is right-handed.

(*b*) Now suppose that the weak interaction is parity-violating and such that left-handed neutrinos produce only μ^- and right-handed neutrinos produce only μ^+.

In that case we do not have to introduce a conserved lepton number! The two neutrinos will represent just two helicity projections of a single Majorana particle - the muon neutrino. Thus we see that there is an intimate relation between the apparent lepton number conservation and the parity violation of the weak interactions, the only interactions neutrinos are supposed to have.

1.3. *Brief formal discussion of discrete symmetries*

As we saw above, discrete symmetries play an important role in neutrino physics. Thus, we digress, and describe briefly formalism involved in their treatment, following conventions and notation of Sakurai (1964) . (The material in subsections 1.3-1.5 is explained nicely in the Kayser, Gibrat-Debu & Perrier (1989))

A free neutrino field ψ is a four-component object which obeys the Dirac equation

$$\left(\gamma_\mu \frac{\partial}{\partial x_\mu} + m \right) \psi = 0 . \tag{1.4}$$

The field $\gamma_5 \psi$ satisfies an equation in which the sign of mass is reversed,

$$\left(\gamma_\mu \frac{\partial}{\partial x_\mu} - m \right) \gamma_5 \psi = 0 . \tag{1.5}$$

The fields with a definite chirality (the name used for the eigenstates of γ_5) are the projections, $\frac{(1 \pm \gamma_5)}{2} \psi$. It is obvious that a free massive particle cannot have a definite chirality at all times. On the other hand, in the ultrarelativistic limit $E/m \to \infty$ (or for massless particles) the chirality becomes a good quantum number.

There is a relation between chirality and helicity (i.e., spin projection on the direction of motion). The helicity operator is $\vec{\Sigma} \cdot \hat{p}$, where $\vec{\Sigma} = -i\gamma_4\gamma_5\vec{\gamma}$. Then $\gamma_5 \to \vec{\Sigma} \cdot \hat{p}$ in the ultrarelativistic limit for positive energy states. There is often used, but not very precise custom, to refer to the chiral projections $\frac{(1 \pm \gamma_5)}{2} \psi$ as right- and left-handed, respectively.

The charge conjugate field ψ^c is defined as

$$\psi^c = \eta_C C \bar{\psi}^T , \tag{1.6}$$

where η_C is a phase factor, C is a 4×4 matrix, (the symbol C is customarily used for two meanings, as the symbol of the charge conjugation operation and as the 4×4 matrix) and the superscript T signifies a transposed matrix ($\bar{\psi} = \psi^+ \gamma_4$). In the Dirac-Pauli representation $C = \gamma_4 \gamma_2$ so that $\psi^c = -\eta_C \gamma_2 \psi^*$. The following relations are independent of representation:

$$C^{-1} \gamma_\mu C = -\gamma_\mu^T \; ; \; C^{-1} \gamma_5 C = \gamma_5^T . \tag{1.7}$$

Therefore, if ψ is a chirality eigenstate ($\gamma_5 \psi = \lambda \psi$), then ψ^c is also a chirality eigenstate with eigenvalue $\lambda^c = -\lambda$.

Under parity transformation $\vec{x}' = -\vec{x}$, $x_4' = x_4$. Requiring that the Dirac equation in both systems describes the same situation leads to the rule

$$\psi(\vec{x}) \to \eta_P \gamma_4 \psi(-\vec{x}) , \tag{1.8}$$

where η_P is a phase factor. The charge conjugate field ψ^c transforms under the parity transformation as

$$\psi^c \to \eta_C \eta_P^* C((\gamma_4 \psi)^+ \gamma_4)^T = -\eta_P^* \gamma_4 \psi^c . \tag{1.9}$$

Thus, for real η_P, we obtain the well-known result that the intrinsic parity of the antifermion is opposite to that of a fermion. (This can be experimentally verified, for example, by a study of the positronium decay.) If, however, ψ should describe a Majorana particle (charge conjugation eigenstate), one can satisfy the above equation only with pure imaginary η_P. Hence we come to important conclusion: *Majorana particles have imaginary intrinsic parity.*

The Majorana field can be defined as

$$\chi(x) = \frac{1}{\sqrt{2}} \left[\psi(x) + \eta_C \psi^c(x) \right] . \tag{1.10}$$

By the appropriate choice of phase we obtain a field which is an eigenstate of charge conjugation with the eigenvalue $\lambda_C = \pm 1$. The above Majorana field has the CP phase $\pm i$.

1.4. *Mass terms*

In the field theory of neutrinos the mass is determined by the mass term of the neutrino Lagrangian. This mass term must be Lorentz invariant and hermitian. This requirement restricts the possible mass terms to two groups: $\bar{\psi}\psi$ and $\bar{\psi}^c\psi^c$, as well as $\bar{\psi}\psi^c$ and its hermitian conjugate $\bar{\psi}^c\psi$. The first group represents so called Dirac mass terms which are invariant under the global phase transformation

$$\psi \to e^{i\alpha}\psi \; ; \;\; \psi^c \to e^{-i\alpha}\psi^c . \tag{1.11}$$

The second possible Lorentz invariant mass term containing $\bar{\psi}\psi^c$ and $\bar{\psi}^c\psi$ is not invariant under this phase transformation.

Thus, the Dirac mass term can be associated with a conserved quantum number (usually called the "lepton number" mentioned earlier), while the second Lorentz invariant mass term violates conservation of the lepton number by two units. It will become apparent shortly why this second mass term is known as the Majorana mass term.

Let us consider the most general mass Lagrangian which is hermitian and Lorentz invariant. It depends on three real parameters m_D, m_1, m_2 ($m_M = m_1 + im_2$, i.e. it is generally complex), and can be rewritten in the matrix form

$$-\mathbf{L}_M = \frac{1}{2}(\bar{\psi}, \bar{\psi}^c) \begin{pmatrix} m_D & m_M \\ m_M^* & m_D \end{pmatrix} \begin{pmatrix} \psi \\ \psi^c \end{pmatrix} . \tag{1.12}$$

To find fields of a definite mass, we have to diagonalize L_M. When this is done, we find that this mass term has two real eigenvalues $m_D \pm |m_M|$, and the corresponding eigenvectors are

$$\begin{pmatrix} \phi_+ \\ \phi_- \end{pmatrix} = \frac{1}{\sqrt{2}} \begin{pmatrix} e^{-i\theta}\psi + e^{i\theta}\psi^c \\ -e^{-i\theta}\psi + e^{i\theta}\psi^c \end{pmatrix} , \tag{1.13}$$

where $\tan 2\theta = m_2/m_1$. Both eigenvectors are Majorana fields, i.e., they are eigenstates of charge conjugation with opposite eigenvalues. (If $m_M = 0$ the mass term is already diagonal with a single eigenvalue m_D; the eigenvectors ψ and ψ^c in that case are not charge conjugation eigenstates and describe a Dirac particle.)

1.5. *See-saw mechanism*

The parameters m_D and m_M are so far undetermined. The only known way by which we can connect their value, i.e. the neutrino mass, to other known masses is the see-saw mechanism introduced independently by Gell-Mann, Ramond & Slansky and by Yanagida in 1979. This mechanism is based on the Majorana mass and is therefore unique to neutrinos. Thus, it explains naturally why neutrinos have a very different mass than the charged leptons.

To see how this situation arises, we will describe in greater detail the steps necessary in order to diagonalize the general mass term $\mathbf{L}_M$. First we have to rewrite it in the basis with chiral projections keeping in mind that the mass term has to connect always left-handed and right-handed components.

It is not important to consider all the steps, but without proof let us state that the mass matrix is transformed to the block diagonal form, with the two blocks equal to M' and $-M'$, where

$$M' = \begin{pmatrix} m_R & m_D \\ m_D & m_L \end{pmatrix} , \tag{1.14}$$

and $m_R = m_1 + |m_2|$, $m_L = m_1 - |m_2|$, where m_2 is the imaginary part of the generally complex Majorana mass parameter m_M.

There are four eigenstates of the mass matrix. The two upper components are charge conjugation eigenstates (Majorana neutrinos) $\psi_R + (\psi_R)^c$ and $\psi_L + (\psi_L)^c$, while the two lower ones are the differences of the same fields $\psi_R - (\psi_R)^c$ and $\psi_L - (\psi_L)^c$. These states are not simultaneous eigenstates of chirality, since e.g. $(\psi_R)^c$ is left-handed, etc.

Now, we can consider the underlying physics. Let us assume that, besides the familiar ν_L, which couples to electrons in weak charged current, there exists a weak interaction singlet neutrino ν_R (i.e. has no weak interaction) of the Majorana mass m_R. This right-handed neutrino must be heavy, otherwise it would have been observed already. The GUTs will make the following assignments

$$m_L \simeq 0 \, , \quad m_R \simeq M_{GUT} \simeq 10^{14} - 10^{16} \text{ GeV} \, , \quad m_D \simeq m_{fermion} \, , \tag{1.15}$$

where $m_{fermion}$ represents either charged lepton or quark masses. The value M_{GUT} is the GUT mass scale. (This is the energy value where the three running coupling constants, corresponding to the groups, $SU(3)$, $SU(2)$, and $U(1)$, come together as a function of the momentum transfer.) The diagonalization of M' is trivial under these circumstances. There are two eigenvalues, a very heavy Majorana neutrino with $m_R \simeq M_{GUT}$, and a light one

$$m_L \simeq m_{fermion}^2 / M_{GUT} . \tag{1.16}$$

Substituting the corresponding fermion masses, we see that the light neutrino, dominantly the Majorana partner of left-handed charged leptons, will have mass as low as 10^{-13} eV and not higher than $\simeq 1$ eV. The light and heavy neutrinos will mix with a very small mixing angle $\tan^2\theta \simeq m_D^2/M_{GUT}^2$.

The see-saw mechanism predicts that the neutrino mass is proportional to the square of the mass of a fermion (lepton or quark) from the same family. Thus, one expects that

$$m_{\nu_e} : m_{\nu_\mu} : m_{\nu_\tau} = m_e^2 : m_\mu^2 : m_\tau^2 \, , \tag{1.17}$$

or, more likely,

$$m_{\nu_e} : m_{\nu_\mu} : m_{\nu_\tau} = m_u^2 : m_c^2 : m_t^2 \, , \tag{1.18}$$

where u, c, t denote the up, charm, and top quarks, respectively. According to this model, the tau neutrino would be the heaviest, and the electron neutrino the lightest. Many modifications of this idea have been proposed, often replacing M_{GUT} by a smaller mass scale and thus making the light neutrinos correspondingly heavier.

1.6. *Cosmological constraint on neutrino mass*

There is an important connection between the problem oı neutrino mass and cosmology. Here we will discuss only one aspect of it, namely the bound on the mass of a light stable neutrino (less than few MeV) based on cosmological arguments. More detailed exposition of the neutrino cosmology can by found in Kolb & Turner (1990) .

1.6.1. *Critical density*

The simplest plausible assumption about the large scale mass and energy distribution in the universe is based on the *cosmological principle*, which states that all positions in the universe are equivalent and hence the universe is homogeneous and isotropic. This principle leads naturally to the uniform expansion of the cosmic fluid with time and is consistent with observational astronomy provided we average over distances of $\sim 10^{26}$ cm ($\sim$ 30 Mpc), a scale larger than typical clusters of galaxies, but significantly smaller than the radius of the visible universe ($\sim 10^{28}$ cm).

The universal expansion is consistent with the observation of the cosmological red shift and leads to the conclusion that the velocity v of distant galaxies is to first approximation proportional to their distance from us,

$$v = H_0 r \ , \tag{1.19}$$

where H_0 is the Hubble's constant.

In the homogeneous universe, a crucial role is played by the critical density ρ_c. If the true density ρ (which is the only important parameter for the homogeneous universe) is equal to the critical density, the universe is "flat"; it is expanding now, but will ultimately, after an infinite time, come to rest. (If $\rho > \rho_c$, the universe is "closed" and will recontract after a finite time, while if $\rho < \rho_c$ the universe is "open" and will expand forever.)

An expression for ρ_c can be obtained using only elementary considerations. Take a sphere of radius R and a probe particle of mass μ on the surface of this sphere. The kinetic and potential energies of this probe particle are,

$$T = \frac{1}{2}\mu\left(\frac{\mathrm{d}R}{\mathrm{d}t}\right)^2 \ , \quad U = -G_N \frac{4\pi}{3}\frac{R^3\rho\mu}{R} \ , \tag{1.20}$$

where G_N is Newton's constant and where we used the fact that the homogeneous matter outside the sphere does not contribute to the potential energy of the probe. Now, a steady state expansion is reached when the total energy of the probe particle vanishes ($T + U = 0$), that is for

$$\left(\frac{1}{R}\frac{\mathrm{d}R}{\mathrm{d}t}\right)^2 = \frac{8}{3}\pi G_N \rho_c \ , \tag{1.21}$$

or, based on the fact that the left-hand side is nothing else than H_0^2, we obtain

$$\rho_c = \frac{3H_0^2}{8\pi G_N} = 1.9 \times 10^{-29}\, h_{100}^2 \text{ g cm}^{-3} = 1.05 \times 10^4\, h_{100}^2 \text{ eVcm}^{-3} \ . \tag{1.22}$$

We have used the notation h_{100} for Hubble's constant in units of 100 km s^{-1} Mpc^{-1} = $(9.8 \times 10^9 \text{ y})^{-1}$. The experimental value of h_{100} is uncertain, but an interval of 0.5 - 1.0 probably sandwiches the correct value. It is customary to express the average density ρ of the universe in units of the critical density ρ_c using the notation $\Omega = \rho/\rho_c$. In the present context we are interested to know how large fraction of Ω can be associated with the massive neutrinos left over from the big bang.

1.6.2. *Density of relic neutrinos*

In the early epochs the universe was much smaller, much hotter, and was "radiation dominated", i.e., photons and other relativistic particles contributed significantly to the energy density. It is quite easy to calculate the number densities of ultrarelativistic Bose and Fermi gases.

$$n_B = \frac{g_s}{(2\pi\hbar)^3} 4\pi \int_0^\infty \frac{p^2 \mathrm{d}p}{e^{pc/kT} - 1} = \frac{g_s}{2\pi^2}\left(\frac{kT}{\hbar c}\right)^3 \times 2 \times \zeta(3) \ , \tag{1.23}$$

$$n_F = \frac{g_s}{(2\pi\hbar)^3} 4\pi \int_0^\infty \frac{p^2 \mathrm{d}p}{e^{pc/kT}+1} = \frac{g_s}{2\pi^2}\left(\frac{kT}{\hbar c}\right)^3 \times \frac{3}{4} \times 2 \times \zeta(3) , \tag{1.24}$$

where g_s is the number of possible spin states, $g_s = 2$ for photons and Majorana neutrinos and $g_s = 4$ for Dirac fermions, and the Riemann zeta function is $\zeta(3) \approx 1.2$.

Thus, while in thermal equilibrium, the number density of each Majorana neutrino flavor is proportional to the photon number density, $n_\nu/n_\gamma = 3/4$. This ratio remains unaltered even when neutrinos "decouple" (at $t \simeq 1$ s and $kT \simeq 1$ MeV), i.e., are no longer in equilibrium. However, at $t \simeq 10$ s the e^+e^- pairs annihilate and contribute their energy to the photons. The entropy (proportional to ρ/T) is conserved and thus the photon number density increases by the factor $1 + 7/4 = 11/4$ (The ratio of energy densities is 7/8). Finally, we conclude that the present number density of each light Majorana neutrino flavor is

$$n_\nu = (4/11) \times (3/4)\, n_\gamma = (3/11) n_\gamma \simeq 110 \text{ particles per cm}^3 , \tag{1.25}$$

since the experimentally determined present temperature of the blackbody background radiation is $kT = 2.7$ K corresponding to the number density of 400 photons per cm^3. (We assume that these neutrinos are stable, or have lifetimes longer than the age of the Universe.)

1.6.3. *Cosmological neutrino mass limit*

The relic neutrinos, if sufficiently massive, will be nonrelativistic at the present time, and their energy density is simply their number density times their mass. Therefore, to close the universe at $\Omega = 1$ we require that

$$m_\nu^M = 10^4\, h_{100}^2/110 \text{ eV} , \tag{1.26}$$

if one Majorana flavor is massive. (Remembering that $0.5 < h_{100} < 1.0$, we see that the required mass is between 20 and 100 eV.) Moreover, we know from observation of the expansion rate that $\Omega \leq 2$ and therefore obtain the often quoted upper limit for the sum of neutrino masses

$$\sum_l \frac{g_s^l}{2} m_\nu \leq 200 h_{100}^2 \text{eV} , \tag{1.27}$$

where the sum is over all light stable neutrino flavors with the corresponding spin degeneracy factor g_s^l.

A somewhat more stringent limit is obtained if one assumes that $\Omega = 1$, i.e. $\rho = \rho_c$. (There are important theoretical and some observational arguments that this is indeed the case.) The age of the Universe is then $t_0 = 2/(3H_0)$ and since observations require that $t_0 \geq 1.1 \times 10^{10}$ years, we conclude that $h_{100} \leq 0.6$. With these assumptions

$$\sum_l \frac{g_s^l}{2} m_\nu \leq 36 \text{ eV} . \tag{1.28}$$

The above equation also shows that if any neutrino flavor has a rest mass of at least a few eV it will contribute significantly to the present energy density of the Universe, and would be an important component of the dark matter.

For completenes let us note that we considered until now only relatively light neutrinos, with masses less than a few MeV. Cosmological arguments (see Kolb & Turner (1990)) alone also allow the existence of relatively heavy stable neutrinos with masses above a few GeV range. Experimentally, we know that none of the three known neutrino flavors belongs to that category.

2. Kinematic determination of neutrino mass. Neutrino Oscillations.

In this second lecture I will first describe the methods that are used in attempts to determine the value, or an upper limit, of the neutrino mass. With one notable exception (observation of supernova neutrinos) these methods are based on an observation of charged particles emitted in weak decays together with neutrinos. One can then use the energy and momentum conservation and determine neutrino mass from the corresponding energy and/or momentum balance. These methods, however, are not at the present time capable to lead to neutrino mass limits below the cosmological bound for the muon and tau neutrinos.

In the second part of the lecture I will describe the phenomenon of neutrino oscillations. If at least one neutrino is massive, and if, in addition, the neutrinos with definite mass are not the same particles as the "flavor" neutrinos, i.e. the particles created or annihilated together with charged leptons in weak decays, neutrino oscillations will be observable. Their observation, in turn, allows to determine the *mass differences* ($\Delta m^2 \equiv m_2^2 - m_1^2$). Search for neutrino oscillations, as we will see, allows one to probe very small values of the parameter Δm^2.

2.1. *Time of flight method*

The conceptually most straightforward way of determining neutrino mass is from the time-of-flight. If one can determine both the velocity and energy of a particle, it is possible to deduce its mass. Indeed, from

$$v = \frac{p}{E} = \frac{(E^2 - m^2)^{1/2}}{E} \approx 1 - \frac{m^2}{2E^2} , \tag{2.29}$$

it is easy to see that, for an ultrarelativistic particle travelling a distance d, the time delay caused by a nonvanishing rest mass equals

$$\Delta t(\mathrm{s}) = 0.026(d/50 \text{ kpc})(m/1\text{eV})^2(10 \text{ MeV}/E)^2 , \tag{2.30}$$

where, in anticipation, we have chosen to measure the time in seconds, the mass in eV, the energy in units of 10 MeV, and the distance in units of 50 kpc = 1.54×10^{21} meters. If a neutrino source emits neutrinos of different energies within a time interval at the source not wider than Δt, the detector at the distance d will detect the most energetic neutrinos first, and the least energetic ones last. The neutrino mass can be calculated from the time spread of the signal. (It is obvious that useful mass information can be obtained only with extraterrestrial neutrino sources.) Mankind had such an opportunity so far only once, in February 1987.

At that time two underground detectors, Kamiokande II and IMB, detected a burst of events (11 events in Kamiokande II and 8 events in IMB) spread over about 12 s time interval. These events, presumably caused by the reaction

$$\bar{\nu}_e + p \to e^+ + n , \tag{2.31}$$

mark the beginning of neutrino astronomy, and confirm that the basic theoretical assumptions (neutrino luminosity, neutrino average energy, time spread of neutrino emission) concerning supernova neutrino emission, are correct. As far as the neutrino mass is concerned, the observed time spread agrees with the expected emission time, and the events are not arranged in an obvious pattern of decreasing energy with arrival time. Many attempts have been made to determine a significant upper limit of the neutrino mass from these data. A conservative upper limit is close to 20 eV (The "official limit" (see PDG (1996)) from such analysis is 23 eV). We will discuss below how the next supernova event could be used to place a limit (or value) on the mass of the heavy flavor (muon and tau) neutrinos.

2.2. *Kinematic tests for neutrino mass*

The above method is quite unique. All other ones use the energy and momentum conservation and are based on the detection of the charged particles created together with neutrinos in weak decays instead of the difficult detection of neutrinos. The decay of the pion into a muon and a muon neutrino, $\pi^+ \to \mu^+ + \nu_\mu$, a weak hadronic process governed by simple two-body kinematics, is the simplest example (particularly for the pions at rest in the laboratory). A beam of pions is stopped in a target where they can decay. The emerging muons and neutrinos are both monochromatic, and the muon momentum, the quantity that can be measured accurately in an experiment, is related to the pion mass m_π, muon mass m_μ, and neutrino mass m_ν by the relation

$$p_\mu^2 = \frac{(m_\pi^2 + m_\mu^2 - m_\nu^2)^2}{4m_\pi^2} - m_\mu^2 \; ; \; m_\nu^2 = m_\pi^2 + m_\mu^2 - 2m_\pi\sqrt{(p_\mu^2 + m_\mu^2)} \; . \tag{2.32}$$

Using the known pion and muon masses, and measuring the muon momentum (which is $\approx$ 29.8MeV/c) we can determine the muon neutrino mass (or its square). The trouble is that the muon momentum is not very sensitive to the mass (28 ppm for the mass of 250 keV) because the mass enters as a difference of two large numbers. Present limit (after resolving troubles with an apparently incorrect determination of the pion mass) is $m_{\nu_\mu} \leq 170$ keV (PDG (1996)).

It is clear that the two-body decay of the pion (or other mesons) will not reach any time soon the sensitivity to neutrino mass in the electronvolt range. For electron neutrinos and antineutrinos the best mass limit is obtained from the study of three-body decays, such as the nuclear beta decay, $(Z,\ A) \to (Z+1,\ A) + e^- + \bar{\nu}_e$. In them the neutrino mass m_ν must be included in the total decay energy E_0, leading to the relation between the maximum energy of the electron, $E_e^{Max}(m_\nu)$, and the masses of the initial atom, M_i, and of the final ion, M_f,

$$E_e^{Max}(m_\nu) = E_0 - m_\nu = M_i - M_f - m_\nu \; . \tag{2.33}$$

(As the electron energy is much smaller than the mass of the nucleus, we have neglected here the recoil energy of the nucleus, which has a maximum value of $(E_0 - m_e)^2/2M_f$.) The atomic masses are rarely known that well, so a better method is to study the beta spectrum shape which for the allowed decay is governed by the phase space factor $E_e p_e E_\nu p_\nu$. More precisely, the electron spectrum (per unit time and unit energy) is given by

$$\frac{\mathrm{d}N}{\mathrm{d}E} = G_F^2 \frac{m_e^5 c^5}{2\pi^3\hbar^7} \cos^2\theta_C |\mathcal{M}|^2 F(Z,E) pE(E_0 - E)[(E_0 - E)^2 - m_\nu^2]^{1/2}, \tag{2.34}$$

where G_F is the Fermi coupling constant, $\mathcal{M}$ contains the nuclear matrix elements, $F(Z,E)$ is the Coulomb function and θ_C is the Cabbibo angle.

We see that appreciable sensitivity to neutrino mass exists only near the endpoint of the spectrum, where the neutrino becomes nonrelativistic. The spectrum terminates abruptly (with infinite slope) at an energy $E_0 - m_\nu$, manifestly distinguishing it from the spectrum for the case of a massless neutrino which has a vanishing slope at the endpoint. In an experiment, the spectrum needs to be studied up to an electron energy E near E_0, so that $E_0 - E \equiv \Delta E$ is of the order of m_ν. However, at this energy E, the number of decays, $\mathrm{d}N/\mathrm{d}E$, becomes extremely small. It is easy to see that for a low energy electron decay with maximum kinetic energy $Q = E_0 - m_e \ll E_0$, the fraction of decays in the interval ΔE is given by $(\Delta E/Q)^3$. In the favorable case of the tritium decay ($Q = 18.6$ keV), the number of electrons in an energy interval of, say, $\Delta E = 9$ eV (near the present neutrino mass limit) is only 10^{-10} of all decays in the spectrum.

The other trouble is that one cannot prepare a source containing simply individual ions (i.e., bare nuclei) of tritium. Even the best experiments use molecular tritium and the presence of the two "spectator" atomic electrons causes serious difficulties in the interpretation of the results. The present situation is unclear, with the best claimed limits of about 5 eV, but the measured spectra do not have quite the expected shape. Since the shape of the measured tritium beta decay spectrum in the most sensitive experiments is not well understood (in fact the best fit typically gives negative neutrino mass squared ($m_\nu^2 < 0$), the official conservative mass limit for $\bar{\nu}_e$ (PDG (1996)) is 15 eV.

2.2.1. *Tau neutrino mass limit*

This third neutrino has never been observed directly, in the sense of being responsible for inducing a reaction. At the present time there is no indication that it has finite mass, however with much less restrictive upper limit of its mass when compared to the muon and, particularly, to the electron neutrino.

Upper limits of the mass of the tau neutrino, ν_τ, have been derived from studies of the decay of the tau lepton. Currently, the best limit is $m_{\nu_\tau} \leq 24$ MeV (PDG (1996)) coming from a two-dimensional fit of the visible energy and invariant mass distribution of $\tau \to 5\pi(\pi^0)\nu_\tau$ decays, performed by the ALEPH collaboration using the data from 1991-93 LEP runs.

2.3. *Muon and tau neutrinos from future galactic supernovae*

Neutrinos play a decisive role in various stages of supernova evolution (Bethe (1990)). In particular, neutrinos generated during the cooling phase of the hot remnant core will be observed in earthbound detectors. Although pairs of all three flavors are expected to be generated with equal luminosity, due to their smaller opacities, ν_μ and ν_τ neutrinos and their antiparticles decouple at smaller radii, and thus higher temperatures in the core, than ν_e and $\bar{\nu}_e$ neutrinos. Thus, it is expected on general grounds that the neutrino spectra after decoupling obey the temperature (and therefore also average energy) hierarchy, $T_{\nu_x} > T_{\bar{\nu}_e} > T_{\nu_e}$, where ν_x stands for ν_μ, ν_τ and their antiparticles, which are assumed to have identical spectra. Is it possible to identify unambiguously the ν_x neutrinos in an existing detector? In a recent paper (Langanke, Vogel & Kolbe (1996)) we have suggested how this could be done in water Čerenkov detectors, such as the Superkamiokande (SK) detector, operational for more than a year now.

The basis of our proposal is the fact that SK can observe photons with energies as low as 5 MeV. Supernova ν_x neutrinos, with average energies of ≈ 25 MeV, will excite the nucleus ^{16}O via the $^{16}\mathrm{O}(\nu_x, \nu_x')^{16}\mathrm{O}^\star$ neutral current reaction. Most final excited states in ^{16}O will be above the particle thresholds and will dominantly decay by proton and neutron emission. Although the proton and neutron decays will be mainly to the ground states of ^{15}N and ^{15}O, respectively, some of them will go to excited states in these nuclei. If these excited states are below the particle thresholds in ^{15}N ($E^\star < 10.2$ MeV) or ^{15}O ($E^\star < 7.3$ MeV), they will decay by γ emission. As the first excited states in both nuclei ($E^\star = 5.27$ MeV in ^{15}N and $E^\star = 5.18$ MeV in ^{15}O) are at energies larger than the SK detection threshold, all of the excited states in ^{15}N and ^{15}O below the respective particle thresholds will emit photons which can be observed in SK.

We calculated the effective $^{16}\mathrm{O}(\nu_x, \nu_x' p\gamma)$ and $^{16}\mathrm{O}(\nu_x, \nu_x' n\gamma)$ cross sections and compared it to the effective "background" cross sections, stemming from the $\bar{\nu}_e + p \to n + e^+$ and $\nu + e \to \nu' + e'$ events with energy release similar to the energy of the photons. SK is expected to detect about 4000 positrons from the $\bar{\nu}_e + p \to n + e^+$ reaction for a supernova going off at 10 kpc (the distance to the galactic center). By scaling the respective effective cross sections, we estimate that such a supernova will produce about 190-360 (depending on the ν_x spectrum) γ events in the energy window $E = 5 - 10$ MeV, to be compared with a smooth background of about 270 positron events from the $\bar{\nu}_e + p \rightarrow n + e^+$ reaction in the same energy window. This number of events produced by supernova ν_x neutrinos is larger than the total number of events expected from ν_x-electron scattering (about 80 events). More importantly, the γ signal can be unambiguously identified from the observed spectrum in the SK detector.

In this scheme, unfortunately, the information about the energy of the incoming ν_x neutrinos is lost. However, one can perhaps observe the time delay, if any, between the positrons from the $\bar{\nu}_e + p \rightarrow n + e^+$ reaction (expected to arrive essentially without delay), and the γ events. If the mass of ν_x is 25 eV, the delay will be about 0.5 s, probably a smallest possible delay observable. On the other hand, if the mass of ν_x is 100 eV, the delay will be 10 s, which should be safely possible to measure. Thus, with a bit of luck (i.e. a galactic supernova should go off in near future), we should be able to improve the mass limits for the heavy flavor neutrinos substantially, and perhaps bring them within reach of the cosmological bound.

2.4. *Number of neutrino flavors from the width of Z*

We have so far discussed mass limits for the three known neutrinos, ν_e, ν_μ and ν_τ. Do we have to worry about possible fourth, etc. neutrinos?

All "standard" neutrinos couple to the Z with the universal coupling

$$\bar{g} \sum_\alpha \bar{\nu}_{\alpha L} \gamma^\mu \nu_{\alpha L} Z_\mu \ , \tag{2.35}$$

where L signifies the left-handed states, α runs over all neutrino flavors, and $\bar{g}^2 = 8G_F m_Z^2/\sqrt{2}$. Therefore, each neutrino contributes to the Z width the amount

$$\Gamma_\nu = \Gamma(Z \rightarrow \nu_l \bar{\nu}_l) \ = \ \frac{\bar{g}^2 m_Z}{96\pi} \ \approx \ 166 \text{ MeV} \ . \tag{2.36}$$

When Z are produced in e^+e^- collision, the resulting resonance can be described by the Breit-Wigner formula (or its relativistic generalization, radiatively corrected)

$$\sigma(E) = \frac{4\pi(2J+1)}{2 \cdot 2m_Z^2} \frac{\Gamma_i \Gamma_f}{(E - m_Z)^2 + \Gamma_{tot}^2/4} \ . \tag{2.37}$$

Thus one gets the m_Z from the resonance energy, Γ_{tot} from the resonance shape, ratios of the partial widths from the ratios of yields of different particles, and the absolute value of the partial width from the peak cross section. After contributions from all "visible" decays of the Z have been added together, the remaining "invisible" width is responsible for the Z decay into neutrinos.

Therefore, to determine the number of neutrino flavors (for the neutrinos lighter that half of m_Z that couple as shown above) we need to determine the "invisible width" i.e.

$$N_\nu = \frac{\Gamma_{inv}}{\Gamma_\nu} = \frac{\Gamma_{tot} - \Gamma_{hadron} - 3\Gamma_l}{\Gamma_\nu} \ . \tag{2.38}$$

The total width was determined (from nearly 8 million visible Z decays) to be 2.4963 ± 0.0032 GeV, and the invisible width is 499.9 ± 2.5 MeV, giving

$$N_\nu = 2.991 \pm 0.016 \ . \tag{2.39}$$

Thus, there are no other neutrinos than the three known flavors, with masses less than half of the Z mass, and with the standard weak couplings.

2.5. *Neutrino oscillations*

In the first lecture we have considered only the mass of one "family" of neutrinos. In reality there are three (see the description of the Z width) neutrino generations. At this point we have to distinguish two kinds of neutrinos. Those, treated so far, have a definite mass. On the other hand, neutrinos observable in an experiment are the "weak interaction" neutrinos, i.e. the weak doublet partners of the electron, muon, and tau. It is by no means obvious that these particles have a definite mass.

The concept of neutrino oscillations, the quantum-mechanical consequence of this distiction (at that time as neutrino-antineutrino oscillations), was introduced by Pontecorvo already in 1957. The flavor oscillation, discussed below, were introduced first by Pontecorvo and Gribov approximately ten years later. To see the physics better let us restrict ourselves to two families for simplicity, and consider what happens if the "mass eigenstates" and "weak eigenstates" are not identical. The mass term expressed through the physical "weak" states is then nondiagonal

$$-\mathbf{L}_M = m_{\nu_e\nu_e}\bar{\nu}_e\nu_e + m_{\nu_\mu\nu_\mu}\,\bar{\nu}_\mu\nu_\mu + m_{\nu_e\nu_\mu}\,(\bar{\nu}_e\nu_\mu + \bar{\nu}_\mu\nu_e)\;. \tag{2.40}$$

Since $\mathbf{L}_M$ is symmetric, we can diagonalize it by the substitution

$$\begin{aligned}\nu_e &= \cos\theta\nu_1 + \sin\theta\nu_2 \\ \nu_\mu &= -\sin\theta\nu_1 + \cos\theta\nu_2\;.\end{aligned} \tag{2.41}$$

The eigenvalues of $\mathbf{L}_M$ are

$$m_{1,2} = \frac{1}{2}\left\{m_{\nu_e\nu_e} + m_{\nu_\mu\nu_\mu} \pm \left[(m_{\nu_e\nu_e} - m_{\nu_\mu\nu_\mu}) + 4m^2_{\nu_e\nu_\mu}\right]^{1/2}\right\}\;, \tag{2.42}$$

and the angle θ is determined by the formula

$$\tan 2\theta = 2m_{\nu_e\nu_\mu}/(m_{\nu_\mu\nu_\mu} - m_{\nu_e\nu_e})\;. \tag{2.43}$$

The particles ν_1 and ν_2 have definite masses m_1 and m_2, and therefore evolve in time as

$$|\nu_1(t)>= e^{-iE_1t}\;|\nu_1(0)\rangle;\;\;|\nu_2(t)>= e^{-iE_2t}\;|\nu_2(0)\rangle\;, \tag{2.44}$$

where $E_i = (p^2+m_i^2)^{1/2}$ is the neutrino energy. We assume that these particles propagate without interactions and are ultrarelativistic ($p \gg m$); the time and distance are then interchangeable. At this point we shall not worry about the complications, largely formal, associated with the correct wave-packet description of the oscillation phenomena. We also assume that all components of a given neutrino state have the same momentum p.

A particle which has been created at $t = 0$ as an electron neutrino ν_e evolves in time as

$$\begin{aligned}|\nu_e(t)> &= \cos\theta e^{-iE_1t}|\nu_1(0)\rangle + \sin\theta e^{-iE_2t}|\nu_2(0)\rangle \\ &= (e^{-iE_1t}\cos^2\theta + e^{-iE_2t}\sin^2\theta)|\nu_e(0)\rangle \\ &+ \cos\theta\sin\theta(e^{-iE_2t} - e^{-iE_1t})|\nu_\mu(0)\rangle\;.\end{aligned} \tag{2.45}$$

The probability that the neutrino, created as ν_e at time $t = 0$, is in the state $|\nu_\mu\rangle$ at time t, is given by the absolute square of the amplitude of $|\nu_\mu(0)\rangle$ above, i.e.,

$$P(\nu_e \to \nu_\mu) = |\langle\nu_\mu|\nu_e(t)\rangle|^2 = \frac{1}{2}\sin^2 2\theta\left(1 - \cos\frac{m_2^2 - m_1^2}{2p}t\right)\;, \tag{2.46}$$

where we have performed a little bit of straightforward algebra. This is the simplest type of neutrino flavor oscillation. The angle θ, the so called mixing angle, describes the amount of mixing. The probability $P(\nu_e \to \nu_\mu)$ varies periodically with time or distance.

This periodicity is characterized by the "oscillation length",

$$L_{osc} = 2\pi \frac{2p}{|m_2^2 - m_1^2|} \ . \tag{2.47}$$

The above equality, rewritten in practical units, means that

$$L_{osc}(\text{meters}) = \frac{2.48 \times E_\nu(\text{MeV})}{|m_2^2 - m_1^2|(\text{eV})^2} \ . \tag{2.48}$$

The oscillation is a typical interference effect caused by the nondiagonal form of the mass term. (For diagonal $\mathbf{L}_M$ one would require $m_{\nu_e \nu_\mu} = 0$ and, consequently, $\theta = 0$ or $\pi/2$, and there are no oscillations.) Thus, we see that flavor oscillations, which are possible only if at least one neutrino is massive, violate conservation of individual lepton numbers (N_e and N_μ in our simplest example). The total lepton number ($N_e + N_\mu$ in our case), however, remains conserved.

In order to observe neutrino oscillations, one can use two strategies. Suppose that one has a beam of a definite neutrino flavor, say a pure ν_μ beam. Observing it at different distances L with a detector which will count the number of muons, one can test whether the neutrino signal scales as the solid angle, i.e. as $1/L^2$, where L is the distance. Deviations from this dependence would be a signal of oscillations. This is so called *disappearance test*. Alternatively, when the initial beam has ν_μ of different energies, and the energy composition is known, one can equivalently search for any change of its energy profile.

The other type of oscillation search would employ a detector capable of distinguishing electrons from muons in the above example. A pure ν_μ beam cannot create electrons in the charged current weak interactions. Thus an *appearance* of the wrong flavor of neutrinos is a signal of oscillations. Again, the effect should depend on the quantity L/E_ν.

Ideally, therefore, one would observe a nonvanishing

$$P(\nu_e \to \nu_\mu) = \sin^2 2\theta \sin^2 \frac{\Delta m^2 L}{4E_\nu} \ , \tag{2.49}$$

in the appearance experiment. (Here $\Delta m^2 = |m_2^2 - m_1^2|$). In the disappearance experiment one would observe the deviation from unity of the quantity

$$P(\nu_\mu \to \nu_\mu) = 1 - P(\nu_e \to \nu_\mu) \equiv 1 - \sin^2 2\theta \sin^2 \frac{\Delta m^2 L}{4E_\nu} \ . \tag{2.50}$$

In both tests the pattern repeats itself when the distance L is changed to $L + L_{osc}$ (at least ideally). Also, the important quantity is not just the distance L, but L/E_ν.

It is easy to generalize the preceding analysis to the case of more than two generations. Again, we encounter periodic behavior of the probability of finding different flavors; however, there are now more mixing angles and oscillation lengths.

Generally, the transformation between the flavor neutrino eigenstates ν_l and neutrino states of the definite mass ν_j is of the form

$$\nu_l = \sum_j U_{l,j} \nu_j \ , \tag{2.51}$$

where $U_{l,j}$ is a unitary matrix. For three flavors this matrix is characterized by three "angles" and one possible CP violating phase. In addition, for three flavors we have, naturally, three masses m_j, i.e. two independent mass differences $\Delta m_{i,j}^2$.

A relatively simple phenomenology is possible if one assumes that one of these three masses is much larger than the other two, e.g. $m_3 \gg m_2 \simeq m_1$. In that case there is only one relevant $\Delta m^2 \cong m_3^3$ and two mixing angles, e.g. $U_{e,3}$ and $U_{\mu,3}$. ($|U_{\tau,3}|$ is then fixed by the unitarity condition.) The relevant oscillation probability is then

$$P(l \to l') = 4|U_{l,3}|^2|U_{l',3}|^2 \sin \frac{m_3^2 L}{4E_\nu} \text{ (for } l \neq l') . \tag{2.52}$$

One can consider now, e.g. oscillations between flavor l which could be dominantly $l = 1$ and l', dominantly $l' = 2$. Nevertheless the oscillation length depends on m_3. This is an example of so-called indirect oscillations. At larger distances the so far neglected much smaller mass difference $\Delta m_{1,2}^2$ can begin to play a role with its own, much larger, oscillation length.

At present, three sets of experimental data suggest that neutrinos are massive and mixed. The solar neutrino deficit has been clearly established in five experiments, using quite different techniques. We will discuss it in more detail in the next lecture. The atmospheric neutrino anomaly, which we will discuss in the fourth lectrure, can be also most readily explained by neutrino mixing. Finaly, the LSND experiment in Los Alamos has found evidence for the appearance channel $\bar{\nu}_\mu \to \bar{\nu}_e$ and more recently for $\nu_\mu \to \nu_e$. That evidence has not been confirmed in any other experiment, and the statistical significance of it is not yet as strong as for the other two. In addition, numerous other attempts to observe neutrino oscillations have seen nothing unusual, and thus exclude certain values of the oscillation parameters from further consideration.

3. Solar neutrinos and matter oscillations

In this lecture we will discuss the so-called "solar neutrino puzzle". There is a vast literature devoted to the subject; many details are summarized in the Bahcall (1989) monograph. The solar model described below is from that book. There have been some modifications since, but they do not change the overall picture. So, we begin by explaining why and how many neutrinos the Sun is expected to produce. Then we describe the observations. The incredibly challenging problems associated with building, testing, calibrating, etc., of the solar neutrino detectors are, unfortunately, beyond the scope of these lectures. An interested reader would have to study the original literature on each of the detectors. The "Neutrino oscillations industry" home page on the World Wide Web (http://www.hep.anl.gov/NDK/Hypertext/nuindustry.html) list links to the individual experiments and can be used for that purpose. The bottom line is that all experiments see a *neutrino deficit.* Moreover, since different experiments have different thresholds, they are sensitive to different parts of the solar neutrino spectrum. It appears that the only plausible explanation of the different ratios between the measured and expected neutrino fluxes is in terms of neutrino oscillations. At that point we change gears and consider how the neutrino oscillation pattern is modified when neutrinos propagate in matter, in particular in matter of varying density, like in the Sun. Finally, we show the preferred sets of oscillation parameters Δm^2 and $\sin^2 2\theta$ that are capable of explaining the existing data.

3.1. *Neutrinos from the Sun*

Exothermic nuclear reactions in the interior of the Sun are the source of energy reaching the surface of the Earth. We believe that the main reactions are those of the proton-proton (pp) chain. The net effect of this chain of reactions is the conversion of four protons and two electrons into an α particle,

$$2e^- + 4p \to {}^4\text{He} + 2\nu_e + 26.7 \text{ MeV} . \tag{3.53}$$

Approximately 97% of the energy released in the *pp* chain is in the form of charged particle and photon energy, which is the source of the Sun's thermal energy. Thus, for each ≈13 MeV of the generated thermal energy we expect one neutrino. Assuming that the Sun is in equilibrium, and has been in stable equilibrium for a sufficiently long time, the rate with which the thermal energy is generated is equal to the rate at which the energy is radiated from the surface. The solar constant, the energy of solar flux reaching the Earth's atmosphere, is $S = 1.4\times10^6$ erg cm^{-2} s^{-1} = 8.5×10^{11} MeV cm^{-2} s^{-1}. Thus, the neutrino flux at the surface of Earth is expected to be $\Phi_\nu \approx S/13 \approx 6\times10^{10}$ ν_e cm^{-2} s^{-1}. Experimental verification of this neutrino flux (or of one or several of its components) would constitute an important test of our understanding of the energy production in the Sun. At the same time, as the path of solar neutrinos to a detector on Earth is much longer than the neutrino path length for any terrestrial neutrino detector, observation of solar neutrinos constitutes a very stringent test for the inclusive oscillations of electron neutrinos. Experiments, which started almost 30 years ago, were motivated by the desire to verify that we indeed understand the energy production in stars (*pure astrophysics*). That goal was achieved, but since now all four running experiments measure neutrino flux somewhat smaller than the predicted one, there is a wide belief that this reduction is a consequence of neutrino oscillations and therefore a window to the "physics beyond the standard model" (*particle physics*).

3.1.1. *The predicted neutrino spectrum from the Sun*

In order to test the neutrino production in the Sun in a quantitative way, we have to evaluate the neutrino flux in greater detail than the crude estimate above, in particular we need to know also its energy distribution. This is so, since the various possible detector reactions have different thresholds and therefore are sensitive to different components of the solar neutrino flux.

The *pp* chain of reactions results in the neutrino spectrum having two continuous components: *pp* (E_{Max} = 420 keV) and ^{8}B decay (E_{Max} = 14 MeV), and three discrete lines: *pep* (E=1.44 MeV), and the two branches of ^{7}Be electron capture (E =862 and 384 keV). In addition, one has to include the weaker contributions of the solar CNO cycle (decays of ^{13}N and ^{15}O).

The branching ratios and the neutrino spectrum are obtained from calculations based on the "standard solar model". The equation of state of the model relates the pressure $P(r)$, the density $\rho(r)$, the temperature $T(r)$, the energy production rate per unit mass $\epsilon(r)$, the luminosity $L(r)$, and the opacity $\kappa(r)$. The equation also depends on the fractional abundances of hydrogen, helium, and the heavier elements. The solution of the equation of state is complicated because the opacity κ is a complex function of density, temperature, and composition, which can be estimated only from detailed knowledge of the atomic transitions involved. In addition, the energy production rate ϵ depends on the detailed knowledge of cross sections for the individual reactions. These cross sections have to be extrapolated from laboratory energies to the much lower solar energies.

Equations of the standard solar model have been carefully solved in a number of papers and the uncertainties in the solar neutrino flux have been estimated. Table I (see Bahcall (1989)) shows neutrino fluxes of individual components. Multiplying the flux by the averaged cross sections, we arrive at the predicted reaction rates for the ^{37}Cl and ^{71}Ga detectors also shown there. Let us note in passing that the fluxes listed in Table I depend, sometimes drastically, on the solar model (e.g. on the central temperature). On the other hand, the spectrum shape of the individual components, in particular the shape of the continuum spectrum from the ^{8}B positron decay, are fully determined by nuclear

TABLE 1. Neutrino fluxes and reaction rates in SNU (1 SNU = 1 capture per second per 10^{36} target atoms) for chlorine and gallium detectors based on the standard solar model. Numbers of ^{37}Ar and ^{71}Ge atoms produced per day in the respective detectors, having 2.2×10^{30} atoms of ^{37}Cl, and 1.0×10^{29} atoms of ^{71}Ga, are given in columns 5 and 7.

Symbol	reaction (decay)	Flux 10^{10} cm^{-2}s^{-1}	^{37}Ar SNU	^{37}Ar d^{-1}	^{71}Ge SNU	^{71}Ge d^{-1}
pp	$p+p \to d+e^+ + \nu_e$	6.0	0	0	70.8	0.61
pep	$p+p+e^- \to d+\nu_e$	0.014	0.2	0.04	3.0	0.03
^{7}Be	^{7}Be+e^- $\to$^{7}Li+ν_e	0.47	1.1	0.21	34.3	0.30
^{8}B	^{8}B$\to$^{8}Be+e^+ + ν_e	0.00058	6.1	1.16	14.0	0.12
^{13}N	^{13}N$\to$^{13}C+e^+ + ν_e	0.06	0.1	0.02	2.8	0.02
^{15}O	^{15}O$\to$^{15}N+e^+ + ν_e	0.05	0.3	0.06	6.1	0.05
	Total	6.6	7.9	1.5	132.0	1.14

physics. Any deviation from the expected (and accurately calculable) shape would be a signal of the "new physics".

3.1.2. *Observations*

I will describe the four (or actually now five) experiments only briefly here. Chlorine, or "Davis" radiochemical experiment, running since 1970, is sensitive primarily to ^{8}B neutrinos. This is a radiochemical experiment, based on the reaction

$$\nu_e + {}^{37}\text{Cl} \to {}^{37}\text{Ar} + e^- , \tag{3.54}$$

with the threshold of 814 keV. The final radioactive ^{37}Ar atoms are periodically extracted, collected, and their decays measured in low-background counters. The expected rates are also listed in Table I. The Davis experiment was historically the first one to show that the solar neutrino flux on Earth is less than the expected one. Compared to the prediction in Table I, the measured flux is $(2.55 \pm 0.17 \pm 0.18)/7.9$ of the expected one. (The first error is statistical, the second systematic, data here and further from PDG (1996) . In calculating the deficit the uncertainty in the expected fluxes is *not* taken into account.)

Kamiokande (Japan), replaced now by SuperKamiokande, are water Čerenkov detectors. In them one can measure the energy and direction of relativistic charged particles. The detection of solar neutrinos is based on the neutrino-electron scattering,

$$\nu_e + e^- \to \nu_e + e^- . \tag{3.55}$$

The final recoil electrons, with kinetic energies above threshold (which is 6.5 MeV in the latest SuperKamiokande available data) are measured. This is the only live-time experiment so far which also clearly shows that the detected neutrinos are indeed coming from the Sun. (Since the recoil electrons point away from the Sun.) The experiment runs since 1986 (stopped now, SK runs since Spring 1996) and, due to its relatively high threshold is sensitive *only* to ^{8}B neutrinos. The measured flux in Kamiokande was $(0.50 \pm 0.04 \pm 0.06$ of the expectation; the preliminary SK result are consistent with this deficit (the preferred value is 0.441).

There are two radiochemical experiments using ^{71}Ga as target, GALLEX (Gran Sasso, Italy) and SAGE (Baksan Valley, Russia). The detection reaction is

$$\nu_e + {}^{71}\text{Ga} \to e^- + {}^{71}\text{Ge} . \tag{3.56}$$

Again, the radioactive daughter atoms of ^{71}Ge are collected and their decays counted in the very low background counters at the same undergound facility where the gallium detectors are. These detectors have a low threshold (236 keV) and therefore are sensitive primarily to the *pp* neutrinos, with a significant contribution of the ^{7}Be neutrinos, but relatively little from the ^{8}B neutrinos (see Table I). Gallex gives $(79 \pm 10 \pm 6)/132$ of the expected flux, and SAGE has a similar value, $(73 \pm 17 \pm 6)/132$. (These are data from PDG (1996) , the experiments are running and new data are published periodically.) The errors on these results are complicated and asymmetric; for orientation I have averaged the upper and lower error bars. Both detectors were successfully calibrated with manmade radioactive sources.

Thus, we see, that all detectors observe a deficit, and the amount of suppression is different depending on the detection scheme. I will just refer to Bahcall (1989) for arguments that possible deficiencies of the solar model cannot explain the observations.

If astrophysical explanations of the solar neutrino deficit fail, it is reasonable to turn to the possibility of neutrino oscillations. Solar neutrinos, produced in the solar center as electron neutrinos, can oscillate into another neutrino flavor. Since the detectors on Earth measure only the electron neutrino component, a reduction in flux results. The observed reduction appears to be substantial, and energy dependent. A relatively consistent picture is obtained if one assumes that the *pp* component is essentially unchanged, the ^{8}B component is reduced to about half of its predicted value, and the ^{7}Be flux is essentially absent. (However, there is no direct verification yet of this last, most dramatic, conclusion.) As stated above, it is very difficult to imagine a scenario based on some flaw in the solar model which would account for such a pattern. Hence most people believe that neutrino oscillations are indeed the true reason for the solar neutrino deficit. Moreover, since there is an apparent energy dependence of the neutrino deficit, it should be possible to determine from the data whether a) the oscillation hypothesis works in the first place and, b) what are the best values of the oscillation parameters Δm^2 and $\sin^2 2\theta$.

In order to explain the observation in terms of vacuum oscillations (so-called "just so" oscillations) one has to make rather fine tuning of the parameters. The oscillation length must be comparable to the Sun - Earth distance (i.e., $\Delta m^2 \simeq 10^{-10}\text{eV}^2$), and the mixing angle must be large. Another possibility is the resonance enhancement of oscillations due to the propagation of neutrinos through solar interior.

3.2. *Neutrino Oscillations in Matter*

Vacuum oscillations, discussed thus far, are an interference effect caused by the phase difference of neutrinos with different masses,

$$\nu(t) = \nu(0)e^{i(px-Et)} \approx \nu(0)e^{-it\frac{m^2}{2p}} \ . \tag{3.57}$$

However, when neutrinos propagate through matter, the phase factor above is changed, $ipx \to ipnx$, where n is the index of refraction. The oscillations of a neutrino beam through matter are then modified, as first recognized by Wolfenstein in 1978. Under favorable conditions a remarkable resonant amplification of oscillations proposed by Mikheyev & Smirnov in 1985 may occur. (This process is often referred to by the acronym MSW.)

The index of refraction deviates from unity owing to the weak interaction of the neutrinos. Thus, for neutrinos of weak interaction flavor l and of momentum p the index of refraction is

$$n_l = 1 + \frac{2\pi N}{p^2} f_l(0) \ , \tag{3.58}$$

where N is the number density of the scatterers and $f_l(0)$ is the forward scattering amplitude. We can neglect the absorption of neutrinos by matter and, therefore, consider only the real part of $f_l(0)$.

Matter is composed of nucleons (or quarks) and electrons. The contribution of nucleons (or quarks) to the forward scattering amplitude is described by the neutral current (Z exchange); it is identical for all neutrino flavors. (We assume that there are no flavor changing neutral currents.) For electrons the situation is different; the electron neutrinos interact with electrons in matter via both the neutral current and the charged current (W exchange). All other neutrino flavors interact only via the neutral current.

That part of the refractive index (or forward scattering amplitude) which is common to all neutrino flavors is of no interest to us, because it modifies the phase of all components of the neutrino beam in the same way. However, the fact that ν_e (and $\bar{\nu}_e$) interact differently has important consequences.

Here I outline the derivation of the additional phase associated with the W exchange. For a neutrino of mass m and momentum $p \gg m$ propagating in matter, the effective energy is

$$E_{eff} \approx p + \frac{m^2}{2p} + \langle e\nu | H_{eff} | e\nu \rangle \ , \tag{3.59}$$

where H_{eff} represents the W exchange:

$$\begin{aligned} H_{eff} &= \frac{G_F}{\sqrt{2}} J^{\mu}_{W} J^{+}_{\mu W} \\ &= \frac{G_F}{\sqrt{2}} \bar{e}\gamma^{\mu}(1-\gamma_5)\nu\bar{\nu}\gamma_{\mu}(1-\gamma_5)e \\ &= \frac{G_F}{\sqrt{2}} \bar{e}\gamma^{\mu}(1-\gamma_5)e\bar{\nu}\gamma_{\mu}(1-\gamma_5)\nu \ . \end{aligned} \tag{3.60}$$

(In the last line we performed the Fierz transformation to the charge retaining form.) For electrons at rest, only the $\mu = 4$ term contributes and we obtain

$$\begin{aligned} \langle e | \bar{e}\gamma^{\mu}(1-\gamma_5)e | e \rangle &= N_e \delta_{\mu 4} \ , \\ \langle \nu | \bar{\nu}\gamma_4(1-\gamma_5)\nu | \nu \rangle &= 2 \ . \end{aligned} \tag{3.61}$$

(The last term above changes sign for the electron antineutrinos.)

Thus,

$$E_{eff} \approx p + \frac{m^2}{2p} + \sqrt{2} G_F N_e \ , \tag{3.62}$$

and the time dependence for a neutrino propagating through matter is

$$\nu(t) = \nu(0) \ e^{-it(\frac{m^2}{2p} + \sqrt{2} G_F N_e)} \ . \tag{3.63}$$

3.2.1. *Formalism for a constant density*

In the simplest case of two neutrino flavors e and α, which are superpositions of two mass eigenstates ν_1, ν_2 we first repeat the derivation for vacuum oscillations:

$$\begin{aligned} \nu_e &= \nu_1 \cos\theta + \nu_2 \sin\theta \ , \\ \nu_\alpha &= -\nu_1 \sin\theta + \nu_2 \cos\theta \ . \end{aligned} \tag{3.64}$$

The time development of a neutrino beam in a vacuum is described in a compact form by the differential equation

$$i\frac{d}{dt}\begin{pmatrix} \nu_1 \\ \nu_2 \end{pmatrix} = \begin{pmatrix} m_1^2/2p & 0 \\ 0 & m_2^2/2p \end{pmatrix} \begin{pmatrix} \nu_1 \\ \nu_2 \end{pmatrix} , \tag{3.65}$$

or, equivalently in terms of the weak eigenstates,

$$i\frac{d}{dt}\begin{pmatrix}\nu_e\\ \nu_\alpha\end{pmatrix} = \begin{pmatrix}Xs^2 & Xcs\\ Xcs & Xc^2\end{pmatrix}\begin{pmatrix}\nu_e\\ \nu_\alpha\end{pmatrix} , \tag{3.66}$$

where $X = 2\pi/L_{osc}$ and $c = \cos\theta, s = \sin\theta$. Both these fully equivalent expressions lead to the familiar expression for vacuum oscillations described by the oscillation length

$$L_{osc} = 2\pi\frac{2p}{m_2^2 - m_1^2} . \tag{3.67}$$

The amplitude of the oscillation is $\sin^2 2\theta$. The oscillation length L_{osc} above in this formalism may now have either sign, depending on the relative magnitude of the masses m_1, m_2. Previously, we used the absolute value of Δm^2, and thus L_{osc} was always positive.

In matter, we have an additional contribution to the time development of the electron neutrino beam (and *only* for the electron neutrinos, except that for the electron antineutrinos the phase is also present, but has an opposite sign)

$$\nu_e(x) = \nu_e(0)e^{ipnx} = \nu_e(0)e^{-i\sqrt{2}G_F N_e x} , \tag{3.68}$$

where in the last term the uninteresting overall phase is omitted. The new phase factor leads to the definition of the "characteristic matter oscillation length" that is, the distance over which the phase changes by 2π,

$$L_o = \frac{2\pi}{\sqrt{2}G_F N_e} \approx \frac{1.7\times 10^7}{\rho(\mathrm{gcm}^{-3})\frac{Z}{A}} \text{ meters} , \tag{3.69}$$

where we used $N_e = \rho N_o Z/A$, Z/A being the average charge-to-mass ratio of the electrically neutral matter, and N_o is the Avogadro number. Unlike the vacuum oscillation length L_{osc}, the matter oscillation length L_o is independent of the neutrino energy and always positive. For ordinary matter (rock) $\rho \approx 3$ g cm^{-3} and $Z/A \approx 0.5$, while at the center of the Sun $\rho \approx 150$ g cm^{-3} and $Z/A \approx 2/3$, thus $L_o \approx 10^4$ km (rock) and $L_o \approx 200$ km (Sun).

Including the forward scattering in matter, the differential equation for the time development of eigenstates ν_e, ν_α is modified

$$i\frac{d}{dt}\begin{pmatrix}\nu_e\\ \nu_\alpha\end{pmatrix} = \begin{pmatrix}Xs^2 + 2\pi/L_0 & Xcs\\ Xcs & Xc^2\end{pmatrix}\begin{pmatrix}\nu_e\\ \nu_\alpha\end{pmatrix} , \tag{3.70}$$

The 2×2 matrix on the right-hand side can be diagonalized as before by the transformation

$$\begin{aligned}\nu_{1m} &= \nu_e\cos\theta_m - \nu_\alpha\sin\theta_m ,\\ \nu_{2m} &= \nu_e\sin\theta_m + \nu_\alpha\cos\theta_m .\end{aligned} \tag{3.71}$$

The combinations ν_{1m} and ν_{2m} represent "particles" which propagate through matter as plane waves. The new mixing angle θ_m depends on the vacuum mixing angle, and on the vacuum and matter oscillation lengths L_{osc} and L_o through the relation

$$\tan 2\theta_m = \tan 2\theta\left(1 - \frac{L_{osc}}{L_o}\sec 2\theta\right)^{-1} . \tag{3.72}$$

(Note that, as stressed already, L_{osc} through its dependence on Δm^2 can in this context have both the positive and negative signs.) The "effective" oscillation length in the presence of matter is obtained from the difference of eigenvalues of the 2×2 matrix and is given by

$$L_m = L_{osc}\frac{\sin 2\theta_m}{\sin 2\theta} = L_{osc}\left[1 + \left(\frac{L_{osc}}{L_o}\right)^2 - \frac{2L_{osc}}{L_o}\cos 2\theta\right]^{-1/2} , \tag{3.73}$$

and the probability of detecting a ν_e at a distance x from the ν_e source is as before

$$P(E_\nu, x, \theta, \Delta m^2) = 1 - \sin^2\theta_m \sin^2\frac{\pi x}{L_m} , \tag{3.74}$$

where θ_m and L_m both depend on the vacuum oscillation angle θ and on the vacuum oscillation length L_{osc} (and through it on $\Delta m^2/E_\nu$).

We can now consider several special cases:

(*a*) $|L_{osc}| \ll L_o$ (low density): matter has virtually no effect on oscillations. In experiments with terrestrial sources of neutrinos, one is at best able to study oscillations with $|L_{osc}| \leq$ Earth diameter, and thus this limiting case is relevant. Matter will have, therefore, only a small effect on experiments with neutrinos of terrestrial origin. (The effect could be observable, however, as day-night or seasonal variation of the neutrino signal in a solar neutrino detector, see discussion below.)

(*b*) $|L_{osc}| \gg L_o$ (high density): the oscillation amplitude is suppressed by the factor $L_o/|L_{osc}|$, while the effective oscillation length $L_m \approx L_o$ is independent of the vacuum oscillation parameters.

(*c*) $|L_{osc}| \approx L_o$: this is, obviously, the most interesting case, since then oscillation effects can be enhanced; in particular, for $L_{osc}/L_o = \cos 2\theta$ one has $\sin^2 2\theta_m = 1$. However, the "effective" oscillation length is correspondingly longer, $L_m = L_{osc}/\sin 2\theta$.

Thus when neutrinos of different energies propagate through matter the resonance condition (c), $L_{osc}/L_o = cos 2\theta$, can be obeyed only for positive Δm^2 ($m_2 > m_1$), and for

$$\frac{E_\nu(\text{MeV})}{|\Delta m^2|(\text{eV}^2)} = \frac{6.8 \times 10^6}{\frac{Z}{A}\rho(\text{gcm}^{-3})}\cos 2\theta \ . \tag{3.75}$$

Therefore, for each value of the density ρ the enhancement affects only certain values of $E_\nu/\Delta m^2$. (For $\bar{\nu}_e$ the resonance would occur if $m_2 < m_1$, corresponding to the so-called inverse hierarchy.)

At resonance the"effective" mixing angle θ_m is enhanced ($\sin^2 2\theta_m \approx 1$) over a finite range of L_0, and hence also over a finite range of matter densities. The width of the resonance region depends on the vacuum mixing angle θ.

The discussion so far has dealt with neutrino propagation in matter of a constant density. The expressions above represent then an exact analytical solution of the coupled first order differential equations. Before discussing further the obviously crucial problem of neutrino propagation when density is not constant, and, in particular, the situations when the resonance enhancement in point (c) above occurs, it is worthwhile stressing that neutrino oscillations in matter do not represent a new physical phenomenon; they are necessary consequences of vacuum oscillations.

3.3. *MSW effect for solar neutrinos*

When neutrinos propagate through a medium of variable density, such as from the center of the Sun, in a collapsing supernova, or through the center of the Earth, one has to solve the system of coupled differential equations. The coefficients in them depend on the electron density N_e and through it on the position, and on time. It is always possible to solve these equations numerically.

The most interesting case is when neutrinos propagate from the central region of the Sun where $L_0 \ll |L_{osc}|$ (i.e. the high density regime above, case (b)), through the region where the resonance condition is obeyed, and to the Sun surface where $L_0 \gg |L_{osc}|$ (i.e. the low density regime, case (a)). In that case the probability that an electron neutrino created in the center remains an electron neutrino rapidly decreases in the resonance region, and remains small for larger radii. For this particular choice of $E_\nu/\Delta m^2$, the neutrinos leaving the Sun are no longer electron neutrinos; indeed, the probability $P(\nu_e \to \nu_e)$ can be very small at the solar edge. Numerical calculations show that if real neutrinos have $|\Delta m^2| \approx 10^{-5}$ eV^2, matter oscillation enhancement would be able to explain the suppression of the neutrino flux as detected by the ^{37}Cl, Kamiokande and Gallium detectors, even if the vacuum mixing angle is as small as $\sin^2 2\theta \simeq 10^{-2}$.

In a vacuum (or at low density), the mass eigenstates ν_1, ν_2 are identical to the effective eigenstates ν_{1m}, ν_{2m}, and are rotated by the vacuum mixing angle θ (which is assumed to be small) with respect to the weak flavor eigenstates ν_e and ν_α. Furthermore, if $m_2 > m_1$ ($\Delta m^2 > 0$), the electron neutrino is dominantly the lighter mass eigenstate ν_1. On the other hand, in the high density region where $L_0 \ll L_{osc}$ the effective mixing angle is such that $\tan 2\theta_m$ is small and negative, i.e. θ_m is near $\pi/2$. An electron neutrino is then dominantly the heavier mass eigenstate ν_{2m}.

Now, let us consider a beam of neutrinos passing through an object of slowly varying density. (Slow density variation means that the density can be regarded as constant over the effective oscillation length L_m, so called adiabatic solution of the MSW effect). In this case neutrinos $\nu_{1m} (\nu_{2m})$ remain in the same state, but the relative orientation of the vectors ν_{1m} and ν_{2m} with respect to the fixed vectors ν_1 and ν_2 (or ν_e and ν_α) will be different at different densities. As the neutrino beam traverses through the star, the pair of vectors ν_{1m}, ν_{2m} slowly rotates with respect to ν_e, ν_α. For positive Δm^2 the rotation is by 90° between the high and low densities. Thus, it is clear that at some intermediate density the full mixing angle $\theta_m = 45^o$ will be reached, i.e., resonance will occur. It is also clear that if ν_e is created dominantly as ν_{2m}, it will remain dominantly ν_{2m} and will emerge from the star as a particle having dominantly weak interaction flavor of the neutrino ν_α.

To find conditions for the neutrino conversion we need to be sure that: a) the resonance occurs at all, and b) the quantity P_x, the probability of transition $1_m \to 2_m$ at the resonance, is small. Condition a) above is obeyed if the ratio $L_{osc}/(L_0 \cos 2\theta) > 1$ is valid, i.e., if

$$1.5 \times 10^{-7} \frac{E_\nu(\mathrm{MeV})}{\Delta m^2(\mathrm{eV}^2)} \rho_{Max}(\mathrm{gcm}^{-3}) \frac{Z}{A} > 1 \ . \tag{3.76}$$

This determines the smallest value of $E_\nu/\Delta m^2$ for which the resonance can occur. Note that the vacuum mixing angle θ does not appear (a horizontal line in the usual Δm^2 vs. $\sin^2 2\theta$ plot). Also, neutrinos of higher energies will obey the condition more easily and will be converted more than the low energy neutrinos.

Condition b) is related to the width of the resonance region which must be wider than the oscillation length L_m at the resonance point. This condition (so called nonadiabatic solution) implies that

$$\frac{\sin^2 2\theta \Delta m^2}{E_\nu} \geq const \ , \tag{3.77}$$

where the *const* above depends on the logarithmic derivative of the solar density distribution (solar density is nearly exponential). This condition represents a downsloping line in the Δm^2 vs. $\sin^2 \theta$ plot. It also affects more the lower energy neutrinos than the higher energy ones, and determines for each θ the largest value of $E_\nu/\Delta m^2$ for which the resonance can occur.

There are two small islands in the Δm^2 vs. $\sin^2 \theta$ plot which are compatible with all the solar neutrino data. Both correspond to $\Delta m^2 \approx 10^{-5}$ eV^2. The preferred one ("small angle solution") has $\sin^2 2\theta \approx 10^{-2}$ while the other one ("large angle solution") has $\sin^2 2\theta \approx 0.5$. New, high statistics experiments, Superkamiokande and SNO, might be able to give a definitive answer by observation of the spectrum distortion and/or neutral current effects.

As we mentioned earlier, an unambiguous observation of the spectrum distortion would be a clear signal of neutrino oscillations. The other possibility is to compare the yield of the charged and neutral current reactions. Only electron neutrinos, ν_e, can cause the charged current reactions in which electrons are created (such as the reactions used in the chlorine and gallium detectors). The neutrino-electron scattering, the reaction used in Kamiokande and SK, is more complicated, since both charged and neutral current reactions contribute. However, if all ν_e would be replaced by other flavor neutrinos, the cross section, and hence the number of observed recoil electrons would be reduced by a factor of about six.

If oscillation involves a conversion to other standard neutrinos (mu or tau), the pure neutral current reaction cross section should not change. In that case, one would observe *full* yield of the reaction products. The planned SNO detector in Sudbury (Canada) will use heavy water D_2O as the target material. Electron neutrinos can disintegrate deuterons either through the charged current reaction

$$\nu_e + d \rightarrow e^- + p + p \ , \tag{3.78}$$

or through the neutral current reaction

$$\nu_e + d \rightarrow \nu_e + p + n \ . \tag{3.79}$$

By measuring the yield of the electrons on one hand, and of neutrons on the other hand, one can determine both reaction rates and make a very convincing case for or contra neutrino oscillations. We have to wait few more years, unfortunately, for this important result.

4. Atmospheric neutrinos, LSND, double beta decay, conclusions

In this last lecture we will first review the other two indications that neutrino oscillate, namely the atmospheric neutrino anomaly and the LSND results. We will then turn to the question of the particle-antiparticle symmetry, i.e. whether neutrinos are Dirac or Majorana particles. We will show that the distinction becomes irrelevant for massless neutrinos. For massive neutrinos the only way to distinguish the two possibilities is by observing a lepton number violating process, such as the neutrinoless double beta decay. At the end of all of that, we will summarize briefly what we have learnt.

4.1. *Atmospheric neutrino anomaly*

When cosmic ray hadrons, dominantly protons and α particles, reach the upper levels of the Earth's atmosphere, they interact with the nuclei of nitrogen and oxygen in air. As a result of these violent strong interactions (see Prof. Gaisser's lectures in this volume) new particles are created, in particular pions. The atmosphere is sufficiently thin that charged pions decay essentially as in the vacuum, after travelling a relatively short distance (the mean life of charged pions is 2.6×10^{-8} s). The dominant decay mode (99.988%) is

$$\pi^+ \rightarrow \mu^+ + \nu_\mu \ \ , \ \ \pi^- \rightarrow \mu^- + \bar{\nu}_\mu \ . \tag{4.80}$$

Most muons also decay before reaching surface by

$$\mu^+ \rightarrow e^+ + \bar{\nu}_\mu + \nu_e \ \ , \ \ \mu^- \rightarrow e^- + \nu_\mu + \bar{\nu}_e \ . \tag{4.81}$$

The muon and electron neutrinos created this way are called "atmospheric neutrinos". They travel generally downwards, and have energies from several hundred MeV up to many GeV (again, Prof. Gaisser will probably cover this subject in detail).

¿From the above decay modes it follows immediately that the composition of atmospheric neutrinos should be governed by

$$\frac{\nu_\mu + \bar{\nu}_\mu}{\nu_e + \bar{\nu}_e} \simeq 2 \ . \tag{4.82}$$

Detailed calculations, taking into account that other mesons, such as kaons, etc. are also produced, confirm the general validity of the above rule.

If one now observes the charged current interactions of these atmospheric neutrinos in underground detectors, one would explore the flight path from about 20 km for neutrinos coming to the detector from above, to 12 000 km for the upward going neutrinos that have traversed the Earth diameter. Experiments with atmospheric neutrinos, therefore, are well suited for exploring neutrino disappearance in the range of $\Delta m^2 \simeq 10^{-2} - 10^{-3}\text{eV}^2$.

Many sources of systematic errors are eliminated if one considers instead of the absolute yield of muons or electrons, the ratio

$$R = \frac{[\nu_\mu/\nu_e]_{data}}{[\nu_\mu/\nu_e]_{MC}} \ , \tag{4.83}$$

where MC stands for the Monte Carlo simulation of the corresponding detector response, and both charges of muons and electrons are added together. (Detectors typically cannot separate e^- from e^+ and similarly for muons.)

Deviations of the double ratio R from unity therefore signals that the actual ratio of muon to electron atmospheric neutrinos deviates from the expectations. There have been several determinations of R. In the Kamioka detector the data were separated into sub-GeV and multi-GeV bins and, see PDG (1996) ,

$$R = 0.60^{+0.06}_{-0.05} \pm 0.05 \ (\text{sub} - \text{GeV}) \ , \ R = 0.57^{+0.08}_{-0.07} \pm 0.07 \ (\text{multi} - \text{GeV}) \ . \tag{4.84}$$

IMB detector, another water Čerenkov detector, present their data somewhat differently, but also find R of about 0.6, with similar statistical significance. The preliminary SK data are in agreement with the published Kamioka result, and actually already surpass them as far as the statistical significance is concerned (Nakamura (1997)):

$$\begin{aligned} R &= 0.641^{+0.043}_{-0.041} \pm 0.031 \pm 0.050 \ (\text{sub} - \text{GeV}) \ , \\ R &= 0.543^{+0.072}_{-0.063} \pm 0.046 \pm 0.050 \ (\text{multi} - \text{GeV}) \ . \end{aligned} \tag{4.85}$$

The data from Soudan detector, an iron calorimeter with very different sources of systematic errors than the water Čerenkov detectors, give (Peterson (1996)) $R = 0.72 \pm 0.19 \pm 0.06$. However, the data from Frejus detector, a no longer existing iron detector, are in agreement with R being unity, $R = 1.00 \pm 0.15 \pm 0.08$.

The ratio R appears to be quite robust, and insensitive to uncertainties in the calculated neutrino fluxes etc. There does not seem to be any other plausible explanation why $R < 1$, except neutrino oscillations. If interpreted as evidence for neutrino oscillations, it implies, as a best fit, $\Delta m^2 \simeq 10^{-2}\text{eV}^2$ and a large mixing angle. Based on these data alone one cannot decide whether the oscillations are $\nu_\mu \to \nu_e$ or $\nu_\mu \to \nu_\tau$.

In fact, only some hint (so far unconfirmed by the SK data) of the zenith angle (i.e. flight path) dependence of R for the higher energy events in Kamiokande points toward "real oscillations". Just the fact that $R < 1$ might be explained as "averaged" oscillations, allowing only to restrict the mass parameter Δm^2 from *below*.

4.2. *LSND evidence*

The team which built the LSND (Liquid scintillator neutrino detector) at LAMPF has been searching for the *appearance* oscillation $\bar{\nu}_\mu \to \bar{\nu}_e$ for several years now, and found evidence that this oscillation indeed occurs with probability of $0.31 \pm 0.12 \pm 0.05\%$, as described in Athanassopoulos et al. (1996).

The neutrino source are neutrinos from the decay at rest of π^+ and μ^+ in the beam dump at LAMPF. The 800 MeV protons of the LAMPF accelerator are stopped in the beam dump, and form pions (dominantly π^+) there. Since the beam dump is dense, almost all pions are stopped before they decay. The π^- mesons, a minority to start with, do not decay. They form first pionic atoms, and finally undergo nuclear capture, a strong process, without producing any neutrinos. Thus, the decay at rest neutrino source contains monoenergetic ν_μ from the two-body π^+ decay, and continuum ν_e and $\bar{\nu}_\mu$ from the three-body decay of μ^+. The continuum has well known Michel-like shape and extends till $E = m_\mu/2$. There are no electron antineutrinos in this source.

In the LSND electron antineutrinos are recognized by their reaction on protons, $\bar{\nu}_e + p \to e^+ + n$. The detector measures the energy and position of the positron track, and searches for the delayed signal of the neutron capture on protons (2.2 MeV γ ray). The positron and neutron signals must be correlated in space and time, and must have energies in the predetermined intervals. Altogether, the collaboration observed $51 \pm 20 \pm 8$ such events with an estimated background of 12.5 ± 2.9. When the criteria for the valid signal are tighten, there are 22 events left with an estimated background of 4.6 ± 0.6 events. There is almost no chance that the signal can be explained by a statistical fluctuation of the background (provided the background rate is calculated reliably). Even though the neutrino energy can be determined for each event, the statistics is so limited that one cannot make a statement about the possible E_ν/L dependence.

This hint of oscillations has not been confirmed independently. But is not excluded by other evidence, either. The region of possible Δm^2 is restricted from below by about 0.1 eV^2 (reactor neutrino experiments) and from above by about 10 eV^2 (accelerator experiments NOMAD and CCFR). An analogous detector KARMEN at the Rutherford laboratory excludes a large fraction, but not all, of the oscillation parameter space compatible with the LSND signal.

Not all pions in the beam dump decay at rest. A small fraction (about 3%) decay in flight, and produce higher energy ν_μ and a smaller amount of $\bar{\nu}_\mu$. A hint of the $\nu_\mu \to \nu_e$ oscillation signal, stemming from these higher energy decay-in-flight neutrinos has been recently announced by the LSND collaboration.

4.3. *Summary of oscillations evidence*

¿From this and the preceding lecture we see, therefore, that there are three groups of data which suggest that neutrinos are indeed massive and mixed.

(*a*) Solar neutrinos signal ν_e disappearance, and suggest that the relevant Δm^2 is 10^{-5} eV^2 (even though the "just-so" oscillations with $\Delta m^2 \simeq 10^{-10}$ eV^2 are also a possibility.) The evidence itself is indisputable; five experiments confirm the solar neutrino deficit. The energy dependence of the deficit, though, is established only indirectly, through the different thresholds of the working detectors. There are several crucial measurements each of which would settle all doubts: i) Establish that the ^{7}Be neutrinos are indeed heavily suppressed. ii) Establish that the ^{8}B neutrinos reaching Earth have a distorted spectrum. iii) Establish that the neutral current reaction rate is not reduced.

(*b*) Atmospheric neutrinos, stripped to the bare bones of evidence, signal ν_μ disappearance. Unless the zenith angle dependence of the ratio R can be reliably confirmed, the only conclusion one can make is that the relevant $\Delta m^2 \geq 10^{-3}$eV^2. Again, there are several measurements that would remove essentially all doubts. They are: i) Firmly establish the zenith angle dependence. ii) See the oscillations in one of the planned long-baseline experiments. The $\nu_\mu \to \nu_e$ channel can be tested in reactor experiments, two of which (Chooz and Palo Verde) are going to begin collect data this year (Chooz already has some data). The $\nu_\mu \to \nu_\tau$ channel can be studied only with accelerator based neutrino beams. Several large long baseline experiments are planned, but will not begin data taking before the turn of the century.

(c) LSND signal suggests the appearance oscillations $\bar{\nu}_\mu \to \bar{\nu}_e$ and $\nu_\mu \to \nu_e$. The energy dependence has not been established, but by combining with other evidence it is clear that $\Delta m^2 \geq 0.1\text{eV}^2$. In order that this claim of oscillations is generally accepted, it should be confirmed by another, independent experiment.

The three sets of data suggest three different scales of Δm^2. That is impossible to accommodate with three neutrino flavors. Thus, if all these data really signal oscillations, we have to either admit that "sterile" neutrinos exist, or we have to "sacrifice" part of the evidence. (For example, suggestions have been made that the zenith angle dependence of the atmospheric neutrino anomaly is absent. One can use then the same value of Δm^2 for atmospheric neutrinos and LSND.) However, only when some, or all, of the experiments enumerated above are performed, can we construct a set of mass differences and mixing angles which accommodates all the data.

4.4. *Dirac versus Majorana neutrinos*

In the first lecture we discussed the possibility that neutrinos are Majorana particles. Suppose that is indeed so, how can we tell whether a neutrino is a Dirac or Majorana particle? (Again, the material in subsections 4.4 and 4.5 is treated in detail in Kayser, Gibrat-Debu & Perrier (1989) .)

4.4.1. *Majorana neutrino in a uniform EM field*

Can we see the difference by studying the behavior of a particle in the electromagnetic field? Consider a Majorana neutrino ν_M with dipole moments μ_{mag} and μ_{el}. The interaction energy is

$$E_{EM} = -\mu_{mag}\langle \vec{s} \cdot \vec{B} \rangle - \mu_{el}\langle \vec{s} \cdot \vec{E} \rangle \ . \tag{4.86}$$

Under CPT transformation $\vec{s} \to -\vec{s}$, $\vec{B} \to \vec{B}$, $\vec{E} \to \vec{E}$ and ν_M gets just a phase which cancels in the matrix element. Therefore E_{EM} changes sign and in an CPT invariant world the magnetic and electric dipole moment of a Majorana neutrino must vanish. (This is intuitively obvious. For a Dirac particle the change of sign of $\langle \vec{s} \cdot \vec{B} \rangle$ is compensated by the change of sign of μ_{mag}, etc.)

Can we use this property to distinguish Dirac and Majorana neutrinos? Unfortunately, not in practice, since the magnetic moment of a Dirac neutrino is proportional to its mass, and therefore very small. For example, in the minimal extension of the standard electroweak model the magnetic moment is predicted to be

$$\mu_{mag} = \frac{3eG_F}{8\pi^2\sqrt{2}} m_\nu \approx 3 \times 10^{-19} \left(\frac{m_\nu}{1\text{eV}} \right) \mu_{Bohr} \ . \tag{4.87}$$

Such a magnetic moment is unobservably small for light neutrinos. However, models have been constructed that have considerable larger neutrino magnetic moments. The present upper limits are about $10^{-10}\mu_{Bohr}$.

Does it mean that a Majorana neutrino cannot interact with EM field at all? Not really. For *any* fermion the interaction with electromagnetic field has the form dictated by Lorentz invariance and current conservation:

$$\langle \vec{p}_f, \vec{s}_f | J_\mu^{EM} | \vec{p}_i, \vec{s}_i \rangle = i\bar{u}(\vec{p}_f, \vec{s}_f)[F\gamma_\mu + M\sigma_{\mu\nu}q_\nu + G(q^2\gamma_\mu - qq_\mu)\gamma_5 + iE\sigma_{\mu\nu}q_\nu\gamma_5]u(\vec{p}_i, \vec{s}_i) \ . \tag{4.88}$$

Here the form factors $F(vector), M(magnetic), G(anapole)$, and E*(tensor or weak electricity)* depend in general on the square of the four-momentum transfer q, and u and $\bar{u}$ are Dirac bispinors. It is simple to show (we have just done that for M in fact) that for a Majorana neutrino F, M and E must vanish. However, G, which is parity violating "anapole" moment (or form factor) can be nonvanishing.

All that is for one neutrino flavor. If there is neutrino mixing, the interaction with EM field can have nondiagonal terms for Majorana neutrinos, e.g. a magnetic moment operator which connects, in the presence of B field, a left handed neutrino of flavor i with a right handed neutrino of another flavor f. This property is sometimes invoked in the discussion of solar neutrinos and a possible effect of the solar magnetic fields.

4.4.2. *Neutral currents of Majorana neutrinos*

There is a wealth of data on the neutral current neutrino scattering,

$$\nu + A \rightarrow \nu + B \ . \tag{4.89}$$

Can we use them to tell whether neutrinos are Dirac or Majorana particles? The neutral current interaction is governed by $\bar{\nu}\gamma_\mu(1 - \gamma_5)\nu$. Consider first the vector part $\bar{\nu}\gamma_\mu\nu$. Remembering the definition $\nu^c = \eta_C C\bar{\nu}^T$ it is easy to check that $\bar{\nu}^c = -\eta_C \nu^T C^{-1}$. Therefore,

$$\bar{\nu}^c\gamma_\mu\nu^c = -\nu^T C^{-1}\gamma_\mu C\bar{\nu}^T = \nu^T\gamma_\mu^T\bar{\nu}^T = -\bar{\nu}\gamma_\mu\nu \ , \tag{4.90}$$

because the fermion fields anticommute. The conclusion that the vector current of an antifermion is opposite to that of an fermion is true for both Dirac and Majorana particles. However, for a Majorana neutrino we also have an additional condition $(\nu^M)^c = phase \times \nu^M$ and therefore

$$(\bar{\nu}^M)^c\gamma_\mu(\nu^M)^c = \bar{\nu}^M\gamma_\mu\nu^M \ . \tag{4.91}$$

And thus *a Majorana particle has no vector current.*

The only contribution to the neutral current scattering is then through the axial current. We have to evaluate then the matrix element

$$\langle \nu_f^M | \bar{\nu}^M\gamma_\mu\gamma_5\nu^M | \nu_i^M \rangle \ . \tag{4.92}$$

The derivation is relatively simple but a bit lengthy and uses the usual expansion

$$\Psi^M(x) = \sum_{\vec{p},s} \left[\frac{M}{E(p)V}\right]^{1/2} [f(\vec{p},s)u(\vec{p},s)e^{ipx} + \lambda[f^+(\vec{p},s)v(\vec{p},s)e^{-ipx}] \ , \tag{4.93}$$

where u and v are bispinors and λ is a phase factor. Using the commutation relations of the fermion fields, and the properties under charge conjugation, we find

$$\langle \nu_f^M | \bar{\nu}^M\gamma_\mu\gamma_5\nu^M | \nu_i^M \rangle = \bar{u}_f\gamma_\mu\gamma_5 u_i - \bar{v}_i\gamma_\mu\gamma_5 v_f = 2\bar{u}_f\gamma_\mu\gamma_5 u_i \ . \tag{4.94}$$

In contrast for a Dirac neutrino we would get

$$\langle \nu_f^D | \bar{\nu}^D\gamma_\mu(1-\gamma_5)\nu^D | \nu_i^D \rangle = \bar{u}_f\gamma_\mu(1-\gamma_5)u_i \ . \tag{4.95}$$

Can we, based on this, detect a difference between these two cases? Not really, since the spinor u_i as we have mentioned previously has the property that

$$\gamma_5 u_i = -u_i + O(m_\nu/E_\nu) \ , \tag{4.96}$$

and hence the Dirac and Majorana matrix elements of the neutral current differ only in the terms which are of the order of neutrino mass divided by neutrino energy, i.e. very small.

This is, in fact a very general property: *Dirac-Majorana confusion theorem states that in the ultrarelativistic limit there is no difference between the Dirac and Majorana neutrinos. The only feasible experiment which might be able to detect the diffence is the neutrinoless double beta decay which would demonstrate that the lepton number is not conserved.*

4.5. *Neutrinoless double beta decay*

Nuclear masses depend on the number of protons (Z) and neutrons (N) in the nucleus. When two nuclei have the same mass number $A = N + Z$, where N is the number of neutrons and Z is the number of protons, the less bound one (the parent) will decay into the more bound nucleus (the daughter) by either the β^- decay (if a neutron is converted into a proton) or by the β^+ decay or electron capture (if a proton is converted into a neutron).

The nuclear mass $M(Z, A)$ is a quadratic function of Z (approximately a parabola). In addition to this smooth dependence on Z there is a small but important "pairing" term in $M(Z, A)$ which does not affect the parabola when A is odd, but splits it into two shifted parabolas when A is even. The even-even nuclei are more bound than the odd-odd nuclei with the same A. Consequently, for every mass number A there is only one stable nucleus for odd A but there could be two stable even-even nuclei for even A. These two stable nuclei do not have equal masses generally, and consequently by energy conservation alone the following decays are possible

$$(Z, A) \rightarrow (Z + 2, A) + 2e^- + 2\bar{\nu} \; , \tag{4.97}$$

which is called two neutrino double beta decay (2ν for short), or

$$(Z, A) \rightarrow (Z + 2, A) + 2e^- \; , \tag{4.98}$$

which is called for obvious reason neutrinoless double beta decay (0ν for short). The important point is that the transmutation of two bound neutrons into two bound protons proceeds in double beta decay "at once" since the intermediate odd-odd nucleus is less bound than both the parent and daughter. This is not a sequential decay. It is also not possible for free nucleons; this is really a purely nuclear effect.

The 2ν decay does not violate anything. You can think of it as a second order weak process, in which the intermediate state is virtual. Since its rate is proportional to G_F^4, it is very slow. Nevertheless, out of some 70 candidate nuclei, the 2ν decay has been observed in about 9 cases, with a typical lifetime of 10^{20} years. It is the slowest radioactive decay ever observed, but has very little to do with our topic of neutrino mass.

On the other hand the neutrinoless decay obviously violates lepton number conservation. We can think of it as a virtual decay of one neutron, in which the emitted neutrino (antineutrino really) is reabsorbed by another neutron, following the so-called "Racah sequence"

$$\begin{aligned} n_1 &\rightarrow p_1 + e^- + \text{“}\nu\text{”} \; , \\ \text{“}\nu\text{”} + n_2 &\rightarrow p_2 + e^- \; . \end{aligned} \tag{4.99}$$

This sequence can occur only if $\bar{\nu}_e$ (i.e. the particle emitted in the first step) and ν_e (i.e. the particle absorbed in the second step) are identical. In other words, the sequence can occur only if the virtual neutrino is a Majorana particle.

But even for a Majorana neutrino the process is suppressed since the particle emitted in the first step is dominantly right-handed, and the particle absorbed in the second step is dominantly left-handed. For massless neutrinos (remember the confusion theorem) the helicity mismatch is absolute and the process does not go at all. However, if the virtual neutrino has mass m_ν, in each vertex there is a "wrong helicity" amplitude proportional to m_ν/E_ν, and the decay rate will be proportional to $(m_\nu/E_\nu)^2$. This is a small quantity since we can crudely estimate the virtual neutrino energy by noting that the neutrino has to be confined in the volume of radius $R_{nucl} \approx$ a few fermis, and its momentum (or energy) is consequently $E_\nu \approx \hbar c/R_{nucl}$ which is something like 50 MeV for a medium size nucleus. We can hope to observe it only because experimentally we are going to use macroscopic amounts of the parent nuclei, and each mol contains $N_A = 6 \times 10^{23}$ nuclei.

4.5.1. *Decay amplitude, purely left-handed currents*

Let us forget the nuclear physics of the problem and just consider the process $W^*W^* \rightarrow e^-e^-$. (The discussion here follows the procedures explained in detail in Kayser, Gibrat-Debu & Perrier (1989) .) The hamiltonian of the charged current interaction will contain (remember the oscillations lecture)

$$W^\mu \bar{e}\gamma_\mu(1-\gamma_5)\nu_e = W^\mu \sum_m \bar{e}\gamma_\mu(1-\gamma_5)U_{e,j}\nu_j \ . \tag{4.100}$$

For the double beta decay this means that we need to evaluate the expression

$$[W^\mu \bar{e}\gamma_\mu(1-\gamma_5)U_{e,m}\nu_m][W^\kappa \bar{e}\gamma_\kappa(1-\gamma_5)U_{e,m}\nu_m] \ . \tag{4.101}$$

Now we would like to contract the neutrino fields to form the usual neutrino propagator. But we cannot do it simply since the expression above does not contain $\bar{\nu}_m$. But there is a trick which uses the identity

$$\bar{e}\gamma_\mu(1-\gamma_5)\nu_m = -\bar{\nu}_m^c\gamma_\mu(1+\gamma_5)e^c \ . \tag{4.102}$$

(Here we used few steps including the anticommutation of two fermion fields.)

Since we assume that the neutrino ν_m is a Majorana field, it is the eigenstate of charge conjugation, i.e.,

$$\nu_m^c = \lambda_m^* \nu_m \ , \tag{4.103}$$

where λ_m is the phase factor. Thus we finally conclude that

$$\bar{e}\gamma_\mu(1-\gamma_5)\nu_m = -\lambda_m\bar{\nu}_m\gamma_\mu(1+\gamma_5)e^c \ . \tag{4.104}$$

So, now we have the neutrino fields in the form where we can make the contraction of $\nu_m\bar{\nu}_m$ which gives the usual fermion propagator

$$\frac{-iq_\mu\gamma^\mu + m_m}{q^2+m_m^2} \ . \tag{4.105}$$

We can certainly safely neglect the m^2 in the denominator. Further, the q part of the propagator will not contribute because $(1-\gamma_5)(1+\gamma_5) = 0$.

So, finally the rate of the 0ν double β decay is given by

$$\frac{1}{\tau} = \left[\sum_m \lambda_m m_m U_{e,m}^2\right]^2 \times \text{[neutrino mass independent factors]} \ . \tag{4.106}$$

It might seem surprising that the decay rate contains the seemingly arbitrary phase factor λ_m. This actually does make sense since we had to invoke the transformation property of the field ν_m under charge conjugation. (Actually, really we need the transformation property under CP, but let us leave it as it is.) More deeply and also more formally, the appearance of the phase is related with our choice that the mixing matrix $U_{e,m}$ is real.

In practice this means that the 0ν decay rate depends on a combination of neutrino masses, so called "effective" mass

$$\langle m_{eff} \rangle = \sum_m \lambda_m |U_{e,m}|^2 m_m \ . \tag{4.107}$$

Due to the phases $\lambda_m = \pm 1$ in the sum $\langle m_{eff} \rangle$ can be smaller than *all* m_m. Thus, the fact that from experimental searches we can conclude that $\langle m_{eff} \rangle$ is less than about 1 eV (see below), we cannot conclude that the actual masses are that small. They can be each arbitrarily large and just almost precisely cancel each other in $\langle m_{eff} \rangle$. On the other hand, naturally, they can be also arbitrarily small, or they may vanish. However, once somebody is lucky enough to actually observe the 0ν double beta decay we would be able to conclude that at least one of the neutrinos is a massive Majorana particle.

4.5.2. *The case of right-handed currents*

For completeness we should also consider the 0ν double beta decay mediated by the right-handed current weak interactions. Let us assume, therefore, that in addition to the known W bosons that couple to the left-handed charged current, there are other vector bosons (let us call them W_R) that couple to leptons via right-handed charged currents. To avoid conflict with experimental evidence these W_R bosons would have to be heavier than the familiar one. What effect would this have on the 0ν double beta decay?

At first blush the decay will simply go, provided the exchanged particle is a Majorana neutrino. We just need a left-handed coupling in one vertex and a right-handed coupling in the other one. In that way we could overcome the helicity suppression, and the fact that the 0ν decay has not been observed so far could be used as a very powerful constraint on the properties of these hypothetical W_R bosons. In addition, if the neutrinoless decay is observed, one would not be able to conclude that neutrinos (at least some) have to be massive. But things are not so simple.

The amplitude corresponding to the process we are considering follows from the propagator (4.105) and is simply

$$\sum_m U_{e,m}^{(L)} \frac{q}{q^2 + m_m^2} U_{e,m}^{(R)} \ , \tag{4.108}$$

where now the unitary matrices $U_{e,m}^{(R)}$ and $U_{e,m}^{(L)}$ are analogous mixing matrices to the $U_{e,m}$ mixing matrix used before, but for the weak eigenstate neutrinos that couple to the right-handed current and left-handed current, respectively. (Until now we considered just the left-handed current, so the superscript L was not needed.) These two matrices are parts of an $2n \times 2n$ (n is the number of neutrino flavors, in our world $n = 3$) unitary matrix which, as before, transforms the mass eigenstates into the weak eigenstates. Since the $2n \times 2n$ matrix is unitary, it must obey the condition

$$\sum_m U_{e,m}^{(L)} U_{e,m}^{(R)} = 0 \ . \tag{4.109}$$

But in that case the double beta decay will not go if all neutrinos are massless since in the above amplitude the q dependent part is then independent of the index m and it vanishes by the unitarity condition. Hence, we have a theorem: *In gauge theories and assuming that the 0ν decay is caused by neutrino exchange, its observation would imply the existence of a massive Majorana neutrino, whether there are right-handed current interactions or not.*

4.5.3. *Search for the 0ν decay*

There have been numerous attempts to observe the 0ν double beta decay and thus show that neutrinos are massive Majorana particles. So far, alas, no evidence have been found.

The principle of the experiment is quite straightforward. One takes certain (as large as possible) number of the parent atoms, and places them into a detector which can detect electrons, and determine their energy. (The best strategy is to use the same atoms as $\beta\beta$ decay source and as detector. That is possible in the so far best studied cases of ^{76}Ge and ^{136}Xe.) In the 0ν decay the two electrons carry all the decay energy. Their sum spectrum is then, ideally, a delta function at the Q value of the decay. In contrast, in the 2ν decay four light particles share the energy, and the sum electron spectrum is continuous, and vanishes at the decay Q value.

However, one tries to observe a very small decay rates, and use as much source nuclei as possible. Thus, the whole problem is reduced to the problem of background reduction. Again, as before, the ingenious experimental techniques are beyond the scope of these lectures (see, however, references in the review by Moe & Vogel 1994).

Then there is still another problem. The decay rate depends not only on the unknown neutrino effective mass $\langle m_{eff} \rangle$, but also on the so-called nuclear matrix elements. These are, in principle, calculable, but in practice they are rather uncertain, by a factor of 2-3. Thus, the resulting limit on $\langle m_{eff} \rangle$ has a similar uncertainty.

It the present time the most stringent limit comes from the Heidelberg-Moscow $\beta\beta$ experiment at Gran-Sasso (Günther et al. (1997)). In it, 11 kg of active mass of ^{76}Ge forms the source-detector system. The published (since surely improved) halflife limit is

$$T_{1/2} > 7.4 \times 10^{24} \text{years } (90\% CL) \ . \tag{4.110}$$

This corresponds to $\langle m_{eff} \rangle < 0.5 - 1.5$ eV, depending on which nuclear matrix element is used. That limit is, as you might have noticed, considerably better than the neutrino mass limit from the tritium beta decay. It excludes, or places serious constraints, therefore, on any scheme in which neutrinos have nearly degenerate masses, and are contributing significantly to the dark matter.

5. Conclusions

The central theme of these lectures was the concept of the neutrino mass. I tried to concentrate on the theoretical aspects of the problem, mentioning experimental results only briefly and not explaining very carefully how they were obtained.

At present there is no clear cut and generally accepted evidence that neutrinos are massive and mixed. But there are several hints, some of them rather longstanding and seen by several experiments like the solar neutrinos. Others are seen by some experiments, but not seen by others, like the atmospheric neutrino anomaly. Yet others, like the recent LSND claim, are seen in only one experiment, and not confirmed by others.

In the first lecture I argued that since neutrino mass is not forbidden, it is likely to be nonvanishing. We do not know what to expect, but a hierarchy of masses, as suggested by the see-saw mechanism, is the most popular assumption. From cosmology there is a hint (at least for an optimist) that the heaviest mass is somewhere near 10 eV - a range accessible to experiments.

In the second lecture I briefly reviewed the methods used to determine neutrino masses, and showed that only for the electron neutrinos we are anywhere near the interesting range. I then introduced the concept of oscillations, which albeit with an additional assumption of neutrino mixing, extends the range of masses (or mass differences) by many orders of magnitude.

In the third lecture I concentrated on the neutrinos from the Sun. It offers the most widely accepted hint of neutrino oscillation. It also spawned the idea of resonant oscillations, a nice theoretical concept.

Finally in the fourth lecture I reviewed the available other evidence for neutrino mass and mixing and I spoke about the attempts to see whether neutrinos could be Majorana particles. I argued that the neutrinoless double beta decay is the only realistic hope of seeing the difference between the Dirac and Majorana particles.

Altogether, this is becoming an industry. I hope that soon one of the hints will turn into an established fact.

Large part of the material in these lectures follows the lines first established in Boehm & Vogel (1992) . The many years of collaboration with Professor Felix Boehm on these and other aspects of neutrino physics are gratefully acknowledged. The work was supported in part by the U.S. Department of Energy under the Grant DE-FG03-88ER40397. Financial support of Academia Mexicanas de Ciencias and of the United States - Mexico Foundation for Sciences is gratefully acknowledged.

REFERENCES

ATHANASSOPOULOS, C. ET AL. 1996 *Phys. Rev. C* **54**, 2685.

BAHCALL, J. N. 1989 *Neutrino Astrophysics.* Cambridge.

BETHE, H. A. 1990 *Rev. Mod. Phys.* **62**, 801.

BOEHM, F. & VOGEL, P. 1992 *Physics of Massive Neutrinos, 2nd ed.*. Cambridge.

GÜNTHER, M. ET AL. 1997 *Phys. Rev. D* **55**, 54-67.

KAYSER, B., GIBRAT-DEBU, F. & PERRIER, F. 1989 *The Physics of Massive Neutrinos.* World Scientific.

KIM, C. W. & PEVSNER, A. 1993 *Neutrinos.* Hardwood.

KLAPDOR-KLEINGROTHAUS, H. V. & STAUDT, A. 1995 *Teilchenphysik ohne Beschleuniger* . Teubner.

KOLB, E. W. & TURNER, M. S. 1990 *The Early Universe*, Addison-Wesley.

LANGANKE, K., VOGEL, P. & KOLBE E. 1996 *Phys. Rev. Lett.* **76**, 2629.

MOE, M. & VOGEL, P. 1994 *Annu. Rev. Nucl. Part. Sci.* **44**, 247-283.

MOHAPATRA, R. & PAL, P. B. 1991 *Massive Neutrinos in Physics and Astrophysics* . World Scientific.

NAKAMURA, K. 1997 *talk at WIN97 conference, to be published in Proceedings.* North Holland.

PETERSON, E. A. 1996 *in Neutrino'96, p. 223.* World Scientific.

SAKURAI, J. J. 1964 *Invariance Principles and Elementary Particles* . Princeton.

PARTICLE DATA GROUP (PDG) 1996 Review of Particle Physics. *Phys. Rev. D* **54**, 1–720.

WINTER, K. (EDITOR) 1991 *Neutrino Physics* . Cambridge.

Cosmic Ray Physics and Astrophysics

By THOMAS K. GAISSER†

Bartol Research Institute, University of Delaware, Newark, DE 19716, USA

This chapter is a review of the background and status of several current problems of interest concerning cosmic rays of very high energy and related signals of photons and neutrinos.

---

1. Introduction

The steeply falling spectrum of cosmic rays extends over many orders of magnitude with only three notable features:

(*a*) The flattened portion below 10 GeV that varies in inverse correlation with solar activity,

(*b*) The "knee" of the spectrum between 10^{15} and 10^{16} eV, and

(*c*) the "ankle" around 10^{19} eV.

For my discussion here I will divide the spectrum into three energy regions that are related to the two high-energy features, the knee and the ankle: I: $E < 10^{14}$ eV, II: $10^{14} < E < 10^{18}$ eV and III: $> 10^{18}$ eV.

In *Region I* (VHE) there are detailed measurements of primary cosmic rays made from detectors carried in balloons and on spacecraft. These observations, and related theoretical work on space plasma physics, form the basis of what might be called the standard model of origin of cosmic rays. Cosmic rays are accelerated by the first order Fermi mechanism at strong shocks driven by supernova remnants (SNR) in the disk of the galaxy. The ionized, accelerated nuclei then diffuse in the turbulent, magnetized plasma of the interstellar medium, eventually escaping into intergalactic space at a rate that depends on their energy. The steady state spectrum that we observe thus reflects a balance between the rate at which particles are accelerated and the rate at which they escape from the galaxy.

Region II (UHE) is the knee region, characterized by a steepening of the spectrum. It is expected that the spectrum in this region should reflect the fact that some accelerators that contribute to the pool of cosmic rays at lower energy have reached their upper limits. Such behavior can also reflect a change in the energy-dependence of diffusion.

The flattening of the spectrum in *Region III* (EHE) may reflect the onset of an extra-galactic component of the cosmic radiation.

Much progress has been made toward understanding the origin of cosmic rays, but there are several problems with our present understanding that are the focus of current research. In this chapter I would like to give some background to the nature of these problems and to current ideas about their solution. The problems include:

Propagation. The most commonly used "leaky box" model of cosmic-ray propagation is inconsistent with the observed isotropy of cosmic rays through the UHE region.

Cosmic accelerators in the Galaxy. Exactly which kinds of SNR accelerate cosmic rays? If all do, why have we not yet seen TeV γ-rays from specific nearby, young SNR produced by collisions of cosmic-rays with gas near the supernova remnant? Is

† I am grateful to the organizing committee for the opportunity to participate in this school and to the Academia Mexicana de Ciencias and to The United States-México Foundation for Science for support that made it possible. The research on which this chapter is based is supported in part by the U.S. Department of Energy.

there a correlation between maximum energy and the type of the SNR? Are there other astrophysical sources in the Galaxy capable of accelerating particles to $\gg 10^{14}$ eV?

Knee of the spectrum. What is the relative composition of the nuclei in the cosmic radiation in this region and how does it reflect the nature of the accelerators and/or the process of diffusion and propagation in the galaxy?

Highest energy cosmic rays. Just on the basis of their large gyroradius in galactic magnetic fields as compared to the size of the galaxy, the highest energy cosmic rays are most likely from extra-galactic sources. What are these sources and how can they accelerate particles to such extremely high energies, $> 10^{20}$ eV? Note also that the interpretation of the observed showers depends on an extrapolation of the physics of hadronic interactions several orders of magnitude beyond the range explored with accelerators on earth.

High energy neutrino astronomy and AGN. The very existence of the highest energy (extragalactic) cosmic radiation suggests that there may also be a diffuse background of $\gg$ TeV neutrinos at a level high enough to be detected with a large ($\sim$ kilometer scale) detector. What, if any, is the relation between the highest energy cosmic rays and observations of >TeV γ-rays from active galactic nuclei (AGN)? Are there other models of origin of EHE (extragalactic) cosmic rays that would also predict detectable levels of neutrinos, and what would their signature be?

2. Standard model of galactic cosmic rays

The processes that lead to the observed spectrum of galactic cosmic rays are illustrated in Fig. 1. Generally, the relation between the power, Q (erg s^{-1}cm^{-3}) of the sources needed to maintain the observed energy density, ρ_E in cosmic rays is

$$Q \sim \rho_E / \tau_e, \tag{2.1}$$

where τ is the typical time that it takes a particle to escape from the galaxy. The median energy of all cosmic rays is about 3 GeV/nucleon. Since particles of this energy typically spend $\sim 10^7$ years in the galaxy, the power required to maintain the supply of cosmic radiation is 5×10^{40} erg/s. Ginzburg & Syrovatskii (1964) emphasized that this is a few percent of the likely power output of supernova explosions. In the 70's the theory of diffusive shock acceleration was developed and shown naturally to give an E^{-2} differential spectrum (Axford, Leer & Skadron, 1977, Bell, 1978, Krymsky, 1977, Blandford & Ostriker, 1978). The relation between this source spectrum and the observed differential spectrum depends on propagation in the galaxy.

2.1. *Propagation and energetics*

The abundance of the various elements in the cosmic radiation are similar to the general abundance of matter in the solar system, but there are several important differences.† The observation that is most relevant for cosmic ray propagation is the fact that the nuclei Li, Be and B, which are virtually absent as the end products of stellar nucleosynthesis, are orders of magnitude more abundant in the cosmic radiation than in the general abundance of matter. The reason is that they are produced mainly in spallation of cosmic-ray carbon and oxygen as these abundant nuclei propagate through the interstellar medium (ISM).

† For a review of all aspects of elemental and isotopic composition of the cosmic radiation see the book edited by Waddington (1988). A recent new development is the work of Meyer, Drury & Ellison, 1997.

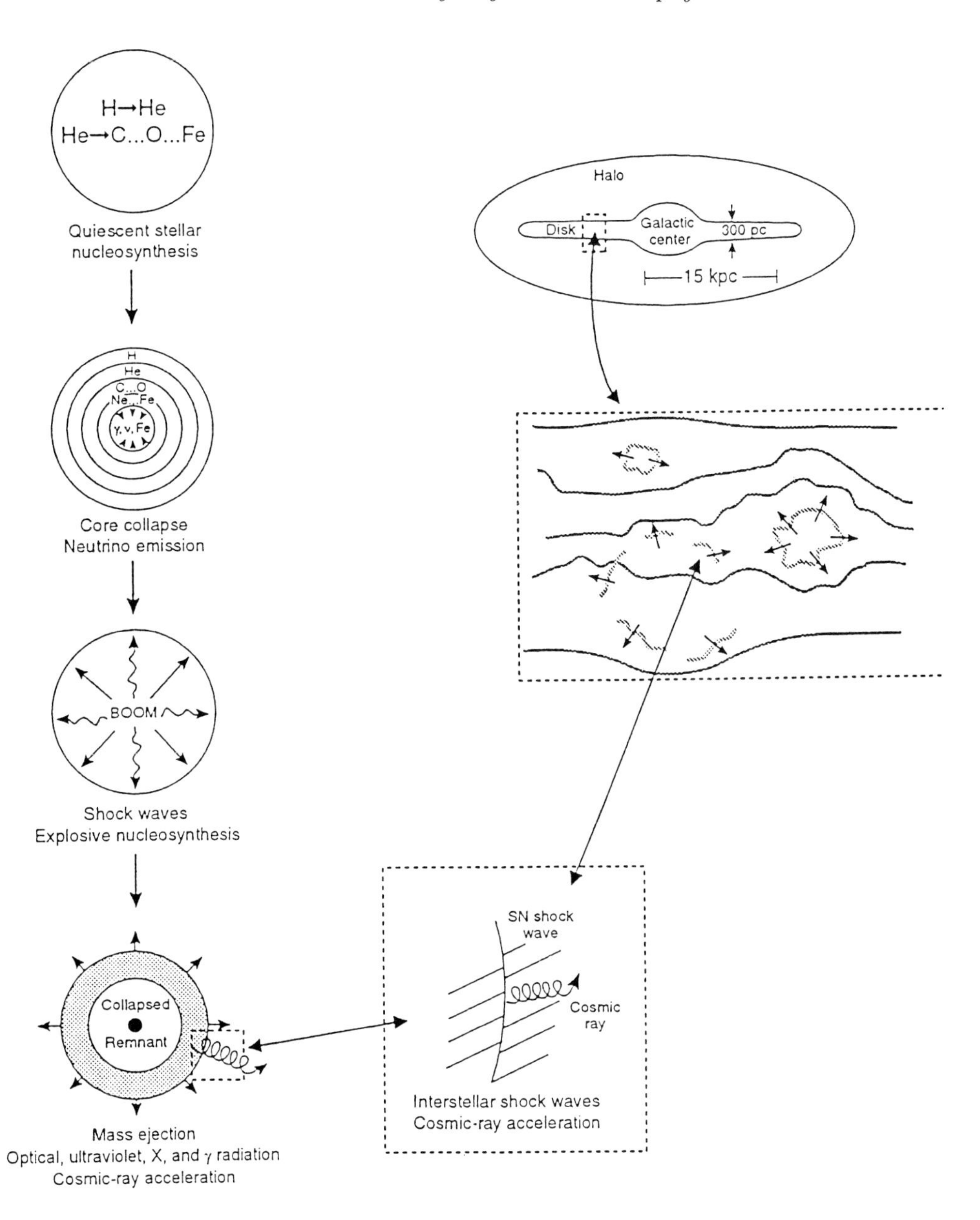

FIGURE 1. Illustration of processes that lead to acceleration of cosmic rays while energizing the interstellar medium. Left side refers to stellar nucleosynthesis leading to core-collapse supernovae in massive stars; right side illustrates dynamical processes in the ISM. From NRC Report, 1995.

The basic equation of cosmic-ray propagation is

$$\dot{n}_i = q_i - n_i \left(\frac{1}{\tau_e} + \frac{1}{\tau_i} + \frac{1}{\gamma \tau_{D,i}} \right) + \sum_{j>i} B_{j\to i} \frac{n_j}{\tau_j}. \qquad (2.2)$$

In equilibrium the density of cosmic rays of type i is constant, so $\dot{n}_i = 0$. The source term, q_i, as well as the cosmic ray densities, n_i, both depend on energy and on position in the galaxy. For purposes of illustration we can neglect the spatial dependence and consider a uniform distribution of cosmic rays in the galaxy. We also assume the sources are distributed uniformly in the disk of the galaxy so that q_i is the number of particles of type i produced in the cosmic ray sources per unit volume per GeV per unit time. The collision rate of nucleus j with gas in the interstellar medium is $1/\tau_j = n_H c\sigma$, where n_H is the density of gas in the interstellar medium, c is the cosmic ray speed and σ the total cross section. The branching ratio for a nucleus j to produce a fragment i is $B_{j\to i}$. To a good approximation the energy per nucleon is conserved during spallation. The rates for loss from the propagation region and for decay are respectively $1/\tau_e$ and $1/\tau_D$.

For a stable secondary nucleus such as ^{11}B the decay term in Eq. 2.2 is absent. Since boron is a secondary nucleus its source term is also absent, and the equation becomes, after some rearrangement,

$$n_{11}\left(1+\frac{1}{\tau_e n_H c\sigma_{11}}\right) = \frac{1}{\sigma_{11}}\sum_{A>11} B_{A\to 11} n_A \sigma_A. \tag{2.3}$$

Using cross sections and branching ratios measured in the lab then allows one to calculate the product $\tau_e n_H$ from the measured densities n_{11} and $n_{A>11}$.† The result is that the characteristic containment time decreases with energy. A fit (Webber *et al.*, 1996) to recent measurements of the ratio of secondary to primary nuclei gives

$$n_H(cm^3)\,\tau_e \sim 10^7\,\text{yr}\times\left(\frac{R}{4.7\,\text{GV}}\right)^{-\delta}, \quad R > 4.7\,\text{GV}, \tag{2.4}$$

with $\delta \approx 0.6$. R in Eq. 2.4 is the magnetic rigidity, defined as

$$R = \frac{P\,c}{Z\,e}, \tag{2.5}$$

where P is the total momentum of a particle of charge $Z\,e$. Thus for a proton of total energy $E \gg m_p$, $R \approx E$, whereas for a high energy nucleus, the rigidity is approximately twice the energy-per-nucleon (since $A \approx 2\times Z$).

The characteristic time $\sim 10^7$ yr from Eq. 2.4 is much larger than the light travel time across the disk of the galaxy. Moreover, the arrival directions of the cosmic rays are observed to be highly isotropic. We conclude, therefore, that propagation of cosmic rays is a diffusive process and that higher energy particles escape from the system more easily than particles of lower energy. Evidence from radioactive decay of ^{10}Be (Garcia-Munoz, Mason & Simpson, 1977) suggests that the propagation volume is larger than the galactic disk and includes also a galactic halo, as suggested by the drawing in Fig. 1. Ginzburg, Khazan & Ptuskin, (1980) describe a simple, instructive disk-halo model of propagation.

Protons are the most abundant component of the cosmic radiation. Although some protons are products of spallation of nuclei, the fraction of secondary protons is small and it is useful to look at Eq. 2.2 neglecting this contribution. Since they are stable, the approximate steady state solution of Eq. 2.2 for protons is

$$q_p(E) = n_p(E)\left(\frac{1}{\tau_e(E)}+\frac{1}{\tau_i}\right) = \frac{n_p(E)}{\tau_e(E)}\left(1+\frac{\tau_e}{\tau_i}\right). \tag{2.6}$$

The inelastic pp cross section is $\sigma_{pp} \approx 30$ mb, so the interaction rate of relativistic protons

† In the full calculation, Garcia-Munoz *et al.* (1987) solve the whole system of equations simultaneously including effects of ionization energy loss in the ISM as well as the correct elemental composition of the interstellar gas, which is the target for the spallation process. My discussion here is only approximate and applies in the multi-GeV range and above.

is

$$\frac{1}{\tau_i} = n_H c \sigma_{pp} \approx \frac{n_H}{(4 \times 10^7 \,\mathrm{yr})},$$

significantly slower than the loss rate inferred from Eq. 2.4. In other words, $\tau_e < \tau_i$ and most protons escape from the galaxy before they interact. Thus Eq. 2.6 for high energy protons simplifies to

$$q_p(E) \approx \frac{n_p(E)}{\tau_e(E)} \sim E^{-(\gamma+1-\delta)}. \tag{2.7}$$

This equation implies a relation between the source spectrum (integral spectral index γ_s) and the observed spectrum (integral spectral index γ) valid at high energy given by

$$\gamma_s = \gamma - \delta. \tag{2.8}$$

In the opposite limit, when $\tau_e > \tau_i$, the relation between the source spectrum and the observed spectrum obtained from Eq. 2.6 would be

$$q_p(E) \approx \frac{n_p(E)}{\tau_i},$$

and the observed spectrum would have the same spectral index as the source (since the interaction cross section which determines τ_i changes relatively slowly with energy). This is the situation for low energy iron. Since $\sigma_{\mathrm{Fe,H}} \approx 725$ mb, the characteristic interaction time for iron is $\tau_{\mathrm{Fe}} n_H \approx 1.5 \times 10^6$ yr cm^{-3}. Thus $\tau_e > \tau_{\mathrm{Fe}}$ for energy less than ~ 40 GeV/nucleon; that is, low energy iron nuclei typically interact on with gas before they escape from the galaxy. As a consequence, if protons and iron both have the same high energy spectrum at the source then we expect the spectrum of cosmic-ray iron to be harder than that of protons in the region of 10 GeV/nucleon, gradually steepening to the same spectrum at high energy when $\tau_e \ll \tau_{\mathrm{Fe}}$. There is some (weak) experimental evidence for this behavior Esposito *et al.*, 1990.

Now that we have an estimate of the rate at which the cosmic rays must be replenished in order to maintain their observed steady state intensity, we can estimate the power required of the sources as

$$P = \int d\mathcal{V} \int Eq(E)dE = \int d\mathcal{V} \int E\frac{n_{CR}(E)}{\tau_e(E)}dE. \tag{2.9}$$

The observed cosmic-ray intensity (including nucleons bound in nuclei) is

$$\frac{c}{4\pi} n_{CR}(E) = \frac{dN}{dE} \approx 1.8\, E^{-2.7} \frac{\mathrm{nucleons}}{\mathrm{cm^2 s\, sr\, GeV}}. \tag{2.10}$$

The corresponding energy density in cosmic rays is

$$\frac{4\pi}{c} \int E\frac{dN}{dE}dE \approx 1.7 \times 10^{-12} \frac{\mathrm{erg}}{\mathrm{cm^3}} \text{ or } \approx 1\frac{\mathrm{eV}}{\mathrm{cm^3}}, \tag{2.11}$$

where the factor $4\pi/c$ converts from intensity to density.

We use Eq. 2.4 to evaluate Eq. 2.9 as

$$P = 1.5 \times 10^{-26} n_H \mathcal{V}_{\mathrm{gal}}, \tag{2.12}$$

where

$$n_H \mathcal{V}_{\mathrm{gal}} \sim 6 \times 10^{66}$$

is the mass of diffuse gas in the galaxy.† The resulting power requirement, $P \sim 10^{41}$ erg/s,

† For example, assume $n_H \sim 1$ in a disk of radius 15 kpc and thickness 300 pc.

can be compared with a rate of injection of kinetic energy of supernova remnants of

$$10^{51}\,\text{erg} \times 3\,\text{SN/century} \;\sim\; 10^{42}\,\text{erg/s}.$$

Thus the power available in kinetic energy of supernova ejecta is well-matched to what is required to maintain the observed cosmic-ray intensity, and an efficiency of $\sim 10\%$ is indicated for the acceleration process. In addition, the source spectrum at high energy, $q \propto E^{-(2+\epsilon)}$, is consistent with what is expected for 1st order Fermi acceleration at a strong shock of the kind produced by a supernova blast wave, as discussed in §2.4 below.

2.2. *Dynamic interstellar medium*

The energy density in cosmic rays is larger than but comparable to the pressure of the diffuse ISM,

$$P_{\text{gas}} \;=\; \langle n_H\, k\, T\rangle\; \frac{1}{2}\rho\langle u^2\rangle \;\sim\; 0.2 \times 10^{-12},$$

and also to the energy density in the galactic magnetic field,

$$P_B \;=\; \frac{B^2}{8\pi} \;\approx\; 0.35 \times 10^{-12},$$

both in c.g.s. units. Parker (1966) developed a model of the diffuse galaxy in which these three components are coupled to each other. The turbulent, partially ionized ISM carries the magnetic field which in turn scatters the high energy charged particles by resonant scattering with those Fourier components of the field with frequencies that match the gyroradius of the particles. This collisionless "scattering" contributes to the processes of cosmic-ray diffusion. The coupling among the three components implies that there should be a correlation among density of cosmic rays, density of thermal plasma and magnetic field strength.

A way to check this model is to use it to estimate the scale-height of the galactic disk. Consider the disk as a plane with a mass per unit area (mostly in stars) of $\sigma \approx 1.5 \times 10^{-2}$ g/cm^2.† The gravitational acceleration above the plane is then

$$g \;=\; \frac{1}{2}\pi\, G\, \sigma \;\approx\; 1.6 \times 10^{-9}\;\text{cm/s}^2. \tag{2.13}$$

The total pressure in the ionized interstellar medium is

$$\mathcal{P} \;=\; P_{\text{gas}} \,+\, P_{\text{B}} \,+\, P_{\text{CR}} \;\equiv\; \frac{1}{2}\rho\langle u^2\rangle\,(1+\alpha+\beta). \tag{2.14}$$

The form of Eq. 2.14 follows from the assumed coupling among the three components. By considering the difference in the gravitational force per unit area at the top and bottom of a small element of volume, we find

$$\frac{d\mathcal{P}}{dz} \;=\; -g\rho(z) \;=\; \frac{1}{2}\langle u^2\rangle\,(1+\alpha+\beta)\,\frac{d\rho(z)}{dz}, \tag{2.15}$$

where the last equality follows from Eq. 2.14. The solution of Eq. 2.15 is

$$\rho(z) \;=\; \rho_0 \exp\left[-z/h\right], \tag{2.16}$$

with the scale height

$$h \;=\; \frac{1}{2}\langle u^2\rangle\,(1+\alpha+\beta)/g \;\approx\; 300\;pc. \tag{2.17}$$

The numerical value follows from Eq. 2.14 with $\mathcal{P} = 2.2 \times 10^{-12}$ and a gas density of 1 hydrogen atom per cm^3, and it is consistent with the usual estimate of 300 pc for the

† This estimate takes a mass of the galaxy as $10^{11}\ M_\odot$, half of which is distributed in the disk, with the remainder in the central bulge.

thickness of the disk of the galaxy. Parker's model also provides a basis for understanding the origin of a galactic halo as a consequence of an instability in the magnetized gas. Injecting a large quantity of cosmic radiation in a local region drives a bubble of magnetic field and cosmic rays out of the disk of the galaxy.‡

2.3. *Diffuse γ-radiation from the galactic disk*

Another feature of cosmic-ray propagation in the galaxy is the γ-radiation produced when cosmic rays collide with interstellar gas. The EGRET map of the gamma-ray sky with point sources subtracted clearly shows this radiation (Hunter *et al.*, 1997). The spectrum consists of three components:

(*a*) $p + H \to \pi^0 + X$, followed by $\pi^0 \to 2\,\gamma$;
(*b*) $e + H \to e + \gamma + X$ (bremsstrahlung) and
(*c*) $e + \gamma_{2.7} \to \gamma + X$ (Inverse Compton scattering).

The proton/electron ratio of cosmic rays is high enough so that the dominant source of photons at high energy is production of π^0 even though Inverse Compton scattering (IC) has a harder spectrum.

Relativistic pions and electrons with power-law spectra both produce secondary spectra with the same power-law shape as the parent spectrum for photon energies much greater than the rest mass of the respective parent particles. However, the spectrum of photons from decay of neutral pions is characterized by a peak at 70 MeV, half the pion rest mass (Stecker, 1971). This kinematic feature is independent of the shape of the spectrum of decaying pions. Thus bremsstrahlung dominates at low energy, and the π^0 peak is reduced to a shoulder in the gamma-ray spectrum that is sometimes difficult to discern.

A detailed calculation of the diffuse flux of photons for cosmic ray interaction with the interstellar gas is described by Hunter *et al.*, 1997. It includes all of the contributions mentioned above and is based on a model of the galaxy which includes expected correlations between density of interstellar gas and cosmic rays. Here I give a much simplified estimate (Gaisser, 1990) of just the π^0 contribution, which is dominant above several hundred MeV. It illustrates some basic features of the calculation of secondary spectra that also occur in estimates of high energy neutrinos and photons from astrophysical point sources, as well as in calculations of muons and neutrinos in the atmosphere.

The production spectrum of neutral pions (π^0 per cm^3 per s) is

$$\frac{dN_\pi}{dE_\pi} = c\, n_H \sigma_{pp} \int_{E_\pi}^{\infty} \frac{dn_\pi(E_\pi, E_p)}{dE_\pi} \times \frac{4\pi}{c}\frac{dN}{dE_p}\, dE_p, \tag{2.18}$$

where the spectrum of nucleons is given by Eq. 2.10.

To evaluate Eq. 2.18, it is useful to take advantage of an approximate scaling of the production spectrum of pions, which is applicable at energies well above the rest masses of the particles involved:

$$E_\pi \frac{dn_\pi(E_\pi, E_p)}{dE_\pi} \approx x\frac{dn_\pi}{dx} \sim F_{p\pi}(x), \tag{2.19}$$

with $x \equiv E_\pi / E_p$. Changing the variable of integration in 2.18 gives

$$\frac{dN_\pi}{dE_\pi} = 4\,\pi\, n_H \sigma_{pp} \left\{ \frac{1.8}{E_\pi^{\gamma+1}} \right\} \int_0^1 F_{p\pi}(x)\, x^{(\gamma-1)}\, dx, \tag{2.20}$$

with $\gamma \approx 1.7$ under the assumption that the cosmic-ray spectrum observed locally is typical of the rest of the galaxy. Note that the factor in curly braces in Eq. 2.20 is

‡ For a discussion of this and many other aspects of the astrophysical aspects of cosmic rays, see the book by Longair (1992).

the primary spectrum of cosmic ray nucleons *evaluated, however, at the energy of the secondary pion.* The integral in 2.20 is simply a number which depends on the spectral index and which can be obtained from accelerator measurements. For production of neutral pions by protons on hydrogen, the spectrum-weighted moment is

$$Z_{p\pi^0} \equiv \int_0^1 F_{p\pi^0}(x)\, x^{(\gamma-1)}\, dx \approx 0.04 \tag{2.21}$$

for $\gamma = 1.7$. Finally, to obtain the production spectrum of photons, the production spectrum of neutral pions must be convolved with the distribution of photons from the process $\pi^0 \to 2\gamma$,

$$\frac{dN_\gamma(E_\gamma, E_\pi)}{dE_\gamma} \approx \frac{2}{E_\pi}$$

at high energy. The convolution of this distribution with Eq. 2.20 gives the production spectrum of photons as

$$q_\gamma = \frac{dN_\gamma}{dE_\gamma} = 4\,\pi\, n_H \sigma_{pp} \left\{ \frac{1.8}{E_\gamma^{\gamma+1}} \right\} Z_{p\pi^0} Z_{\pi^0\gamma}. \tag{2.22}$$

for spectral index $\gamma \approx 1.7$, $Z_{\pi^0\gamma} \approx \frac{2}{\gamma+1} \approx 0.74$.

The observed flux of photons per unit detector area from a given direction in galactic latitude, b and longitude ℓ is given by the integral over the line of site in that direction as

$$\int_0^{Rmax} \frac{q_\gamma}{4\pi r^2} dr^3 = \int_0^{Rmax} \frac{q_\gamma}{4\pi}\, d\Omega\, dr.$$

Thus the integral flux (photons per second per steradian per cm^2) is

$$\phi_\gamma(> E_{min}) = \int_{Emin} \frac{q_\gamma}{4\pi} R_{max}(b, \ell). \tag{2.23}$$

For example, for $b = 4°$, $R_{max} \approx 2.2$ kpc and $\phi_\gamma(> 1\,GeV) \sim 10^{-5}\text{cm}^{-2}\text{s}^{-1}\text{sr}^{-1}$, in qualitative agreement with the general level of the corresponding measurement of EGRET (Hunter *et al.*, 1997).

Although the general level and angular dependence of the diffuse galactic γ-radiation are consistent with the expectation, there is one notable difference from the detailed model calculation of Hunter *et al.*, 1997. The photon spectrum from the inner galaxy is harder above 1 GeV than predicted under the assumption that the parent spectrum of cosmic rays is the same in the inner galaxy as observed locally. This could indicate that propagation of cosmic rays is different there than at Earth (e.g. $\delta \sim 0.3$ rather than ~ 0.6 in Eqs. 2.4 and 2.8). Another possibility is that the observed flux from the inner region could include an admixture of unresolved sources with hard spectra characteristic of the cosmic accelerators. This could occur if some of the gamma-ray production occurs in gas near the source, a possibility discussed further in §2.5 below.

2.4. *Fermi acceleration and maximum energy*

Fermi acceleration is a mechanism for transferring macroscopic kinetic energy of large-scale motion of ionized gas to individual charged particles. Two things are required for it to work: differential motion within the plasma (characterized by speed $V = u_1 - u_2$) and coupling of individual high-energy particles to the plasma in such a way that the resulting diffusion length is much less than the typical distance for collisions with the gas. Electromagnetic coupling through turbulent magnetic fields as described above is sufficient.

A particle of energy E_1 that enters an adjacent region moving with a different velocity $\vec{V}$ will diffuse in that region until its average velocity vector is equal to $\vec{V}$. If it then re-enters region 1 it will have a different energy, $E_2 = E_1 + \Delta E$, with

$$\Delta E = f(\theta_1, \theta_2) \frac{V}{c} E_1.$$

In general, ΔE can be either positive or negative, and the magnitude and sign of the coefficient f depend on the angles θ_1 and θ_2 that describe the orientation of the particle as it enters and leaves region 2. A simplified derivation of the geometry following the treatment of Bell, 1978 is given in Gaisser, 1990. The result is

$$\langle \Delta E \rangle \propto \xi E, \tag{2.24}$$

with $\xi \propto (V/c)^2$ for the original second order Fermi mechanism (Fermi, 1949) and $\xi = (4/3) \times (V/c)$ for first order Fermi acceleration (Bell, 1978). Here c is the speed of the particle being accelerated.

From Eq. 2.24 it is straightforward to derive the basic results of Fermi acceleration. Suppose a test particle is injected into the acceleration region with energy E_0 (assumed here to be fully relativistic for simplicity) then after n transitions from region 1 to 2 and back the particle has on average

$$E_n = (1 + \xi)^n \times E_0.$$

Therefore the number of encounters needed to reach energy E is

$$n = \frac{1}{\xi} \ln \left(\frac{E}{E_0} \right). \tag{2.25}$$

If $P_{\rm esc}$ is the probability per transition that the particle escapes from the accelerator, then the number of particles with energy $> E$ is

$$N(\geq E) = N_0 \sum_{m=n}^{\infty} (1 - P_{\rm esc})^m = N_0 \frac{(1 - P_{\rm esc})^n}{P_{\rm esc}}, \tag{2.26}$$

with n given by Eq. 2.25. Thus

$$N(\geq E) \propto E^{-\gamma_s}, \ \gamma_s = \frac{P_{\rm esc}}{\xi}.$$

In the rest frame of a large, plane shock, upstream (unshocked) gas is flowing into the shock with speed u_1 and out with $u_2 < u_1$. The relation between the upstream and downstream velocities is

$$\frac{u_1}{u_2} = 4 - \frac{12}{M^2}, \tag{2.27}$$

where M is the Mach number of the shock (Landau & Lifshitz, 1982). In the lab coordinate system the shock advances into the upstream region with velocity $-\vec{u}_1$. As long as the radius of curvature of a cosmic-ray particle is small compared to that of the shock, particles cannot escape upstream because the shock will eventually catch up. Particles can escape on the downstream side because they are being convected away from the shock at an average rate u_2. Therefore $P_{\rm esc}$ can be estimated by comparing the rate at which particles in the downstream region cross back upstream to the rate at which they are convected downstream ($n_{CR} u_2$). The crossing rate is the projection of the downstream cosmic ray density, n_{CR} assumed isotropic, onto a plane:

$$\int_0^1 d\cos\theta \int_0^{2\pi} d\phi \frac{c\, n_{CR}}{4\pi} \cos\theta = \frac{c\, n_{CR}}{4}. \tag{2.28}$$

Note that the angle average in Eq. 2.28 is only over the half of phase space moving toward the shock. Thus

$$P_{\rm esc} \;=\; (u_2 n_{CR}) \times 4/(c\, n_{CR}) \;=\; 4u_2/c,$$

and

$$\gamma_s \;=\; \frac{4u_2}{c\xi} \;=\; \frac{4u_2}{\frac{4}{3}(u_1 - u_2)} \;\approx\; 1 + \frac{4}{M^2}. \tag{2.29}$$

The rate of acceleration is

$$\frac{dE}{dt} \;=\; \frac{\Delta E}{T_{\rm cycle}} \;=\; \frac{\frac{4}{3}\frac{V}{c}E}{\lambda/u_1}, \tag{2.30}$$

where the cycle time is estimated as the characteristic scale of the diffusion process, λ, divided by the shock speed, u_1. Since the diffusion involves scattering in the turbulent magnetic field,

$$\lambda > r_{\rm gyro} \;=\; \frac{Pc}{Ze\langle B\rangle},$$

where P is the total momentum of a particle of charge Ze, and $pc = E$ in the relativistic limit. Therefore

$$\frac{dE}{dt} \;\le\; \frac{u_1}{c}\, Ze\langle B\rangle u_1. \tag{2.31}$$

At a time T after injection, the maximum energy of a particle is

$$E_{\rm max} \le \frac{u_1}{c}\, Ze\langle B\rangle\, u_1 T. \tag{2.32}$$

Note the factor of nuclear charge in the expression for the maximum total energy per nucleus in Eq. 2.32. It implies that, for a particular cosmic accelerator, the particles with the greatest mass should achieve the highest total energy.

It is important to estimate E_{max} for parameters typical of a supernova remnant. Suppose the shock expands with a speed $u_1 \approx 10^9$ cm/s into a uniform external medium with density $n_H = 1$ atom per cm^3. Swept up material behind the shock has kinetic energy $\frac{1}{2}m_H V^2 \sim 5\times10^{-7}$ erg/particle. By energy conservation the expansion will have to slow down when the shock has transferred a significant fraction of the initial kinetic energy of the ejected stellar envelope to the swept up gas. If the initial kinetic energy of the shell driving the shock is 10^{51} erg, this will occur when the shock encloses a volume $\sim 10^{57}$cm^3 or at a radius of $u_1 T \sim 2$ pc. Inserting these values into Eq. 2.32 leads to an estimate of $E_{max} \sim 200 \times Z$ TeV, similar to the early result of Lagage & Cesarsky, 1983. (A more recent calculation by Berezhko, 1996, gives a higher estimate.)

After several hundred to 1000 years, the supernova begins to slow down as more matter is swept up. Although the shock may remain strong ($M \gg 1$) for tens of thousands of years, the maximum energy accessible to particles injected during the later phases decreases slowly because the expansion velocity (u_1 in Eq. 2.32) decreases as $\dot{R} \propto t^{-3/5}$, a result which follows from energy conservation. For a helpful discussion of explosions in the ISM, see the paper of McKee & Truelove (1995).

The characteristic power-law index of the integral spectrum of particles accelerated by a strong shock according to Eq. 2.29 is $\gamma_s \approx 1$. This result is obtained neglecting the possible non-linear effects of the cosmic-rays themselves acting on the shock, the "test-particle" approximation. If a significant fraction of the energy of the SNR is converted into high energy cosmic rays, however, this back-reaction may not be negligible. According to the discussion following Eq. 2.9 above, if SNR are indeed the main source of galactic cosmic rays, then the acceleration efficiency must be quite high, of order 10%.

In the theory of cosmic-ray modified shocks,† the accelerated particles streaming into the upstream region generate the turbulence from which they scatter. This coupling, which is an integral part of the acceleration mechanism, slows down the upstream (unshocked interstellar medium) as it flows into the shock. (Viewed from the lab system, the shock expands into the external (upstream) medium and the coupling imparts a velocity to material in front of the shock.) The net effect is to spread out the velocity discontinuity over a scale given by the gyroradius of the highest energy particles in the accelerated spectrum. Lower energy particles do not see the full discontinuity ($u_1/u_2 = 4$), but only experience a smaller ratio, which, according to Eq. 2.29, corresponds to a steeper spectrum. On the other hand, the contribution to the gas pressure of a relativistic component (the cosmic rays) changes the properties of the shock so that the velocity discontinuity in Eq. 2.27 increases and the spectral index γ_s in Eq. 2.29 decreases. The qualitative effect is to give a concave spectrum that is softer at low energy and harder at high energy. Examples of such concave spectra are given by Berezhko & Völk, 1997.

There are several other considerations that would lead one to expect the effective cosmic ray source spectrum integrated over time to be somewhat steeper than the ideal $\gamma_s = 1$ of the test particle limit of first order Fermi acceleration:‡ Particles accelerated in the later phases of expansion of an SNR may have steeper spectra because the shock weakens. Adiabatic deceleration as the shock expands and disintegrates may also affect the spectrum. In addition, if many different SNR contribute to the cosmic ray pool, each with its own parameters and characteristic E_{max}, the overall source spectrum will be steepened if SNR with small E_{max} are the most numerous. In the remainder of this chapter, I will characterize a power-law fit to the observed spectrum by an integral spectral index

$$\gamma = \gamma_s + \delta = 1 + \epsilon + \delta, \tag{2.33}$$

where ϵ refers to effects in the sources and δ to effects of propagation. In certain conditions, as discussed above, ϵ may be negative.

2.5. *SNR as γ-ray sources*

A supernova remnant that puts a significant fraction of its energy into cosmic rays is a potential source of high energy photons. For significant acceleration to occur, there must necessarily be enough gas present to slow down the expansion of the SNR. This will provide target material in which the accelerated cosmic rays can interact to produce neutral pions and hence photons. Accelerated electrons will radiate lower energy photons by synchrotron radiation and γ-rays by bremsstrahlung. Synchrotron radiation has long been used as a tracer of acceleration of electrons in SNR (Reynolds, 1994). Recently, the ASCA satellite detected non-thermal x-rays from the rim of SN1006 (Koyama *et al.*, 1995), which have been interpreted as synchrotron radiation from an electron spectrum extending up to ~ 100 TeV (Reynolds, 1996; Mastichiadis, 1996).

There are several calculations of photon production by cosmic rays in SNR (for example, Drury, Aharonian & Völk, 1994; Berezhko & Völk, 1997). The following sketch is a very qualitative discussion, and the reader is strongly encouraged to consult the recent work of Berezhko & Völk, 1997 for an informative and detailed treatment as well as for

† See the review of Jones & Ellison, 1991 for a discussion of cosmic-ray modified shocks and references to other work on the subject.

‡ Note that the energy content of a power-law spectrum with $\gamma = 1$ diverges logarithmically as $E_{max} \to \infty$. Such a spectrum contains equal amounts of energy per logarithmic interval of energy.

references to earlier work on the theory of shock acceleration and γ-ray production in SNR.

The source spectrum of photons is a convolution of the cosmic ray spectrum and the cross section for nucleon$\rightarrow \pi^0 \rightarrow 2\gamma$ integrated over the volume $\mathcal{V}$ of the cosmic ray source. The analysis is completely parallel to the calculation of diffuse gamma-radiation from the galaxy except that here we use the cosmic ray source spectrum and the density and distribution of gas in the source rather than those of the interstellar medium. The production spectrum of γ-rays (photons/s/GeV) is

$$\frac{dq_\gamma}{dE_\gamma} = c\int d\mathcal{V}\, n_{\rm gas} \int_{E_\gamma}^{\infty} \frac{d\sigma(E_\gamma, E)}{dE_\gamma}\, n_{CR}(E)\, dE. \tag{2.34}$$

For purposes of an estimate we can take n_{CR} = constant behind the shock with a spectrum given by

$$\mathcal{V}\, n_{CR}(E) = AE^{-(\gamma_s+1)}, \quad E < E_{max}.$$

We estimate $n_{\rm gas} \approx 4n_H$ because of the compression of the surrounding interstellar medium by the shock. The integration over the inclusive π^0 distribution and the π^0 decay spectrum is the same as before except that the spectrum of cosmic rays in the source is harder ($\propto E^{-(2+\epsilon)}$, differential). The result is

$$\frac{dq_\gamma}{dE_\gamma} = c\,\sigma_{pp}^{\rm inel} n_{\rm gas}\, \mathcal{V} \times Z_{\pi\gamma} Z_{p\pi^0}(r) \times n_{CR}(E_\gamma), \tag{2.35}$$

where $r = E_\gamma/E_{max}$. For $E_\gamma \ll E_{max}$, $Z_{p\pi^0}$ is given by the integral expression of Eq. 2.21, but as $E_\gamma \rightarrow E_{max}$ it approaches zero Harding & Gaisser, 1990.

If Θ is the efficiency of cosmic-ray acceleration in the SNR then the total energy content in the cosmic rays at time t is

$$\mathcal{V}\int_{E_{min}} E\, n_{CR}(E)\, dE = \Theta\left(\frac{t}{T}\right)^3 \mathcal{E}_{SN} \rightarrow \Theta\,\mathcal{E}_{SN}, \quad t \geq T, \tag{2.36}$$

where T is the characteristic time for the SNR to begin slowing down and $E_{min} \approx 1$ GeV. With the normalization provided by Eq. 2.36, we can calculate, for example, the integral spectrum of photons at a distance d from the source at time $t \gtrsim T$. For a power-law spectrum with integral index $\gamma_s = 1+\epsilon$,

$$\begin{aligned} n_\gamma(>E_\gamma) &\sim Z_{\pi^0\gamma} Z_{p\pi^0} c\, \sigma_{pp} n_{\rm gas} \frac{\gamma-1}{\gamma} (E_{min})^{(\gamma-1)} \frac{\Theta \mathcal{E}_{SN}}{4\pi\, d^2} \\ &= Z_{p\pi^0} \frac{2(\gamma-1)}{(\gamma+1)\gamma}\, \Theta \left(\frac{\mathcal{E}_{SN}}{10^{51} erg}\right) \left(\frac{1\ \rm kpc}{d}\right)^2 \times \frac{4.8\times 10^{-6}}{\rm cm^2\, s}. \end{aligned} \tag{2.37}$$

For $\gamma = 1.1$ and 1.4, $Z_p\pi^0 \approx 0.12$ and 0.066 respectively (Harding & Gaisser, 1990). Inserting these values into Eq. 2.37 gives the estimates shown in Table 1 for nominal parameters of $\mathcal{E}_{SN} = 10^{51}$ erg, $d = 1$ kpc and $\Theta = 0.1$. For comparison, the values at $t = T$ from Berezhko & Völk, 1997 are shown in the last column of TABLE 1. Note that Berezhko & Völk, 1997 have $\Theta \sim 50\%$ and $n_H = 0.3$ as compared to 10% and 1.0 used here (2nd and 3rd columns).

The numbers for $\gamma = 1.1$ in TABLE 1 are in remarkably good agreement with the results of the detailed calculation of Berezhko & Völk, 1997, considering the extremely rough approximations made here, for example neglecting the spatial and energy-dependence of the overlap of the cosmic rays with the gas behind the shock. In addition, the structure of the supernova ejecta, its velocity profile and the possibility of the interaction of cosmic rays with the ejecta have been neglected here, and this leads to an underestimate of the

	$\gamma = 1.4$	$\gamma = 1.1$	Berezhko & Völk, 1997
$E_\gamma > 1$ GeV	4×10^{-8}	2×10^{-8}	$\sim 2 \times 10^{-8}$
$E_\gamma > 1$ TeV	3×10^{-12}	1×10^{-11}	$\sim 5 \times 10^{-11}$

TABLE 1. Estimates of γ-ray luminosity of SNR at 1 kpc (photons $\text{cm}^{-3}\text{s}^{-1}$).

cosmic-ray production for $t << T$ in Eq. 2.36. Berezhko & Völk, 1997 have included these effects and have calculated the evolution of the shock and the cosmic-ray spectrum self-consistently. They find an efficiency of $\sim$ 50% for cosmic-ray acceleration, with the remaining energy going into kinetic energy and heating of the shocked gas for $t \gtrsim T$. Depending on details of the parameters used, they find average differential spectra slightly harder than E^{-2} for the relativistic part of the spectrum as a consequence of the non-linear feedback of the accelerated particles on the structure of the shock. They note that the final contribution of the SNR to the cosmic ray pool will be reduced from 50% as a consequence of adiabatic cooling as the old SNR expands.

The EGRET Collaboration (Esposito, J.A. *et al.*, 1996) has reported signals from two SNR (γ-Cygni and IC433 at distances somewhat greater than 1 kpc) at a level that is a factor 3 to 10 higher than the nominal numbers in Table 1 for $E > 1$ GeV. This could happen, for example, if the density surrounding these remnants is higher than the nominal value of $n_H = 1\ \text{cm}^{-3}$ used here. Before following up this possibility, let us check what fraction of the produced cosmic rays interact to make sure that most are not absorbed in the source, which would rule them out as sources of galactic cosmic rays. We compare

$$L_\gamma = \int E_\gamma \frac{dq_\gamma}{dE_\gamma} dE_\gamma$$

to the average rate at which the SNR produces cosmic rays, $\Theta\, \mathcal{E}_{SN}/T$. The integrals are similar to those in Eqs. 2.35 and 2.37. The result is that for a nominal ISM with $n_H = 1\ \text{cm}^{-3}$ only $\sim$ 0.1% of the accelerated cosmic rays interact per thousand years.

Upper limits on $\sim$TeV photons from several supernova remnants including γ-Cygni and IC433 from the Whipple telescope are at the level of several times $10^{-11}\text{cm}^{-2}\text{s}^{-1}$ and are significantly below a simple power-law extrapolation from the level of the EGRET signals with a spectral index of $\gamma = 1.0$ or $\gamma = 1.1$ (Buckley, *et al.*, 1997). One possibility is that the source spectra are steeper. For example, an extrapolation from the EGRET data with $\gamma_s \approx 1.4$ would be consistent with the Whipple upper limits. When the contributions of bremsstrahlung and inverse Compton photons are included, the Egret data itself is also consistent with source spectrum of $\gamma_s = 1.3$ or 1.4 (Gaisser, Protheroe & Stanev, 1997). On the other hand, the theoretical calculations (Berezhko & Völk, 1997) predict spectra even harder than $\gamma_s = 1$.

The search for TeV γ-rays from SNR is a rapidly evolving field at present and the final word is not in. On the one hand, if SNR are indeed the sources of galactic cosmic rays up to the knee region, then they should be point sources of photons at some level. On the other hand, the level of the signal depends on the details of the environment of each particular supernova remnant. For the remainder of this chapter I continue to assume that SNR are the principal sources of galactic cosmic rays.

2.6. *Summary*

Several of the results described in this section are rather general and apply on other size and distance scales. These include

- Power law spectra of accelerated particles with integral spectral index $\gamma_s \sim 1$;
- A maximum energy of accelerated particles characterized by the magnetic field, the scale and the velocity of a strong shock,

$$E_{\max} \sim Z\,e \times \frac{V}{c}\,B\,R, \tag{2.38}$$

where the scale R may in some cases be expressed as velocity multiplied by a characteristic time;

- Frozen in magnetic fields in turbulent plasma;
- Equipartition among energy in accelerated particles, in magnetic fields and in kinetic energy of the diffuse gas.

3. The knee region and beyond

The upper limit for acceleration by SNR estimated in Eq. 2.32 is of order 10^{14} to 10^{15} eV for protons and proportionately higher for nuclei with $Z > 1$. Although it is possible that all cosmic rays above this energy are from outside our galaxy (see e.g. Protheroe & Szabo, 1992), there is evidence (to be discussed below) that the transition to extragalactic (EG) cosmic rays occurs at much higher energy, above 10^{18} eV. Therefore I adopt as a working hypothesis the assumption illustrated in Fig. 2. The three regions indicated by the lines in Fig. 2 correspond to the three regions defined in the introduction. The terminology of Axford (1994) suggests the physical origins: I (VHE)=GCRI; II (UHE)=GCRII and III (EHE)=EG. Region I corresponds to standard supernova accelerated particles as discussed above. The knee region is Galactic Cosmic Rays-II.

A useful starting point is to estimate the power required to maintain the observed spectrum in region II. To do this we must evaluate an integral like that in Eq. 2.9 for the spectrum shown by the dashed line in Fig. 2. We immediately run into a problem if we use the expression 2.4 for the containment time because τ_e becomes so short at high energy that the cosmic rays would be highly anisotropic, which is not observed. For example, Eq. 2.4 gives an escape time corresponding to $\sim$ 2 kpc for a proton of 10^{15} eV and an escape length not much bigger than the thickness of the galactic disk at 10^{16} eV. The weak re-acceleration model for cosmic-ray propagation (Seo & Ptuskin, 1994, Heinbach & Simon, 1995) offers a way to avoid this problem. In that model, $\delta = 0.33$, which is preferred in turbulence theory. Re-acceleration refers to the recognition in these models that particles undergo second-order Fermi acceleration as they propagate, and that the amount of redistribution of energy is more important for lower energy particles because they spend more time in the galaxy. Re-acceleration affects also the ratios of secondary/primary nuclei, and the data are described well by a fit with $\epsilon = 0.37$ and $\delta = 0.33$ (Seo & Ptuskin, 1994, Heinbach & Simon, 1995).

The slower dependence of $\tau_e \propto E^{-0.33}$ is consistent with the low anisotropy of cosmic rays up to the highest energies of the galactic component. Using a fit of this form in Eq. 2.9 leads to a similar estimate of $\sim 10^{41}$ erg/s for GCRI, which are dominated by the low energy end of the spectrum. The estimate for GCRII (inserting the dashed line in Fig. 2 into Eq. 2.9 leads to an estimate of $\sim 10^{40}$ erg/s.† This requirement could be satisfied, for example, by one in ten SNR having 10% efficiency for accelerating particles

† This estimate is very rough because it depends on how the spectrum is extrapolated to low energy.

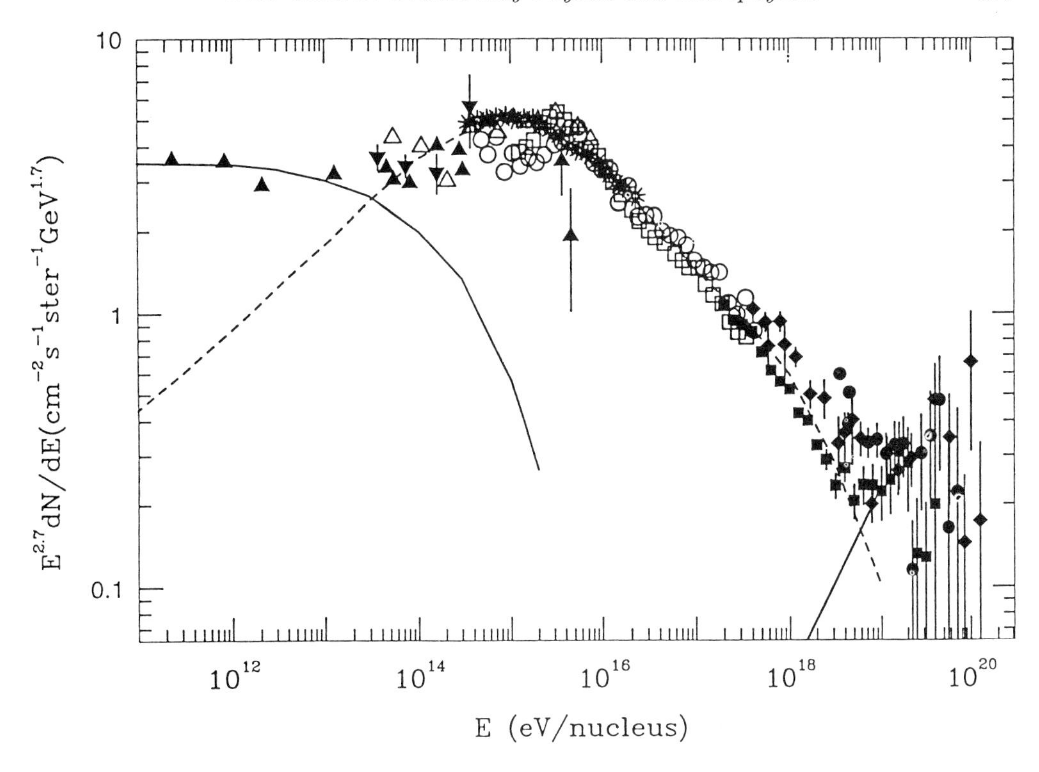

FIGURE 2. The all-particle cosmic-ray spectrum. The lines indicate the three components of the spectrum discussed in the text.

to $\sim 3 \times 10^{17}$ eV$\times Z$. Erlykin & Wolfendale, 1997a recently made the interesting point that the shape of the spectrum in the knee region resembles the concave spectrum of a single SNR, as in the work of Berezhko & Völk, 1997. They suggest that the knee is the result of acceleration at a single, nearby, recent supernova. Because of the dependence on nuclear charge in Eq. 2.38, this idea has rather specific consequences for composition (Erlykin & Wolfendale, 1997b).

3.1. *Models for higher energy galactic sources*

There are several ideas for how expanding SNR might produce cosmic-rays with energy beyond the knee. Jokipii (1987) points out the possible importance of magnetic field configuration. The angle between the expansion velocity of an approximately spherical supernova shock and the average field in the interstellar medium will vary from parallel to perpendicular around the explosion. Since cross-field diffusion is slower than diffusion along the field, the acceleration rate could be higher in the perpendicular configuration, leading to a higher E_{max}. Axford (1994) and collaborators consider the acceleration of high energy cosmic rays injected from individual SNR to higher energy when they interact with the converging flow of several SNR. In both of these cases there is nothing intrinsic to the ideas that would lead to different spectra for different elements.

Another possibility for achieving higher energy explicitly recognizes that different kinds of supernovae may lead to different accelerated spectra (Völk & Biermann, 1988) . Type II supernovae are core-collapse supernovae with hydrogen lines in their spectra. Their progenitors are usually red giant stars with mass $\sim 10\,M_{\odot}$. Much more massive giant stars also lead to core-collapse supernovae, but they lack hydrogen lines and are therefore classified as Type I. To distinguish them from supernovae from explosion of carbon-oxygen white dwarves, they are labelled as Type Ib, Ic,.... The progenitors are massive ($\gtrsim 20\, M_\odot$) stars with fast winds that have blown away the outer hydrogen layer before the supernova. For this reason hydrogen lines are not seen in the supernova outburst.†

The progenitor stellar wind will establish a cavity analogous to the heliosphere into which the supernova explodes. The boundary of the cavity (observed as a ring nebula–Lozinskaya, 1992) occurs where the pressure of the ISM balances the ram pressure of the stellar wind. There is a termination shock in the wind inside an "astropause" (by analogy with the heliopause). The astropause is the contact discontinuity with the external ISM, which itself may be shocked. The fast wind expands freely with velocity $V_w \sim$ 1000 km/s up to the termination shock. The region between the termination shock and the astropause, called the stagnation zone, contains the shocked stellar wind.

Interior to the termination shock the fast ionized wind carries the magnetic field from the surface of the star with the characteristic Parker spiral configuration (Parker, 1958). The Parker field has a near zone defined by $r < R_L = V_w/\Omega_*$, where Ω_* is the angular velocity of the progenitor. Outside R_L the magnetic field in the equatorial region is

$$B = B_* \left(\frac{R_*}{R_L}\right)^2 \frac{R_L}{R} \equiv B_L \frac{R_L}{R}, \tag{3.39}$$

where B_* is the magnetic field at the surface of the star. For $V_w = 2000$ km/s and $\Omega = 10^{-6}\mathrm{s}^{-1}$, $R_L = 2\times 10^{14}$ cm as compared to $R_* \approx 3\times 10^{12}$ cm for a typical massive main sequence O-star. (I use parameters from an example discussed by Berezhko & Völk, 1997 in a related context.) For a surface magnetic field of $B_* = 50$ Gauss, $B_L \approx 10^{-2}$ Gauss.

This is the environment in which the supernova occurs. As the SNR expands into the wind cavity of its progenitor, it accelerates particles out of the gas swept up by the shock at the rate given by Eq. 2.30. In the present case, however, the magnetic field at the shock is not constant but decreases with time as the shock sweeps out through the decreasing field.

A prediction of this model for GCRII would be that this component would have a different composition, reflecting the material eroded from the surface of the progenitor star rather than the general ISM. In addition, if type Ib, Ic, etc. supernovae accelerate to higher E_{max}, then there would be a characteristic signature of composition in the knee region. For example, if a particular star had blown away all its hydrogen and helium and had a surface mostly of carbon at the time of the supernova explosion, it could produce a composition relatively rich in helium and carbon, depending on the details of nucleosynthesis in the progenitor. Specific predictions along these lines are made by Silberberg *et al.*, 1990.

3.2. *Evidence from composition*

Because the maximum energy for a particular cosmic accelerator is proportional to the charge Z of the nucleus, the top end of a particular acceleration mechanism should manifest itself as a change toward a relatively higher abundance of heavy nuclei when the spectra are classified by energy per nucleus. A related statement is that if all nuclei come from the same distribution of sources and all have the same propagation history, then the spectrum of each component should look the same when plotted as a function

† The book *Supernovae and Stellar Winds in the Interstellar Medium* by Lozinskaya, 1992 is an excellent source for background on these subjects.

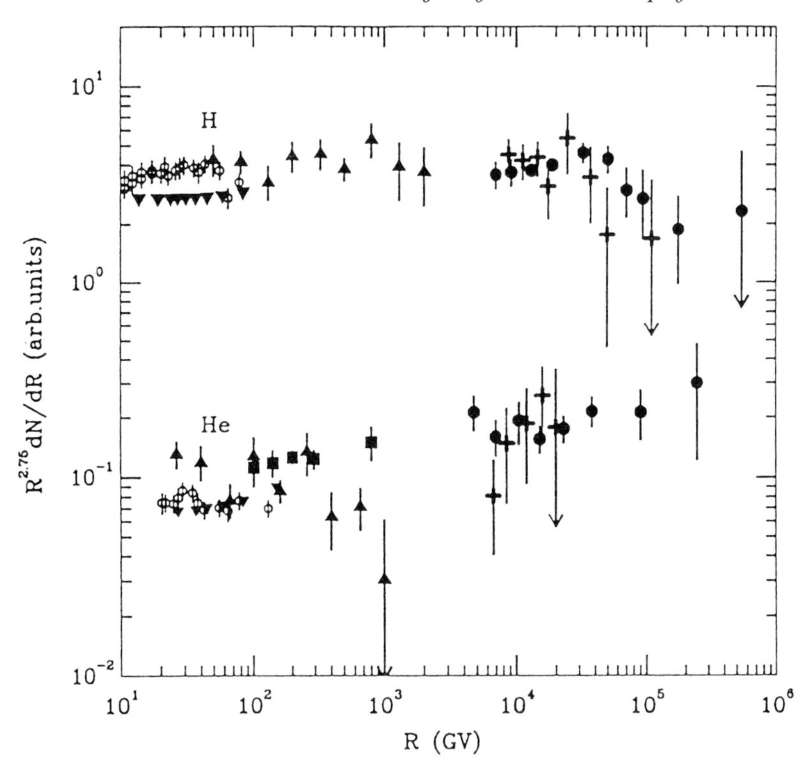

FIGURE 3. Rigidity spectra of hydrogen and helium above 10 GV. Key to the data points is given by Biermann, Gaisser & Stanev (1995).

of magnetic rigidity. This statement is valid provided all acceleration and propagation processes depend only on coupling to electromagnetic fields.‡

Fig. 3 (Biermann, Gaisser & Stanev, 1995) is a compilation of direct measurements of the spectra of hydrogen and helium as a function of magnetic rigidity. It seems that protons have somewhat steeper spectra than helium. Fits to major groups of heavy nuclei also give spectra that are somewhat steeper than for protons (Swordy, 1993). Such differences suggest that different kinds of sources are at work. In a qualitative sense, the data of Fig. 3 are not inconsistent with the model discussed in the previous section (Völk & Biermann, 1988; Silberberg *et al.*, 1990), in which helium and carbon might be relatively more abundant at high energy. The most recent report (Cherry *et al.*, 1997) from the JACEE balloon-borne experiment is also consistent with this picture. It gives power-law fits to measured energy spectra of hydrogen and helium in the energy range from 2 to 800 TeV that correspond to differential spectral indices of $\gamma + 1 = 2.80 \pm 0.04$ and 2.68 ± 0.06 respectively.

Unfortunately, direct measurement of the primary cosmic radiation with detectors above the atmosphere (in high altitude balloons or on spacecraft) run out of statistics approaching the knee region. Because of the steeply falling energy spectrum, indirect experiments with ground based detectors must be used in the knee region and beyond to collect a big enough sample of events. Ground-based air shower experiments overcome the problem of low flux by their large area and long exposure times. The price is that

‡ Recall the earlier discussion of the propagation of iron: for low energies, collisions with interstellar gas become important, and the spectrum of iron differs from that of protons even if they have the same source spectra.

they do not detect the primary cosmic rays but only the secondary cascades they generate in the atmosphere. The interpretation is complicated and at present still inconclusive. (See Watson, 1997 for a recent review.)

3.3. *Hybrid air shower detectors*

There are several fairly new air shower experiments with promise for improving our knowledge of composition in the knee region. They are designed to measure three or more different components of showers to provide redundancy and thus to resolve ambiguities in interpretation of pervious experiments.

Air showers produce both neutral and charged pions. The π^0 decay promptly to a pair of photons, each of which initiates an electromagnetic sub-shower. The resulting electrons and positrons are the most abundant charged particles in the cascades. Charged pions either interact or decay with a relative probability that depends on their energy and the depth in the atmosphere. High energy nucleons and charged pions from the interaction of the incident nucleus are concentrated in a beam parallel to the axis of the shower, and they generate further high-energy interactions along the trajectory of the shower. The electrons and low energy muons spread over a large area with a density that on average decreases with distance from the shower core, but is still detectable out to hundreds or even thousands of meters from the core, depending on the initial energy, E_0 of the shower. Generally higher energy showers produce more particles of all types than showers of lower energy, but the relation between shower size and primary energy also depends on the identity of the primary nucleus. Basic features of air showers are reviewed in more detail in Gaisser, 1990.

Because of the competition between decay and interaction of charged pions in the atmosphere, showers generated by heavy nuclei produce more muons than showers of the same energy initiated by single protons. This is basically because the showers generated by heavy nuclei develop higher in the atmosphere where the atmosphere is less dense and it is therefore relatively easier for the charged pions to decay to muons before they interact. Conversely, proton showers are more penetrating, reaching cascade maximum deeper in the atmosphere than showers generated by heavy primaries of the same energy. This is because the energy in a nucleus of mass A is already subdivided into A packets at the top of the atmosphere. Thus a large ratio of muons to electrons suggests heavy primaries and *vice versa*.

The problem is that intrinsic fluctuations in shower development are large. In addition, a conventional air shower array only samples the shower front, and sampling errors are also significant. Several experiments that measure at least three shower components are in progress.

One possibility is to measure the Cherenkov light generated by charged particles high in the atmosphere. This technique provides a measure of the shower near maximum development, where intrinsic fluctuations are at a minimum. Detectors are spaced so that the light density can be sampled at distances from $\sim$ 20 to $\sim$ 120 m or somewhat greater from the shower core. Thus for each shower there is a measurement of the lateral distribution of the Cherenkov light as well as a measure of the density at $>$ 100 m from the core. It is known (Patterson & Hillas, 1983, Karle *et al.*, 1995) that deeply penetrating showers have steep Cherenkov lateral distributions while showers that develop high in the atmosphere have flatter lateral distributions. The Cherenkov light density at $R >$ 100 m is proportional to the total energy in the shower. Since, for a given total energy, protons give showers that are more penetrating than showers initiated by heavy nuclei, we expect the correlations summarized in Table 2.

There are several air shower arrays which include a sub-array of open photomultipli-

	Protons, α	heavy nuclei
SIZE	large	small
Muon signal	small	large
Cherenkov density (> 100m)	$\propto E_0$	$\propto E_0$
Cherenkov lateral distribution	steep	flat

TABLE 2. Dependence of shower components to composition.

ers to measure the intensity and lateral distribution of the Cherenkov light. The most well-developed is the HEGRA array, which samples the shower front with an array of scintillators. The sub-array AIROBICC measures the Cherenkov light (Karle *et al.*, 1995). The array also has 17 detectors for low energy muons. The BLANCA air Cherenkov sub-array (144 detectors) inside the CASA-MIA array at Dugway, Utah began operation in January 1997 (Cassidy *et al.*, 1997). This very large hybrid array also samples the shower front with scintillators and measures electron, low-energy muon and Cherenkov components of showers in the region of the knee. The SPASE-AMANDA coincidence experiment (Dickinson, J.E. *et al.*, 1997) at the South Pole is much smaller. The South Pole Air Shower Experiment (SPASE) is a conventional air shower array with a sub-array (VULCAN) for Cherenkov detection. The distinguishing feature of this hybrid experiment is that the muon detector is the Antarctic Muon and Neutrino Detector Array (AMANDA), which is deep enough so that the ice overburden filters out muons produced with less than several hundred GeV. It is also at somewhat higher altitude and therefore observes showers somewhat closer to maximum.

Another major new effort aimed primarily at an improved determination of composition in the knee region is the KASCADE experiment at Karlsruhe (Klages, 1997). This detector has a hadron calorimeter in the middle of an array of electron and muon counters. With the calorimeter it is possible to measure the energy spectra of high energy hadrons in the cores of showers. There is a useful discussion of issues associated with interpretation of air shower experiments and determination of the composition in the region of the knee of the spectrum in the workshop proceedings edited by Rebel, Schatz & Knapp, 1997.

3.4. *Fly's Eye experiment*

The great advantage of the Fly's Eye technique Sokolsky, Sommers & Dawson, 1992 is that it traces out the longitudinal development of a shower as it crosses the atmosphere rather than sampling the shower at one depth only. The shower energy (apart from the fraction lost to neutrinos) can be obtained rather directly from the convolution of the number of ionizing particles along the trajectory with the energy lost per g/cm^2 of the atmosphere. Each shower can be fitted with a curve to obtain parameters such as the depth at which the shower reaches maximum development and the depth at which it reaches, say, 20% of maximum on the rising and falling portions. The magnitudes and distributions of these quantities are sensitive to the nature of the incident nuclei and to the differential and total cross sections for production of various secondary particles.

The photons detected by Fly's Eye are from the atmospheric fluorescence generated by ionization of nitrogen in the atmosphere. The fluorescence radiation is emitted isotropically. The yield per charged particle in the shower is well known, and the total number of shower particles on a short segment of the event can be inferred from the pulse-height in each pixel of the detector provided the clarity of the atmosphere is monitored and the distance to the trajectory is known. In the stereo version of the Fly's Eye detector, each composite eye determines a plane, and the intersection of the planes determines the trajectory.

An important accomplishment of the Fly's Eye experiment is the simultaneous measurement of the energy spectrum and the depth of shower maximum ($\langle X_{max} \rangle$) as a function of energy (Bird *et al.*, 1993). The spectrum shows a change of slope (from steep to less steep) around 3×10^{18} eV (Bird *et al.*, 1994). A similar hardening of the spectrum (the "ankle") in the same energy region appears in other experiments (see for example Yoshida *et al.*, 1995). Such a feature is characteristic of a transition from a lower energy component, which dies out, to an emerging higher energy component. Because the gyroradii of particles in this energy region are comparable with galactic scales (1 kpc for a proton of 3×10^{18} eV, for example) it is natural to associate the ankle with a transition from galactic cosmic rays to particles of extra-galactic origin.

The results for $\langle X_{max} \rangle$ from the same data appear to show a transition from nearly all heavy nuclei at 10^{17} eV to mostly protons at 10^{19} eV. Unlike the spectrum result, which depends only on a simulation of detector response to charged particles, this result also depends on the assumptions made about hadronic interactions in the cores of the showers (see for example Gaisser, 1997). The energies involved are equivalent to as much as two orders of magnitude higher than present $\bar{p}p$ colliders. Moreover, they require a knowledge of properties of the fragmentation region of nuclear collisions whereas colliders explore the central regions of hadron-hadron collisions. As an illustration, Fig. 4 shows $\langle X_{max} \rangle$ calculated by Gaisser, Lipari & Stanev, 1997 with the SIBYLL interaction model (Fletcher *et al.*, 1994) and compares it with the result of a different model that was used by Bird *et al.*, 1993. A calculation with the quark-gluon string model (Kalmykov & Khristiansen, 1995; Kalmykov, Ostapchenko & Pavlov, 1997) shows a result indicating some transition from heavy toward light primaries as energy increases, but not as strong as in the work of Bird *et al.*, 1993. The width of the distribution of depth of maximum is less model dependent than its absolute value. Comparison with the Fly's Eye data clearly shows that in the region of 3×10^{18} eV there are both heavy and light components in the incident cosmic radiation (see Fig. 3 of Gaisser, 1997).

4. Extragalactic cosmic rays

Assuming that the observed cosmic rays above $\sim 3 \times 10^{18}$ eV are indeed of extragalactic origin we can estimate the power required to generate the intensity. As before we use Eq. 2.9, this time with the line indicated for region III in Fig. 3 as the observed spectrum and with τ_e replaced by the age of the Universe. The numerical value of the result depends in an important way on how the extragalactic spectrum is extrapolated to low energy. Assuming an extrapolation with $\gamma = \gamma_s = 1$ for the extragalactic component, we find an energy density in the extragalactic component of 3×10^{-19} erg cm^{-3}, with a corresponding power of 10^{-36} erg cm^{-3} s^{-1}. If the spectrum is steeper, the power requirement will be correspondingly greater. Whether and how the spectrum cuts off above $\approx 5 \times 10^{-19}$ does not have a big effect on this estimate because most of the integral comes from the energy content at lower energy.

It is useful to express the power requirement for extragalactic cosmic rays on astronomical scales. It corresponds to

- $\sim 3 \times 10^{39}$ erg/s per galaxy
- $\sim 3 \times 10^{42}$ erg/s per cluster of galaxies
- $\sim 2 \times 10^{44}$ erg/s per active galaxy
- $\sim 10^{53}$ erg per cosmological gamma-ray burst (GRB).

The interesting thing about these numbers is that they correspond roughly to power

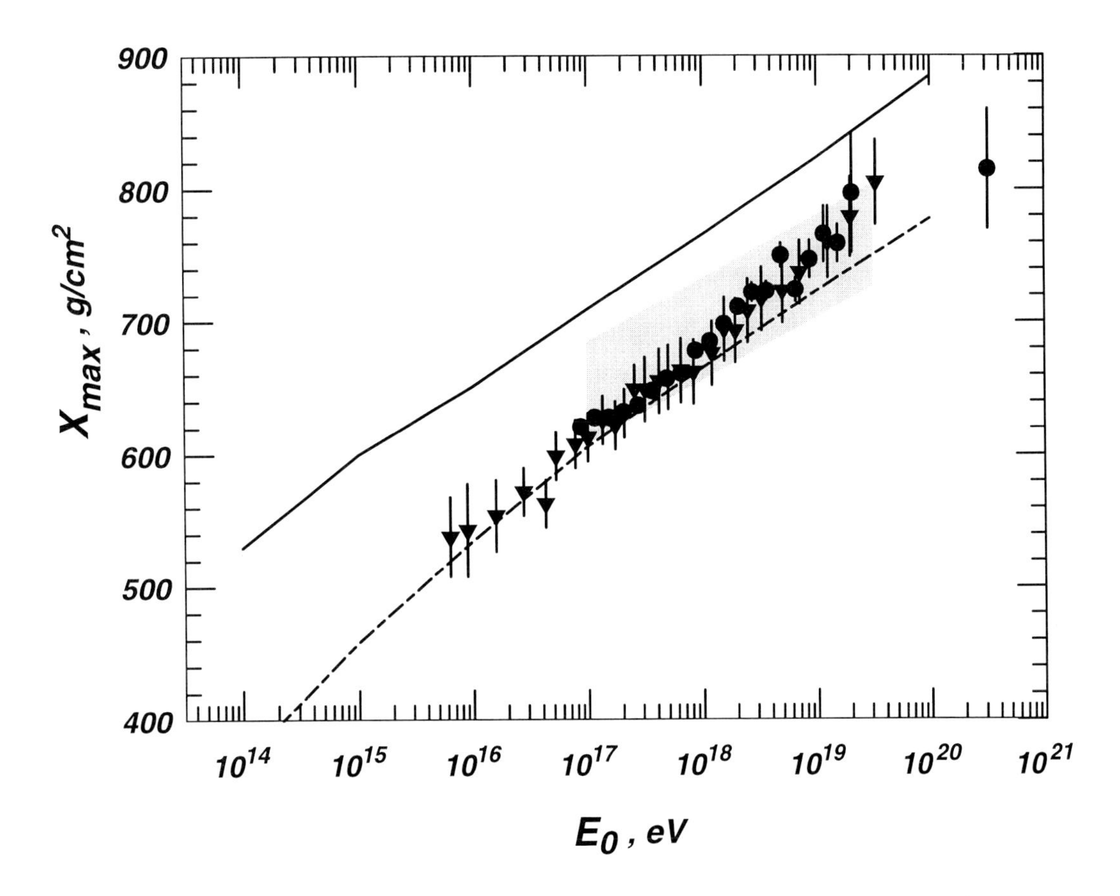

FIGURE 4. Mean depth of shower maximum vs. primary energy from Gaisser, Lipari & Stanev (1997). Dashed and solid lines are for primary iron and primary protons, respectively. Shaded region is bounded below by calculated iron and above by calculated protons in the paper of Bird *al.* (1993). Solid circles are stereo Fly's Eye data (Bird *et al.*, 1993) and solid triangles from Yakutsk (Kalmykov & Khristiansen, 1995).

output estimated from other phenomena associated with these objects. For example, 3×10^{39} erg/s is close to the power radiated in magnetic dipole radiation from a single 10 ms pulsar with a magnetic field of 10^{12} Gauss at the surface of a neutron star with an appreciable angle between the magnetic and rotational axes. It is interesting that highly magnetized neutron stars are among the few objects that nearly meet Hillas' criterion (Hillas, 1984) for acceleration to 10^{20} eV.

4.1. *Some possible accelerators*

Luminosities associated with cooling flows in clusters of galaxies are $> 10^{42}$ erg/s, and X-ray luminosities of rich clusters have $L_X \sim 10^{44}$ erg/s (Fabian, 1994). It has been suggested (Norman, Melrose & Achterberg, 1995) that shocks associated with accretion during formation of large structures such as clusters of galaxies could accelerate protons to energies of $\sim 5 \times 10^{19}$ eV and possibly higher (Kang, Rachen & Biermann, 1997).

Estimates of maximum energy are made by inserting estimates of shock velocity and scale into Eq. 2.38.

Active galaxies (Blandford, Netzer & Woltjer, 1990) typically have total luminosity $> 10^{44}$ erg/s. Shocks near the termination of relativistic jets of active galaxies are a particularly promising source of the highest energy cosmic rays (Rachen & Biermann, 1993). The assumed parameters of the system are shock velocity of 0.3 c, $R \sim 1$ kpc and $B \sim 5 \times 10^{-4}$Gauss. A straightforward application Eq. 2.38 then gives $E_{max} \sim Z \times 10^{20}$ eV. With such high fields and large distances, it is also necessary to check the limit that comes from equating the acceleration rate (2.30) to the rate of losses due to radiative processes. In this case both synchrotron losses and photopion production on photon fields in the source play a role. This consideration gives a similar limit (Rachen & Biermann, 1993). Comparison of this model with the observed cosmic-ray spectrum requires an integration over the cosmological distribution of the appropriate class of radio galaxies with extended jets, followed by modification of the injected spectrum by photoproduction ($p + \gamma_{2.7} \rightarrow \pi + X$) on the microwave background radiation. Since the radio luminosity of the jets is due to synchrotron radiation from electrons, normalization of the spectrum also requires an assumption about the ratio of accelerated protons to electrons in the jets.

Another possibility (Miralda-Escudé & Waxman, 1996; Vietri, 1995) is that cosmological GRB sources may also accelerate cosmic rays to $> 10^{19}$ eV. Given the rate of observed bursts (Fishman & Mccgan, 1995), the total rate of bursts in the Universe should be of order 1000 per year. Comparing this with the required power of 10^{-36} erg cm^{-3} s^{-1} gives the above estimate of 10^{53} erg/GRB in cosmic rays. This is to be compared with the luminosity per burst in the γ-ray band of $\sim 2 \times 10^{51}$ erg. Recall that the estimate of the power requirement of the extragalactic cosmic rays is made assuming the spectrum extrapolates as E^{-2} (differential) down to ~ 1 GeV. The energy requirement per decade would be a factor of 10 lower, still somewhat high compared to the flux of γ-rays. On the other hand, if the total power is available, a ratio of 10/1 for protons/photons is not unreasonable, especially if the photons are secondaries of the protons.

4.2. *The highest energy cosmic ray*

An event detected with the stereo Fly's Eye (Bird *et al.*, 1993) currently holds the record as the highest energy cosmic ray. This single event is so remarkable that it is the subject of a separate paper to discuss it (Bird *et al.*, 1995). The energy is determined as $3.2 \pm 0.9 \times 10^{20}$ eV. The Akeno group have also reported an extremely high energy event estimated as $\sim 2 \times 10^{20}$ eV. As discussed in §3.4 above, the Fly's Eye technique has the advantage for energy determination of seeing the entire shower profile (see Fig. 5).

Photoproduction by protons (Greisen, 1966; Zatsepin & Kuz'min, 1966) and photodisintegration of nuclei (Puget, Stecker & Bredekamp, 1976) makes the universe opaque to cosmic rays above approximately 5×10^{19} eV. The high energy cosmic rays emerge from interaction with the photons with reduced energy and/or reduced mass. These energy loss processes only occur as long as the particle has high enough energy so that a microwave photon is above the threshold for production of a pion (or for photodisintegration) in the rest frame of the incident cosmic ray particle. This leads to some degree of "pile-up" just below the cutoff as a result of accumulation of particles from higher energy falling below the threshold. This feature, which is expected for a cosmological distribution of cosmic ray sources, is known as the GZK cutoff.

The exact nature of the GZK cutoff depends on the distribution of sources as a function of redshift and on the injection spectrum at the sources. In particular, the acceleration mechanisms suggested for extragalactic cosmic rays typically have maximum energy in the same energy region as expected for the cutoff. Therefore the spectrum could be quite complicated in this energy region, perhaps revealing a contribution from a nearby high-energy source (to account for the highest energy events) superimposed on contributions from a variety of sources, some with cutoffs intrinsic to the acceleration mechanism and some cutoff by the GZK mechanism. Aharonian & Cronin, 1994 give a discussion of the evolution of extragalactic cosmic ray spectra through the microwave background motivated by the highest energy events. They emphasize the point that, no matter how high its initial energy, a particle observed with 3×10^{20} eV, must have originated within 50 Mpc.

Sigl, Schramm & Bhattacharjee (1994) have emphasized the possibility that one manifestation of grand unification could be the production of high energy particles when topological defects (TD) left over from the big bang annihilate (Sigl, 1996). The radiation would be at the scale of Grand Unified Theories (GUT-scale $\sim 10^{24}$ to $\sim 10^{25}$ eV). The spectrum would be quite hard (Bhattacharjee, Hill & Schramm, 1992). The most numerous secondaries would be photons and neutrinos from pions produced in the hadronic, parton-induced jets from the GUT-scale annihilation. Clearly, the high energy of the $> 10^{20}$ eV events is no problem for this model, although the rate of annihilations is uncertain. A potential problem is the radiation produced at lower energy by photon cascading in the diffuse radio background (Protheroe & Stanev, 1997). If the annihilation rate is adjusted to fit the observed intensity of cosmic rays with energy $> 10^{20}$ eV, then there is a danger of over-producing diffuse γ-rays in the sub-GeV energy range (Protheroe & Stanev, 1997). This question is still under investigation.

Figure 5 shows the measured profile of the Fly's Eye event compared to simulated atmospheric showers initiated by protons and by iron nuclei (Gaisser, Lipari & Stanev, 1997). The shower size is shown normalized to the energy in GeV and is plotted as a function of atmospheric depth along the shower trajectory in g/cm^2. There are approximately 2×10^{11} particles in the shower near maximum! The simulations shown here are based on the SIBYLL model (Fletcher *et al.*, 1994). In this model, this particular shower looks more like a heavy nucleus than a proton, but other models of hadronic interactions (e.g. the quark-gluon string model, Kalmykov, Ostapchenko & Pavlov, 1997) may give showers at this energy that develop somewhat higher in the atmosphere. The fitted depth of shower maximum is ≈ 815 g/cm^2 (Bird *et al.*, 1995). In contrast, the depth of maximum of an electromagnetic cascade initiated by a photon of 3×10^{20} eV is 1070 g/cm^2. Accounting for the Landau-Pomeranchuk-Migdal (LPM) effect (Landau & Pomeranchuk, 1953; Migdal, 1956) would make a calculated electromagnetic shower penetrate deeper (Stanev, 1982). On the other hand, there could be significant cascading in the geomagnetic field before the photon reaches the atmosphere (Stanev & Vankov, 1997), and this would have the effect of making the shower develop higher in the atmosphere.

In summary, the highest energy shower has a profile that looks like a nucleus or proton of extraordinarily high energy, but a photon cannot be completely ruled out as the primary of this event. Whether there is structure in the spectrum around 10^{20} eV, perhaps indicating a separate source for the particles beyond the GZK cutoff, cannot be answered without significantly more events. The Auger project (Auger Collaboration, 1996), as well as proposals such as OWL (Ormes *et al.*, 1996) and Airwatch (Linsley, *et al.*, 1997) to observe $> 10^{20}$ eV events from space, indicate the great interest in the nature and origin of the highest energy cosmic rays.

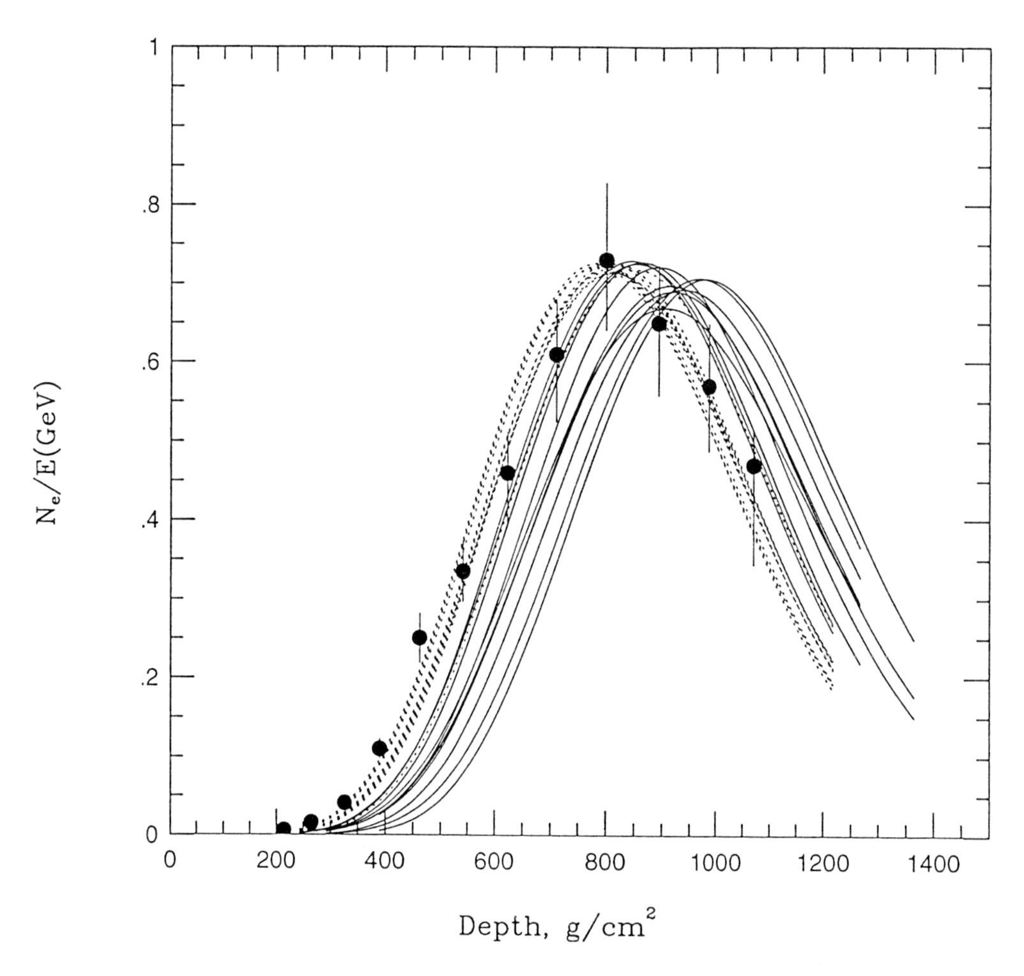

FIGURE 5. Profile of 3×10^{20} eV event (Bird *et al.*, 1995). Dashed lines show 10 simulated iron events; solid lines show 10 simulated proton events calculated with the version of SIBYLL described by Gaisser, Lipari & Stanev (1997).

5. High energy neutrino astronomy?

The detection of high energy neutrinos would open a new window on the study of energetic astrophysical processes because the neutrinos can emerge from deep inside their sources. In principle, neutrino astronomy is complementary to γ-ray astronomy. Neutrinos would be expected whenever accelerated hadrons interact, either with gas or with photons, to produce pions. The neutral pions decay to photons and the charged pions to ν_μ and $\mu \rightarrow \bar{\nu}_\mu + \nu_e + e$. There could be circumstances in which high energy photons are absorbed in the source and only the neutrinos escape. Gamma-ray sources involving only accelerated electrons would not be neutrino sources.

Most of the theoretical ideas and experimental techniques for high energy neutrino astronomy were first developed in connection with the DUMAND project (Learned, 1993). The basic idea is to instrument a large volume of water (or ice) with photomultipliers and look for the characteristic signal of Cherenkov radiation from energetic upward or horizontal muons produced in charged current interactions of ν_μ. Large depth combined with good up-down discrimination is needed to identify signals in a large background of downward muons from cosmic-ray interactions in the atmosphere. Effective volumes much larger than the instrumented volume can be obtained for ν_μ with energies in the TeV range and above because of the long range of high energy muons. Charged current interactions of ν_e (and ν_τ)† can also be detected, but their interactions would produce an electromagnetic (or hadronic) cascade with a typical length of 5 to 10 meters inside the instrumented volume. Another approach, which becomes efficient for $E_\nu >> 1000$ TeV is to search for the radio signals produced when neutrinos interact in ice (Frichter, Ralston, & McKay, 1995). Horizontal air showers may also be used to search for extremely energetic neutrinos. Price (1996) compares the effective volumes of the different techniques as a function of neutrino energy, E_ν.

At present there are two operating arrays, Lake Baikal (Belolaptikov, *et al.*, 1997) and AMANDA (Barwick, *et al.*, 1997) There are also two proposals for large neutrino detectors in the deep ocean, Nestor (Resvanis *et al.*, 1992; Capone *et al.*, 1997) and Antares (de Botton *et al.*, 1997). It is generally agreed that a kilometer-scale detector is needed to study high energy astrophysical neutrinos (see Gaisser, Halzen & Stanev, 1995 for a review). The most promising sources may be associated with extragalactic sources, such as AGN (Mannheim, 1995; Stecker & Salamon, 1996; Szabo & Protheroe, 1994) or GRB (Waxman & Bahcall). The most likely sources in fact correspond to possible models for the origin of extragalactic cosmic rays (Gaisser, Halzen & Stanev, 1995).

Rather than discuss the predictions of individual models here, I will discuss a simplified flux which illustrates the features of signals expected in more realistic models. As noted in the previous section, the energy density in extragalactic cosmic rays is likely to be at least 3×10^{-19} erg/cm^3. Since high energy neutrinos require the presence in the sources of energetic hadrons, I use this estimate of energy density to normalize a trial neutrino spectrum. Such a situation would correspond to sources of extragalactic cosmic rays in which the accelerated protons were magnetically trapped in the source until they interact with ambient photons or gas to produce fragment nucleons and pions. Neutral particles (neutrinos and high energy neutrons) could escape and could lead to similar energy densities and spectra for neutrinos and cosmic ray protons (from decay of neutrons outside the source).

If the neutrino spectrum is expressed as

$$E_\nu \frac{dN}{dE_\nu} = A\, E_\nu^{-\gamma}, \tag{5.40}$$

then the normalization condition is

$$\frac{c}{4\pi} \int A\, E_\nu^{-\gamma} = 3 \times 10^{-19}\ \mathrm{erg/cm}^3. \tag{5.41}$$

For example, for $\gamma = 1$

$$A = \frac{4 \times 10^{-7}}{\ln(E_{max}/E_{min})} \frac{\mathrm{GeV}}{\mathrm{cm}^2\,\mathrm{sr\,s}}. \tag{5.42}$$

We now estimate the signal of neutrino-induced muons that would result from a flux of ν_μ normalized as in Eq. 5.42.

† Learned & Pakvasa (1995) point out that a ν_τ of sufficiently high energy would produce two resolvable bursts, one when the ν_τ is produced and one when it decays. For $E_\nu \sim 1000$ TeV the two events would typically be separated by 100 m.

5.1. *Neutrino-induced muons*

The rate of neutrino-induced muons is related to the flux of ν_μ (and $\bar{\nu}_\mu$) by

$$\frac{d\,Rate}{d\ln E_\nu} = Area(\theta) \times \frac{dN_\nu}{d\ln E_\nu} \times P_{\nu\to\mu}(E_\nu, E_{\mu,min}), \tag{5.43}$$

where

$$\begin{aligned} P_{\nu\to\mu} &= \int_{E[\mu,min]}^{E[\nu]} N_A \frac{d\sigma_\nu(E_\nu)}{dE_\mu} R(E_\mu)\,d\,E_\mu \\ &\sim N_A\,\sigma_\nu(E_\nu)\,R\,[(1-\langle y\rangle)E_\nu] \end{aligned} \tag{5.44}$$

is the probability that a neutrino on a trajectory through the detector produces a muon that reaches the detector. N_A is Avogadro's number, y is the fraction of energy to hadrons in a charged current interaction of a ν_μ (or $\bar{\nu}_\mu$) and $R(E_\mu)$ is the range in g/cm^2 of a muon with energy E_μ. The last line of this equation allows one to make reasonably accurate estimates easily which display the dependence of the signal on neutrino energy. To display the dependence on muon energy it is straightforward to include the convolution of the differential cross section with the neutrino range as obtained, for example, from Lipari & Stanev, 1991.

The neutrino cross sections at high energy require extrapolation beyond the energies at which structure functions are known (Gandhi *et al.*, 1996). The neutrino-nucleon cross section is nearly proportional to E_ν up to $E_\nu \approx \frac{1}{2}M_W^2/m_p \approx 3400$ GeV, after which the increase with energy is less rapid. The corresponding interaction length decreases with energy and becomes comparable to the diameter of the Earth for $E_\nu \approx 40$ TeV, so that higher energy upward neutrinos are increasingly obscured. Note that this effect leads to an angular dependence for neutrinos with $E > 100$ TeV because the vertically upward neutrinos are absorbed while the more nearly horizontal neutrinos can penetrate the Earth.

For illustrative purposes, I use a simplified treatment of the range in Eq. 5.44 that displays the essential physical effects of muon propagation. The energy-loss rate is

$$\frac{d\,E_\mu}{d\,X} = -\alpha - E/\xi, \tag{5.45}$$

with $\epsilon_{critical} = \alpha\xi \sim 500$ GeV. The first term in Eq. 5.45, which varies slowly with energy, represents ionization loss, while the last term represents radiative losses due to pair production, bremsstrahlung and hadronic interactions. At the critical energy ϵ the two energy loss terms are equal. The solution for the average energy is

$$E(X) \sim (E_0 + \epsilon)\,exp[-X/\xi] - \epsilon, \tag{5.46}$$

and the range R (neglecting fluctuations) is obtained by setting $E(R) = 0$:

$$R \approx \xi\,\ln(1 + E_0/\epsilon), \tag{5.47}$$

with $\xi \approx 2$ km water equivalent. (See Lipari & Stanev, 1991 for a nice discussion of the effects of fluctuations, which depend on the neutrino spectrum. Formulas and tables for muon energy loss in various media are given by Lohmann, Kopp & Voss, 1985.)

The muon range is shown in Fig. 6 as a function of its initial energy. It is interesting that for $E_{\mu,0} > 0.5$ TeV $R > 1$ km. Thus for TeV (and higher energy) neutrinos the effective volume for neutrino-induced muons is larger that the physical volume of the detector even for a kilometer-cubed detector. This is the reason that this mode of detection is of such great interest.

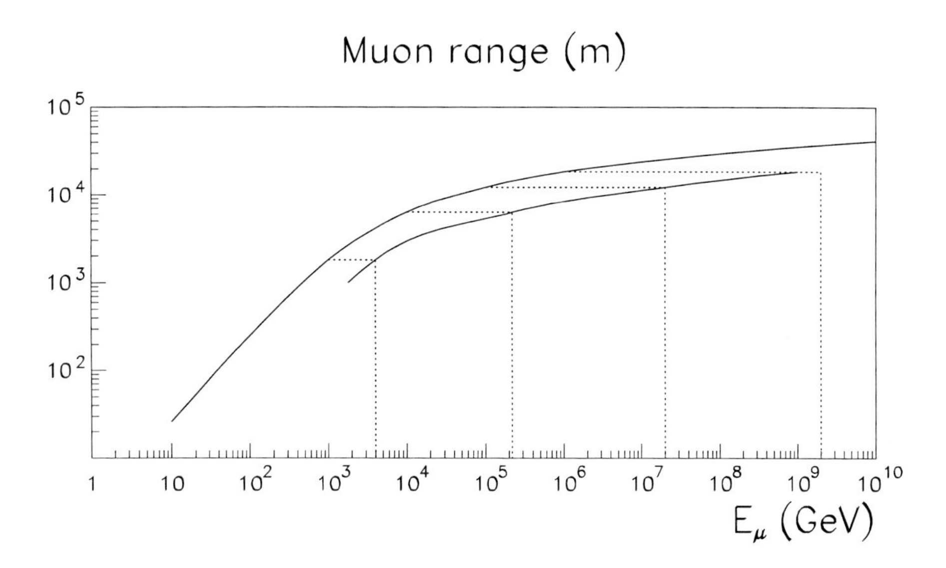

FIGURE 6. Muon range in meters of water equivalent (m.w.e.) vs. muon energy E in GeV is shown as the upper solid line. The lower solid line shows the relation between the initial muon energy and the median muon energy at the detector (see text).

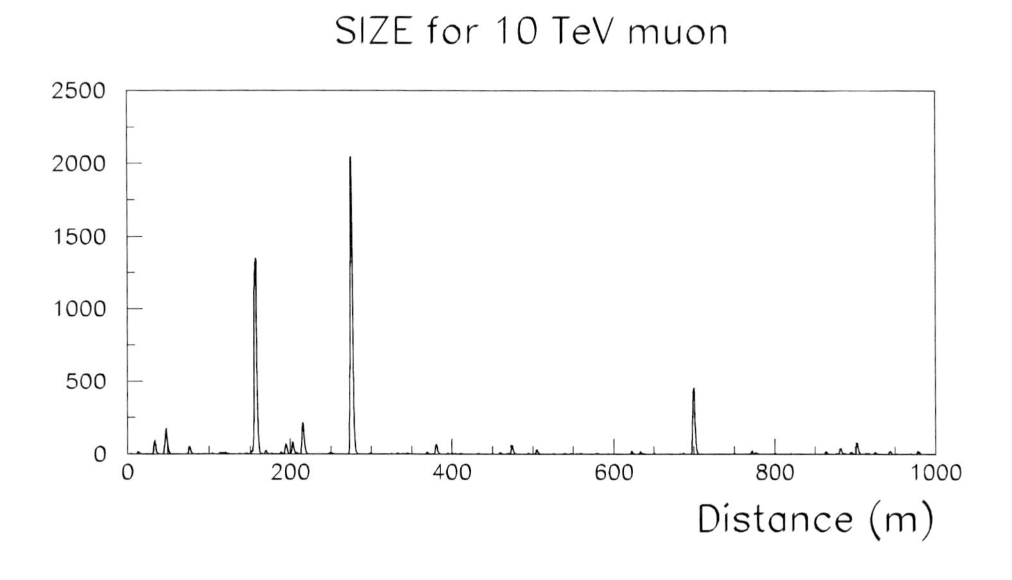

FIGURE 7. Energy loss profile of a 10 TeV muon propagating through 1 km of water or ice. Left hand scale is approximately GeV of energy deposited.

5.2. *Signature of astrophysical high energy neutrinos*

Returning to the illustrative example of a diffuse flux of neutrinos with an E^{-1} integral spectrum normalized as in Eq. 5.42, it is straightforward to us Eq. 5.47 together with the charged current cross sections tabulated by Gandhi *et al.*, 1996 to evaluate Eq. 5.43 for this example. The rate 5.43 peaks around 30 TeV at a level that is still below the diffuse flux of atmospheric neutrinos (Agrawal *et al.*, 1996). Because the assumed astrophysical spectrum is harder than the atmospheric flux, however, the rate exceeds the atmospheric neutrino background for $E_\nu >\sim 2 \times 10^5$ GeV.† For this example, the total rate of neutrino-induced muons in the upward hemisphere with $E_\mu > 10^5$ GeV in a km^3 detector is approximately 100 per year. In a situation with a uniform distribution of weak sources, it is unlikely that a single source will be strong enough to be detected if the total signal from all directions is as low as 100 per year.

This somewhat artificial example therefore illustrates one of the principal challenges of neutrino astronomy: Some degree of energy resolution is desirable to help discriminate a diffuse astrophysical signal from the atmospheric background. Fortunately, the nature of the muon energy loss at high energy provides this possibility. To see this, we first need to look at the relation between the energy of the muon at production and its energy at the detector.

A charged-current interaction of a ν_μ (or $\bar{\nu}_\mu$) can occur anywhere between the full range away from the detector to just outside (or perhaps inside) the detector volume. Thus the distance a detected muon has traveled before reaching the detector has a median

$$R_{\frac{1}{2}} \approx \frac{1}{2} \ln(1 + E_0/\epsilon), \tag{5.48}$$

with a flat distribution. The corresponding distribution of energies at the detector is therefore very broad with a median

$$E_\mu(median) \approx \epsilon \sqrt{(1 + E_0/\epsilon)} - \epsilon. \tag{5.49}$$

The right-hand curve in Fig. 6 represents this relation. For example, the median energy at the detector of a muon that starts with 4 TeV is 1 TeV, while for 200 TeV it is 10 TeV. For $E_\mu \gg \epsilon$, $E_\mu(\text{median}) \approx \sqrt{\epsilon\, E_0}$.

In the example under discussion, we therefore want to distinguish muons with E_μ more than $\sim$ 10 TeV at the detector. Fig. 7, made from calculations by Stanev, shows the propagation history of a single 10 TeV muon through 1 km of water (or ice). The upper panel is a linear scale and the lower panel the same event on a logarithmic scale. What is plotted is the total number of electrons and positrons along the track. Each burst is a cascade with a typical length in water of 5 m. Since for each sub-shower

$$E_{shower} \approx 1\, GeV \times Size\ at\ maximum \tag{5.50}$$

for showers in this energy range, the vertical axis can also be read as number of GeV deposited per burst.

The radiative processes approximately scale with energy in this region, so we can summarize the situation by noting that a muon with $E_\mu \gg$TeV will typically generate a couple of bursts per kilometer in which it radiates about 10 per cent of its energy. This is the characteristic signature that a km^3 detector should be sensitive to.

† There is significant uncertainty in the crossover energy because of the uncertainty in the level of atmospheric neutrinos from decay of charm.

REFERENCES

Aharonian, F.A. & Cronin, J.W. 1994 *Phys. Rev.* **D50**, 1892.

Agrawal, V., Gaisser, T.K., Lipari, P. & Stanev, T. 1996 *Phys. Rev. D* **53**, 1314.

Auger Collaboration 1996 *The Pierre Auger Project Design Report* http://www.fnal.gov. See also *Cosmic Rays Above* 10^{19} *eV Nucl. Phys. (Proc. Suppl.)* **28B**.

Axford, W.I. 1994 *Ap.J. (Suppl.)* **90**, 937.

Axford, W.I., Leer, E. & Skadron, G. 1977 *Proc. 15th Int. Cosmic Ray Conf.* (Plovdiv) **11**, 132.

Barwick, S.W. *et al.* 1997 *Proc. 25th Int. Cosmic Ray Conf.* (Durban) **7** 1.

Bell, A.R. 1978 *Mon. Not. R. Astr. Soc.* **182**, 147 and 443.

Belolaptikov, I.A. *et al.* 1997 *Astroparticle Physics* **7**, 263.

Berezhko, E.G. 1996 *Astroparticle Physics* **5**, 367.

Berezhko, E.G. & Völk, H.J. 1997 *Astroparticle Physics* **7**, 183.

Bhattacharjee, P., Hill, C.T. & Schramm, D.N. 1992 **Phys. Rev. Letters 69**, 567.

Biermann, P.L., Gaisser, T.K. & Stanev, T. 1995 *Phys. Rev. D* **51**, 3450.

Bird, D.J. *et al.* 1993 *Phys. Rev. Letters* **71** 3401.

Bird, D.J. *et al.* 1994 *Ap.J.* **424**, 491.

Bird, D.J. *et al.* 1995 *Ap.J.* **441**, 144.

Blandford, R.D. & Ostriker, J.P. 1978 *Ap.J.* **221**, L29.

Blandford, R.D., Netzer, H. & Woltjer, L. 1990 *Active Galactic Nuclei* Springer Verlag.

Buckley, J.H., *et al.* 1997 *Astron. Astrophys.* (to be published).

Capone, A. *et al.* 1997 *Proc. 25th Int. Cosmic Ray Conf.* (Durban) **7**, 49.

Cassidy, M. *et al.* 1997 *Proc. 25th Int. Cosmic Ray Conf.* (Durban) **5**, 189.

Cherry, M. 1997 *Proc. 25th Int. Cosmic Ray Conf.* (Durban) **4**, 1.

de Botton, N. *et al.* 1997 *Proc. 25th Int. Cosmic Ray Conf.* (Durban) **7**, 17.

Dickinson, J.E. *et al.* 1997 *Proc. 25th Int. Cosmic Ray Conf.* (Durban) **5**, 229.

Drury, L.O'C., Aharonian, F.A. & Völk, H.J. 1994 *Astron. Astrophys.* **287**, 959.

Erlykin, A.D. & Wolfendale, A.W. 1997a *Astroparticle Physics* **7**, 1.

Erlykin, A.D. & Wolfendale, A.W. 1997b *Astroparticle Physics* **7**, 203.

Esposito, J.A., Streitmatter, R.E., Balasubrahmanyan, V.K. & Ormes, J.F. 1990 *Ap.J.* **351**, 459.

Esposito, J.A., Hunter, S.D., Kanbach, G. & Sreekumar, P. 1996 *Ap.J.* **461**, 820.

Fabian, A.C. 1994 *Ann. Rev. Astron. Astrophys.* **32**, 277.

Fermi, E. 1949 *Phys. Rev.* **75**, 1169.

Fishman, G.J. & Meegan, C.A. 1995 *Ann. Rev. Astron. Astrophys.* **33**, 415.

Fletcher, R.S., Gaisser, T.K., Lipari, P. & Stanev, T. 1994 *Phys. Rev. D* **50**, 5710.

Frichter, G.M., Ralston, J.P. & McKay, D.W. 1995 *Phys. Rev. D* **53**, 1684.

Gaisser, T.K. 1990 *Cosmic Rays and Particle Physics*, Cambridge University Press.

Gaisser, T.K., Halzen, F. & Stanev, T. 1995 *Physics Reports* **258**, 173 (erratum: 1996 **271**, 355). See also astro-ph/9707283 by T.K. Gaisser.

Gaisser, T.K. 1997 *Nucl. Phys. B (Proc. Suppl.)* **52B**, 17. See also Gaisser, T.K. *et al.* in Snowmass 1994: *Particle and Nuclear Astrophysics in the next Millennium* (ed. Kolb, E.W. & Peccei, R.D.; World Scientific, 1995) p. 371.

Gaisser, T.K., Lipari, P. & Stanev, T. 1997 *Proc. 25th Int. Cosmic Ray Conf.* (Durban) **6**, 281.

Gaisser, T.K., Protheroe, R.J. & Stanev, T. 1997 *Ap.J.* (to be published).

Gandhi, R., Quigg, C., Reno, M.H. & Sarcevic, I. 1996 *Astroparticle Physics* **5**, 81.

GARCIA-MUNOZ, M., G.M. MASON & J.A. SIMPSON 1977 *Ap.J.* **217**, 859

GARCIA-MUNOZ, M. *et al.*, 1987 Ap.J. (Suppl.), 64, 269

GINZBURG, V.L. & SYROVATSKII, S.I. 1964 *The Origin of Cosmic Rays*, Pergamon Press.

GINZBURG, V.L., KHAZAN, YA.M. & PTUSKIN, V.S. 1980 *Astrophys. Space Sci.* **68**, 295.

GREISEN, K. 1966 *Phys. Rev. Letters* **16**, 748.

HARDING, A.K. & GAISSER, T.K. 1990 *Ap.J.* **358**, 561.

HEINBACH & SIMON, M. 1995 *Ap.J.* **441**, 209.

HILLAS, A.M. 1984 *Ann. Rev. Astron. Astrophys.* **22**, 425.

HUNTER, S.D., *et al.* 1997 *Ap. J.* (to be published).

JOKIPII, J.R. 1987 *Ap.J.* **313** 842 and references therein.

JONES, F.C. & ELLISON, D.C. 1991 *Space Sci. Rev.* **58**, 259. See also the short paper, "Some non-linear effects in diffusive shock acceleration", Ellison, D.C. in *Proc. of Gamow Seminar* St. Petersburg, Sept. 12-14, 1994.

KALMYKOV, N.N. & KHRISTIANSEN, G.B. 1995 *J. Phys. G* **21**, 1279.

KALMYKOV, N.N., OSTAPCHENKO, S.S. & PAVLOV, A.I. 1997 *Nucl. Phys. B (Proc. Suppl.)* **52B**, 17.

KANG, H., RACHEN, J.P. & BIERMANN, P.L. 1997 *Mon. Not. R. Astron. Soc.* **286**, 257.

KARLE, A. *et al.* 1995 *Astroparticle Physics* **3**, 321.

KLAGES, H.O. *et al.* 1997 *Nucl. Phys. B (Proc. Suppl.)* **52B**, 92.

KOYAMA, K. *et al.* 1995 *Nature* **378**, 255.

KRYMSKY, G.F. 1977 *Soviet Physics Doklady* **23**, 327.

LAGAGE, P.O. & CESARSKY, C.J. *Astron. Astrophys.* **118**, 223 and **125**, 249.

LANDAU, L.D. & LIFSHITZ, E.M. 1982 *Fluid Mechanics* (Pergamon Press, Oxford), pp. 315, 331.

LANDAU, L.D. & POMERANTCHUK, I. 1953 *Dokl. Akad. Nauk. SSSR* **92**, 535.

LEARNED, J.G. 1993 *Nucl. Phys. B (Proc. Suppl.)* **33A,B**, 77.

LEARNED, J.G. & PAKVASA, S. 1995 *Astroparticle Physics* **3**, 267.

LINSLEY, J. *et* 1997 *Proc. 25th Int. Cosmic Ray Conf.* (Durban) **5**, 381, 385.

LIPARI, P. & STANEV, T. 1991 *Phys. Rev. D* **44**, 3543.

LOHMANN, W., KOPP, R. & VOSS, R. 1985 *CERN Yellow Report* No. EP/85-03 (unpublished).

LONGAIR, M.S. *High Energy Astrophysics* 1992 Cambridge University Press.

LOZINSKAYA, T.A. 1992 *Supernovae and Stellar Wind in the Interstellar Medium* (American Institute of Physics).

MANNHEIM, K. 1995 *Astroparticle Physics* **3**, 295.

MASTICHIADIS, A. 1996 *Astron. Astrophys.* **305**, 53.

MCKEE, C.F. & TRUELOVE, J.K. 1995 *Physics Reports* **256**, 157.

MIGDAL, A.B 1956 *Phys. Rev.* **103**, 1811.

MIRALDA-ESCUDÉ, J. & WAXMAN, E. 1996 *Ap.J.* **462**, L59 and WAXMAN, E. 1995 *Phys. Rev. Letters* **75**, 386.

MEYER, J.-P., DRURY, L.O'C. & ELLISON, D.C. and ELLISON, D.C., DRURY, L.O.'C. & MEYER, J.-P. 1997 *Ap.J.* (to be published).

NRC REPORT *Opportunities in Cosmic-Ray Physics and Astrophysics*, Report of the Committee on Cosmic-Ray Physics of the Board on Physics and Astronomy of the National Research Council (National Academy Press, 1995).

NORMAN, C.A., MELROSE, D.B. & ACHTERBERG, A. 1995 *Ap.J.* **454**, 60.

ORMES, J. *et al.* 1996 *Orbiting array of Wide angle Light collectors* GSFC PROPOSAL, HTTP://LHEAWWW.GSFC.NASA.GOV/DOCS/GAMCOSRAY/HECR/OWL.

PARKER, E.H. 1966 *Ap.J.* **145**, 811. SEE ALSO *Space Sci. Rev.* **9**, 654.

PARKER, E.H. 1958 *Ap.J.* **128**, 664.

PATTERSON, J.R. & HILLAS, A.M. 1983 *J. Phys. G* **9**, 1433.

PRICE, P.B. 1996 *Astroparticle Physics* **5**, 43.

PROTHEROE, R.J. & SZABO, A.P. 1992 *Phys. Rev. Letters* **69**, 2885.

PROTHEROE, R.J. & STANEV, T. 1997 *Phys. Rev. Letters* **77**, 3708.

PUGET, J.L., STECKER, F.W. & BREDEKAMP, J.H. 1976 *Ap.J.* **205**, 638.

RACHEN, J.P. & BIERMANN, P.L. 1993 *Astron. Astrophys.* **272**, 161.

REBEL, H. SCHATZ, G. & KNAPP, J. (EDS.) 1997 *Very High energy Cosmic Ray Interactions Nuclear Physics B (Proc. Suppl.)* **52B**.

RESVANIS, L.K. 1996 *Space Science Reviews* **75**, 213 (AND REFERENCES THEREIN).

REYNOLDS, S.P. 1994 *Ap.J. (Suppl.)* **90**, 845.

REYNOLDS, S.P. 1996 *Ap.J.* **459**, L13.

SEO, E.S. & PTUSKIN, V.S. 1994 *Ap.J.* **431**, 705.

SIGL, G. 1996 *Space Sci. Rev.* **75**, 375.

SIGL, G., SCHRAMM, D.N. & BHATTACHARJEE 1994 *Astroparticle Physics* **2**, 401.

SILBERBERG, R. *et al.* 1990 *Ap.J.* **363**, 265.

SOKOLSKY, P., SOMMERS, P. & DAWSON, B.R. 1992 *Phys. Reports* **217**, 225.

STANEV, T. 1982 *Phys. Rev. D* **25**, 1291.

STANEV, T. & VANKOV, H.P. 1997 *Phys. Rev.* **D55**, 1365.

STECKER, F W.. 1971 *Cosmic Gamma Rays*, (NASA SCIENTIFIC AND TECHNICAL INFORMATION OFFICE) NASA SP-249.

STECKER, F.W. & SALAMON, M.H. 1996 *Space Sci. Rev.* **75**, 341 AND REFERENCES THEREIN.

SWORDY, S. 1993 *Proc. 23rd Int. Cosmic Ray Conf.: Invited, Rapporteur & Highlight Papers* (WORLD SCIENTIFIC, ED. LEAHY, D.A., HICKS, R.B. & VENKATESAN, D., 1994), 243.

SZABO, A.P. & PROTHEROE, R.J. 1994 *Astroparticle Physics* **2**, 375. SEE ALSO ASTRO-PH/9607165 BY R.J. PROTHEROE.

VIETRI, M. 1995 *Ap.J.* **453**, 883.

VÖLK, H.J. & BIERMANN, P.L., 1988 *Ap.J.* **333**, L65.

WADDINGTON, C.J. (ED.) 1989 *Cosmic Abundances of Matter*, A.I.P. CONF. PROCEEDINGS 183 (A.I.P., NEW YORK).

WATSON, A.A. 1997 *Proc. 25th Int. Cosmic Ray Conf.* (DURBAN), RAPPORTEUR TALK (TO BE PUBLISHED).

WAXMAN, E. & BAHCALL, J. 1997 *Phys. Rev. Letters* **78**, 2292.

WEBBER, W.R., A. LUKASIAK, F.B. MCDONALD & P. FERRANDO 1996 *Ap.J.* **457**, 435.

YOSHIDA, S. *et al.* 1995 *Astroparticle Physics* **3**, 105.

ZATSEPIN, G.T. & KUZ'MIN, V.A. 1966 *JETP Letters* **4**, 78.

Physical Cosmology for Nuclear Astrophysicists

By DAVID N. SCHRAMM

Department of Astronomy and Astrophysics, University of Chicago, Chicago, IL 60637, USA

This lecture series provides an overview of modern physical cosmology with an emphasis on nuclear arguments and their role in the larger framework. In particular, the current situation on the age of the universe and the Hubble constant are reviewed and shown now to be in reasonable agreement once realistic systematic uncertainties are included in the estimates. Big bang nucleosynthesis is mentioned as one of the pillars of the big bang along with the microwave background radiation. It is shown that the big bang nucleosynthesis constraints on the cosmological baryon density, when compared with dynamical and gravitational lensing arguments, demonstrate that the bulk of the baryons are dark and also that the bulk of the matter in the universe is non-baryonic. The recent extragalactic deuterium observations as well as the other light element abundances are examined in detail. Comparison of nucleosynthesis baryonic density arguments with other baryon density arguments is made.

---

1. Introduction

Modern physical cosmology has entered a "golden period" where a multitude of observations and experiments are guiding and constraining the theory in a heretofore unimagined manner. Many of these constraints involve nuclear physics arguments, so the interface with nuclear astrophysics is extemely active. This review opens with a discussion of the three pillar of the big bang: the Hubble expansion, the cosmic microwave background, and big bang nucleosynthesis (BBN). It will then focus on one of the most exciting topics in all the physical sciences, namely, the apparent fact that the bulk of the matter of our universe is not only not seen but seems to be made out of a different substance than baryons. This paper reviews the key arguments regarding the existence of this "dark matter." The conclusion that we need the bulk of the universe to be in the form of non-baryonic dark matter will be emphasized since this seems to require new physics beyond the standard particle model. The dark matter issue will also be related to the problem of forming large scale structure in the universe. The review will close with a discussion of some of the new experiments currently being built and what they will do.

2. Age of the Universe and the Hubble Expansion

The problem of estimating the age of the universe is longstanding. For example, in 1650, Bishop James Ussher (1658) determined, by a technique of summing the Biblical begats and making other corrections and connections based on the then available historical and astronomical records, that the universe began in 4004 BC, at the moment that would correspond to sunset in Jerusalem on the evening before October 23. This would correspond to 4 PM U.T. on October 22.

This early determination illustrates a key point which we will also apply to more modern techniques. Namely, while Bishop Ussher was able to obtain a result with reported accuracy of about 8 significant figures, his systematic errors are considerably larger. (Even his intrinsic error is larger than the accuracy of his result indicates, since the Jewish calendar, using essentially the same technique, obtains an age that is over 200 years off from Ussher's.)

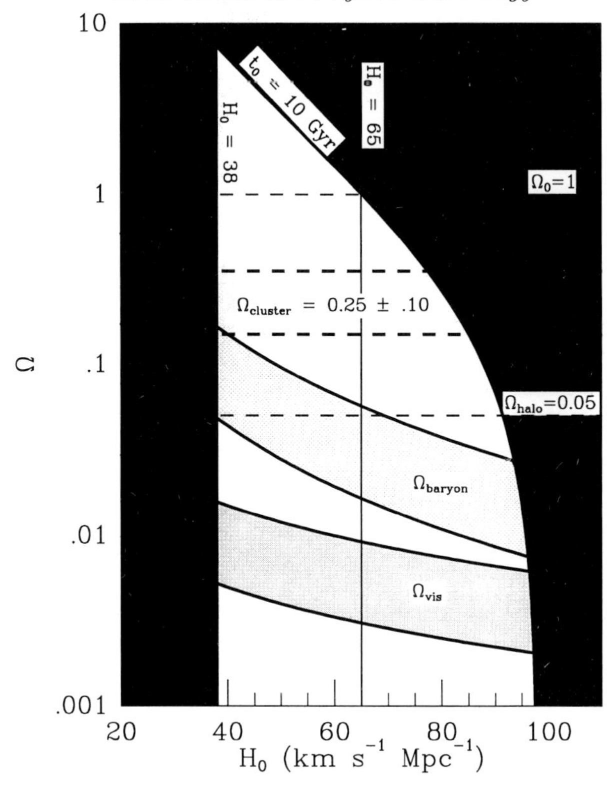

FIGURE 1. An updated version of $H_0 - \Omega$ diagram of Gott, Gunn, Schramm & Tinsley (1974) showing that Ω_b does not intersect $\Omega_{VISIBLE}$ for any value of H_0 and that $\Omega_{TOTAL} > 0.1$, so non-baryonic dark matter is also needed (Shi *et al.* 1995).

Today the age of the universe can be estimated by three independent means:

- 1. Dynamics (Hubble age and deceleration)
- 2. Oldest Stars (globular clusters)
- 3. Radioactive Dating (nucleocosmochronology)

We will see that despite much activity on the dynamical technique, the best age bounds are still those derived from nuclear arguments - namely #2 and #3. Each of these gives a lower bound of $t \gtrsim 10$ Gyr as plotted in Figure 1. Furthermore, the age of the disk also bounds the age of the universe ($t_{disk} \sim 10Gyr$), as does radioactive dating of the Earth-Meteorite system at $t_{ss} = 4.6 \pm 0.1Gyr$.

2.1. *The Age From Dynamics*

The use of the Hubble constant to determine an age is the most quoted and still least accurate of all the age determination methods. The point is that it is not really determining an age but only a dynamic timescale. For perspective let us note that in the past decade astronomers have published values ranging from $H_0 \sim 100$ km/sec/Mpc down to values near $H_0 \sim 40$ km/sec/Mpc. The higher values tended to come from people using empirical techniques like Tulley-Fisher, whereas the smaller values came primarily from people using supernovae. In principle, supernovae are better understood physically, but some astronomical calibrations inevitably creep in. However, few hidden-variables should creep in since the physics is in reasonable shape, unlike the empirical technology. A critical question tends to be the accuracy of intermediate distance calibrators and the correction for infall into the Virgo cluster. Most of us can't see anything wrong at face value with the Tulley-Fisher techniques other than a possible susceptibility to the so-called Malmquist bias. However, many physicists have a certain fondness for the use of Type-I supernovae as standard candles. Type I's seem to be due to the detonation of a C-O white dwarf star converting its C-O to Fe. Such a model has a physical relationship between its luminosity and basic nuclear quantities that can be measured in the lab. Current best-fit models (c.f. Nomoto) tend to convert about $0.7M_\odot$ of C-O, which yields $H_0 \sim 60$ km/sec/Mpc. However, even in the extreme where the entire $1.4M_\odot$ Chandrasekhar mass is burned, H_0 is never below ~ 40 km/sec/Mpc (see also Nugent et al. 1995). Sandage and Tammann's (1995) empirical calibrations, which ignore the nuclear mechanism, now yield $H_0 \sim 58 \pm 7$ km/sec/Mpc after using HST-measured cepheids to calibrate M101, which fall within the theoretically allowed range and correspond to almost complete burning of a Chandrasekhar core. Recently, Riess, Press and Kirshner (1995) have argued that there may be some variation in type IA light curves which shifts H_0 up to $\sim 66 \pm 3$. Kirshner (1992) also argues that the expanding photosphere of type II supernova implies $H_0 \sim 73$. While selecting between 40 and 75 is still a matter of choice, it does seem that values less than 40 can be reliably excluded. Why these numbers tend to be systematically lower than the Tulley-Fisher numbers remains to be fully understood.

Most recently there has been much publicity about the Hubble Space Telescope (HST) seeing individual cepheid variable stars in Virgo Cluster (Freedman *et al.* 1994) galaxies as well as other potential calibrator galaxies out to about 20 Mpc. Over the next few years, HST will find many more cepheids in other galaxies in Virgo so that part of the uncertainty will decrease. Freedman et al. (1997) have been very conscientious in listing both statistical and systematic errors in their recently quoted valuc of $H_0 = 73 \pm 7 \pm 8$ km/sec/Mpc. Although most astronomers add the errors in the quadrature, it is probably more realistic not to add systematic errors in quadrature since the second derivatives of these systematic errors are probably not well behaved. In fact, in some cases, the distributions may even be bi-modal. Hence, a better estimate is 73 ± 15. Even this large error does not include the possibility that cepheids themselves may have a systematic shift in luminosity between the LMC (where the calibration is done) and other galaxies. This latter point has now been supported with the new Hipparcus observations of Feist and Catchpole (1997) which seem to increase the cepheid scale by $\sim$10%. Historically, Hubble got $H_0 \sim 500$ due to using cepheids calibrated from Pop II objects and applied to Pop I in other galaxies. While much of the metalicity effect is now taken into account, the observed trends of cepheids in M31 may hint that there is still some residual effect. Thus, while some systematic errors will be reduced with more HST detections of cepheids in other galaxies, some systematic errors will remain (including potential differential reddening between the southern LMC direction and the northern Virgo direction).

With all of these systematics, it is clear at the present time that the SN technique and the Tulley-Fisher techniques are not really in conflict. Recently, the first principles technique of using gravitational lensing, and the time delay between two images have entered the fray. In particular, Kundic et al. (1997), using the Q50 0957$\pm$561 A, B system found a time delay of 417 ± 3 days which yields $H_0 = 63 \pm 12$ at 95% confidence where the uncertainly now is primarily due to the mass distribution in the lens. Furthermore, Schechter et al. (1996) using PG1115 + 080 found a 24 ± 3 days time delay, which Keeton and Kochonek (1997) said yields $H_0 = 60 \pm 17$. Another first principles technique that has undergone a revolution is the Sunyaev-Zeldovich technique of observing the microwave background spectral distortion due to Compton scattering off hot x-ray gas in clusters. Carlstrom (1997) and other have now observed this effect in over a dozen clusters. The best fit value is in the mid-50's with the dominant uncertainty now coming from the x-ray data which should be remedied with AXAF. Perhaps we are finally seeing a convergance at

$$H_0 = 65 \pm 10 \text{ km/sec/Mpc.}$$

This range would include all recent determinators. Converting to years, this yields

$$\frac{1}{H_0} = 15 \pm^3_2 \text{ Gyr.}$$

While such a convergance may be a bit premature, it is, nonetheless, not a driving factor in the age discussion.

Age, t_u, is related to H_0 by:

$$t_u = \frac{f(\Omega)}{H_0}$$

where for standard matter-dominated models with cosmological constant $\Lambda = 0$,

$$f(\Omega) = \begin{cases} 1 & \Omega = 0 \\ 2/3 & \Omega = 1 \\ \sim 0.5 & \Omega = 4 \end{cases} .$$

From dynamics alone we can put an upper limit on Ω by limiting the deceleration parameter q_o. From limits on the deviations of the redshift-magnitude diagrams at high redshift, we know that $q_o \lesssim 2$ (for zero cosmological constant $\Omega = 2q_o$). Thus, we can argue that $\Omega \lesssim 4$ or that $f(\Omega) \gtrsim 0.5$. Therefore, from dynamics alone, with no further input, we can conclude only that

$$6 \lesssim t_u \; (Gyr) \; \lesssim 18$$

Since the lower bound here could almost be obtained from the age of the earth, it is clear that the dynamical technique is not overly restrictive, even with our convergence assumption.

Even high values of H_0 can be consistent with high ages by invoking the cosmological constant, $\lambda = \Lambda/3H_0^2$. Even $\lambda \sim 0.4$ allows high H_0 to be consistent with a flat universe. It is interesting to speculate that $\Lambda \neq 0$ can be produced by a late-time vaccum phase transition of the type proposed by Hill, Schramm and Fry (1989). However, such models do require a fair degree of tuning.

2.2. *The Age From the Oldest Stars*

Globular cluster dating is an ancient and honorable profession. The basic age comes from determining how long it takes for low mass stars to burn their core hydrogen and thereby move off the main sequence. The central temperature of such stars is determined by their composition and the degree of mixing. While there has certainly been some static as to what is the dispersion between the age of the youngest versus the oldest globular cluster in a given calculation, there is a surprising convergence on the age of the oldest clusters. Since the age of the very oldest cluster is the critical cosmological question, it is really somewhat of a red herring as to how much less the youngest cluster may be. The convergence on the age of the oldest does require a consistency of assumptions about primordial helium and metalicity (including O/Fe). Some differences between different groups can be explained away once agreement is made on these assumptions. For example, Sandage's(1993) oldest ages of $\sim$ 18 Gyr and Iben and Renzini's (1984) of $\sim$ 16 Gyr are consistent if the same helium is used. (Lower helium yields higher ages.

Iben assumed the Pagel (Pagel 1991) value of mass fraction, Y, = 0.23, whereas BBN favors Y$\simeq$ 0.245.) Another decrease of a billion years occurs if O/Fe is assumed high as current observations show for extreme Pop II.

Another effect is the fact that these old stable stars will have some gravitational settling of their helium which will also shorten the ages about 1 Gyr relative to calculations where core helium enrichment is purely due to nuclear burning. All of these assumptions give a standard model (Pagel & Jimenez 1996, Chaboyer *et al.* 1996, Sandage 1993, Mazzitelli *et al.* 1997)) for the oldest globulars of $\sim 14 \pm 1$ Gyr where the ± 1 is only the difference between different groups using the same standard assumptions. Additionally, Feist and Catchpole (1997) have argued from Hipparcus data that the RR Lyrae stars in the LMC should be recalibrated, which shifts the age downward to a central value of around 12 Gyr. In addition to the calculational errors, there are also uncertainties in composition/opacity, uncertainties in distance/turnoff luminosity, and uncertainties in reddening/surface temperature at turnoff which increase the statistical error from ± 1 to ± 2 Gyr. Then, there are systematic uncertainties due to model assumptions: the helium abundance, settling, O/Fe, etc. For example, helium abundances might even be enhanced from the big bang (BBN) value due to helium production accompanying the extreme Pop II metal production and perhaps preferential helium in cluster formation (Shi, Schramm & Dearborn 1995). Also note that the current best fit BBN helium is actually closer to 0.25 than 0.23. Shi, Schramm and Dearborn (1995) showed that assumptions about He could lower the best fit age by as much as 2 Gyr without violating any other constraint (e.g. Y must be ≤ 0.28 to fit RR Lyrae blue edge). Furthermore, there are recent suggestions from the first Keck spectroscopic temperature determinations of globular cluster stars that the true temperatures are as much as 200 K hotter than the photometric determinations. This could also shift the age downward by as much as 2 Gyr. Furthermore, Shi, Schramm and Dearborn (1995; see also Shi 1995) have shown that mass loss due to the variable strip crossing the main sequence near the cluster turnoff could also shift the age down by 1 to 2 Gyr. However, these combined effects do not add linearly. No matter what, low mass stars can burn their hydrogen only so fast. We estimate that systematics add an additional ± 2 Gyr which should not be added in quadrature with the ± 2 Gyr statistical uncertainty, since most of the systematic effects are binary assumptions rather than selections from smooth, well behaved distributions. Thus, we conclude that $t_{GC} = 12 \pm 2 \pm 2$ Gyr, which is in good agreement with the recent Chaboyer and Krauss (1997) estimate of 11.5 ± 1.5, but with larger allowed uncertainties.

One can use the standard solar model to get a quick estimate of an extreme globular age. The main line pp-chain is the main energy generation mechanism for the Sun and the globular clusters. The basic pp part of the solar model is now well confirmed by the calibrated GALLEX and SAGE solar neutrino experiments. Since the Sun has a much higher metalicity than the oldest globular clusters, and presumably has higher helium content and is at least as massive, if not more massive, it is paramount that the calculated main sequence lifetime of 10 Gyr for our Sun will always be a lower bound on the oldest globular cluster lifetimes. This 10 Gyr is also consistent with Shi, Schramm and Dearborn (1995) and with an independant study by Chaboyer (1995). Thus, it is reasonable to conclude that fting the "best" fit age for the oldest globulars down to 11 Gyr cannot be excluded. But an extreme lower bound at 10 Gyr is not able to be broken.

Note that the time delay for cluster formation does not change this limit, since it is certainly possible to hypothesize an isocurvature model where globular clusters are the first objects (Lee, Schramm & Mathews 1995) to form after recombination (their Jeans mass at that time is the globular cluster mass). Their Kelvin-Helmholtz time is only

Globular Clusters $t_{GC} = 13 \pm 2 \pm 2Gyr$
Long Lived Radioactive Isotopes (Nucleocosmochronology) $t_{NC} \gtrsim 10$ Gyr
Solar System $t_{SS} = 4.6 \pm 0.1$ Gyr

TABLE 1. AGE OF OLD THINGS IN THE UNIVERSE
(Age of Universe is Greater Than Age of Oldest Things)

$\sim 10^7$ yr, so in principle, they could be present as early as 10^8 yr after the big bang. (Of course, standard CDM models extend this to several Gyr.)

2.3. *Nucleocosmochronology*

Nucleocosmochronology is the use of abundance and production ratios of radioactive nuclides coupled with information on the chemical evolution of the Galaxy to obtain information about time scales over which the solar system elements were formed. Typical estimates for the Galaxy's (and Universe's) minimum age as determined from cosmochronology are of the order of 9.6 Gyr (e.g. Meyer and Schramm (1986)), using the model independent "mean age" technique correcting for second order effects. In recent years questions about the role of β-delayed fission in estimating actinide production ratios as well as uncertainties in ^{187}Re decay due to thermal enhancement and the discussion of Th/Nd abundances in stars have obfuscated some of the limits one can obtain. In particular, we note that the formalism of Schramm and Wasserburg (1970) as modified by Meyer and Schramm (1986) continues to provide firm bounds on the mean age of the heavy elements (see also recent reprint of Wasserburg and Busso 1996). In fact, Th/U provides a firm lower limit to the age and Re/Os, a firm upper limit. These limits are based solely on nuclear physics inputs and abundance determinations. To extend these mean age limits to a total age limit requires some galactic evolution input. However, as Reeves and Johns (1976) first showed, and as Meyer and Schramm (1986) developed further, one can use chronometers to constrain Galactic evolution models and thereby further restrict the age from the simple mean age limits of Schramm and Wasserburg (1970). To try to push further on such ranges and give ages to ± 1 Gyr accuracy, as some authors have done, always necessitates making some very explicit assumptions about Galactic evolution beyond the pure chronometric arguments. At the present time such model-dependent ages are not fully justified and should probably not be used as arguments to question (or support) cosmological models, but pure, nuclear derived lower bounds are very useful. In particular, the Meyer and Schramm (1986) lower bound of $t_{NC} > 9.6$ Gyr which involves the mean age and the nuclear constrains on maximal evolutionary effects is a very firm bound.

2.4. *Age Summary*

The age situation at the present time can be summarized by Figure 1 and by Table 1. We see there that an $\Omega = 1$ universe is consistent with $t > 10$ Gyr as long as $H_0 \lesssim 65$

km/sec/Mpc. If uncertainties on H_0 (including bounds on systematics) ever exclude 65, then one would require $\Lambda_o \neq 0$ to achieve the flat universe favored by inflation models.

Naively, we expect gravitational microphysics on the Planck scale, M_p to determine the scale of Λ_o. An effective $\lambda_o \sim 1$ requires $\rho_\Lambda \sim 10^{-121} M_p^4$. This seems like remarkable tuning. Of course, some late-time transition on the fraction of an eV scale could substitute for M_p if the early $\rho_\Lambda \sim M_p^4$ effects could be surpressed to more than 121 orders of magnitude. Because these problems seem awkward to avoid, most physicists think $\Lambda = 0$.

As an anthropic aside, if it were ever shown that $\Lambda_o \neq 0$, then we may have to appeal to the following anthropic argument (ugh!). While particle physics prefers a large value for $\Lambda_o \sim M_p^4$, the only values consistent with an old universe have to have $\Lambda_o < 10^{-121} M_p^4$. Thus, our existence plus particle theory would make the maximum value consistent with our existence the most likely value. (Hopefully, a better motivated physics explanation for Λ_o will eventually be found.)

To repeat the main conclusion: at present there is no age problem, even for $\Omega = 1, \Lambda = 0$ models, since the real uncertainties including systematics allow completely consistent age values.

3. The Cosmic Microwave Background (CMB)

Since non-big bang models like the "steady state" also have a Hubble expansion, the linear Hubble expansion is only a necessary but not a sufficient condition for the big bang. However, the other two pillars, the cosmic microwave background, described in this section, and big bang nucleosynthesis, described in in the next, are sufficient arguments and push our observational knowledge of the Universe back to much earlier times.

Much has been written about the Penzias and Wilson discovery of the CMB in the mid 1960s, so it will not be repeated here. Let us merely note that the COBE satellite has now measured the CMB to higher precision than any thermal blackbody spectrum has ever been measured (including laboratory measurements). The temperature of 2.728 ± 0.004 K is a remarkable achievement and the spectrum is thermal to better than parts in 10^4. See Figure 2 for a current summary of spectral data on the CMB. Thus, we can conclude that the early Universe was hot and dense enough to be in thermal equalibrium. Such a hot-dense Universe we call the "big bang." Later, we will turn to anisotropies in the CMB which COBE detected at a level of 10^{-5} and are an important new toll for probing the Universe.

4. Big Bang Nucleosynthesis

The study of the light element abundances has undergone a recent burst of activity on many fronts. New results on each of the cosmologically significant abundances have sparked renewed interest and new studies. The bottom line remains: primordial nucleosynthesis has joined the Hubble expansion and the microwave background radiation as one of the three pillars of big bang cosmology. Of the three, big bang nucleosynthesis probes the universe to far earlier times ($\sim$ 1 sec) than the other two and led to the interplay of cosmology with nuclear and particle physics.

Recent heroic observations of ^{6}Li, Be and B, as well as ^{2}D, ^{3}He and new ^{4}He determinations, have all gone in the direction of strengthening the basic picture of cosmological nucleosynthesis. Theoretical calculations of cosmic ray production of ^{6}Li, Be and B have fit the observations remarkably well, thus preventing these measurements from disturbing the standard scenario (Olive 1990). The recent reports of D/H

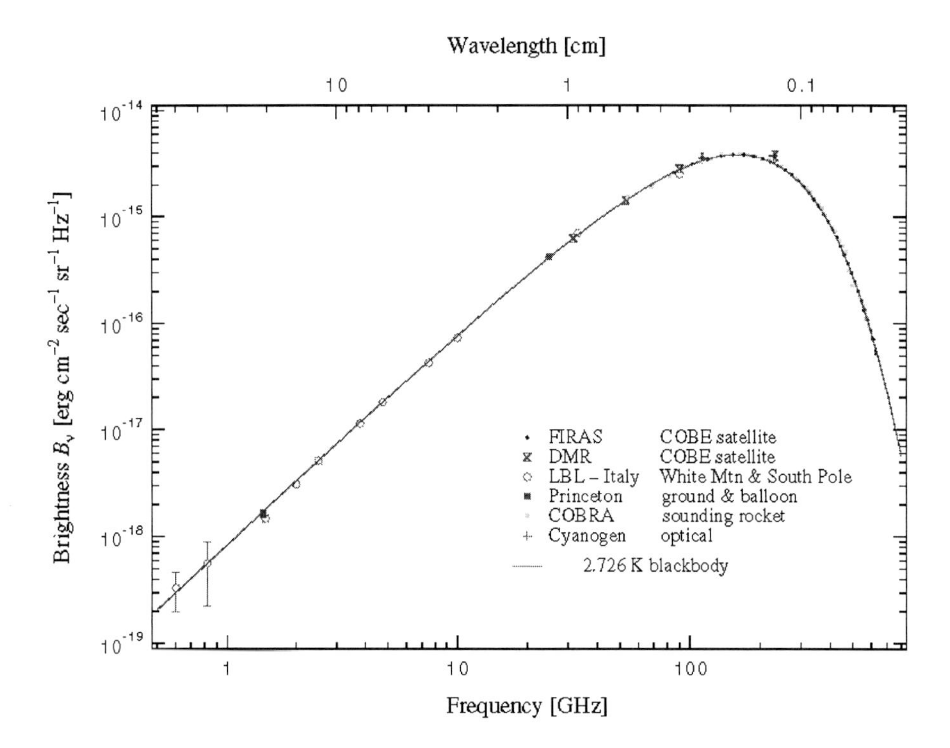

FIGURE 2. Summary of spectral data on the CMB

in quasar absorption systems at redshift $Z \sim 3$ are particularly interesting and will be discussed (Songaila *et al.* 1990, Carswell *et al.* 1994, Tytler *et al.* 1996) since BBN requires that fragile deuterium be found in primitive material. However, the possible variation of D/H in different lines of sight at $Z \sim 3$ argues that perhaps hydrogen cloud interlopers may be immitating deuterium, at least in lines of sight having higher apparent D/H. New work by Tytler, Burles, and Kirkman(1996) on a previous high D/H system seems to support this point of view. Furthermore, recent theoretical calculations have confirmed that quark-hadron inspired inhomogenous Big Bang Nucleosyntheis does not significantly alter the basic conclusions of standard BBN. We will also briefly discuss the possible impact on BBN of the recent ROSAT and ASCA x-ray satellite results on clusters of galaxies and the recent halo microlensing results. This summary will attempt to put it all together within an historical framework. The bottom line that emerges is how dramatically robust BBN is and how it gives a dramatically tight constraint on the density of baryons in the universe.

Let us now briefly review the history. This agreement works only if the baryon density is well below the cosmological critical value. This summary draws on the reviews of Walker et al. (1991); Schramm (1994); Copi, Schramm and Turner (1994); and Turner and Schramm (1998).

It should be noted that there is a symbiotic connection between BBN and the 3K background dating back to Gamow and his associates, Alpher and Herman. The initial BBN calculations of Gamow's group (1948) assumed pure neutrons as an initial condition and thus were not particularly accurate, but their inaccuracies had little effect on the group's predictions for a background radiation.

Once Hayashi (1950) recognized the role of neutron-proton equilibration, the framework for BBN calculations themselves has not varied significantly. The work of Alpher, Follin and Herman (1964) and Tayler and Hoyle (1964), preceeding the discovery of the 3K background, and of Peebles (1966) and Wagoner, Fowler and Hoyle (1967), immediately following the discovery, and the more recent work of our group of collaborators (Walker *et al.* 1991, Copi *et al.* 1994, Olive *et al.* 1990, Schramm & Wagoner 1977, Olive *et al.* 1981, Boesgard & Steigman 1985, Yang *et al.* 1984, Kawano *et al.* 1988) all do essentially the same basic calculation, the results of which are shown in Figure 1. As far as the calculation itself goes, solving the reaction network is relatively simple by the standards of explosive nucleosynthesis calculations in supernovae, with the changes over the last 25 years being mainly in terms of more recent nuclear reaction rates as input, not as any great calculational insight, although the current Kawano code (Kawano *et al.* 1988) is somewhat streamlined relative to the earlier Wagoner code (1967). In fact, the earlier Wagoner code is, in some sense, a special adaptation of the larger nuclear network calculation developed by Truran (1965) for work on explosive nucleosyntheis in supernovae. With the exception of Li yields and non-yields of Be and B, to which we will return, the reaction rate changes over the past 25 years have not had any major affect (see Yang *et al.* 1984 and Krauss and his collaborators (Krauss & Romanelli 1990, Kernan & Krauss 1994) or Copi, Schramm and Turner 1994 for discussion of uncertainties). The one key improved input is a better neutron lifetime determination (Mampe *et al.* 1996). There has been much improvement in the $t(\alpha,\gamma)$ ^{7}Li reaction rate (Schramm & Mathews 1995) but as the width of the curves in Figure 3 shows, the ^{7}Li yields are still the poorest determined, both because of this reaction and even more because of the poorly measured ^{3}He (α,γ) ^{7}Be.

With the exception of the effects of elementary particle assumptions, to which we will also return, the real excitement for BBN over the last 25 years has not really been in redoing the basic calculation. Instead, the true action is focused on understanding the evolution of the light element abundances and using that information to make powerful conclusions. In the 1960's, the main focus was on ^{4}He which is very insensitive to the baryon density. The agreement between BBN predictions and observations helped support the basic big bang model but gave no significant information, at that time, with regard to density. In fact, in the mid-1960's, the other light isotopes (which are, in principle, capable of giving density information) were generally assumed to have been made during the T-Tauri phase of stellar evolution (Fowler *et al.* 1962), and so, were not then taken to have cosmological significance. It was during the 1970's that BBN fully developed as a tool for probing the universe. This possibility was in part stimulated by Ryter et al. (1970) who showed that the T-Tauri mechanism for light element synthesis failed. Furthermore, ^{2}D abundance determinations improved significantly with solar wind measurements (Geiss & Reeves 1971, Black 1971) and the interstellar work from the Copernicus satellite (Rogerson & York 1973). [Recent HST observations reported by Linsky et al. (1993) have compressed the local ISM ^{2}D error bars considerably.] Reeves, Audouze, Fowler and Schramm (1973) argued for cosmological ^{2}D and were able to place a constraint on the baryon density excluding a universe closed with baryons. Subsequently, the ^{2}D arguments were cemented when Epstein, Lattimer and Schramm (1976) proved that no realistic astrophysical process other than the Big Bang could produce significant ^{2}D. This baryon density was compared with dynamical determinations of density by Gott, Gunn, Schramm and Tinsley (1974). See Figure 1 for an updated $H_0 - \Omega$ diagram.

In the late 1970's, it appeared that a complimentary argument to ^{2}D could be developed using ^{3}He. In particular, it was argued (Rood *et al.* 1976) that, unlike ^{2}D, ^{3}He was

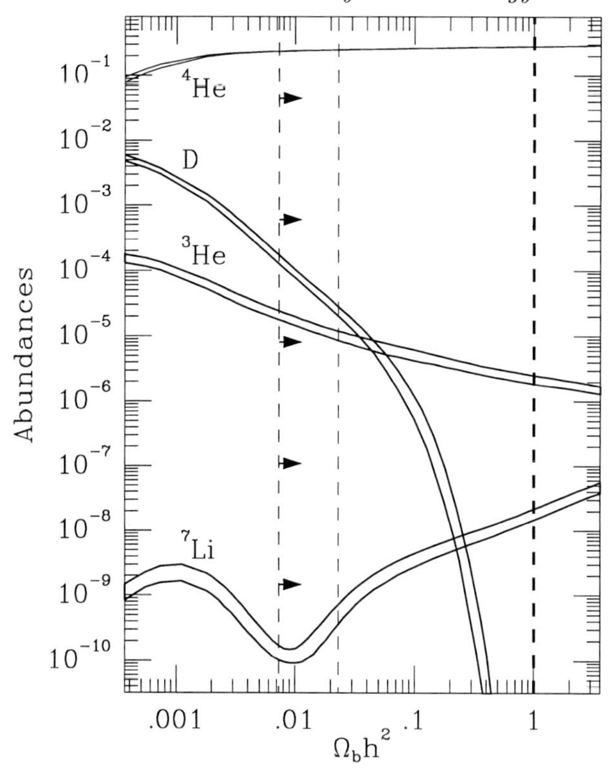

FIGURE 3. Big Bang Nucleosynthesis abundance yields versus baryon density (Ω_b) and $\eta \equiv \frac{n_b}{n_\gamma}$ for a homogeneous universe. ($h \equiv H_0/100$ km/sec/Mpc; thus, the concordant region of $\Omega_b h^2 \sim 0.015$ corresponds to $\Omega_b \sim 0.06$ for $H_0 = 50$ km/sec/Mpc.) Figure is from Copi, Schramm and Turner (1994). Note concordance region is slightly larger than Walker et al. (1991) due primarily to inclusion of possible systematic errors on Li/H. The width of the curves represents the uncertainty due to input of nuclear physics in the calculation.

made in stars; thus, its abundance would increase with time. Unfortunately, recent data on ^{3}He in the interstellar medium (Gloeckler & Geiss 1996) has shown that ^{3}He has been constant for the last 5 Gyr. Thus, low mass stars are not making a significant addition, contrary to these previous theroetical ideas. Furthermore, Rood, Bania and Wilson (1992) have shown that interstellar ^{3}He is quite variable in the Galaxy, contrary to expectations for a low mass star-dominated nucelus. However, the work on planetary nebulae shows that at least some low mass stars do produce ^{3}He. Nonetheless, the current observational situation clearly shows that arguments based on theoretical ideas about ^{3}He evolution should be avoided (c.f. Hata *et al.* 1996). Since ^{3}He now seems not to have a well behaved history, simple ^{3}He or ^{3}He + D inventory arguments are misleading at best. One is not free to go to arbitrary low baryon densities and high primordial D and ^{3}He, since processing of D and ^{3}He in massive stars also produces metals which are constrained (Copi *et al.* 1995, Scully *et al.* 1996) by the metals in the hot intra-cluster gas, if not the Galaxy.

It was interesting that other light elements led to the requirement that ^{7}Li be near its minimum of $^7\mathrm{Li/H} \sim 10^{-10}$, which was verified by the Pop II Li measurements of Spite and Spite (Spite & Spite 1982, Rebolo *et al.* 1988, Hobbs & Pilachowski 1988), hence yielding the situation emphasized by Yang *et al.*(1984) that the light element abundances are consistent over nine orders of magnitude with BBN, but only if the cosmological baryon density, Ω_b, is constrained to be around 6% of the critical value (for $H_0 \simeq 50$ km/sec/Mpc). The Li plateau argument was further strengthened with the observation of ^{6}Li in a Pop II star by Smith, Lambert and Nissen (1982). Since ^{6}Li is much more fragile than ^{7}Li, and yet it survived, no significant nuclear depletion of ^{7}Li is possible (Olive & Schramm 1992, Steigman *et al.* 1993, Lemoine *et al.* 1997). This observation of ^{6}Li has now been verified by Hobbs and Thorburn (1994). Lithium depletion mechanisms are also severly constrained by the recent work of Spite et al. (1996) showing that the lithium plateau also is found in Pop II tidally locked binaries. Thus, meridonal mixing is not causing lithium depletion. Recently Nollett *et al.* (1997) have discussed how ^{6}Li itself might eventually become another direct probe of BBN depending on the eventual low energy measurement of the ^{2}D $(\alpha\gamma)$ ^{6}Li cross section on spectroscipy improvements for extreme metal-poor dwarfs.

Another development back in the 70's for BBN was the explicit calculation of Steigman, Schramm and Gunn (1977) showing that the number of neutrino generations, N_ν, had to be small to avoid overproduction of ^{4}He. [Earlier work (Tayler and Hoyle 1964, Schvartzman 1969, Peebles 1971) had commented about a dependence on the energy density of exotic particles but had not done an explicit calculation probing N_ν.] This will subsequently be referred to as the SSG limit. To put this in perspective, one should remember that the mid-1970's also saw the discovery of charm, bottom and tau, so that it almost seemed as if each new detector produced new particle discoveries, and yet, cosmology was arguing against this "conventional" wisdom. Over the years, the SSG limit on N_ν improved with ^{4}He abundance measurements, neutron lifetime measurements, and with limits on the lower bound to the baryon density, hovering at $N_\nu \lesssim 4$ for most of the 1980's and dropping to slightly lower than 4 just before LEP and SLC turned on (Walker *et al.* 1991, Olive 1990, Schramm & Kawano 1989, Pagel 1991). This was verified by the LEP results (1993) where now the overall average is $N_\nu = 2.987 \pm 0.02$. A recent examination of the cosmological neutrino limit by Copi et al. (1997) in the light of the recent ^{3}He and D/H work shows that the BBN limit remains between 3 and 4 for all reasonable assumption options.

The recent apparent convergence of the extra-galactic D/H measurements towards the lower values (Tytler 1997, Hogan 1997) D/H $\sim 3\times10^{-5}$ may eventually collapse the Ω_B band in Figure 3 and Figure 1 to a relatively narrow strip on the high Ω_B side. However, such a collapse at present is probably a bit premature. It should be noted that this low D/H value also has important implications for galactic evolution since the present ISM value is within a factor of 2 fo the primordial value. This would seem to favor infall models or models with variable initial mass functions to explain heavy element production in our Galaxy.

It might be noted that the previous tension between the low D/H of Tytler and the primordial helium value seems to have eased. The previous Y of ~ 0.23 seems to be moving up to ~ 0.24 due to a re-evaluation of IZW 18, the old low point and the other recent observational work by Skillman (1997) and by Thuan *et al.* (1996).

The power of homogeneous BBN comes from the fact that essentially all of the physics input is well determined in the terrestrial laboratory. The appropriate temperature regimes, 0.1 to 1 MeV, are well explored in nuclear physics laboratories. Thus, what nuclei do under such conditions is not a matter of guesswork, but is precisely known. In fact, it is known for these temperatures far better than it is for the centers of stars like our sun. The center of the sun is only a little over 1 keV, thus, below the energy where nuclear reaction rates yield significant results in laboratory experiments, and only the long times and higher densities available in stars enable anything to take place.

5. Dark Matter and Visible Matter

The success and robustness of BBN have given renewed confidence to the limits on the baryon density constraints. Let us convert this density regime into units of the critical cosmological density for the allowed range of Hubble expansion rates. This is shown in Figure 1. In particular, $\Omega_b = (0.08 \pm 0.02)h_{50}^{-2}$ where h_{50} is the Hubble constant in units of 50km/sec/Mpc. Figure 1 also shows the lower bound on the age of the Universe of 10 Gyr from both nucleochronology and from globular cluster dating (Shi *et al.* 1995) and a lower bound on H_0 of 40 from extreme type IA supernova models with pure 1.4 $M_\odot$ carbon white dwarfs being converted to ^{56}Fe. The constraint on Ω_b means that the universe *cannot be closed with baryonic matter.* [This point was made over twenty years ago (Reeves *et al.* 1973) and has proven to be remarkably strong.] If the Universe is truly at its critical density, then nonbaryonic matter is required. This argument has led to one of the major areas of research at the particle-cosmology interface, namely, the search for non-baryonic dark matter. In fact, from the lower bound on Ω_{TOTAL} from cluster dynamics of $\Omega_{TOTAL} > 0.15$, it is clear that non-baryonic dark matter is required unless $H_0 < 40$. The need for non-baryonic matter is strengthened on even larger scales (Davis & Nusser 1995). Figure 1 also shows the range of $\Omega_{VISIBLE}$ and shows that there is no overlap between Ω_b and $\Omega_{VISIBLE}$. Hence, the bulk of the baryons are dark.

The estimate of the Ω_{VIS} in Figure 1 can be obtained from noting that stellar material tends to have mass-to-light ratios in solar units as shown below

$$\frac{M}{L}|_* \sim 2 - 8\ \frac{M_\odot}{L_\odot}$$

for blue band light. (The number is greater than one since mass is dominated by low mass stars and light by high mass ones, and initial mass functions favor low mass ones.) The uncertainty comes from the location of the low mass cutoff. Recent observations from HST (Flynn *et al.* 1996) argue that the cutoff is above the red dwarf limit of several tenths $M_\odot$ since these objects were not found to be ubiquitous. To obtain a density from an M/L one needs to multiply by the average luminosity density, $\mathcal{L}_B$, in that wave band. Kirshner et al. (1979) have determined :

$$\mathcal{L}_B \simeq 1.8 \pm 0.2 \times 10^8 h \frac{L_\odot}{Mpc^3}$$

where h is the Hubble constant in units of 100 km/sec/Mpc. One can then obtain Ω_{VIS} by dividing by

$$\rho_{crit} \equiv \frac{3H_0^2}{8\pi G} = 2.8 \times 10^{11} h^2 \frac{M_\odot}{Mpc^3} = 2 \times 10^{-29} h^2 g/cm^3$$

Hence,

$$\Omega_{VIS} = \frac{\frac{M}{L}|_B \cdot \mathcal{L}_B}{\rho_{crit}} = \frac{0.003 \pm 0.002}{h}.$$

Note that Ω_{VIS} obtained in this way is inversely proportional to H_0. (One can also estimate Ω_{VIS} from the dynamics of the shining regions of galaxies and obtain a similar value but independent of H_0.)

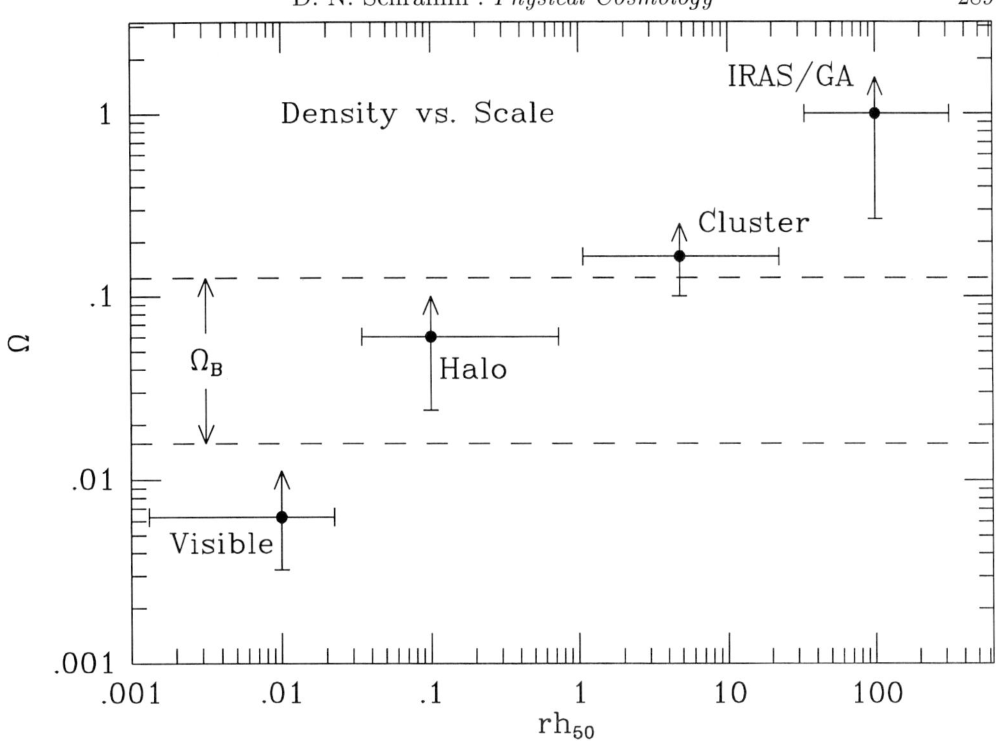

FIGURE 4. Implied densitites versus the scale of the measurements.

6. The Need for Non-Baryonic Dark Matter

The arguments requiring some sort of dark matter fall into separate and quite distinct areas. These arguments are summarized in Figures 1 and 4. First are the arguments using Newtonian mechanics applied to various astronomical systems that show that there is more matter present than the amount that is shining. It should be noted that these arguments reliably demonstrate that galactic halos seem to have a mass ~ 10 times the visible mass.

For dynamical estimates of Ω one estimates the mass from

$$M \sim \frac{v^2 r}{G}$$

where v is the relative velocity of the objects being studied, r is their separation distance, and G is Newton's constant. The proportionality constant out front depends on orientation, relative mass, etc. For large systems such as clusters, one uses averaged quantities. For single galaxies v would represent the rotational velocity and r the radius of the star or gas cloud. Note that since v is measured by relative redshift, it has no H_0 dependence, whereas distances will scale with $1/H_0$. Also note that the luminosity of a system will scale as the surface area and hence is proportional to $1/H_0^2$. Hence, for dynamically measured systems,

$$\frac{M}{L}|_{DYN} \propto H_0.$$

Again,

$$\Omega_{DYN} = \frac{\frac{M}{L}|_B^{DYN} \cdot \mathcal{L}_B}{\rho_{crit}}$$

which now is independent of H_0. It is this technique which yields the halo and cluster points shown in Figure 1 and Figure 4. It should be noted that the value of $\Omega_{CLUSTER} \sim 0.25 \pm 0.10$ is also obtained in those few cases where alignment produces giant gravitational-lens arc. Recent work using weak gravitational lensing by Kaiser (Fahlmer *et al.* 1995) also supports large Ω. Furthermore, the temperature of the hot x-ray emitting gas in large clusters can be used to determine their gravitational potential owell and hence their mass, and it too gives values consistent with the lensing and dynamics arguments. The IRAS/Great Attractor point on Figure 4 comes from the large scale flow arguments. As Davis and Nusser(1995) show, if the large scale velocity flows measured from the IRAS survey are due to gravity, then $\Omega_{IRAS} \gtrsim 0.2$. Similar arguments have been obtained using the Great Attractor study or the Potent (Bertschinger *et al.* 1990, Dekel *et al.* 1993) technique. All imply $\Omega > \Omega_{BARYON}$, hence the need for non-baryonic dark matter. However, there is still considerable uncertainty of the exact value of Ω determined in this way as discussed by Szalay (1995). But all groups agree it is greater than 0.2. However, as Figure 1 illustrates, except for $H_0 < 40$, $\Omega_{CLUSTER}$ already required $\Omega_{TOTAL} > \Omega_{BARYON}$ and hence the need for non-baryonic dark matter.

An Ω of unity is, of course, preferred on theoretical grounds since that is the only long-lived natural value for Ω, and inflation (Guth 1981, Linde 1990) or something like it provided the early universe with the mechanism to achieve that value and thereby solve the flatness and smoothness problems. Note that our need for exotica is not dependent on the existence of dark galatic halos and that high values of H_0 (see Figure 1) increase the need for non-baryonic dark matter.

Non-baryonic matter can be divided into two major categories (Bond 1992) for cosmological purposes: hot dark matter (HDM) and cold dark matter (CDM). Hot dark matter is matter that is relativistic until just before the epoch of galaxy formation, the best example being low mass neutrinos with $m_\nu \sim$ 20eV. Cold dark matter is matter that is moving slowly at the epoch of galaxy formation. Because it is moving slowly, it can clump on very small scales, whereas HDM tends to have more difficulty in being confined on small scales and be CDM as well. Examples of CDM could be lightest super-symmetric particle which is presumed to be stable and might have a mass of several GeV. Following Michael Turner, all such weakly interacting massive particles are called "WIMPS." Axions, while very light, would also be moving very slowly (Turner *et al.* 1983) and thus would clump on small scales and be CDM as well. Or, one could also go to non-elementary particle candidates, such as primordial blackholes or quark nuggets of strange quark matter, possibly produced at the quark-hadron transition (Crawford & Schramm 1982), (Alcock & Olinto 1988), as we will discuss below. Another possibility would be any sort of massive toplogical remnant left over from some early phase transition. Note that CDM would clump in halos, thus requiring the dark baryonic matter to be out between galaxies, whereas HDM would allow baryonic halos. The MACHO and EROS events may eventually require at least some CDM to fill out the halo. Obviously, mixed models with some HDM and some CDM have even more flexibility and have thus become quite popular as data constraints increase.

In discussing primordial black holes as dark matter, it is worth noting that the recent MACHO results (Alcock *et al.* 1997) are now best fit by MACHOS of mass $\sim 0.5\mathrm{M}_\odot$. However, loopholes do exist. For example, it may be that the microlensing events are not truly in the halo but rather all near the large Magellanic Cloud, in which case they would not have this mass. If we acept the $\sim 0.5\mathrm{M}_\odot$ at face value, it leads to a serious problem for the previous MACHO brown dwarf hypothesis, since stars above $\sim 0.1\mathrm{M}_\odot$ produce enough light that they should have been seen in previous halo searches. While it may be possible to populate the halo with $\sim 0.5\mathrm{M}_\odot$ white dwarfs which would be too dim to be picked up in these previous surveys, the Galactic evolution necessary to make so many white dwarfs without over producing heavy elements requires several ad hoc assumptions (Fields *et al.* 1997). While these cannot be excluded, they are nonetheless rather ad hoc.

It is interesting that the mass within the horizon M_H at the time of the Q-H transition is

$$M_H \sim \frac{0.01\mathrm{M}_\odot}{T_{QH}(\mathrm{Gev})^2}$$

which for $T_{QH} \sim 140$ MeV (a reasonable Q-H transition value) yields $M_H \sim 0.5\mathrm{M}_\odot$. Since the larger the black hole mass, M_{BH}, the smaller the density, ρ_{BH}, required to achieve an event horizon

$$\rho_{BH} \propto \frac{1}{M_{BH}^2}.$$

Thus, it may be easier to make horizon scales black holes than smaller ones.

It was noted in the early work on making black holes at the Q-H transition (Crawford & Schramm 1982) that any black hole production must be very inefficient since the density of black holes will scale like matter ($\propto T^3$) rather than radiation ($\propto T^4$) as the Universe expands and cools. If most horizons became black holes at the Q-H transiton, it would imply $\Omega \sim 10^9$. That is a very short-lived Universe. Thus, only about 1 horizon in 10^9 could have become a black hole. Note that since these black holes form prior to BBN, they circumvent the BBN limit on Ω_b. Thus, $\Omega \sim 1$ in such black holes is possible. Also, since these would be slow moving, they behave like cold dark matter (see Freese, Price & Schramm 1983).

Getting about 1 horizon in 10^9 to go over the Schvarzschild limit at the Q-H transition is still not trivial. There are some positive speculative points. A weak first order transition (as now is favored since the mass of the strange quark is so close to the transition temperature) yields a phase transition which is almost scale free, hence yielding horizon scale (km) fluctuations even for nucelar interactions, while still releasing some entropy to enable a fluctuation to be "fixed." Keeping the fluctuation from continuing to expand with the Universe does require density variations of the order unity so that the black hole forms prior to BBN. Jedamsik (1997) and others (e.g. Olinto, Schramm and Fuller 1997) have investigated possible Q-H scenarios to form black holes, but much work remains to be done.

Before closing this dark matter section, let us remember that some baryonic dark matter must exist since we know that the lower bound from Big Bang Nucleosynthesis is greater than the upper limits on the amount of visible matter in the universe. If the baryonic dark matter is not in the halo, it could be in hot intergalactic gas, hot enough not to show absorption lines in the Gunn-Peterson test, but not so hot as to be seen in the x-rays. The exciting report by Jakobsen *et al.* (1994; see also Davidsen1997) of a Gunn-Peterson effect observed with HST for He-II at high Z showed that at least some hot IGM exists (and verifies that He seems primordial). Recent analyses by Davidsen

Density Fluctions	Topological Seeds
Quantum Fluctuations From Inflation	Top
Gaussian	Non-Gaussian
Adiabatic	Iso-Curvature or almost equivalently isothermal

TABLE 2. Two Generic Seed Families

(1997) show that the universe at high Z probably had the bulk of its baryons in the Lyman-α gas clouds.

Furthermore, hot gas has been found in clusters of galaxies by ROSAT and ASCA. As mentioned earlier, temperature of the gas can be used to estimate the gravitational potential of the clusters if it is assumed that the gas is vivialized and purely supported by thermal pressure. Similarly, the intensity of the emission can be used to estimate the density of the gas. White *et al.* (1993) have shown that the typical values for x-ray clusters yield a hot gas to total mass ratio M_{HOT}/M_{TOT} of about 0.2. Cluster masses can be estimated either from the temperature of the hot gas or from dynamics or from gravitational lensing. All yield the cluster implied density, $\Omega_{CLUSTER}$, of $\sim 0.25 + 0.10$. Thus, the implied baryon density from x-ray gas in clusters, $\frac{\Omega_{B,\ CLUSTER} \simeq M_{HOT}}{M_{TOT}} \times \Omega_{CLUSTER} \sim 0.05$ in good agreement with the BBN value for H = 50 and for the Tytler D/H.

7. Seeds and Large Scale Structure

To form structure in the universe requires some sort of "seed" in addition to the baryonic matter and the non-baryonic dark matter. The types of seeds tend to fall into two broad categories:

1. density fluctuations;
2. topological seeds.

Density fluctuations are presumably produced via quantum fluctuations near the end of the inflation epoch and topological seeds such as cosmic strings or textures are topological defects again left over from some vacuum phase transition. Thus, in either case a cosmological phase transition seems to be required. Table 2 shows other names and properties these two seed classifications sometimes go by.

Since COBE only was able to observe anisotropies on angular scales $\gtrsim 7^\circ$, it was not able to directly probe the scales that eventually led to the observed large scale structure.

The horizon size at the decoupling of the microwave radiation corresponds only to an angular size of less than 2°, so COBE could not probe the causal processes that led to structure. Many new microwave experiments are now probing the angular scales of $\sim 30'$ to $\sim 2^\circ$ to try to differentiate between the models of Table 1 and various CDM/HDM assumptions. These experiments are being done in balloons and at the South Pole and at various other remote terrestrial locations such as Saskatoon, Canada. At present, there is much ambiguity among the different experiments, so no single picture emerges. Non-gaussian models may even claim to predict that different directions should see different results, but the error bars are too large to claim this has been proven yet. Many experiments do find a few times higher signal at $\sim 1^\circ$ than COBE finds for larger scale (COBE $\frac{\Delta T}{T} \sim 10^{-5}$) which is what the fluctuation model predicts. Detailed analysis by Jungman, Kosowsky, Kamionkowski and Spergel (1996) and by Scott and White (1995) show that if the fluctuations are gaussian then the details of the accoustic peaks near 1° and less will determine H_0, Ω_T, Ω_B, Λ_0, etc. Thus, the next generation satellites, MAP and PLANCK (formerly called COBRAS/SAMBA), may be able to give us precise answers to most of the traditional cosmological questions. The intercomparison of statistically significant galaxy distribution data and microwave anisotropy data, plus direct dark matter searches and accelerator searches will hopefully resolve the origin of structure in the universe by the end of this decade.

8. Alternative Measures of Ω_B

In addition to the BBN arguments presented above in section 4, it is interesting that we are now obtaining new alternative probes of Ω_B. In particular, the height of the first doppler peak in the CMB anisotropy, mentioned in section 7 above, is directly related to $\Omega_B h^2$. Since H_0 is also related to the ratio of the first and third peak heights, one might eventually be able to deconvolve the two quantities using just the CMB anisotropy. The correct preliminary analyses based on Sasketoon, South Pole and balloon data overlap with the BBN $\Omega_B h^2$ values on the high side; thus, they are more consistent with the low primordial D/H values. That this overlap exists is remarkable independent support for BBN arguments.

Eventually, with MAP and PLANCK and the new South Pole and balloon results, we should have a very definitive test here.

Another way of probing Ω_B is via the Lyman-α clouds. The recent work of Davidsen (1997), showing that the presence of ionized intergalactic helium and clouds at high redshift, is well interpreted by baryonic gas with Ω_B again overlapping with the high side of the BBN range.

The x-ray gas in clusters might also provide some clues here, and, as mentioned in the previous section, it too supports the Ω_B from BBN.

9. Conclusion

It seems clear that we need the bulk of the matter in the universe to be in some non-baryonic form. A possible mixed dark matter solution may be motivated by the need for neutrino masses from the solar neutrino problem. The best fit solar ν mass hints at $m_{\nu_\mu} \sim 10^{-3}$ eV which in a simple see-saw model may imply $m_{\nu_\tau} \sim 10$ eV, that is, great hot dark matter.

Particularly exciting are the experimental searches for non-baryonic dark matter. These searches include accelerator searches for supersymmetry (or other Weakly Interactiong Massive Particles, WIMPS) and for neutrino oscillations. They also include direct underground searches for WIMPS and axions and even satellite searches for WIMP annihilation products.

Not only will the new CMB measurement revolutionize physical cosmology, but we also have other astronimcal projects of great significance. In particular, the Sloan Digital Sky Survey will map the 2D positions of 10^8 galaxies, the 3D position of 10^6 galaxies, and the 3D positions of 10^5 QSOs. This should settle once and for all what is the large scale structure we need to get the seeds and the dark matter to make.

Hopefully, with all this activity, we will find the dark matter within a decade.

I would like to thank my recent collaborators, David Dearborn, Martin Lemoine, Brian Fields, Dave Thomas, Gary Steigman, Brad Meyer, Keith Olive, Angela Olinto, Bob

Rosner, Michael Turner, George Fuller, Karsten Jedamzik, Rocky Kolb, Grant Mathews, Bob Rood, Jim Truran and Terry Walker for many useful discussions. I would further like to thank Poul Nissen, Jeff Linsky, David Tytler, Scott Burles, Julie Thorburn, Doug Duncan, Lew Hobbs, Evan Skillman, Bernard Pagel and Don York for valuable discussion regarding the astronomical observations.

This work is supported by the NASA and the DoE(nuclear) at the University of Chicago, and by the DoE and NASA grant NAG5-2788 at Fermilab.

REFERENCES

Alcock, C. *et al.* 1997 *Astrophys. J.* **486**, 697-726.

Alcock, C. & Olinto, A. 1988 *Ann. Rev. Nuc. Part. Phys.* **38**, 161-184.

Alpher, R.A., Bethe, H., & Gamow, G. 1948 *Phys. Rev.* **73**, 803-804.

Alpher, R.A., Follin, J.W., & Herman, R.C. 1953 *Phys. Rev.* **92**, 1347-1361.

Bertschinger, E., Dekel, A., Faber, S.M., Dressler, A. & Burstein, D. 1990 *Astrophys. J.* **364**, 370-395;

Black, D. 1971 *Nature* **234**, 148-149.

Boesgaard, A.M. & Steigman, G. 1985 *Ann. Rev. of Astron. and Astrophys.* **23**, 319-378.

Bond, R. & Szalay, A. 1992 in Proc. Texas Relativistic Astrophysical Symposium, Austin, Texas.

Carlstrom, J. E., Grego, L., Holzapfel, W. L. & Joy, M. (1997) in *Procceding of the 18th Texas Symposium on Relativistic Astrophysics*, ed. A. Olinto, in press, World Scientific.

Carswell, R.F. *et al.* 1994 *MNRAS* **268**, L1-L12.

Chaboyer, B. 1995 *Astrophys. J.* **444**, L9-L12.

Chaboyer, B., Demarque, P., Kernan, P.J., & Krauss, L. 1996 *Science* **271**, 957-961.

Chaboyer, B. & Krauss, L. 1997 Case Western Reserve University preprint.

Copi, C., Schramm, D.N. & Turner, M.S. 1994 *Science* **267**, 192-199.

Copi, C.J., Schramm, D.N., & Turner, M.S. 1995 *Astrophys. J.* **455**, L95-L98.

Copi, C., Schramm, D.N., & Turner, M.S. 1997 *Phys. Rev. D* **55**, 3389-3393.

Crawford, M. & Schramm, D.N. 1982 *Nature* **298**, 538-540.

Davidsen, A. 1997 in *Proceeding of the 18th Texas Symposium on Relativistic Astrophysics*, ed. A. Olinto, in press, World Scientific.

Davis, M. & Nusser, A. 1995 in Proc. Maryland Symposium on Dark Matter, in press; see also Davis, M. 1997 in *Proceeding of the 18th Texas Symposium on Relativistic Astrophysics*, ed. A. Olinto, in press, World Scientific.

Dekel, A., Bertschinger, E., Yahil, A., Strauss, M., Davis, M., & Huchra, J. 1993 *Astrophys. J.* **412**, 1-21.

Epstein, R., Lattimer, J. & Schramm, D.N. 1976 *Nature* **263**, 198-202.

Fahlmer, G., Kaiser, N., Squires, G. & Woods, D. 1994 *Astrophys. J*, **437**, 56-62.

Feast, M.W., Carter, B.S., Roberts, G., Marang, F. & Catchpole, R.M. 1997 *MNRAS* **285**, 317-338.

Fields, B.D., Mathews, G.J., & Schramm, D.N. 1997 *Astrophys. J.* **483**, 625-637.

Flynn, C., Gould, A. & Bahcall, J. 1996 *Astrophys. J. Lett.* **466**, L55-L58.

Fowler, W.A., Greenstein, J. & Hoyle, F. 1962 *Geophys. R.A.S.* **6**, 148-220.

Freedman, W.L. *et al.* 1994 *Nature* **371**, 757-762.

Freedman, W.L. *et al.* 1997 *Critical Dialogues in Cosmology*, ed. N. Turok, 92-129, World Scientific.

Freese, K., Price, R. & Schramm, D.N. 1983 *Astrophys. J.* **275**, 405-412.

Geiss, J. & Reeves, H. 1971 *Astron. and Astrophys.* **18**, 126-132.

Gloeckler, G. & Geiss, J. 1996 *Nature* **381**, 210-212.

Gott, J.R., Gunn, J., Schramm, D.N., & Tinsley, B.M. 1974 *Astrophys. J.* **194**, 543-553.

Guth, A. 1981 *Phys. Rev. D* **23**, 347-356.

Hata, N., Scherrer, R.J., Steigman, G., Thomas, D., & Walker, T.P. 1996 *Astrophys. J.* **458**, 637-640.

Hayashi, C. 1950 *Prog. Theor. Phys.* **55**, 224-235.

Hobbs, L. & Pilachowski 1988 *Astrophys. J.* **326**, L23-L26.

Hobbs, L. & Thorburn, J. 1994 *Astrophys. J. Lett.* **428**, L25-L28.

Hill, C., Schramm, D.N., & Fry, J. 1989 *Comm. Nuc. Part. Phys.* **19**, 25-39.

Hogan, C. 1997 in *Proceeding of the 18th Texas Symposium on Relativistic Astrophysics*, ed. A. Olinto, in press, World Scientific.

Iben, I. & Renzini, A. 1984 *Physics Reports* **105**, 330-406.

Jakobsen, P., Boksenberg, A., Deharveng, J.M., Greenfield, P., Jedrzewski, R. & Paresce, F. 1994 *Nature* **370**, 35-39.

Jedamzik, K. 1997 Max-Planck-Insitute-Munich preprint.

Jungman, G., Kosowsky, A., Kamionkowski, M. & Spergel, D.N. 1996 *Phys. Rev. Lett.* **76**, 1007-1010; *Phys. Rev. D* **54**, 1332-1344.

Kawano, L., Schramm, D.N., & Steigman, G. (1988) *Astrophys. J.* **327**, 750-754.

Keeton, C.R. & Kochonek, C.S. (1997) astro-ph/9611216, *Astrophys. J.* **482**, 604-620.

Kernan, P. & Krauss, L. 1994 *Phys. Rev. Lett.* **72**, 3309-3312.

Kirshner, R., Oemler, A. & Schechter, P.L. 1979 *Astronom. J.* **84**, 951-959.

Krauss, L.M. & Romanelli, P. 1990 *Astrophys. J.* **358**, 47-59.

Kundic, T. *et al.* 1997 *Astrophys. J.* **482**, 631-635.

Lee, S., Schramm, D.N., & Mathews, G. 1995 *Astrophys. J.* **449**, 616-622.

Lemoine, M., Schramm, D.N., Truran, J.W. & Copi, C.J. 1997 *Astrophys. J.* **478**, 554-562.

The LEP Collaboration: ALEPH, DELPHI, L3, & OPAL 1992 *Phys. Lett. B* **276**, 247-253.

Linde, A. 1990 *Particle Physics and Inflationary Cosmology*, Harwood Academic Publishers, N.Y.

Linsky, J., Brown, A., Gayley, K., Diplas, A., Savage, B., Ayres, T., Landsman, W., Shore, S. & Heap, S. 1993 *Astrophys. J.* **402**, 694-709.

Mampe, W., Ageron, P., Bates, C., Pendlebury, J.M. & Steyerl, A. 1996 *Phys. Rev. Lett.* **63A**, 593-596; Mampe, W. *et al.* 1993 *JETP Lett.*57, 82-87.

Mazzitelli, I & d'Antonna, F. 1997 *Astron. and Astrophys.* **302**, 382-400.

Meyer, B.S. & Schramm, D.N. 1986 *Astrophys. J.* **311**, 406-417.

Nollett, K., Lemoine, M. & Schramm, D.N. 1997 *Phys. Rev. C* **56**, 1144-1151.

Nugent, P., Branch, D., Boran, E., Fisher, A., Vaughn, T, & Hauschildt, D. 1995 *Phys. Rev. Lett.* **75**, 394-397.

Olinto, A., Schramm, D.N. & Fuller, G. 1997, in preparation.

Olive, K. & Schramm, D.N. 1992 *Nature* **360**, 4349-442.

Olive, K., Schramm, D.N., Steigman, G., Turner, M.S., & Yang, J. 1981 *Astrophys. J.* **246**, 557-568.

Olive, K., Schramm, D.N., Steigman, G. and Walker, T. 1990 *Phys. Lett. B* **236**, 454-460.

Pagel, B. 1991 *Physica Scripta* **T36**, 7-15.

Pagel, B. & Jimenez, R. 1997 *Physics Reports*, in preparation.

Peebles, P.J.E. 1966 *Phys. Rev. Lett.* **16**, 410-413.

PEEBLES, P.J.E. 1971 *Physical Cosmology.* PRINCETON UNIVERSITY PRESS.

REBOLO, R., MOLARO, P. & BECKMAN, J. 1988 *Astron. and Astrophys.* **192**, 192-205.

REEVES, H., AUDOUZE, J., FOWLER, W.A., & SCHRAMM, D.N. 1973 *Astrophys. J.* **179**, 909-930.

REEVES, H. & JOHNS, O. 1976 *Astrophys. J.* **206**, 958-962.

RIESS, A.G., PRESS, W.H., & KIRSHNER, R.P. 1995 *Astrophys. J. Lett.* **438**, L17-L20.

ROGERSON, J. & YORK, D. 1973 *Astrophys. J.* **186**, L95-L98.

ROOD, R.T., BANIA, T. & WILSON, J. 1992 *Nature* **355**, 618-620.

ROOD, R.T., STEIGMAN, G., & TINSLEY, B.M. 1976 *Astrophys. J.* **207**, L57-L60.

RYTER, C., REEVES, H., GRADSTAJN, E. & AUDOUZE, J. 1970 *Astron. and Astrophys.* **8**, 389-397.

SANDAGE, A. 1993 *Astrophys. J.* **106**, 719-723.

SANDAGE, A. & TAMMAN, G. 1995. IN *Current Topics in Astrofundamental Physics: The Early Universe*, EDS. SANCHEZ, N. & ZICHICHI, A. PP. 403-443. KLUWER, DORDRECHT.

SCHECHTER, P.L. 1996 *IAU Symposia* **173**, 263-264.

SCHMIDT, B., KIRSHNER, R., & EASTMAN, R. 1992 *Astrophys. J.* **395**, 366-386.

SCHRAMM, D.N. 1994 IN *Evolution of the Universe and Its Observational Quest*, PROC. OF YAMADA CONF. XXXVII, TOKYO, JUNE 1993, ED. SATO, K., PP. 61-74. UNIVERSAL ACADEMIC PRESS, TOKYO. SEE ALSO SCHRAMM, D.N. 1995 IN *The Light Element Abundances*, PROC. OF ESO/EIPC WORKSHOP, ELBA, MAY 1994, ED. CRANE, P., 51-72. SPRINGER-VERLAG, HEIDELBERG.

SCHRAMM, D.N. & KAWANO, L. 1989 *Nuc. Inst. and Methods A* **284**, 84-88.

SCHRAMM, D.N. AND MATHEWS, G. (1995) IN *Particle and Nuclear Astrophysics in the Next Millenium: Proc. Snowmass Summer Study*, EDS. KOLB, E.W. & PECCEI, R.D., 479-497. WORLD SCIENTIFIC, SINGAPORE.

SCHRAMM, D.N. & WAGONER, R.V. 1977 *Ann. Rev. of Nuc. Sci.* **27**, 37-74.

SCHRAMM, D.N. & WASSERBURG, G.J. 1970 *Astrophys. J.* **162**, 57-69.

SCHVARTZMAN, V.F. 1969 *JETP Letters* **9**, 184-186.

SCOTT, D. & WHITE, M. 1995 *Gen. Rel. and Grav.* **27**, 1023-1030.

SCULLY, S.T., CASSÉ, M., OLIVE, K.A., SCHRAMM, D.N., TRURAN, J. & VANGIONI-FLAM, E. 1996 *Astrophys. J.* **462**, 960-968.

SHI, X. 1995 *Astrophys. J.* **446**, 637-645.

SHI, X., SCHRAMM, D.N., & DEARBORN, D. 1995 *Phys. Rev. D* **50**, 2414-2420.

SHI, X., SCHRAMM, D.N., DEARBORN, D., & TRURAN, J.W. 1995 *Comments on Astrophysics* **17**, 343-360.

SKILLMAN, E. 1997, PRIVATE CONVERSATION.

SMITH, V.V., LAMBERT, D.L. & NISSEN, P. 1982 *Astrophys. J.* **408**, 262-276.

SONGAILA, A., COWIE, L.L., HOGAN, C.J., & RUGERS, M. 1990 *Nature* **368**, 599-604.

SPITE, M., NISSEN, P.E. & SPITE, F. (1996) *Astron. and Astrophys.* **307**, 172-183.

SPITE, J. & SPITE, M. 1982 *Astron. and Astrophys.* **115**, 357-366.

STEIGMAN, G., FIELDS, B., OLIVE, K., SCHRAMM, D.N., & WALKER, T. 1993 *Astrophys. J.* **415**, L35-L38.

STEIGMAN, G., SCHRAMM, D.N., & GUNN, T. 1977 *Phys. Lett. B* **66**, 202-204.

SZALAY, A. 1995 IN PROC. OF THE AUSTRALIAN NATIONAL UNIVERSITY SUMMER SCHOOL ON COSMOLOGY, CANBERRA, AUSTRALIA, JANUARY 1994, WORLD SCIENTIFIC, SINGAPORE.

TAYLER, R. & HOYLE, F. 1964 *Nature* **203**, 1108-1110

THUAN, T.X., IZOTOV, Y.I. & LIPOVETSKY, V.A. 1996 *Astrophys. J.* **463**, 120-133.

TRURAN, J. 1965, DOCTORAL THESIS, YALE UNIVERSITY; J.W. TRURAN, A.G.W. CAMERON, & GILBERT, A. 1966 *Can. Jour. of Phys.* **44**, 563-5900000002.

Turner, M.S. & Schramm, D.N 1998 *Review of Modern Physics*, in press (January).

Turner, M.S., Wilczek, F. & Zee, A. 1983 *Phys. Lett. B***125**, 35-38; **125**, 519-522.

Tytler, D. 1997 in *Proc. of the 18th Texas Symposium on Relativisitic Astrophysics*, ed. A. Olinto, in press, World Scientific.

Tytler, D., Burles, S., & Kirkman, D. (1996) astro-ph/9612121.

Ussher, J. 1658 *The Annals of the World.* E. Tyler for J. Crook, London.

Wagoner, R., Fowler, W.A., & Hoyle, F. 1967 *Astrophys. J.* **148**, 3-49.

Walker, T., Steigman, G., Schramm, D.N., Olive, K., & Kang, H.S. 1991 *Astrophys. J.* **376**, 51-69.

Wasserburg, G. & Busso, M. 1996 Caltech abstract, LPSXXVIIz.

White, S.D.M., Navarro, J.F., Evrard, A.E. & Frenck, C.S. 1993 *Nature* **366**, 429-433.

Yang, J., Turner, M.S., Steigman, G., Schramm, D.N., & Olive, K. 1984 *Astrophys. J.* **281**, 493-511.